Circulating Fluidized Beds

JOIN US ON THE INTERNET VIA WWW, GOPHER, FTP OR EMAIL:

WWW: http://www.thomson.com
GOPHER: gopher.thomson.com
FTP: ftp.thomson.com
EMAIL: findit@kiosk.thomson.com

A service of I(T)P®

Circulating Fluidized Beds

Edited by

J.R. GRACE
University of British Columbia
Vancouver
Canada

A.A. AVIDAN
Mobil Research and Development Corporation
Paulsboro
USA

and

T.M. KNOWLTON
Particulate Solid Research Incorporated
and formerly of Institute of Gas Technology
Chicago
USA

BLACKIE ACADEMIC & PROFESSIONAL
An Imprint of Chapman & Hall
London · Weinheim · New York · Tokyo · Melbourne · Madras

Published by Blackie Academic and Professional,
an imprint of Chapman & Hall, 2–6 Boundary Row, London SE1 8HN, UK

Chapman & Hall, 2–6 Boundary Row, London SE1 8HN, UK

Chapman & Hall GmbH, Pappelallee 3, 69469 Weinheim, Germany

Chapman & Hall USA, 115 Fifth Avenue, New York, NY 10003, USA

Chapman & Hall Japan, ITP-Japan, Kyowa Building, 3F, 2-2-1 Hirakawacho, Chiyoda-ku, Tokyo 102, Japan

DA Book (Aust.) Pty Ltd, 648 Whitehorse Road, Mitcham 3132, Victoria, Australia

Chapman & Hall India, R. Seshadri, 32 Second Main Road, CIT East, Madras 600 035, India

First edition 1997
© 1997 Chapman & Hall

Typeset in 10/12pt Times by Academic & Technical Typesetting, Bristol
Printed in Great Britain by St Edmundsbury Press, Bury St Edmunds, Suffolk

ISBN 0 7514 0271 0

Apart from any fair dealing for the purposes of research or private study, or criticism or review, as permitted under the UK Copyright Designs and Patents Act, 1988, this publication may not be reproduced, stored, or transmitted, in any form or by any means, without the prior permission in writing of the publishers, or in the case of reprographic reproduction only in accordance with the terms of the licences issued by the Copyright Licensing Agency in the UK, or in accordance with the terms of licences issued by the appropriate Reproduction Rights Organization outside the UK. Enquiries concerning reproduction outside the terms stated here should be sent to the publishers at the London address printed on this page.

The publisher makes no representation, express or implied, with regard to the accuracy of the information contained in this book and cannot accept any legal responsibility or liability for any errors or omissions that may be made.

A catalogue record for this book is available from the British Library

Library of Congress Catalog Card Number: 96-83012

∞ Printed on acid-free text paper, manufactured in accordance with ANSI/NISO Z39.48-1992 (Permanence of Paper).

Contents

List of contributors		xiii
Preface		xv
1	**Introduction to circulating fluidized beds** JOHN R. GRACE and HSIAOTAO BI	**1**
	1.1 Introduction	1
	1.2 Distinguishing characteristics	1
	1.3 Advantages and disadvantages of CFB systems	4
	1.4 CFB applications	4
	1.5 Regimes of fluidization	6
	1.5.1 Regime transitions	6
	1.5.2 Regime diagrams	10
	1.6 Particle properties	15
	1.7 Terminology	16
	Nomenclature	17
	References	18
2	**Hydrodynamics** MASAYUKI HORIO	**21**
	2.1 Background	21
	2.2 Fundamental nature and meso-scale flow structure of gas–solid suspensions	25
	2.2.1 Governing equations	25
	2.2.2 Aggregating tendency of suspensions	27
	2.2.3 Meso-scale suspension structure in each flow regime	32
	2.3 Pressure profiles and macroscopic flow structure	42
	2.3.1 Boundary effects and macroscopic flow structure	42
	2.3.2 Formulation of core-annulus flow structure	50
	2.3.3 Total pressure loop prediction	61
	2.4 Scaling relationships	65
	2.4.1 Scale-up and hydrodynamic scaling law	65
	2.4.2 Derivation of scaling law and experimental validations	66
	2.5 Closing remarks	72
	Acknowledgements	74
	Nomenclature	74
	References	78
3	**Gas mixing** UMBERTO ARENA	**86**
	3.1 Introduction	86
	3.2 Experimental studies	88
	3.2.1 Experimental procedures	88
	3.2.2 Mixing coefficients	92

	3.2.3	Axial mixing	93
	3.2.4	Lateral mixing	101
	3.2.5	Effect of design and operating variables	103
	3.2.6	Lateral injection of a gas stream	105
3.3	Mixing models		107
	3.3.1	Single-phase approach	109
	3.3.2	Two-region approach	110
3.4	Concluding remarks		114
Acknowledgements			114
Nomenclature			114
References			115

4 Solids motion and mixing — 119
JOACHIM WERTHER and BERND HIRSCHBERG

4.1	Introduction		119
4.2	Particle motion and solids mixing mechanisms		120
	4.2.1	Particle motion in the bottom zone	120
	4.2.2	Particle motion in the dilute zone	121
	4.2.3	Particle motion in the transition zone	125
	4.2.4	Particle motion in the exit zone	125
4.3	Axial solids mixing		126
	4.3.1	Experimental techniques	127
	4.3.2	Axial solids dispersion model	127
	4.3.3	Core-annulus interchange model	134
4.4	Lateral solids mixing		134
	4.4.1	Experimental techniques	134
	4.4.2	Evaluation of lateral solids mixing experiments	135
4.5	Fluidization of dissimilar particles/solids segregation		139
	4.5.1	Experimental findings	139
	4.5.2	Segregation mechanisms	142
Nomenclature			144
References			146

5 Hydrodynamic modeling — 149
JENNIFER L. SINCLAIR

5.1	Introduction	149
5.2	Governing equations	151
5.3	Summary of governing equations in various two-fluid models	152
5.4	Earlier two-fluid models	158
5.5	Kinetic theory	159
5.6	Recent two-fluid models	161
5.7	Boundary conditions	167
5.8	Other two-fluid models	170
5.9	Semi-empirical models	172
5.10	Computer simulations	173
Nomenclature		176
References		177

6 Cyclones and other gas–solids separators — 181
EDGAR MUSCHELKNAUTZ and VOLKER GREIF

6.1	Introduction	181
6.2	Particle size distribution	183
6.3	Cyclones with and without a vortex tube	184
6.4	Entrance duct and entrance velocities	187

	6.5	Pressure drop and separation efficiency	191
		6.5.1 General flow pattern	191
		6.5.2 Separation efficiency of a cyclone according to the model of Barth/Muschelknautz	193
		6.5.3 Cyclone pressure drop	203
	6.6	Downcomer tube and fluidized bed seal with valve at its end	205
	6.7	Inserts in the separation zone	207
	6.8	Other separators	208
	6.9	Closure	210
		Nomenclature	210
		References	213

7 Standpipes and return systems — 214
TED M. KNOWLTON

	7.1	Introduction	214
	7.2	Standpipes	214
	7.3	Standpipes in CFB systems	228
	7.4	Standpipes in CFBCs	229
		7.4.1 Automatic solids recirculation systems in CFBCs	230
		7.4.2 Controlled solids recirculation systems in CFBCs	230
	7.5	Standpipes in FCC units	232
	7.6	Laboratory CFB systems	235
	7.7	Non-mechanical solids flow devices	240
		7.7.1 Non-mechanical valve mode	242
		7.7.2 Automatic solids flow devices	249
	7.8	Cyclone diplegs and trickle valves	254
		Nomenclature	258
		References	259

8 Heat transfer in circulating fluidized beds — 261
LEON R. GLICKSMAN

	8.1	Introduction	261
		8.1.1 General observations	261
	8.2	Hydrodynamics	263
	8.3	Heat transfer fundamentals	269
		8.3.1 Particle convection	270
		8.3.2 Gas convection	275
		8.3.3 Radiation heat transfer	276
	8.4	Heat transfer models	279
		8.4.1 Parameter values	279
		8.4.2 Wall resistance	280
		8.4.3 Cluster wall coverage	280
		8.4.4 Cluster solids concentration	280
		8.4.5 Contact time	280
		8.4.6 Parametric trends	282
	8.5	Advanced considerations	285
		8.5.1 Radiation heat transfer	285
		8.5.2 Wall coverage	287
		8.5.3 Deposition rate to the wall	288
	8.6	Thermal and dynamic scaling	290
		8.6.1 Hydrodynamic scaling	290
		8.6.2 Thermal scaling	291
	8.7	Heat transfer measurement techniques	293
	8.8	Heat transfer results – laboratory-scale beds	293
		8.8.1 Elevated temperature	296

	8.9	Large units	300
	8.10	Fins and heat transfer augmentation	304
	8.11	Conclusions and recommendations	305
		Acknowledgements	305
		Nomenclature	306
		References	307

9 Experimental techniques — 312
MICHEL LOUGE

	9.1	Introduction	312
	9.2	Visualization	313
	9.3	Pressure and stresses	314
	9.4	Solids volume fraction	317
		9.4.1 Capacitance instruments	317
		9.4.2 Optical fibers	321
		9.4.3 Transmission densitometry	325
		9.4.4 Capacitance tomography	329
	9.5	Local particle flux	332
	9.6	Solids mass flow rate	336
	9.7	Particle velocity	337
		9.7.1 Cross-correlation	338
		9.7.2 Laser–Doppler anemometry	340
		9.7.3 Other velocimetry	343
	9.8	Heat and mass transfer	344
		9.8.1 Forced convection in cold risers	344
		9.8.2 High temperatures	348
		9.8.3 Mass transfer	350
	9.9	Tracers and sampling	351
		9.9.1 Particle swarms	351
		9.9.2 Individual particles	353
		9.9.3 Transient gas injection	354
		9.9.4 Continuous gas injection	354
		9.9.5 Gas and solids chemical sampling	355
	9.10	Closure	355
		Nomenclature	356
		References	358

10 Combustion performance — 369
CLIVE BRERETON

	10.1	Introduction – history and status	369
	10.2	Clean fossil fuel combustion in fluid bed systems	370
	10.3	Fluid bed versus alternative combustors	372
	10.4	Selection of operating temperature	373
	10.5	Circulating bed combustion fluid mechanics and comparisons with FCC units	377
	10.6	Turndown and control strategies	379
	10.7	Temperature profiles and the effect of turndown	384
	10.8	Relationship between combustion fundamentals and heat release profiles	386
	10.9	Generation and destruction of pollutants	392
		10.9.1 Hydrocarbons and carbon monoxide	392
		10.9.2 Sulfur capture	394
		10.9.3 Nitrogen oxide emissions	402
	10.10	Summary	410
		References	411

11 Design considerations for CFB boilers — 417
YAM Y. LEE

11.1 Introduction	417
11.2 Boiler configuration	421
11.3 Combustor design	422
11.4 Fuel characteristics	424
11.5 Refractory	425
11.6 Air and solids feed system	426
11.7 Separator and return system features	428
11.8 Ash handling system	428
11.9 Control system	430
11.10 Scale-up	430
11.10.1 Heat transfer surfaces	431
11.10.2 Air distribution	431
11.10.3 Fuel distribution	431
11.10.4 Cyclone design	433
11.10.5 Research needs for scale-up	433
11.11 Pressurized circulating fluidized beds (PCFB) technology	434
11.12 Summary	437
References	438

12 Applications of CFB technology to gas–solid reactions — 441
RODNEY J. DRY and COLIN J. BEEBY

12.1 Introduction	441
12.2 Process considerations	442
12.3 CFB gasification of coal	443
12.4 CFB calcination of alumina	446
12.5 CFB roasting of sulfide ores	449
12.6 CFB treatment of hot smelter offgas	452
12.7 CFB pre-reduction of iron ore for direct smelting	456
12.8 Rotary kiln metallization of ilmenite	462
12.9 The future	463
References	464

13 Fluid catalytic cracking — 466
AMOS A. AVIDAN

13.1 Introduction	466
13.2 Brief history	467
13.3 Catalyst flow in modern FCC units	470
13.4 Catalyst regeneration	473
13.5 FCC process basics	476
13.6 FCC feed atomization and mixing	479
13.7 FCC catalyst–product separation	484
Nomenclature	487
References	487

14 Design and scale-up of CFB catalytic reactors — 489
JOHN M. MATSEN

14.1 Introduction	489
14.2 Scale up issues	489
14.2.1 Study design	489

	14.2.2	Reactor engineering	490
	14.2.3	Experimental work	491
14.3	Commercial components in CFB systems		492
	14.3.1	Gas and particle introduction	492
	14.3.2	The riser	492
	14.3.3	Riser termination	493
	14.3.4	Solids separation	493
	14.3.5	Standpipe	494
	14.3.6	Circulation control	494
14.4	Development of specific processes		495
	14.4.1	Fluid catalytic cracking	495
	14.4.2	Maleic anhydride	497
	14.4.3	The Synthol process	498
14.5	Closure		500
Nomenclature			501
References			502

15 Reactor modeling for high-velocity fluidized beds   504
JOHN R. GRACE and K. SENG LIM

15.1	Introduction		504
15.2	Turbulent regime		505
15.3	Fast fluidization: single-region one-dimensional models		506
	15.3.1	Without allowance for hydrodynamic axial gradients	506
	15.3.2	With allowance for hydrodynamic axial gradients	507
15.4	Fast fluidization: core/annulus models		508
	15.4.1	Without allowance for hydrodynamic axial gradients	509
	15.4.2	With allowance for hydrodynamic axial gradients	512
15.5	Some experimental findings		515
15.6	Other models		516
	15.6.1	Cluster/gas two-phase model	516
	15.6.2	Co-existing upflow/downflow model	517
	15.6.3	Monte Carlo model	517
	15.6.4	Models based on solving fundamental equations	518
	15.6.5	Transient models	519
15.7	Concluding remarks		521
Nomenclature			521
References			522

16 Novel configurations and variants   525
YONG JIN, JING-XU ZHU and ZHI-QING YU

16.1	Introduction		525
16.2	Internals		525
16.3	End configurations		531
	16.3.1	Bottom sections	531
	16.3.2	Exit configuration	536
16.4	Other bed configurations		538
16.5	Gas–solids co-current downflow systems		541
	16.5.1	Existing and potential applications of downflow systems	542
	16.5.2	Typical structure of downers	544
	16.5.3	Hydrodynamics of downer	546
	16.5.4	Gas and solids mixing	551
16.6	Liquid–solids and gas–liquid–solids systems		554
	16.6.1	Liquid–solids (L–S) circulating fluidized bed	554
	16.6.2	Gas–liquid–solids (G–L–S) three-phase circulating fluidized bed	557

		Acknowledgements	561
		Nomenclature	561
		References	562
17	**Future prospects**		**568**
	AMOS A. AVIDAN		
	17.1	Introduction	568
	17.2	Fischer–Tropsch synthesis	570
	17.3	Ultra-short contact time fluid–particle reactors	572
		Nomenclature	576
		References	577

Index   **579**

Contributors

Professor Umberto Arena Institute for Combustion Research, National Research Council, Naples, Italy

Dr Amos A. Avidan Mobil Research and Development Corporation, Paulsboro, New Jersey 08066-0480, USA

Dr Colin J. Beeby Comalco Research Centre, Thomastown, Melbourne, Victoria, Australia 3074

Dr Hsiaotao Bi Département de génie chimique, Ecole Polytechnique, Monteal, Canada H3C 3A7

Professor Clive Brereton Department of Chemical Engineering, University of British Columbia, Vancouver, B.C., Canada, V6T 1Z4

Dr Rodney J. Dry CRA Advanced Technical Development, 1 Research Avenue, Bundoora 3083, Victoria, Australia

Professor Leon R. Glicksman Building Technology Group, Department of Architecture, Massachusetts Institute of Technology, Cambridge, Massachusetts 02139, USA

Professor John R. Grace Department of Chemical Engineering, University of British Columbia, Vancouver, B.C., Canada, V6T 1Z4

Volker Greif Department of Applied Chemistry, Technical University of Stuttgart, Germany

Dr Bernd Hirschberg Technical University of Hamburg–Harburg, VTI, Denickstr. 15, 21071 Hamburg, Germany

CONTRIBUTORS

Professor Masayuki Horio	Department of Chemical Engineering, Tokyo University of Agriculture and Technology, Tokyo 184, Japan
Professor Yong Jin	Fluidization Laboratory, Department of Chemical Engineering, Tsinghua University, Beijing 100084, P.R. China
Dr Ted M. Knowlton	Particulate Solid Research Inc., 3424 South State Street, Chicago, Illinois 60616, USA
Dr Yam Y. Lee*	Ahlstrom Pyropower Inc., San Diego, California, USA
Dr K. Seng Lim	Department of Chemical Engineering, University of British Columbia, Vancouver, B.C., Canada, V6T 1Z4
Professor Michel Louge	School of Mechanical and Aerospace Engineering, Cornell University, Ithaca, New York 14853-7501, USA
Dr John M. Matsen	Exxon Research and Engineering Company, P.O. Box 101, Florham Park, New Jersey 07932-0101, USA
Prof. Dr-Ing. Edgar Muschelknautz	Department of Applied Chemistry, Technical University of Stuttgart, Germany
Professor Jennifer Sinclair	Department of Chemical Engineering, University of Arizona, Tucson, Arizona 85721, USA
Professor Joachim Werther	Technical University of Hamburg Harburg, VTI, Denickstr. 15, 21071 Hamburg, Germany
Professor Zhi-Qing Yu	Fluidization Laboratory, Department of Chemical Engineering, Tsinghua University, Beijing 100084, P.R. China
Professor Jing-Xu Zhu	Department of Chemical and Biochemical Engineering, University of Western Ontario, London, Canada, N6A 5B9

*Currently with Y. Y. Lee Consulting Services, 11212 Corte Playa Azteca, San Diego, CA 92124, USA

Preface

Since the late 1970s there has been an explosion of industrial and academic interest in circulating fluidized beds. In part, the attention has arisen due to the environmental advantages associated with CFB (circulating fluidized bed) combustion systems, the incorporation of riser reactors employing circulating fluidized bed technology in petroleum refineries for fluid catalytic cracking and, to a lesser extent, the successes of CFB technology for calcination reactions and Fischer–Tropsch synthesis. In part, it was also the case that too much attention had been devoted to bubbling fluidized beds and it was time to move on to more complex and more advantageous regimes of operation.

Since 1980 a number of CFB processes have been commercialized. There have been five successful International Circulating Fluidized Bed Conferences beginning in 1985, the most recent taking place in Beijing in May 1996. In addition, we have witnessed a host of other papers on CFB fundamentals and applications in journals and other archival publications. There have also been several review papers and books on specific CFB topics. However, there has been no comprehensive book reviewing the field and attempting to provide an overview of both fundamentals and applications. The purpose of this book is to fill this vacuum.

When we began planning the book, it was to have been a comprehensive volume written by four authors. However, the explosion of work which then appeared over the next few years led us to reconceptualize the book as an edited multi-authored volume. However, we have attempted to retain some elements of the original plan by exercising more editorial direction and intervention than one normally finds in multi-authored volumes of this general nature.

The choice of topics has been made to provide coverage of fundamentals (Chapters 2–5, 8 and 15), applications (Chapters 10–14) and specific topics which are essential to the operation (Chapter 7) and understanding (Chapter 9) of CFB systems. Chapter 16 deals with novel geometries, including the downflow (or 'downer') reactor. Chapters 1 and 17 bracket the book, the former providing some background and contextual material, mostly for the novice, and the latter attempting to look at future prospects for CFB technology.

We have assembled a first-class group of authors for the volume, providing a good mixture of industrial and academic authors, a blend of ages, and widely international geographic coverage. We are grateful to the authors for responding to our invitation to prepare chapters and for accepting our proddings and editorial actions with equanimity.

We are also pleased to acknowledge the role of Dr Lesley Anderson in enthusiastically backing the project on behalf of the publisher. Finally, we wish to record our gratitude to those on home territory who have helped with this project and many others, in particular Grace Lee (JRG), June Cavallo (AAA) and Sharon Solar (TMK).

John R. Grace
Amos A. Avidan
Ted M. Knowlton
June 1996

1 Introduction to circulating fluidized beds
JOHN R. GRACE AND HSIAOTAO BI

1.1 Introduction

Solid particles are often of great interest in the chemical process industry, mineral processing, pharmaceutical production, energy-related processes, etc. In some cases the particles serve as catalysts for reacting gases and/or liquids. In other cases, as in ore processing, the particles must be chemically converted. In still other processes the particles must undergo physical transformation, as in drying of particulate solids.

A number of possible configurations are available for carrying out such reactions and contacting operations. For example, in industrial calcination and combustion processes, there have long been competing technologies based on fixed beds (or moving packed beds where the particles travel slowly downward in contact with each other), fluidized beds (where the particles are supported by gas or liquid introduced through a distributor at the bottom of a vessel) and dilute-phase transport systems (where the particles are conveyed through a duct or pipe). The circulating fluidized bed (often abbreviated CFB) has come to prominence in the past two decades in terms of major applications. A typical configuration for a CFB reactor is shown schematically in Figure 1.1. Required are a tall vessel, a means of introducing particles (often simply called 'solids') usually near the bottom, a sufficient upwards flow of fluid (generally a gas, but a liquid or a gas–liquid mixture is also possible – see Chapter 16) to cause substantial entrainment of particles from the top of the vessel, and a means of capturing a substantial majority of these particles and returning them continuously to the bottom. The term 'circulating' signifies that the particle separation and return systems are integral and essential components of the overall reactor configuration. The words 'fluidized bed' denote the fact that the particles are supported by the fluid, while there is still a substantial suspension density. Note that there is unlikely to be a true 'bed' in the normal sense; in particular, most circulating fluidized beds operate in the so-called 'fast fluidization' hydrodynamic regime where there is no distinct or recognizable upper bed surface.

1.2 Distinguishing characteristics

A gas–solids two-phase vertical flow system without mechanical restraint

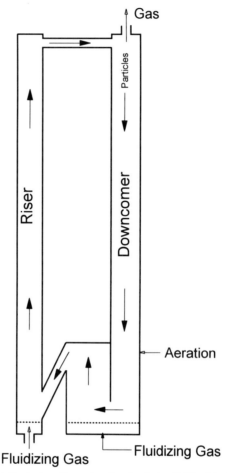

Figure 1.1 Typical configuration for circulating fluidized bed system.

can be operated in three modes: co-current upflow, co-current downflow and counter-current flow with gas flowing upward as shown in Figure 1.2. A circulating fluidized bed shown in Figure 1.1 resembles a bottom restraint system in which solids are prevented from escaping from the bottom. At low solids feed rates, all injected particles are carried upward giving co-current upward flow. When the solids feed rate is increased to such an extent that the upward flow collapses due to saturation of solids entrainment, excess particles fall downward and a dense region forms in the bottom section of the riser. A circulating fluidized bed can thus be operated in either a co-current upward flow mode or fast fluidization mode, depending on the gas velocity and solids circulation rate.

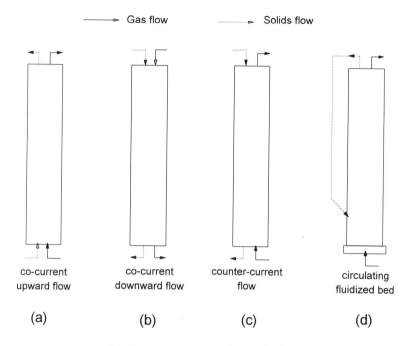

Figure 1.2 Flow modes of gas–solids vertical flow systems.

Both a configuration and a suitable mode of operation as described above are required if one wishes to establish a circulating fluidized bed contactor. As discussed in Chapter 16, there can be considerable variants in the physical geometry. Chapter 16 even contains a description of co-current downflow (or 'downer') reactors [Figure 1.2(b)], which clearly do not represent fluidized beds in any normal sense, but which are gaining favour for certain applications, e.g. where reactions are extremely quick and any variations in particle residence time would result in a decrease in efficiency. Leaving aside such major variants, one can distinguish typical CFB systems from low-velocity fluidized beds and dilute-phase transport systems, as indicated in Table 1.1.

Note that the values in the table are typical operational ranges as currently practised in industry, but not limits of operation. For example, there appears to be no reason why particles somewhat smaller or substantially larger than the 50 to 500 μm range indicated in the table could not be used if this turned out to be advantageous for some processes. It is seen in Table 1.1 that CFB systems are generally intermediate between low-velocity fluidized beds and dilute-phase systems. In recent years they have found a niche for some processes where the combination of mixing characteristics, gas–solid contacting, residence time and heat transfer properties make them advantageous.

Table 1.1 Key features that distinguish circulating fluidized bed reactors from low-velocity fluidized beds and from dilute-phase transport reactors

	Low-velocity fluidized bed reactors	Circulating fluidized bed reactors	Dilute-phase transport reactors
Particle histories	Particles spend substantial time (minutes or hours) in main reactor vessel. Occasional excursions through cyclone and standpipe	Particles pass repeatedly through the recirculating system; residence time in the main vessel for each circuit is counted in seconds	Once-through system
Hydrodynamic regime (see section 1.5)	Bubbling, slugging or turbulent fluidization, with a distinct upper interface	Usually fast fluidization, though bottom of the reactor may correspond to turbulent fluidization conditions or even bubbling	Dilute transport conditions
Superficial gas velocity	Generally below 2 m/s	Usually 3 to 16 m/s	Usually 15 to 20 m/s
Mean particle diameter	0.03 to 3 mm	Usually 0.05 to 0.5 mm	Typically 0.02 to 0.08 mm
Net circulation flux of solids	Low, generally 0.1 to 5 kg/m^{2}s	Substantial, e.g. 15 to 1000 kg/m^{2}s	Up to ~20 kg/m^{2}s
Voidage	Typically 0.6 to 0.8 in bed. Much higher in freeboard above bed	Typically 0.8 to 0.98 averaged over riser	Generally >0.99
Gas mixing	Substantial axial dispersion; complex two-phase behaviour	Some gas downflow near walls typically results in intermediate gas mixing	Very little axial dispersion

1.3 Advantages and disadvantages of CFB systems

Now that the general features of both CFB equipment and their hydrodynamics have been covered, we are in a position to delineate advantages and disadvantages. A comparison with conventional low velocity fluidized beds is given in Table 1.2. Note that each application requires careful weighing and cost-benefit analysis of the factors that affect the choice of equipment in that case. Generally speaking, it is more likely that the additional capital cost of CFB systems can be afforded in large systems than in small ones.

1.4 CFB applications

The unique features and advantages of CFB reactors have resulted in them being used in a number of chemical processes. They have not yet found any significant application for physical processes such as drying.

Table 1.2 Typical advantages and disadvantages of CFB reactors relative to conventional low-velocity fluidized bed reactors without baffles

Advantages:
1. Improved gas–solid contacting given the lack of bubbles;
2. Reduced axial dispersion of gas;
3. Reduced cross-sectional area given the higher superficial velocities;
4. Potentially more control over suspension-to-wall heat transfer because of the ability to use the solids circulation flux as an additional variable;
5. No region like the freeboard region of low-velocity beds where there can be substantial temperature gradients;
6. Less tendency to show particle segregation and agglomeration;
7. Recirculation loop provides a location where a separate operation (e.g. regeneration or heat transfer) can be carried out;
8. Easier to have staged processes;
9. Because of superior radial mixing, fewer solids feed-points are needed;
10. Higher solids flux through the reactor.

Disadvantages:
1. Increased overall reactor height;
2. Higher capital cost;
3. Decreased suspension-to-wall heat transfer coefficients for given particles;
4. Somewhat more restricted range of particle properties;
5. Do not lend themselves to horizontal surfaces due to erosion of in-bed surfaces;
6. Added complexity in designing and operating recirculating loop;
7. Increased particle attrition.

The principal applications, fluid catalytic cracking (FCC) and circulating fluidized bed combustion (CFBC), are so important that they are given their own chapters in this book, and their histories, as well as key features of the two technologies, are covered there (FCC in Chapter 13, CFBC in Chapters 10 and 11). An account of early work on high velocity fluidization, chiefly in hydrocarbon processing, has been published recently by Squires (1994). FCC attempted to exploit CFB conditions in the 1940s, but operational problems and inferior catalysts then postponed further development for several decades. In the case of CFBC, it was the pioneering work of Reh (1971), drawing on his experience with respect to CFB calcining, that led to the rapid development of the technology. For a brief historical statement on the development of calcination and combustion processes by Lurgi, see Reh (1986). In the mid 1990s, there are approximately 250 FCC units in operation worldwide using CFB risers and approximately 400 commercial CFBC units. In addition, there are more than 70 laboratory scale CFB cold models in operation in a host of countries.

While FCC and CFBC are by far the major technologies for circulating fluidized beds, a number of other applications have been investigated and some have been developed commercially or are under development for industrial processes. Table 1.3 give a list of applications and key references. Further details are given in Chapter 12 for gas–solid reactions and Chapter 14 for catalytic reactions.

Table 1.3 Applications of CFB reactors and key references

I. Gas–solids reactions (see also Chapters 10 to 12)	
Combustion of coal, wood and shale	Reh, 1986, 1995; Yerushalmi, 1986
Incineration of solid waste	Chang et al., 1987
	Hallstrom and Karlsson, 1991
Synthesis of AlF_3 and SiC	Reh, 1995
Recovery/cleaning of off-gases	Reh, 1986
Desulphurization of flue gas	Graf, 1986
Gasification of coal, biomass, etc.	Blackadder et al., 1991; Hirsch et al., 1986; Reh, 1995
Calcination of alumina, phosphate rock, clay, etc.	Reh, 1971, 1986, 1995
Reduction of iron ore, lateritic nickel ore, etc.	Hirsch et al., 1986; Suzuki et al., 1990
Roasting of sulphidic ores (ZnS, Cu_2S, gold ores)	Reh, 1995
Dehydration of boric acid	Li et al., 1990
Decomposition of sulphate, chloride and carbonate	Reh, 1995
Cement production	Deng, 1993
II. Solid-catalysed gas-phase reactions (see also Chapters 13 and 14)	
Fluid catalytic cracking (FCC)	Avidan et al., 1990; King, 1992
Fischer–Tropsch synthesis	Shingles and McDonald, 1988
Butane oxidation to maleic anhydride	Contractor, 1988
Oxidation of o-xylene/naphthalene to phthalic anhydride	Wainwright and Hoffman, 1974
Ethylene epoxidation	Park and Gau, 1986
Oxidative dehydrogenation of butene to butadiene	Liu et al., 1989
Oxidative coupling of methane to ethylene and ethane	Baerns et al., 1994
Methanol to olefins	Schoenfelder et al., 1994
Simultaneous NO_x and SO_2 removal from off-gases	Reh, 1995

1.5 Regimes of fluidization

1.5.1 Regime transitions

The introduction of gas from the bottom of a column containing solid particles via a gas distributor can cause the particles to be fluidized. As shown in Figure 1.3 and Table 1.4, several flow patterns/regimes have been identified. With increasing gas velocity, these are the fixed bed, delayed bubbling or bubble-free fluidization, bubbling fluidization, slugging fluidization, turbulent fluidization, fast fluidization and dilute pneumatic conveying regimes.

The transition from a fixed bed to fluidization is delineated by the minimum fluidization velocity, U_{mf}, which corresponds to the lowest gas velocity at which all bed particles are suspended by the gas. Experimentally, U_{mf} is determined from the levelling off of the pressure drop across the bed with increasing superficial gas velocity after a stage of initial increase in the fixed bed regime. U_{mf} has been extensively studied, and a number of equations are available to predict this transition velocity (e.g. Lippens and Mulder, 1993). One of these equations is due to Grace (1982) modified from the well-known correlation of Wen and Yu (1966):

$$Re_{mf} = \sqrt{27.2^2 + 0.0408 Ar} - 27.2 \qquad (1.1)$$

INTRODUCTION TO CIRCULATING FLUIDIZED BEDS

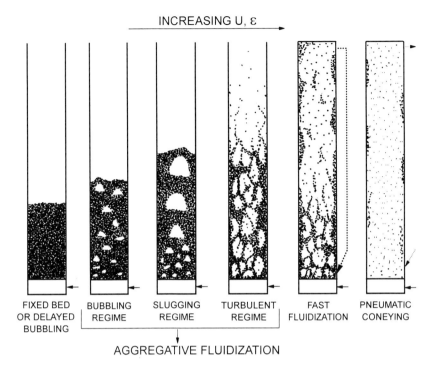

Figure 1.3 Flow patterns in gas–solids fluidized beds (Grace, 1986).

The onset of bubbling is indicated by the minimum bubbling velocity, U_{mb}, the gas velocity at which the bubbles first appear in the bed (Abrahamsen and Geldart, 1980). The minimum bubbling velocity has been found to be a strong function of particle properties. It is higher than U_{mf} for fine particles in Group A of the Geldart (1973) particle classification and equal to U_{mf} for Group B and Group D particles. A bubble-free fluidization regime between U_{mf} and U_{mb} thus exists only for Group A particles, these being small enough that interparticle forces play a significant role. In that case, U_{mb} can be estimated by the Geldart and Abrahamsen (1978) dimensional correlation,

$$U_{mb} = 33 d_p \left(\frac{\rho_g}{\mu_g}\right)^{0.1} \quad \text{(SI units)} \qquad (1.2)$$

For Groups B and D particles, where U_{mb} predicted from equation (1.2) should be less than U_{mf}, U_{mb} must be taken as equal to U_{mf}.

When the superficial gas velocity is increased further, gas bubbles become larger. Slugging is said to occur when the bubbles grow to sizes comparable with the column diameter (Clift *et al.*, 1978). The minimum

Table 1.4 Major characteristics of gas–solid flow regimes

Velocity range	Regime	Appearance and principal features
$0 < U < U_{mf}$	Fixed bed	Particles are stationary; gas flows through interstices
$U_{mf} < U < U_{mb}$	Bubble-free fluidization	Bed expands smoothly and uniformly; top surface is well defined; some small-scale particle motion; little tendency for particles to aggregate; very little pressure fluctuation
$U_{mb} < U < U_{ms}$	Bubbling fluidization	Voids form near the distributor, grow mostly by coalescence, and rise to the surface; top surface is well defined with bubbles breaking through periodically; irregular pressure fluctuations of appreciable amplitude. Bubble size increases as U increases
$U_{ms} < U < U_c$	Slugging fluidization	Voids fill most of the column cross-section; top surface rises and collapses periodically with a reasonably regular frequency; large and regular pressure fluctuations
$U_c < U < U_{se}$	Turbulent fluidization	Small voids and particle clusters dart to and fro; top surface difficult to distinguish; small amplitude pressure fluctuations only
$U_{se} < U$ and $\max(V_{CB}, V_{CC}, V_C) < U < V_{CA}$	Fast fluidization	No distinguishable upper bed surface; particles are transported out at the top and must be replaced by adding solids near the bottom. Clusters or strands of particles move downward, mostly near the wall, while gas and entrained widely dispersed particles move upward in the interior. Increasingly dilute as U is increased at a fixed solid feed rate
$V_{CA} < U$	Dilute-phase transport	No axial variation of solids concentration except in the bottom acceleration section. Some particle strands may still be identified near the wall

slugging velocity, U_{ms}, can be estimated by an equation due to Stewart and Davidson (1967):

$$U_{ms} = U_{mf} + 0.07\sqrt{gD} \tag{1.3}$$

However, slugging is not encountered for shallow beds (e.g. $H/D < 1$), in columns of very large diameter or for fine particles (e.g. $d_p < 60\,\mu m$) because bubbles are then unable to grow to be of comparable size to the column diameter. Several criteria to distinguish slugging from non-slugging systems have been identified by Bi *et al.* (1993). Note that there are no satisfactory techniques that account for the influence of particle size distribution.

The turbulent and fast fluidized regimes are considered to be high-velocity fluidization regimes. Two different definitions are commonly used to distinguish between the bubbling regime and the turbulent regime. The first defines

U_c, the superficial gas velocity at which the standard deviation of the pressure fluctuations reaches a maximum, as the onset of the turbulent regime (Yerushalmi and Cankurt, 1979). U_c is believed to reflect the condition at which bubble coalescence and break-up reach a dynamic balance, with bubble break-up becoming predominant if the gas velocity is increased further. The experimental results obtained by different investigators, however, tend to be inconsistent with each other, with much of the scatter attributable to the effects of measurement method (absolute versus differential pressure fluctuations), signal interpretation method (dimensional versus dimensionless standard deviation) and the measurement location, as well as different particle size distributions (Grace and Sun, 1991; Brereton and Grace, 1992). Data based on differential pressure fluctuation measurements are preferred and can be predicted (Bi and Grace, 1995a) by

$$Re_c = 1.24 Ar^{0.45} \qquad (2 < Ar < 1 \times 10^8) \qquad (1.4)$$

Efforts have also been made to determine U_c using other techniques, such as bed expansion (Avidan and Yerushalmi, 1980) and capacitance signals (Lancia et al., 1988), but they are generally less satisfactory and do not necessarily lead to the same values (Bi and Grace, 1995a).

A second definition used to indicate the transition from bubbling to turbulent fluidization is based on U_k, the superficial gas velocity at which the root-mean-square standard deviation of pressure fluctuation starts to level off with increasing U (Yerushalmi and Cankurt, 1979). This implies that bubble coalescence and break-up have become stabilized, with only small dispersed voids identifiable. Recent experimental findings suggest that U_k is not always a determinable parameter and that it is affected by the measurement method, solids return system, solids recirculation rate, data interpretation method and particle properties (Rhodes and Geldart, 1986b; Brereton and Grace, 1992; Mei et al., 1994; Bi and Grace, 1995a). For large Group B and Group D particles, levelling off of the standard deviation is primarily caused by blow-out of bed particles (Mei et al., 1994), while for Group A particle systems it can be caused by insufficient return of entrained particles (Bi and Grace, 1995a) or by the approach to constant bed density (Schnitzlein and Weinstein, 1988; Perales et al., 1991). Further discussion of U_c and U_k is given in Chapter 2.

Transition from turbulent to fast fluidization is said to occur at the transport velocity, U_{tr}, where significant numbers of particles are carried out from the top of the column (Yerushalmi, 1986). According to Yerushalmi and Cankurt (1979), a sudden change of pressure drop with increasing solids flow rate disappears when the superficial gas velocity exceeds U_{tr}. See Chapter 2 for more detailed discussion of U_{tr}. Rhodes and Geldart (1986a) and Schnitzlein and Weinstein (1988), using the same method as Yerushalmi and Cankurt (1979), were unable to identify a transport velocity. The choice of U_{tr} appears to depend on the location of the two taps across which the pressure drop is measured and the distance between them (Bi, 1994). A more practical way to define this transition is based on the critical velocity, U_{se}, earlier (Lewis and Gilliland, 1950; Squires, 1986) called the 'blow-out velocity', deduced from the entrainment versus superficial gas velocity curve (Bi et al., 1995). U_{se} behaves like U_{tr}, but is more reliable. Based on extensive literature data, the following correlation has been developed for estimating U_{se} (Bi et al., 1995)

$$Re_{se} = 1.53 Ar^{0.50} \qquad (2 < Ar < 4 \times 10^6) \qquad (1.5)$$

For Type D particles, when U_{se} predicted from equation (1.5) is less than the individual particle terminal settling velocity, v_t, U_{se} should be taken as v_t rather than the lower value that results from equation (1.5).

The transition from fast fluidization to pneumatic transport is marked by the disappearance of a dense-phase region of relatively high density and large amplitude pressure fluctuations in the bottom sector of the riser (Bi et al., 1993, 1995). The transition velocity can be measured most readily by decreasing U at a constant solids circulation rate until type A choking (Bi et al., 1993) occurs. The suspension collapse process has been postulated to be caused by the formation of particle clusters (Matsen, 1982), an increase of particle–wall friction (Yang, 1975) and the particle weight overcoming gas shear in the near wall region (Louge et al., 1991). The equations of Yang (1983),

$$\frac{2gD(\epsilon_{CA}^{-4.7} - 1)}{\left(\dfrac{U_{CA}}{\epsilon_{CA}} - v_t\right)^2} = 6.81 \times 10^5 \left(\frac{\rho_g}{\rho_p}\right)^{2.2} \qquad (1.6)$$

and

$$G_s = \rho_p(1 - \epsilon_{CA})\left(\frac{U_{CA}}{\epsilon_{CA}} - v_t\right) \qquad (1.7)$$

can be used to estimate U_{CA} (Bi et al., 1993).

1.5.2 Regime diagrams

Various attempts have been made to plot flow regime maps for gas–solid suspensions. Zenz (1949) proposed an early flow diagram in which both the dense fluidization and co-current pneumatic flow regimes are indicated, but the 'turbulent' region was not delineated. A similar flow regime map was proposed by Yerushalmi et al. (1976), with bed voidage or pressure gradient plotted against superficial gas velocity to show the transitions among the packed bed, bubbling bed, turbulent fluidization and fast fluidization regimes. The regime map developed by Li and Kwauk (1980) also plots voidage against superficial gas velocity. Squires et al.

(1985) expanded such a map to include the pneumatic transport regime and choking points, and this was further modified by Rhodes (1989). The transition from low velocity to high velocity fluidization is, however, still poorly characterized. Grace (1986) extended and modified the approach of Reh (1971) to propose a unified regime diagram based on literature data to show the operating ranges of conventional fluidized beds, spouted beds, circulating beds and transport systems.

Following another approach, Leung (1980), Klinzing (1981) and Yang (1983) proposed flow regime maps of gas–solids transport in which superficial gas velocity was plotted against solids flux, with gas–solid transport divided into dense-phase flow and dilute-phase flow regimes. The transition between pneumatic transport and dense-phase fluidization was again unclear. Takeuchi *et al.* (1986) proposed a flow map based on their experimental findings to define the boundaries of fast fluidization. This flow regime map was modified by Bi and Fan (1991) to include the transition from heterogeneous dilute flow to homogeneous dilute flow. Hirama *et al.* (1992) tried to extend such a diagram to the transition from high-velocity to low-velocity fluidization, but the transition was again not fully defined.

For conventional gas–solids fluidization, the flow regimes include the fixed bed, bubbling fluidization, slugging fluidization and turbulent fluidization. A flow regime diagram consistent with the above picture for cases where the overflow is small enough that the inventory of the column is constant or nearly so is shown in Figure 1.4. In this diagram, the dimensionless gas velocity $U^*(=Re/Ar^{1/3})$ is plotted against the dimensionless particle diameter $d_p^*(=Ar^{1/3})$ as suggested by Grace (1986). U_{mf}^* is based on equation (1.1), U_c^* on equation (1.4), and U_{se}^*, which sets an upper limit on conventional fluidized bed operation, is based on equation (1.5). Since U_{ms} depends on the column diameter [see equation (1.3)], a variable that is not included in any of the dimensionless groups, U_{ms} cannot be plotted on this diagram. U_{mb} based on equation (1.2) also cannot be included in the diagram because this equation is dimensional.

As with gas–liquid vertical transport lines, a gas–solid system can be operated in transport modes when both gas and solids are supplied at sufficient rates to the bottom of the column, with gas and solids also leaving continuously at the top. Ideally, the flow patterns are completely determined by the relative velocity between the gas and the particle phase (i.e. by the slip velocity) rather than by the superficial gas velocity alone. Analogous to gas–liquid upward transport, at a fixed solids flux, a gas–solids transport line may experience bubbly flow, slug flow and fast fluidization or turbulent fluidization before achieving dilute-phase transport. A bubble-free dense-phase transport regime may also exist for fine Group A particles. The slug flow regime may again be bypassed for large diameter columns or for fine particles.

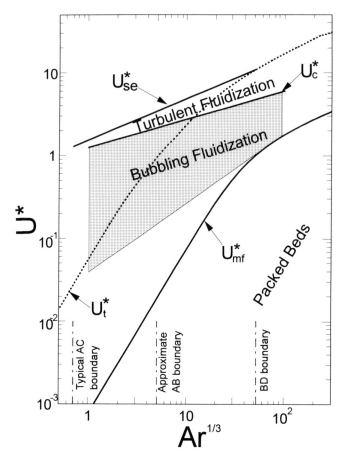

Figure 1.4 Flow regime map for gas–solids fluidization. Heavy lines indicate transition velocities, while the shaded region is the typical operating range of bubbling fluidized beds (Bi and Grace, 1995b).

In solids transport systems, the transition velocity depends on the relative velocity between the two phases. Ideally, the transition velocity in transport lines, V_i, can be predicted by

$$V_i = U_i + \frac{G_s \epsilon_i}{\rho_p (1 - \epsilon_i)} \qquad (1.8)$$

where $i = mf, mb, ms, c$ or se.

It is seen that $i = mf$, mb and ms correspond to minimum fluidization, minimum bubbling and minimum slug flow. The velocity V_c marks the onset of fast fluidization or turbulent flow. The voidage, ϵ_c, at this transition point is approximately 0.65 (Bi and Grace, 1994). The minimum transport velocity, V_{se}, is also called type A choking velocity (Bi et al., 1993). The bed voidage, ϵ_{se}, at this transition point ranges from 0.96 to 0.99, depending on particle properties (Bi et al., 1995). For fine Group A particles, $\epsilon_{CA} \approx 0.96$, while for large Group B and Group D particles, $\epsilon_{CA} \approx 0.99$.

Based on the above considerations, a flow regime diagram, Figure 1.5, similar to Figure 1.4, can be produced with Ar as the abscissa axis and V^*, defined as

$$V^* = \left[\frac{\rho_g^2}{g\mu_g(\rho_p - \rho_g)}\right]^{1/3}\left[U - \frac{G_s\epsilon}{\rho_p(1-\epsilon)}\right] \quad (1.9)$$

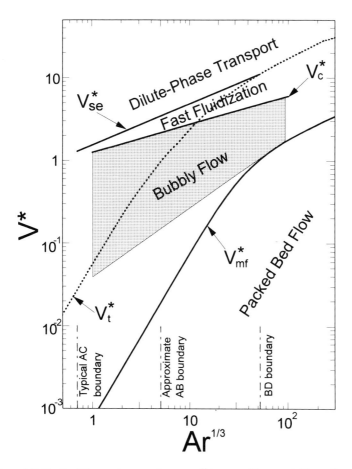

Figure 1.5 Idealized flow regime map for gas–solids upward transport. Heavy lines indicate transition velocities, while the shaded region is the typical operating range of bubbly flow with solids flux maintained constant (Bi and Grace, 1995b).

as the ordinate. Again, neither V_{mb}^* nor V_{ms}^* can be included in this diagram. For a batch operated fluidized bed with given particles, the flow pattern is determined by the superficial gas velocity only. For $G_s = 0$, Figure 1.5 becomes the same as Figure 1.4 with $V^* = U^*$. Figure 1.5 can thus be considered as a generalized flow regime diagram. In a solids transport system with given particles, the flow pattern depends on both the superficial gas velocity and the solids circulation rate. To determine the flow pattern under given operating conditions, both Ar and V^* need to be calculated and then located on Figure 1.5.

The dilute-phase transport regime has been studied extensively (Marcus et al., 1990), while relatively few studies have been reported on dense-phase transport (Konrad, 1986), possibly because it is difficult to maintain dense-phase flow under stable operation in transport lines. Unlike pneumatic transport above U_{CA}, where particles are fully suspended in the gas, particles in dense-phase transport are pushed up the column, requiring a relatively high gas pressure and a high feed rate of particles. When the gas blower is unable to provide sufficient pressure head or when solids cannot be fed to the riser at the required rate, stable dense-phase operation becomes impossible due to Type B choking (Bi et al., 1993). In some cases, even though sufficient blower pressure and solids feeding are provided, it is impossible to achieve dense-phase transport due to severe slugging, i.e. due to Type C or 'classical choking' (Bi et al., 1993). This is clearly demonstrated in Figure 1.6 where three possible flow transition routes with decreasing superficial gas velocity at a fixed solids circulation rate are included. The ideal transition route in Figure 1.5 can be realized only in systems with no equipment-related restrictions and where no classical choking occurs.

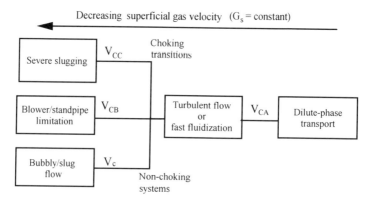

Figure 1.6 Flow chart showing regime transitions in circulating fluidized beds and transport risers with decreasing gas flow (Bi et al., 1993).

Table 1.5 Key characteristics of the turbulent fluidization, fast fluidization and dilute phase transport regimes

Characteristic	Turbulent fluidization	Fast fluidization	Dilute-phase transport
Gas velocity range	$U_c < U < U_{se}$	$U_{se} < U < V_{CA}$	$V_{CA} < U < V_{mp}$
Solids flux range	$G_s \leq G_{s,CA}$	$G_s > G_{s,CA}$	$G_s < G_{s,CA}$
Overall voidage	$\epsilon = 0.6$–0.8	$\epsilon = 0.8$–0.98	$\epsilon > 0.98$
Axial voidage gradients	High	High	Low
Radial voidage gradients	Moderate	High	Low to moderate
Gas–solids slip velocity	Low	High	Low
Particle backmixing	High	High	Low

A circulating fluidized bed is generally operated in the region between the Type A and the Type B or C choking velocities and with the gas velocity close to the minimum pressure gradient point. It can be considered to cover both fast fluidization and core-annular dilute-phase transport, as indicated in Figure 1.6. In most cases, it is impossible to operate a circulating fluidized bed under dense-phase transport conditions because insufficient solids can be provided from the standpipe owing to pressure imbalance between the riser and downcomer (Bi, 1994). Key characteristics of different high velocity fluidization regimes commonly encountered in circulating fluidized bed risers are summarized in Table 1.5.

1.6 Particle properties

While there is no exact prescription for particles that will give good fluidization behaviour including fast fluidization behaviour, there are some 'rules of thumb'. In CFB processes, there appear to be no restrictions on particle density. The surface/volume mean particle diameter, defined as

$$d_p = 1 \bigg/ \sum_{i=1}^{n} (x_i/d_{pi}) \tag{1.10}$$

should normally be between 50 and 500 μm as already noted. It is conventional to determine the weight fractions, x_i, and corresponding interval characteristic particle dimensions, d_{pi}, by sieving. The particle size distribution should not be too narrow, with it being desirable to have at least a tenfold variation between the sizes of the 5% and 95% cumulative particle sizes. Angular particles are usually acceptable, but not if the shapes are extreme, e.g. needle-like or flaky. Surface moisture capable of causing particle agglomeration should also be avoided, although a circulating fluidized bed tends to be less sensitive to this property than low velocity systems. Because of the high velocities and circulation rates in CFB systems, the particles should be as resistant as possible to attrition.

1.7 Terminology

This section is intended for the novice, unfamiliar with the jargon in common usage in the fluidization community. The goal is to provide some basic definitions that will be helpful to the uninitiated in addressing later chapters:

Agglomerate: A group of particles held together by interparticle forces.

Attrition: Breaking of particles due to collisions or other stresses.

Choking: Collapse of dilute gas–solids suspension into a dense-phase flow with decrease in gas velocity at constant solids flow; for a discussion of different modes of choking, see Bi *et al.* (1993).

Cluster: A group of particles travelling together due to hydrodynamic forces.

Dense phase: Gas–solid suspension that is sufficiently concentrated that there are significant particle–particle interactions and inter-particle contacts.

Dilute phase: Suspension sufficiently sparse that there are relatively few interparticle contacts.

Distributor: Perforated plate or other device at the bottom for introducing the fluidizing gas and for supporting the particles during shut-downs.

Entrainment: The physical transport or carrying of particles out of the vessel by the fluidizing fluid.

Fast fluidization: Hydrodynamic regime in which most circulating fluidized beds operate, characterized by dilute upflow in core and downwards movement of strands and streamers near the outer wall (see section 1.2).

Fines: Particles smaller than about 37 to 44 μm in diameter.

Fluoseal: Alternative name for loop seal.

Geldart powder classification: Scheme introduced by Geldart (1973) to distinguish four types of particles with characteristic properties. In order of increasing size with words that help the memory, they are C (cohesive), A (aeratable), B (bubble readily) and D. The original Geldart classification is for systems with air at room temperature and pressure. For extension to other gases and to non-atmospheric pressures and temperatures, see Grace (1986).

Grid: Alternate term used instead of distributor.

H-valve, J-valve, L-valve, V-valve: Configurations shaped like the letter without moving parts that permit solids to be returned to the bottom of a riser against a pressure gradient while not permitting gas to travel up the valve; see Chapter 7 for details.

Loop seal: Common configuration for providing return of solids without reverse flow of gases; see Chapter 7 for details.

Membrane walls: Containing walls employed in combustion equipment composed of parallel (usually vertical) tubes connected by longitudinal fins (see Chapter 8).

Riser: Tall reactor vessel or column used to provide principal reaction zone. On average, particles travel upwards in the riser, though the motion at the wall may be downwards.

Secondary/Tertiary air: Air added at some height above the distributor.
Segregation: Tendency for particles to become separated from each other in different zones based on different physical characteristics such as density or particle size.
Shaft: Alternate name sometimes used for the riser.
Solids: Generic term referring to solid particles.
Strand/Streamer: Alternate name sometimes used for clusters, especially for clusters in the near wall region, which tend to be vertically elongated.
Superficial velocity: Fluid volumetric flow rate divided by total riser (or vessel) cross-sectional area.
Transport Disengagement Height (TDH): Height from surface to bubbling or turbulent bed to level above which there is negligible variation in suspension density with vertical position.
Voidage: Fraction by volume of a fluid–solid suspension occupied by the fluid.

Nomenclature

Ar Archimedes number, $\rho_g(\rho_p - \rho_g)d_p^3 g/\mu_g^2$
D column diameter, m
d_p mean particle diameter, μm
d_p^* dimensionless particle diameter, $(= Ar^{1/3})$
g acceleration due to gravity, m/s^2
G_s solids net circulation rate or solids entrainment flux, kg/m^{2}s
Re Reynolds number, $\rho_g U d_p/\mu_g$
U superficial gas velocity, m/s
U^* dimensionless superficial velocity, $(= Re/Ar^{1/3})$
V_i transition velocity in transport lines, $(i = mf, mb, ms, c \text{ or } se)$
V^* dimensionless net superficial gas velocity,
$$\left(= \left[\frac{\rho_g^2}{g\mu_g(\rho_p - \rho_g)}\right]^{1/3} \left[U - \frac{G_s \epsilon}{\rho_p(1-\epsilon)}\right]\right)$$
v particle velocity, m/s
ϵ voidage
μ_g gas viscosity, kg/m.s
ρ_g gas density, kg/m^3
ρ_p particle density, kg/m^3

Subscripts

c transition from bubbling to turbulent fluidization
CA type A or accumulative choking
CB type B or blower-induced choking

CC   type C or classical choking
mb   minimum bubbling
mf   minimum fluidization
ms   minimum slugging
se   onset of significant solids entrainment
t   terminal settling of single particles

References

Abrahamsen, A.R. and Geldart, D. (1980) Behaviour of gas-fluidized beds of fine powders, Part I. Homogeneous expansion. *Powder Technol.*, **26**, 35–46.
Avidan, A.A. and Yerushalmi, J. (1980) Bed expansion in high velocity fluidization. *Powder Technol.*, **32**, 223–232.
Avidan, A.A., Edwards, E. and Owen, H. (1990) Innovative improvements highlight FCC's past and future. *Oil and Gas J.*, Jan. 33–58.
Baerns, M., Mleczko, L., Tjiatjopoulos, G.J. and Vadalos, I.A. (1994) Comparative simulation studies on the performance of bubbling and turbulent bed reactors for the oxidative coupling of methane in circulating fluidized bed, in *Circulating Fluidized Bed Technology IV* (ed. A.A. Avidan), AIChE, New York, pp. 414–421.
Bi, H.T. (1994) Flow regime transitions in gas–solid fluidization and vertical transport. Ph.D. thesis, University of British Columbia, Vancouver, Canada.
Bi, H.T. and Fan, L.-S. (1991) Regime transitions in gas–solid circulating fluidized beds. AIChE Annual Meeting, Los Angeles, Nov. 17–22.
Bi, H.T. and Grace, J.R. (1994) Transition from bubbling to turbulent fluidization. AIChE Annual Meeting, San Francisco, Nov. 13–18.
Bi, H.T. and Grace, J.R. (1995a) Effects of measurement methods on velocities used to demarcate the transition to turbulent fluidization. *Chem. Eng. J.*, **57**, 261–271.
Bi, H.T. and Grace, J.R. (1995b) Flow regime diagrams for gas–solids fluidization and upward transport. *Int. J. Multiphase Flow*, **21**, 1229–1236.
Bi, H.T., Grace, J.R. and Zhu, J.X. (1993) On types of choking in pneumatic systems. *Int. J. Multiphase Flow*, **19**, 1077–1092.
Bi, H.T., Grace, J.R. and Zhu, J.X. (1995) Regime transitions affecting gas–solids suspensions and fluidized beds. *Trans. I. Chem. Engr.*, **73**, 154–161.
Blackadder, W., Morris, M., Rensfelt, E. and Waldheim, L. (1991) Development of an integrated gasification and hot gas cleaning process using circulating fluidized bed technology, in *Circulating Fluidized Bed Technology III* (eds P. Basu, M. Horio and M. Hasatani), Pergamon Press, Toronto, pp. 511–517.
Brereton, C.M.H. and Grace, J.R. (1992) The transition to turbulent fluidization. *Chem. Eng. Res. Des.*, **70**, 246–251.
Chang, D.P.Y., Sorbo, N.W., Murchison, G.S., Adrian, R.C. and Simeroth, D.C. (1987) Evaluation of a pilot-scale circulating fluidized bed combustor as a potential hazardous waste incinerator. *J. Air Pollution Control Assoc.*, **37**, 266–274.
Clift, R., Grace, J.R. and Weber, M.E. (1978) *Bubbles, Drops and Particles*, Academic Press, New York.
Contractor, R.M. (1988) Butane oxidation to maleic anhydride in a circulating solids riser reactor, in *Circulating Fluidized Bed Technology II* (eds P. Basu and J.F. Large), Pergamon Press, Toronto, pp. 467–477.
Deng, X.J. (1993) CFPC process, in *Preprint for CFB-IV conference* (ed. A.A. Avidan), AIChE, New York, pp. 472–477.
Geldart, D. (1973) Types of gas fluidization. *Powder Technol.*, **7**, 185–195.
Geldart, D. and Abrahamsen, A.R. (1978) Homogeneous fluidization of fine powders using various gases and pressures. *Powder Technol.*, **19**, 133–136.
Grace, J.R. (1982) Fluidized bed hydrodynamics. Chapter 8.1 in *Handbook of Multiphase Flow* (ed. G. Hetsroni), Hemisphere, Washington.

Grace, J.R. (1986) Contacting modes and behaviour classification of gas–solid and other two-phase suspensions. *Can. J. Chem. Eng.*, **64**, 353–363.

Grace, J.R. and Sun, G. (1991) Influence of particle size distribution on the performance of fluidized bed reactors. *Can. J. Chem. Eng.*, **69**, 1126–1134.

Graf, R. (1986) First operating experience with a dry flue gas desulphurization process using a circulating fluidized bed, in *Circulating Fluidized Bed Technology* (ed. P. Basu), Pergamon Press, Oxford, pp. 317–328.

Hallstrom, C. and Karlsson, R. (1991) Waste incineration in circulating fluidized bed boilers test results and operating experiences, in *Circulating Fluidized Bed Technology III* (eds P. Basu, M. Horio and M. Hasatani), Pergamon Press, Toronto, pp. 417–422.

Hirama, T., Takeuchi, T. and Chiba, T. (1992) Regime classification of macroscopic gas–solid flow in a circulating fluidized-bed riser. *Powder Technol.*, **70**, 215–222.

Hirsch, M., Janssen, K. and Serbent, H. (1986) The circulating fluidized bed as reactor for chemical and metallurgical processes, in *Circulating Fluidized Bed Technology* (ed. P. Basu), Pergamon Press, Oxford, pp. 329–340.

King, D. (1992) Fluidized catalytic crackers. An engineering review, in *Fluidization VII* (eds O.E. Potter and D.J. Nicklin), Engineering Foundation, New York.

Klinzing, G.E. (1981) *Gas–Solid Transport.* McGraw-Hill, New York.

Konrad, K. (1986) Dense-phase pneumatic conveying: a review. *Powder Technol.*, **49**, 1–35.

Lancia, A., Nigro, R., Volpicelli, G. and Santoro, L. (1988) Transition from slugging to turbulent flow regimes in fluidized beds detected by means of capacitance probes. *Powder Technol.*, **56**, 49–56.

Leung, L.S. (1980) Vertical pneumatic conveying: a flow regime diagram and a review of choking versus non-choking systems. *Powder Technol.*, **25**, 185–190.

Lewis, W.K. and Gilliland, E.R. (1950) US patent No. 2,498,088.

Li, Y. and Kwauk, M. (1980) The dynamics of fast fluidization, in *Fluidization* (eds J.R. Grace and J.M. Matsen), Plenum, New York, pp. 537–544.

Li, Y., Wang, F. and Tseng, Q. (1990) A new process of preparing anhydrous boric oxide by dehydration of boric acid in a fast fluidized bed. *Chem. Reaction Eng. and Technol.*, **6**(2), 43–48.

Lippens, B.C. and Mulder, T. (1993) Prediction of minimum fluidization. *Powder Technol.*, **75**, 67–78.

Liu, J., Zhang, R., Luo, G. and Yang, G.L. (1989) The macrokinetic study on the oxidative dehydrogenation of butene to produce butadiene. *Chem. Reaction Eng. and Technol.*, **5**(1), 1–8.

Louge, M.Y., Mastorakos, E. and Jenkins, J.T. (1991) The role of particle collisions in pneumatic transport. *J. Fluid Mech.*, **231**, 345–356.

Marcus, R.D., Leung, L.S., Klinzing, G.E. and Rizk, F. (1990) Flow regimes in vertical and horizontal conveying. Chapter 5 in *Pneumatic Conveying of Solids* (eds R.D. Marcus, L.S. Leung, G.E. Klinzing and F. Rizk), Chapman and Hall, New York, pp. 159–191.

Matsen, T.M. (1982) Mechanisms of choking and entrainment. *Powder Technol.*, **32**, 21–33.

Mei, J.S., Rockey, J.M. and Robey, E.H. (1994) Effects of particle properties on fluidization characteristics of coarse particles, in *Circulating Fluidized Bed Technology IV* (ed. A.A. Avidan), AIChE, New York, pp. 600–608.

Park, D.W. and Gau, G. (1986) Simulation of ethylene epoxidation in a multitubular transport reactor. *Chem. Eng. Sci.*, **41**, 143–150.

Perales, J.F., Coll, T., Llop, M.F., Puigjaner, L., Arnaldos, J. and Casal, J. (1991) On the transition from bubbling to fast fluidization regimes, in *Circulating Fluidized Bed Technology III* (eds P. Basu, M. Horio and M. Hasatani), Pergamon Press, Toronto, pp. 73–78.

Reh, L. (1971) Fluid bed processing. *Chem. Eng. Progr.*, **67**, 58–63.

Reh, L. (1986) The circulating fluid bed reactor – a key to efficient gas/solid processing, in *Circulating Fluidized Bed Technology* (ed. P. Basu), Pergamon Press, Oxford, pp. 105–118.

Reh, L. (1995) New and efficient high-temperature processes with circulating fluidized bed reactors. *Chem. Eng. Technol.*, **18**, 75–89.

Rhodes, M.J. (1989) The upward flow of gas/solid suspensions. Part 2: a practical quantitative flow regime diagram for the upward flow of gas/solid suspensions. *Chem. Eng. Res. Des.*, **67**, 30–37.

Rhodes, M.J. and Geldart, D. (1986a) The hydrodynamics of re-circulating fluidized beds, in *Circulating Fluidized Bed Technology* (ed. P. Basu), Pergamon Press, Oxford, pp. 193–200.

Rhodes, M.J. and Geldart, D. (1986b) Transition to turbulence? in *Fluidization V* (eds K. Ostergaard and A. Sorensen), Engineering Foundation, New York, pp. 281–288.

Schnitzlein, M.G. and Weinstein, H. (1988) Flow characterization in high-velocity fluidized beds using pressure fluctuations. *Chem. Eng. Sci.*, **43**, 2605–2614.

Schoenfelder, H., Hinderer, J., Werther, J. and Keil, F. (1994) Methanol to olefins, prediction of the performance of a circulating fluidized bed reactor on the basis of kinetic experiments in a fixed bed reactor. *Chem. Eng. Sci.*, **49**, 5377–5390.

Shingles, T. and McDonald, A.F. (1988) Commercial experience with Synthol CFB reactors, in *Circulating Fluidized Bed Technology II* (eds P. Basu and J.F. Large), Pergamon Press, Toronto, pp. 43–50.

Squires, A.M. (1986) The story of fluid catalytic cracking: The first circulating fluid bed, in *Circulating Fluidized Bed Technology* (ed. P. Basu), Pergamon Press, Oxford, pp. 1–19.

Squires, A.M. (1994) Origins of the fast fluid bed. *Adv. Chem. Eng.*, **20**, 1–37.

Squires, A.M., Kwauk, M. and Avidan, A.A. (1985) Fluid beds: at last, challenging two entrenched practices. *Science*, **230**, 1329–1337.

Stewart, P.S.B. and Davidson, J.F. (1967) Slug flow in fluidized beds. *Powder Technol.*, **1**, 61–80.

Suzuki, S., Kunitomo, K., Hayashi, Y., Egashira, T. and Yamamoto, T. (1990) Iron ore reduction in a circulating fluidized bed. *Proceedings 2nd Asian Conference on Fluidized-Bed and Three-Phase Reactors*, pp. 118–125.

Takeuchi, H., Hirama, L., Chiba, T., Biswas, J. and Leung, L.S. (1986) A quantitative regime diagram for fast fluidization. *Powder Technol.*, **47**, 195–199.

Wainwright, M.S. and Hoffman, T.W. (1974) The oxidation of *o*-xylene in a transported bed reactor. *Chem. Reaction Eng. II, Advances in Chem. Sciences* (ed. H.M. Hulburt), American Chemical Society, Washington DC, pp. 669–685.

Wen, C.Y. and Yu, Y.H. (1966) A generalized method for predicting the minimum fluidization velocity. *AIChE J.*, **12**, 610–612.

Yang, W.C. (1975) A mathematical definition of choking phenomenon and a mathematical model for predicting choking velocity and choking voidage. *AIChE J.*, **21**, 1013–1021.

Yang, W.C. (1983) Criteria for choking in vertical pneumatic conveying lines. *Powder Technol.*, **35**, 143–150.

Yerushalmi, J. (1986) High velocity fluidized beds, Chapter 7 in *Gas Fluidization Technology* (ed. D. Geldart), John Wiley & Sons, Chichester, UK, pp. 155–196.

Yerushalmi, J. and Cankurt, N.T. (1979) Further studies of the regimes of fluidization. *Powder Technol.*, **24**, 187–205.

Yerushalmi, J., Turner, D.H. and Squires, A.M. (1976) The fast fluidized bed. *Ind. Eng. Chem. Process Des. Dev.*, **15**, 47–51.

Zenz, F.A. (1949) Two-phase fluidized-solid flow. *Ind. Eng. Chem.*, **41**, 2801–2806.

2 Hydrodynamics
MASAYUKI HORIO

2.1 Background

Hydrodynamics of circulating fluidized beds deal, on the one hand, with the dynamics of gas–solid suspensions over a certain solid concentration range (voidage: 0.7–0.999) and, on the other, with the hydrodynamic characteristics of particular types of gas–solid contacting devices. From a scientific viewpoint, the clustering nature of dilute suspensions, which was first detected from the large gas–solid slip (i.e. relative) velocity, should be the essential point of interest. From an engineering viewpoint, the major hydrodynamic issues are the effects of such design factors as column diameter, wall shape, gas distributor design, exit structure, solid separation and recycling devices, as well as operating conditions, on the performance of circulating systems. The engineering and scientific aspects are closely interrelated. For instance, the behavior of dilute gas–solid suspensions could not have been found so easily and studied so systematically without the development of circulating devices (i.e. riser and downcomer systems) and the hydrodynamic behavior of gas and solids in a circulating apparatus is not independent of the essential nature of suspensions. However, in the development of our hydrodynamic understanding of circulating fluidized beds, paradigm shifts have occurred between the above two aspects, causing controversy.

If we look back to the historical development of fluidization science, we find definite stepwise development of paradigms, leading to stepwise progress of our knowledge in each decade (Horio *et al.*, 1986a).

During the first stage of fluidization research, initiated in the late 1930s by the development of the first fluid catalytic cracking (FCC) process, researchers sought in vain for the causes of disappointingly low process conversions. Systematic structural examination became possible after the two-phase hypothesis of Toomey and Johnstone (1952) because, without such a flow allotment, even the evaluation of the gas interchange coefficient from experimental conversion was not possible. From a pure hydrodynamic approach, a comprehensive theory was established by Davidson (1961) to explain quantitatively gas bypassing and gas interchange. During the 1960s, most hydrodynamic phenomena were systematically analyzed based on the essential bubble hydrodynamics.

This three-step paradigm, shifting from chaotic to structural and then to essential, appears to be repeating itself again in circulating fluidized bed research. Since the first FCC plant was designed based on the upflow concept, circulating fluidized bed issues had been encountered in the FCC development process (e.g. see Squires, 1982; Squires *et al.*, 1985). Nevertheless, knowledge was only fragmental at that time. It took several decades before the circulating fluidized bed was recognized as unique, differing from the bubbling fluidized bed and pneumatic conveying. While most scientific efforts had been focused on the bubbling bed, some endeavors existed to revive the upflow concept. The circulating fluidized bed roaster (Reh, 1971), which appeared towards the end of the 1960s, was one of them, which then led to CFB combustors (see Chapters 10 and 11). Thus, early in the 1970s the need arose to study the suspension characteristics in circulating fluidized beds. This issue was first raised systematically in 1975 at the International Fluidization Conference by Yerushalmi *et al.* (1976b) together with the new terminology 'fast fluidized bed'.

The hydrodynamic study of circulating fluidized beds thus started from the chaotic stage was, however, accompanied from its beginning by a basic confusion concerning the reactor concepts, which differ very much between combustors and catalytic reactors. In CFB combustors, whose historical appearance and success activated CFB research, the pressure drop limits the height of the lower dense region. On the contrary, in catalytic reactors, one seeks an extended dense region of constant voidage. In CFB combustors the internal solids circulation in the upper dilute region is favorable for heat transfer to membrane walls and to carry the heat generated by combustion of volatiles and char back to the dense region. The external solid circulation flux in combustors is said to be $10-30\,\text{kg/m}^2\text{s}$. In catalytic reactors the internal solids circulation due to the downflow along the wall is not favorable because it increases gas backmixing and decreases yields. In most cases CFB catalysts provide rapid reactions but they have to be regenerated. Accordingly, the external solid circulation flux can be as high as $800\,\text{kg/m}^2\text{s}$.

Macroscopic differences in the design and flow structure of combustors and catalytic reactors are illustrated in Figure 2.1. Most of the pioneers of CFB research (e.g. Squires, Yerushalmi, Kwauk, etc.) were under the strong influence of catalytic reactor developments as reviewed by Squires (1982). Accordingly, early experiments were done in tall columns of a large length to diameter ratio, L/D_t, with smooth recycle of solids into their risers (cf. Yerushalmi *et al.*, 1976b). This tended to focus attention on the nature of the long dense region formed in a vessel of a large L/D_t ratio, which might not be appreciated by people working on combustor designs.

The differences between catalytic reactors and boilers are not only limited to the riser conditions, but a big difference may also exist in the downcomer. In processes of rapid catalytic reactions with catalyst regeneration, regenerators contain most of the catalyst inventory since regeneration is done rather slowly in the bubbling-bed mode. In combustors, the downcomer may have

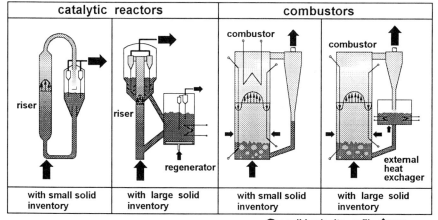

Figure 2.1 Designs and riser flow conditions of catalytic reactors and combustors.

a much smaller volume compared to the riser (combustion chamber) unless it has an external heat exchanger.

The chaotic stage of the CFB hydrodynamic study lasted until the first half of the 1980s. Li and Kwauk (1980) presented the experimental data shown in Figure 2.2, in which is evident the S-shaped axial voidage distribution, with a gradual transition over several meters from the 'dense region' to the 'dilute region' of an 8 m high column (see Figure 2.3(a) for terminology). Previously, the uniqueness of the flow regime 'fast fluidized bed' was not convincing. Li and Kwauk's data were, however, sufficiently impressive to negate almost all objections to the claim of the uniqueness of the fast fluidized bed regime. Yerushalmi *et al.* (1976a) had already presented some data on axial voidage distribution. Even earlier observations were made by Reh (1971), who recognized in Figure 2.3(b)

> the gradient in the solids concentration with height, and the occurrence of gas–solid-flow demixing.... The solids move mostly in strands. The gas–solids demixing seen... is undoubtedly the reason for the high slip velocities observed by Lewis and others (1949) in circulating fluid beds having a fixed inventory of fine particles.

By 1980 all the fragmental findings came together, such as the presence, without choking of the transition, in highly loaded pneumatic transport (Yousfi and Gau, 1974; Yang, 1976), the operating flexibility over a wide voidage range (Yerushalmi and Squires, 1977) and the indistinct boundary between denser region and more dilute region in the axial voidage distribution (Li and Kwauk, 1980). The unique nature of the suspension in circulating fluidized beds became evident. Circulating fluidized bed combustion technology quickly emerged onto the market. It did not take long

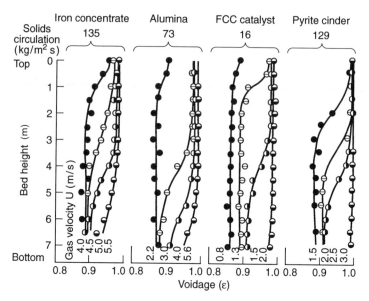

Figure 2.2 Typical S-shaped voidage distributions in a riser ($D_t = 0.09$ m) (Li and Kwauk, 1980).

before the flow structures in circulating fluidized beds were investigated in detail.

Among fluidization researchers there has been an intention to understand the nature of particulate systems, as general as possible, inspired by the analogy between particulate systems and molecular systems, i.e. fixed beds versus solids, fluidized beds versus liquids and entrained beds versus gases (cf. Gelperin and Einstein, 1971). Now, with the fast fluidization regime we have the whole spectrum of gas–particle systems ranging from fixed bed to entrained bed. The flow regimes of circulating fluidization should also be studied in the light of the above analogy. However, the wall, entrance and exit effects in realistic circulating fluidized bed systems often make it difficult to separate the suspension nature from these boundary effects. For instance, although the turbulent fluidization regime (cf. Section 2.2) has been claimed as a unique regime, the distinction between bubbling and turbulent regimes is not yet clear.

Previous understanding of flow structures also needs further review. For instance, the typical S-shaped axial voidage distribution, obtained from most fundamental experiments done with large L/D_t columns, never exists in atmospheric circulating fluidized bed combustors. More precise information is needed concerning the bottom regions for both CFB combustors and catalytic reactors.

Nevertheless, the decade of structural investigation seems to be ending. Treating several issues of controversy as open subjects for future research

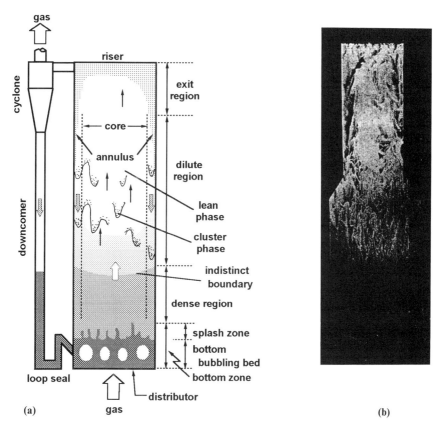

Figure 2.3 Illustration of a typical flow structure in circulating fluidized beds. (a) Major regions and phases in a CFB. (b) Flow visualization with a two-dimensional fluidized bed (Reh, 1971).

and stressing the more essential way of thinking, this chapter addresses the key hydrodynamic factors, including governing equations, different modes of suspension dynamics, meso-scale and micro-scale suspension structures, macroscopic suspension structures in an apparatus, criteria for phase transition, scaling relationships and design methodologies.

2.2 Fundamental nature and meso-scale flow structure of gas–solid suspensions

2.2.1 Governing equations

Anderson and Jackson (1967) presented a rigorous derivation of the two-fluid model for solid suspensions through local averaging of the point equation of motion for gas and the equation of motion for a single solid particle. Pritchett *et al.* (1978) first applied similar two-fluid equations to direct numerical simulation of bubbling fluidized beds. The first numerical simulation for circulating fluidized bed conditions was presented by Tsuo and Gidaspow (1990).

With the progress of computers it became possible to treat individual particle motion, e.g. using the discrete element method (DEM) and a direct simulation Monte Carlo method (DSMC method) (Tanaka et al., 1993). Discussions on suspension models are dealt with in detail in Chapter 5. The present chapter starts with the Anderson and Jackson two-fluid model, given by:

equation of continuity for gas:

$$\frac{\partial \epsilon}{\partial t} + \nabla \cdot (\epsilon u) = 0 \qquad (2.1)$$

equation of motion for gas:

$$\rho_f \epsilon \left[\frac{\partial u}{\partial t} + (u \cdot \nabla)u \right] = \epsilon \rho_f \vec{g} - \epsilon \nabla p - R \qquad (2.2)$$

equation of continuity for solid particles:

$$\frac{\partial (1-\epsilon)}{\partial t} + \nabla \cdot [(1-\epsilon)v] = 0 \qquad (2.3)$$

equation of motion for solids:

$$\rho_p (1-\epsilon) \left[\frac{\partial v}{\partial t} + (v \cdot \nabla)v \right] = (1-\epsilon)\rho_p \vec{g} - (1-\epsilon)\nabla p + R + \nabla \cdot P_s \qquad (2.4)$$

where R denotes

$$R = \beta(u-v) + (1-\epsilon)M\rho_f \left[\frac{D(u-v)}{Dt} \right] \qquad (2.5)$$

Applying equations (2.2) and (2.4) and the Richardson and Zaki (1954) relationship $u_z - v_z = u_T \epsilon^{n-1}$ for homogeneous dilute suspensions, we obtain:

$$\beta = (\rho_p - \rho_f)g(1-\epsilon)/u_T \epsilon^{n-2} \qquad (\epsilon_{mf} \ll \epsilon \le 1) \qquad (2.6a)$$

For homogeneous dense suspensions, substitution of the Ergun (1952) correlation for ∇p into equation (2.2) leads to:

$$\beta = \frac{1-\epsilon}{\phi_s d_p \epsilon} \left[\frac{150(1-\epsilon)\mu}{\phi_s d_p} + 1.75 \rho_f \epsilon |u-v| \right] \qquad (\epsilon \ll 1) \qquad (2.6b)$$

Furthermore, in the same manner as the derivation of equation (2.6a), we obtain the following expression for incipiently fluidized suspensions:

$$\beta = (\rho_p - \rho_f)g(1-\epsilon_{mf})\epsilon_{mf}^2 / U_{mf} \qquad (\epsilon \simeq \epsilon_{mf}) \qquad (2.6c)$$

2.2.2 Aggregating tendency of suspensions

(a) Theoretical and numerical analyses

Homogeneous suspensions are almost always unstable. To quantify the aggregating tendency of stationary dense suspensions, Jackson (1963) examined the local stability of a steady-state flow based on equations (2.1) to (2.4) with the perturbation method. Although the perturbation analysis for a non-linear system only gives information about the vicinity of a steady state, it was found that dense suspensions are unstable so that the amplitude of a density wave always grows with height. For an entrained homogeneous suspension introduced continuously into a column, Grace and Tuot (1979) performed a similar perturbation analysis relaxing the condition of zero superficial velocity of the solid phase. They found that even in dilute suspensions, non-homogeneity in voidage grows, so that the real part of s_1 in the following solution of voidage perturbation was positive for physically possible cases:

$$\epsilon = f(t)\exp[j(k_1 x + k_2 y + k_3 z)] \qquad (2.7)$$

$$f(t) = \exp(s_1 t) \text{ or } \exp(s_2 t) \qquad (2.8)$$

where s_1 and s_2 are the roots of the characteristic equation.

To add some physical understanding of the instability, Jackson (1963) defined the growth distance (the vertical distance in which the amplitude of voidage wave grows by a factor of 2.718) as:

$$\Delta x = -\frac{\text{Im}(s_1)}{|k|\,\text{Re}(s_1)} \qquad (2.9)$$

Here, a larger growth distance indicates a less aggregating suspension. Figure 2.4 shows the growth distance for suspensions of interest. It is evident that, as for bubbling fluidized beds, the dilute suspensions tend to be more homogeneous for larger fluid-to-solid density ratios.

The origin of the instability and the aggregating tendency is related to the nature of particle–particle and fluid–particle interactions. It is widely known that either an attractive or a repulsive force exists between particles, depending on the direction of their alignment. A repulsive force is created between two particles aligned perpendicular to the flow direction and an attractive force between those aligned parallel with the flow (cf. Quadflieg, 1977; Moriya et al., 1983 for two-dimensional cases). Accordingly, two sedimenting particles tend to vertically align. However, as demonstrated by Fortes et al. (1987) and Joseph et al. (1991), the alignment is not stable and particles collide with kissing and tumbling motions (see Figure 2.5). When we have numerous particles, such particle-to-particle and particle-to-fluid interactions make a suspension behave like a continuum. The collision between particles mostly accompanies energy dissipation due to the presence of fluid friction, particle–particle friction or cohesion, plastic

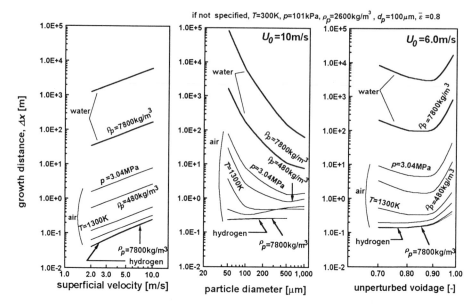

Figure 2.4 Growth distances for different particles and operating conditions (data are taken from Grace and Tuot, 1979).

deformation of solids and attrition, which results in non-elastic collisions. The particle–particle attractive force and non-elastic collisions give particles a definite tendency to gather. This kind of particle gathering can be called 'clustering', in contrast with 'agglomeration' in which particle gathering is maintained by surface forces (Horio and Clift, 1992).

The hydrodynamic attractive and repulsive forces should exist not only between individual particles but also between their assemblies, i.e. structures of higher orders, such as clusters. Furthermore, the buoyancy force acting on

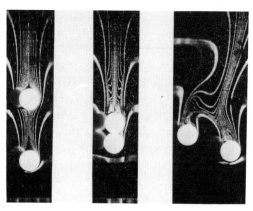

Figure 2.5 Interaction of two sedimenting spheres showing drafting, kissing and tumbling (sphere diameter: 11.113 mm, $Re_T = 8.5$, Joseph et al., 1991).

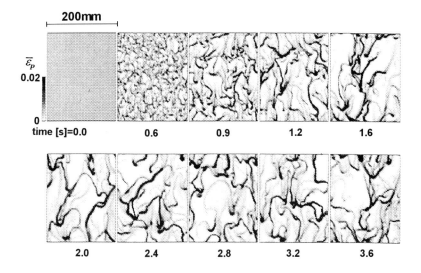

Figure 2.6 DSMC simulation showing cluster formation from a homogeneous suspension ($U_0 = 0.6$ m/s, $d_p = 61.3\,\mu$m, $\epsilon_p = 0.004$, Yonemura et al., 1995).

a gas-rich region, together with the gravity force acting on a solid-rich region, causes relative motion and shear between clusters and gas-rich regions. These forces are supposed to control the development of particle assemblies and to form a steady but turbulent meso-scale suspension structure with a certain mean length scale.

Figure 2.6 shows a result of a computer simulation using the DSMC method for a two-dimensional suspension of solids in gas. Due to the limitation of computer capacity, the simulation is restricted to a small region where the boundary conditions are periodic. Even for these restricted conditions, it can be clearly seen that the initially homogeneous suspension evolves into a complicated cluster structure.

(b) Experimental information

The real suspension behavior can be investigated with point probes, mostly based on optical fiber techniques or needle-type capacitance probes. To obtain cluster velocity and pierced cluster length simultaneously, two-point optical fiber probes can be applied.

Figure 2.7 shows measurements of light reflection intensity corresponding to voidage fluctuations. Figure 2.8 shows the radial distribution of vertically pierced cluster lengths (c) in different levels of a CFB column (see (a) and (b)). It can be seen in Figure 2.8(c) that the length of clusters over the wall increases as they fall, while the cluster length remains rather constant in the core region.

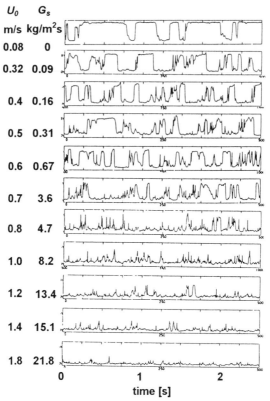

Figure 2.7 Change in the condition of fluidized FCC bed with superficial gas velocity measured by the light reflection from the bed ($d_p = 60\,\mu$m, $\rho_p = 1000\,\text{kg/m}^3$, probe height $= 0.61$ m, column: see Figure 2.8) (Horio and Morishita, 1988).

Takeuchi *et al.* (1986a), Takeuchi and Hirama (1991), Li *et al.* (1991a) and Hatano and Kido (1991) applied an optical fiber scope system. However, it appears that the view area of the image scope was too small to obtain images larger than the length scale of the meso-scale suspension structure. Visual observation has been attempted with two-dimensional circulating fluidized beds (Arena *et al.*, 1989; Bai *et al.*, 1991).

Much wider images of suspension behavior can be obtained using the laser sheet technique, as shown in Figures 2.9 and 2.10 for a very dilute system and for a 'realistic' circulating fluidized bed, respectively. Some interesting details shown in these figures are the round-nosed clusters (e.g. cluster A in Figures 2.9 and 2.10), the film-like cluster tails shed upward (e.g. B in Figure 2.9) and the gas pockets or channels (e.g. C in Figure 2.9). These observations are similar to those available from two-dimensional experiments and computer simulations. However, in the three-dimensional systems quite a rapid horizontal motion exists, as can be inferred from Figure 2.11.

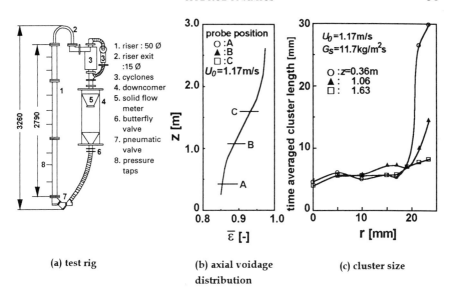

(a) test rig (b) axial voidage distribution (c) cluster size

Figure 2.8 Radial distribution of cluster size (Horio et al., 1988).

There have been several attempts to observe and analyze cluster behavior and to obtain information on the slip velocity versus drag force relationship and gas–solid contact mode based on simplified cluster concepts. Wirth (1991) developed a drag expression for a strand. Mueller and Reh (1994) measured the drag force on a columnar strand. Marzocchella (1991) observed the transient behavior of a circular cluster placed in a two-dimensional gas stream. Bobkov and Gupalo (1986) derived gas and solid streamlines for

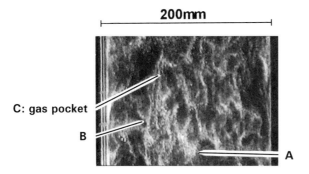

Figure 2.9 Vertical laser sheet image of dilute suspension in a circular riser showing clusters (brighter part) and gas pockets (FCC, $d_p = 61.3\,\mu m$, $\rho_p = 1780\,\text{kg/m}^3$, $U_0 = 0.67\,\text{m/s}$, $G_s = 0.018\,\text{kg/m}^2\text{s}$, $z = 780\,\text{mm}$, shutter speed: 1/1000 s, column: CFBA in Figure 2.35) (Tsukada, 1995).

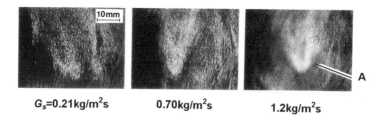

$G_s=0.21$ kg/m²s   0.70 kg/m²s   1.2 kg/m²s

Figure 2.10 Cluster images taken by laser sheet technique with internal picturing for high solid mass flux operation (FCC, $d_p = 61.3\,\mu\text{m}$, $\rho_p = 1780\,\text{kg/m}^3$, $U_0 = 0.67\,\text{m/s}$, $G_s = 0.21$–$1.2\,\text{kg/m}^2\text{s}$, $z = 1150\,\text{mm}$, shutter speed: 1/250 s, column: CFBA in Figure 2.35) (Kuroki and Horio, 1994).

a spherical cluster. However, further investigation is necessary for more realistic evaluation.

2.2.3 Meso-scale suspension structure in each flow regime

The major fluidization regimes so far claimed to exist are homogeneous, bubbling, turbulent and fast fluidization (Figure 2.12). Here, slugging is omitted because it is a limiting case of bubbling fluidization with a dominant effect of the column wall. The regime map provides information concerning the denser section of a suspension vertically distributed in a vessel or a column.

As discussed in Chapter 1, correlations of the following form have been proposed for transition velocities U_c, U_k and U_{tr} (Horio, 1986; Hashimoto et al., 1988; Bi and Fan, 1992; and Bi et al., 1993):

$$Re_i \equiv \frac{d_p \rho_g U_i}{\mu} = c_i Ar_i^{n_i} \qquad (i = c, k, tr) \qquad (2.10)$$

Values of c_i and n_i from recent works are listed in Table 2.1.

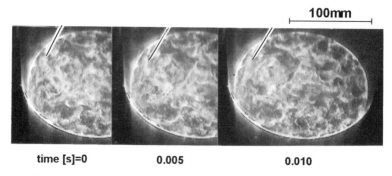

time [s]=0   0.005   0.010

Figure 2.11 Horizontal laser sheet image of dilute suspension in a circular riser (FCC, $d_p = 61.3\,\mu\text{m}$, $\rho_p = 1780\,\text{kg/m}^3$, $U_0 = 0.74\,\text{m/s}$, $G_s = 0.019\,\text{kg/m}^2\text{s}$, $z = 740\,\text{mm}$, shutter speed: 1/1000 s, column: CFBA in Figure 2.35) (Tsukada, 1995).

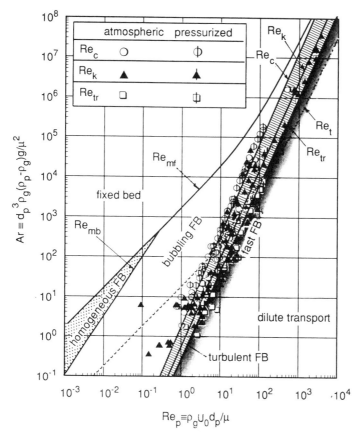

Figure 2.12 Flow regime map for fluidized suspensions (Adánez *et al.*, 1993; Avidan and Yerushalmi, 1982; Bi *et al.*, 1991; Cai *et al.*, 1989; Canada and McLaughlin, 1978; Canada *et al.*, 1978; Carotenuto *et al.*, 1974; Chen *et al.*, 1980; Geldart and Rhodes, 1986; Han *et al.*, 1985; Hashimoto, 1989; Hashimoto *et al.*, 1988; Hirama *et al.*, 1992; Horio *et al.*, 1992a; Ishii and Murakami, 1991; Jiang and Fan, 1991; Kehoe and Davidson, 1971; Kwauk *et al.*, 1986; Lanneau, 1960; Lee and Kim, 1982; Le Palud and Zenz, 1989; Leu *et al.*, 1990; Li and Kwauk, 1980; Li *et al.*, 1988; Massimilla, 1973; Perales *et al.*, 1991; Satija and Fan, 1985; Shin *et al.*, 1984; Son *et al.*, 1988; Thiel and Potter, 1977; Tsukada *et al.*, 1993, 1994; Yang and Chitester, 1988; Yang *et al.*, 1990; Yerushalmi and Cankurt, 1979).

In all the correlations n_i is close to 0.5, which is equivalent to:

$$Fr_i \equiv \frac{U_i^2}{gd_p} = \frac{\rho_g}{\rho_p - \rho_g} \frac{Ar}{Re_i^2} = \frac{c_i^2 \rho_g}{\rho_p - \rho_g} \quad (2.11)$$

It is therefore evident that the gas–solid slip condition in the turbulent and fast regimes is affected little by the gas viscosity but rather by the balance between inertia and gravity forces. The fact that the transition criteria contain little viscosity effect, even for fine particles whose terminal velocity

Table 2.1 $Re_i = c_i Ar^{n_i}$ type correlations for regime transitions

Author	c_i	n_i	Ar range	Comment
Onset of bubbling to turbulent fluidization ($i = c$)				
Horio (1986)	0.936	0.472		
Lee and Kim (1988)	0.700	0.485		
Tsukada (1995)*	0.791	0.435		Re_c indicated by line in Fig. 2.12
Offset of bubbling to turbulent fluidization ($i = k$)				
Horio (1986)	1.41	0.56	$<10^4$	
	1.46	0.472	$>10^4$	
Bi and Fan (1992)	0.601	0.695	<125	
	2.28	0.419	>125	
Bi and Fan (1992)	16.31Φ	0.136	<125	$\Phi = (U_T/\sqrt{gD_t})^{0.941}$
D_t effect considered	2.274Φ	0.419	>125	$\Phi = (U_T/\sqrt{gD_t})^{0.0015}$
Tsukada (1995)*	1.31	0.450		Re_k indicated by line in Fig. 2.12
Onset of fast fluidization ($i = tr$)				
Lee and Kim (1990)	2.916	0.354		
Perales et al. (1991)	1.415	0.483		
Bi and Fan (1992)	2.28	0.419		
Adánez et al. (1993)	2.078	0.463		
Tsukada (1995)*	1.806	0.458		Re_{tr} indicated by line in Fig. 2.12

* Determined from all data of Re_c, Re_k or Re_{tr} in Figure 2.12 except for Kehoe and Davidson (1971), and Thiel and Potter (1977).

is in the Stokes range, implies that clusters are sufficiently large to be in the inertia dominant region.

From equation (2.10) these transition velocities should be proportional to p^{n_i-1}. The experimental results deviate slightly from this. The effect can be expressed by the following correlation (Tsukada et al., 1993, 1994):

$$U_i/U_{i0} = (p/p_0)^{-0.3} \quad (i = c, k, tr) \tag{2.12}$$

Accordingly, with increasing pressure, operation in the circulating fluidized bed mode becomes possible at lower gas velocities.

With respect to the flow structure, each regime is supposed to correspond to a unique meso-scale suspension structure. Here, the wall effects as well as the entrance effects are assumed to be negligible and the dense region to be sufficiently high that the suspension structure is well developed. In regime mapping experiments there are at least three possible situations for the denser section, as illustrated in Figure 2.13, which are analogous to the three situations in molecular systems (A′–C′). First, no denser section may exist if there are no solids in the column or if solids supply is insufficient. Second, the denser section may coexist with a dilute section above it in the column. This stratification is of course due to gravity. Third, the whole column can be occupied by the dense suspension.

The onset velocities for turbulent and fast fluidization have been introduced based on different aggregation indices of suspensions, i.e. (1) pressure fluctuation for the bubbling-to-turbulent regime transition and (2) for the

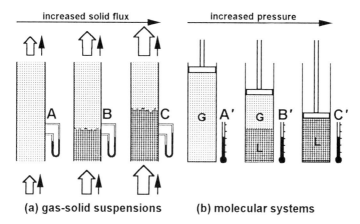

Figure 2.13 Vertical stratification and possible coexisting situation of gas–solid suspensions in comparison with molecular systems.

turbulent-to-fast regime transition the abruptness of density change caused by changes in gas velocity or solid circulation rate. In some measurements the reduction of pressure fluctuation takes place with increasing gas velocity merely because of the entrainment loss of bed materials in the riser (Geldart and Rhodes, 1986). As shown in Figure 2.14 the typical pressure fluctuation decrease for $U_0 > U_c$ takes place even when the probe is located carefully in the lower dense region.

The turbulent-to-fast phase change detected by a sensor at a certain height can also be related to the solids distribution in the column. The abrupt transition between dense and dilute phases corresponds to a sharp dense/dilute boundary in the axial voidage distribution, and the gradual transition above U_{tr} to an indistinct boundary between the phases. Figure 2.15 is an example of the $|dp/dz|, G_s, U_0$ phase diagram, where $|dp/dz|$ is the axial pressure gradient measured at a lower section of a riser, which is equivalent to the cross-sectional average bed weight per unit volume, $\rho_p g(1-\epsilon)$ (Arena et al., 1986), and G_s is the cross-sectional average solid circulation flux. Figure 2.15 shows that when solid circulation flux is small, $|dp/dz|$ remains small, indicating that the suspension in the measured section is very dilute. At a certain value of G_s, $|dp/dz|$ takes off from its low values and starts increasing to much higher ones, which correspond to dense suspensions. Above a critical gas velocity, named the transport velocity, U_{tr}, by Yerushalmi and Cankurt (1979), the switching takes place gradually over a wide range of G_s. This corresponds to the operation named 'fast fluidization' in which we have a dense region called 'fast fluidized bed' in its narrow sense.

The axial voidage profiles in Figure 2.2 are typical of fast fluidization. As discussed in section 2.3.2, the volume fraction of solids, as well as the upward flux of solids entrained in the dilute region, decreases exponentially with

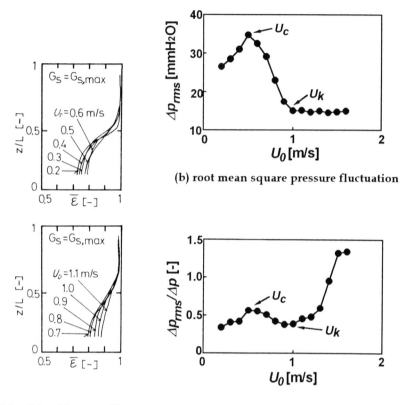

(a) axial voidage profiles

(b) root mean square pressure fluctuation

(c) pressure fluctuation relative to the static pressure difference in the dense part between the measuring sections

Figure 2.14 Pressure fluctuation for detecting the transition from bubbling regime to turbulent regime ($d_p = 61.3 \,\mu$m, $\rho_p = 1780 \,$kg/m^3, column: CFBA in Figure 2.35) (Ishii and Horio, 1991).

height. It has been said that above a certain height called the TDH (transport disengaging height) the solid volume fraction as well as the solid mass flux becomes constant. The limiting mass flux has been called 'saturation carrying capacity'. The denser and dilute regions exist for solid circulation flux between the saturation carrying capacity and the maximum upward solid flux at the surface of the denser region.

Modifying the Zenz and Weil (1958) expression to cover both upper dilute and lower dense regions, Li and Kwauk (1980) proposed the following expression for axial voidage profiles:

$$\frac{\bar{\epsilon}_p - \bar{\epsilon}_{p\infty}}{\bar{\epsilon}_{pd} - \bar{\epsilon}_p} = \exp(-a(z - z_i)) \qquad (2.13)$$

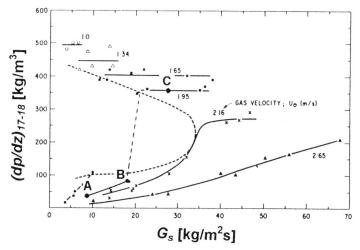

Figure 2.15 Pressure gradient – G_s–U_0 phase diagram (particles: HFZ – 20, $D_t = 0.152$ m) (Yerushalmi and Cankurt, 1979).

In this expression the second derivative of $\bar{\epsilon}_p$ with respect to z equals zero at $z = z_i$. They, therefore, called height z_i the inflection height.

Hirama *et al.* (1992) showed that even when the downcomer was disconnected and a forced solids feeding system was attached to the riser, axial solids distributions similar to those in circulating fluidized beds can be obtained, as shown in Figure 2.16.

In the early definitions of fast fluidized beds, the coexistence of dense and dilute regions was thought to be a necessary condition (Li and Kwauk, 1980; Takeuchi *et al.*, 1986b). However, as discussed above, the definition of a regime should be relevant to the dense region in the lower part, if two regions coexist, so that we can provide a basis for investigation of the characteristics of each suspension. The appropriate definition of 'fast fluidized bed' can be, therefore, 'the dense region which typically appears over the range of superficial gas velocity above the transport velocity U_{tr}' (Horio, 1991). Both Kwauk and Takeuchi later agreed that coexistence is not necessary for fast fluidization (Kwauk *et al.*, 1986; Hirama *et al.*, 1992).

Taking the Hirama *et al.* (1992) data, the voidage in the lower dense region $\bar{\epsilon}_{pd}$ is shown on the $U_0 - G_s$ plane in Figure 2.17. As in the case of supercritical fluid, the density changes gradually with U_0 and G_s. Moreover, with increasing $\bar{\epsilon}_{pd}$ the contour lines become parallel to the G_s axis. The bubbling-to-turbulent and turbulent-to-fast transitions take place at $\bar{\epsilon}_{pd} \simeq 0.25$ and $\bar{\epsilon}_{pd} \simeq 0.1$, respectively. This implies that the boundary between fast fluidized and bubbling regimes should also be almost parallel with the G_s axis, i.e. G_s does not enter into the criteria for the above regime transitions, in agreement with the early prediction of Matsen (1982). For $\bar{\epsilon}_{pd}$ Bai and Kato (1994)

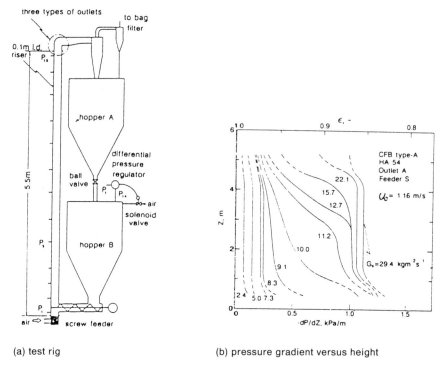

(a) test rig   (b) pressure gradient versus height

Figure 2.16 Axial pressure gradient (i.e. solid hold up) distribution in a special riser with forced solids feeding and without solids recycle ($D_t = 0.1$ m, particles: high alumina, $d_p = 54\,\mu$m) (Hirama et al., 1992).

proposed the following correlations for $\bar{\epsilon}_{pd}$ and $\bar{\epsilon}_{p\infty}$ ($U_0 = 0.8$–9 m/s, $G_s = 4$–220 kg/m²s, $d_p = 49$–280 μm, $\rho_p = 706$–4510 kg/m³):

$$\bar{\epsilon}_{pd} = \frac{v_0}{U_0 - u_T}\left[1 + 0.103\left(\frac{U_0}{v_0}\right)^{1.13}\left(\frac{\rho_p - \rho_g}{\rho_g}\right)^{-0.013}\right] \quad (2.14)$$

$$\bar{\epsilon}_{p\infty} = \frac{v_0}{U_0 - u_T}\left[1 + 0.208\left(\frac{U_0}{v_0}\right)^{0.5}\left(\frac{\rho_p - \rho_g}{\rho_g}\right)^{-0.082}\right] \quad (2.15)$$

Here, to avoid semantic confusion between turbulent and fast regimes it is necessary to examine the uniqueness of each fluidization regime based on more detailed measurements of flow structures. Figure 2.18 shows the signals obtained with an optical fiber probe from different locations in the column with careful determination of flow regimes based on the Yerushalmi and Cankurt (1979) criteria for both turbulent and fast fluidization (Ishii and Horio, 1991). Comparison of signals from the dense regions of fast and turbulent fluidized beds clearly shows that in turbulent fluidized beds bubbles or gas slugs are often observed while in fast fluidized beds they are not.

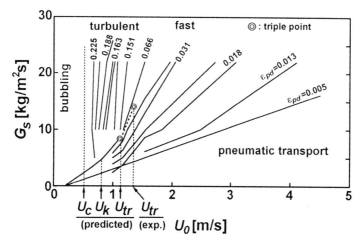

Figure 2.17 G_s–U_0–ϵ_{pd} (particle volume fraction in the lower dense region) phase diagram drawn from the data of Figure 2.16.

To diagnose the slip conditions both in turbulent and fast fluidizations, let us estimate the minimum fluidization velocity U_{mf} and the terminal velocity u_T from measured cluster size and density. The following correlations can be used for the inertia dominant region, so that we can compare the suspension in a turbulent fluidized bed with that in a fast fluidized bed.

$$U_{mf,cl} = [g(\rho_{cl} - \rho_{lean})d_{cl} \cdot \epsilon^{*3}/1.75 \cdot \rho_{lean}]^{1/2} \quad (2.16)$$

$$u_{T,cl} = [3g(\rho_{cl} - \rho_{lean})d_{cl}/\rho_{lean}]^{1/2} \quad (2.17)$$

where the densities of cluster and lean phases, ρ_{cl} and ρ_{lean}, respectively, can be related to voidages by:

$$\rho_i = (1 - \epsilon_i)\rho_p + \epsilon_i \rho_g \quad (i = cl, lean) \quad (2.18)$$

Figure 2.19 shows a sample calculation. Since the intersections of the $U_{mf,cl}$ curves with the diagonal for both FCC and silica sand are close to the corresponding transport velocities, the turbulent and fast regimes are essentially different because in a turbulent fluidized bed clusters in the dense phase are still too large and too heavy to be suspended in the gas stream, while in a fast fluidized bed they are sufficiently light to be macroscopically fluidized by the gas.

The turbulent regime may be understood to have a transition structure that can be included in bubbling beds. Werther and Wein (1994) presented an elaborate set of data for comparing the two regimes for ash ($d_{p50} \simeq 200$ μm) and sand ($d_{p50} \simeq 400$ μm). Some of their results are shown in Figure 2.20. There it is evident that (1) the axial voidage profile around the bed surface, (2) the fraction of dense phase and (3) the radial voidage profiles of the turbulent fluidized bed are clearly different from those of the bubbling fluidized bed.

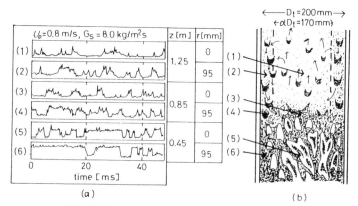

(a) bubbling - turbulent intermediate fluidization

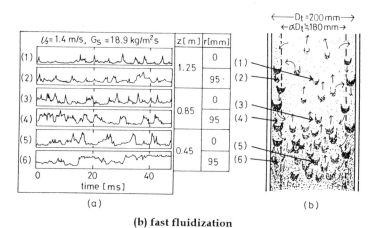

(b) fast fluidization

Figure 2.18 Conceptual sketches of flow structures of turbulent and fast regimes based on light reflection signals from different locations of the bed ($D_t = 0.2$ m, particles: FCC, $d_p = 61.3\,\mu$m, $\rho_p = 1780$ kg/m^3, column: CFBA in Figure 2.35) (Ishii and Horio, 1991).

They further showed that the structure of a turbulent fluidized bed can be well described using conventional bubbling bed concepts by:

$$\bar{\epsilon}_b = \dot{V}_b/u_b \qquad (2.19)$$

$$\dot{V}_b = \varphi(U_0 - U_{mf}) \quad (\varphi = 1.45 Ar^{-0.18}, 10^2 < Ar < 10^4) \qquad (2.20)$$

$$u_b = \dot{V}_b + \gamma\sqrt{gD_b} \qquad (2.21)$$

$$\gamma/0.71 = \begin{cases} 0.63 & (D_t < 0.1\,\text{m}) \\ 2.0\sqrt{D_t} & (0.1\,\text{m} < D_t \le 1.0\,\text{m}) \\ 2.0 & (1.0\,\text{m} < D_t) \end{cases} \qquad (2.22)$$

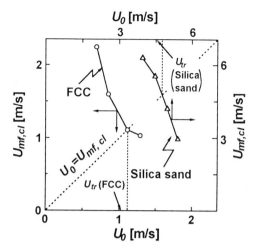

Figure 2.19 Comparison of transport velocity U_{tr} with $U_{mf,cl}$ calculated by equation (2.16) from observed cluster sizes ($D_t = 0.2$ m, particles: FCC, $d_p = 61.3\,\mu$m, $\rho_p = 1780$ kg/m^3 and silica sand, $d_p = 106\,\mu$m, $\rho_p = 2600$ kg/m^3) (Horio et al., 1992a).

Even for gas velocities corresponding to a high velocity fluidization regime, there exists a possibility of transition to denser regimes if the solid circulation rate is high enough. Nevertheless, the coexistence of two well-developed dense zones of different regimes has not yet been reported.

When the ratio of the dense region height to the equivalent vessel diameter is small as in large-scale CFB boilers, the bottom zone can dominate the whole dense region. As pointed out by Johnsson et al. (1992), the dense

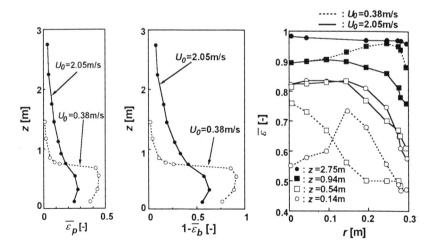

Figure 2.20 Comparison of flow structure of turbulent regime with that of bubbling regime (Werther and Wein, 1994).

42   CIRCULATING FLUIDIZED BEDS

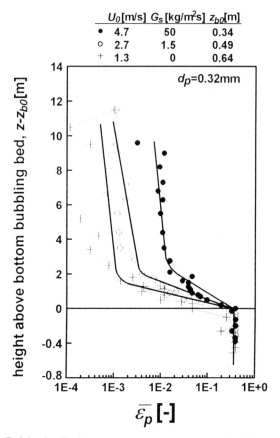

Figure 2.21 Bed density distribution in the bottom region of a 12 MWth circulating fluidized bed boiler (Johnsson and Leckner, 1995).

region of CFB boilers can be characterized as a bubbling bed of an exploding nature since the initial bubble diameter already equals or exceeds the dense region height. Figure 2.21 is an example of such situations, where the bed voidage is about 0.6 and a clear boundary exists between the bottom zone and the dilute region. In this respect, the flow regime classification is for a general understanding of circulating fluidized beds that may not be directly applied to the characterization of the dense region in CFB boilers.

2.3 Pressure profiles and macroscopic flow structure

2.3.1 Boundary effects and macroscopic flow structure

The major features of macroscopic flow structure in ordinary circulating fluidized beds are, (1) dense/dilute coexistence in the axial voidage profile,

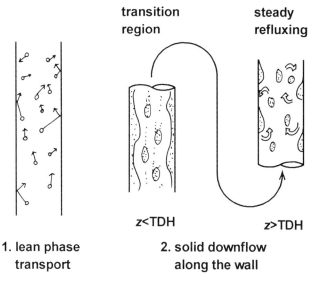

Figure 2.22 The role of wall: solid holding and refluxing.

(2) core-annulus flow and solid downflow along the wall, (3) solid and gas maldistribution and solid acceleration in the bottom zone, (4) variable solid refluxing at the column top depending on the exit design and (5) gas and solid flows between the riser and downcomer depending on the total solids inventory, downcomer aeration and the valve design.

The effect of the vertical wall on the structure of riser flow is illustrated in Figure 2.22. The boundary layer along the wall has the capability to hold more particles than the central part. This is because of the solid downflow formation due to the low gas velocity in the boundary layer and, further, because of the particle capturing capability of the solid downflow over the wall. In lean phase transport, particles may rebound back from the wall. In the region of axially constant suspension density, the intermittent refluxing of the dense phase from the wall keeps their macroscopic density profiles in equilibrium.

In large-scale CFB combustors of small L/D_t ratios, the dilute region height can be well below the TDH. However, in catalytic reactors of large L/D_t ratios, radial profiles can be well developed in the major part of their risers. Early measurements for catalytic risers were reported by van Breugel et al. (1969) and Saxton and Worley (1970), although their experimental conditions may not represent commercial scale FCC risers (see Zenz, 1994).

Figure 2.23 shows some examples of recent measurements on radial voidage distributions as well as radial distributions of vertical solid velocity for columns of various sizes and shapes. These data are presented together with the axial voidage distribution data. This is important because the

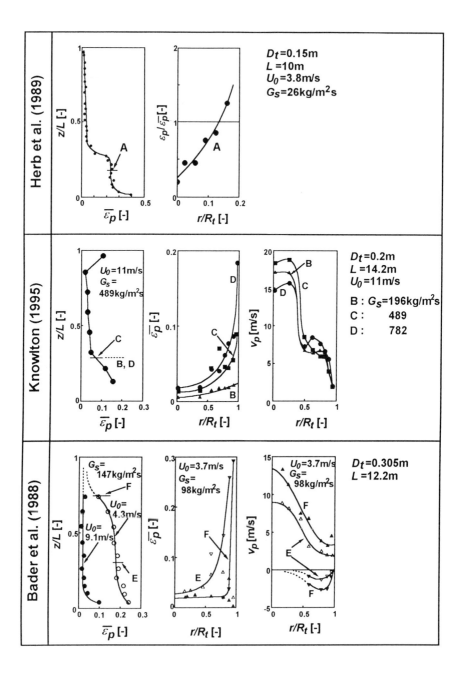

Figure 2.23 Comparison of radial structure of suspensions in circulating fluidized beds.

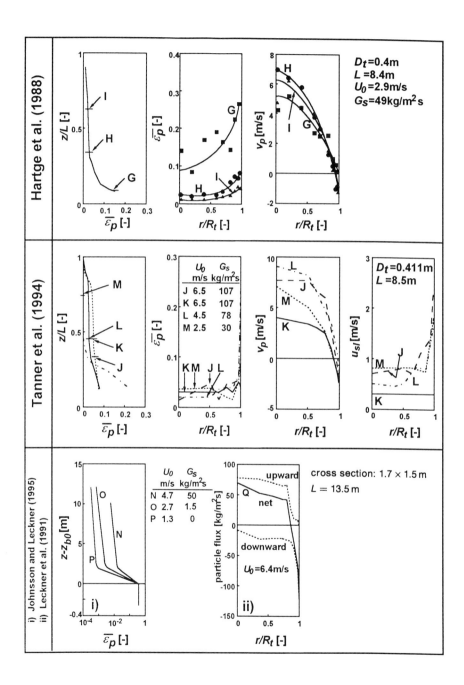

Figure 2.23 Continued.

dense region height and, accordingly, the radial voidage and velocity distributions can vary depending on column design and operating conditions. From Figure 2.23 the formation of the downflow over the wall (see H–Q) and the intermittent upflow and downflow (E, F, Q) can be confirmed. For convenience, the time-averaged solid downflowing area is called the annulus and the rest is the core. In this chapter these terms are used for risers of either circular or rectangular cross-section.

Measurements on the core-annulus structure have been done mostly for the upper dilute region, presumably for experimental convenience. Based on an optical fiber probe technique, Ishii et al. (1989) and Tanner et al. (1994) (see Figure 2.23 (J, K, L)) reported that core-annulus structures exist not only in the upper dilute region but also in the lower dense region in the fast fluidization. However, some data (see Figure 2.23(B–G)) show the absence of downflow in the time-averaged sense. No clear downflow was found in the lower dense region of a turbulent fluidization mode (Ishii and Horio, 1991). However, the wall effect exists even when there is no net downflow. In this respect, it may be worth defining the boundary layer thickness in addition to the annulus thickness to treat the wall effects more quantitatively.

It has been reported that the time-averaged radial profiles of both local voidage ϵ and local solids mass flux G'_s have similar shapes. Although this may not be valid for very dilute cases ($\bar{\epsilon}_p < 0.005$) and for cases of small L/D_t, the following correlations for ϵ are available in the literature:

$$\epsilon = \bar{\epsilon}^{(0.191 + (r/R_t)^{2.5} + 3(r/R_t)^{11})}$$

$$(D_t = 0.03\text{–}0.3\,\text{m}, d_p = 34\text{–}75\,\mu\text{m}, \rho_p = 607\text{–}2003\,\text{kg/m}^3)$$

(Zhang et al., 1991)   (2.23a)

$$\epsilon_p/\bar{\epsilon}_p = 1 - \beta'/2 + \beta'(r/R_t)^2 \quad (\beta' = 1.3\text{–}1.9)$$

$$(\rho_p = 2456\,\text{kg/m}^3, d_p = 75\,\mu\text{m}, D_t = 0.15\text{–}0.30\,\text{m}, U_0 = 3\text{–}5\,\text{m/s}$$

$$\text{and } G_s = 2\text{–}111\,\text{kg/m}^2\text{s})$$ (Rhodes et al., 1992)   (2.23b)

$$(\bar{\epsilon}^{0.4} - \epsilon)/(\bar{\epsilon}^{0.4} - \bar{\epsilon}) = 4(r/R_t)^6$$

(covering both Zhang et al., 1991 and Rhodes et al., 1992 conditions)

(Patience and Chaouki, 1995)   (2.23c)

For time-averaged local solid velocity, Patience and Chaouki (1995) proposed:

$$v/v_{r=0} = 1 - (r/\alpha R_t)^2 \quad (2.24)$$

However, as mentioned by Horio and Ito (1996), the parameters of the above correlations need more careful review to make them satisfy the material balance.

For the dimensionless downflow thickness $1 - \alpha$, Werther (1993) proposed the following correlation, which was further confirmed by data from large-scale boilers, i.e. Duisburg plant ($D_t = 8$ m, Werther, 1993) and EDF E. Huchet plant ($D_t = 9.65$ m, 125 MWe, Lafanechere and Jestin, 1995):

$$1 - \alpha = 1.1(U_0 D_t \rho_f / \mu)^{-0.22}(L/D_t)^{0.21}(1 - z/L)^{0.73} \qquad (2.25)$$

However, the above correlation does not contain the effect of G_s on $1 - \alpha$ (Bi et al., 1996). More data are needed for high G_s conditions.

In real risers the radial profiles cannot be completely symmetric due to the effects of non-symmetric solids recycling and exit design (Rhodes et al., 1989; Zhou et al., 1994). The non-symmetric effect may affect the horizontal swirl motion of solids. Wang and Gibbs (1991) and Brereton and Grace (1994) reported that the secondary air injection and swirl increase solids hold-up in the column. Leckner and Andersson (1992) showed the thickening of the solid downflow region in the corner of a column of square cross-section.

The effects of top and bottom design on the structure of riser flow are illustrated in Figure 2.24. In the dilute region the exponential decay of suspension density takes place up to TDH level. Above TDH, the suspension density may remain constant. However, the exponential decay to a constant density above TDH is not usually true, especially in the upper part of the dilute region, as investigated by Brereton et al. (1988), Jin et al. (1988), Bai et al. (1992), Brereton and Grace (1994) and Zheng and Zhang (1994). Figure 2.25 is an example of the exit effect. The exit design can affect the density profile over several meters in the upper region of the riser.

In the bottom zone both pressure gradient and solids volume fraction are much higher than those in the upper part of the dense region. The height of the bottom zone can be roughly determined in Figures 2.16 and 2.23.

The contribution of acceleration effect in the bottom zone can be evaluated here. Summing equations (2.2) and (2.4), neglecting the gas density effects

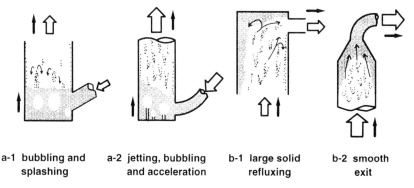

Figure 2.24 Entrance and exit regions typical for combustors (a-1, b-1) and catalytic reactors (a-2, b-2).

48 CIRCULATING FLUIDIZED BEDS

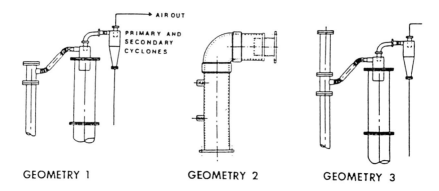

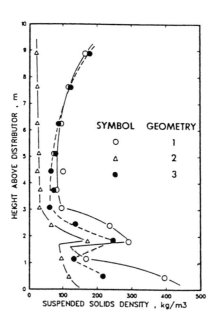

Figure 2.25 Effect of exit structure on the axial bed density profile ($U_0 = 7.1$ m/s, $G_s = 73$ kg/m^{2}s, solids re-entry: 1.98 m above distributor, no secondary air) (Brereton and Grace, 1994).

as well as the particle phase stress, and using ϵ_p instead of $(1-\epsilon)$, we can write the following equation for the particle velocity in a steady state:

$$\rho_p \epsilon_p (v \cdot \nabla) v = \epsilon_p \rho_p \vec{g} - \nabla p \qquad (2.26)$$

The vertical component of equation (2.26) is then area-averaged over the cross-section to give:

$$G_s \frac{d\bar{v}_z}{dz} = -\bar{\epsilon}_p \rho_p g - \frac{dp}{dz} \qquad (2.27)$$

Equation (2.26) can be integrated over the bottom zone (i.e. $z=0-z_b$, $\bar{\epsilon}_p = \bar{\epsilon}_p(0) - \bar{\epsilon}_{pd}$, $\bar{\epsilon}_{pb}$ = average solid volume fraction in the bottom zone, $\bar{v}_z = 0 - G_s/\rho_p \bar{\epsilon}_{pd}$) to give the pressure drop Δp_b across the bottom zone. Then we obtain:

$$\Delta p_b = \bar{\epsilon}_{pb} \rho_b g z_b + G_s^2 / \rho_p \bar{\epsilon}_{pd} \qquad (2.28)$$

$$\underbrace{\phantom{\bar{\epsilon}_{pb}\rho_b g z_b}}_{\text{gravity term}} \quad \underbrace{\phantom{G_s^2/\rho_p\bar{\epsilon}_{pd}}}_{\text{acceleration term}}$$

or, in other form:

$$\Delta p_b = (1/\bar{\epsilon}_{pd}\rho_p)(G_{sb}^{*2} + G_s^2) \qquad (2.28)'$$

where G_{sb}^*, i.e. the G_s equivalent of gravity contribution, is defined as:

$$G_{sb}^* \equiv (\bar{\epsilon}_{pb}/\bar{\epsilon}_{pd})^{1/2} \rho_p \bar{\epsilon}_{pd} (gz_b)^{1/2} = (\bar{\epsilon}_{pb}\bar{\epsilon}_{pd})^{1/2} \rho_p (gz_b)^{1/2} \qquad (2.29)$$

Accordingly, if G_s is sufficiently smaller than G_{sb}^*, the acceleration effect can be neglected.

Table 2.2 shows an example of G_{sb}^* for $\rho_p = 1000 \text{ kg/m}^3$, $(\bar{\epsilon}_{pb}\bar{\epsilon}_{pd})^{1/2} \simeq \bar{\epsilon}_{pd} = 0.1-0.3$ and $z_b = 0.25-4 \text{ m}$. In the case of combustors, G_s ranges from 10 to 30 kg/m²s and the acceleration term $G_s^2/\bar{\epsilon}_{pd}\rho_p$ remains less than 1% of the gravity term $G_{sb}^{*2}/\bar{\epsilon}_{pd}\rho_p$ (i.e. $(G_s/G_{sb}^*)^2 < 0.01$). This is consistent with the observation of Johnsson et al. (1992), which has already been shown in Figure 2.21. However, in the case of catalytic reactors with catalyst regeneration, G_s can be 300–1000 kg/m²s and the solid acceleration contributes to the pressure drop to the same extent as the gravity ($G_{sb}^* \simeq 800$ kg/m²s, $z_b \simeq 6$ m, $\bar{\epsilon}_{pd} \simeq 0.04$, $\bar{\epsilon}_{pb} \simeq 0.1$, and $\rho_p \simeq 1700$ kg/m³ for Knowlton, 1995).

Table 2.2 Gravity contribution, G_{sb}^* [kg/m²s], to obtain acceleration/gravity ratio, $(G_s/G_{sb}^*)^2$

$(\bar{\epsilon}_{pb}\bar{\epsilon}_{pd})^{1/2}$	z_b [m]: 0.25	1	2.25	4
0.1	157	300	469	626
0.2	313	626	939	1250
0.3	469	939	1410	1880

Assumption: $\rho_p = 1000$ kg/m³

The contribution of acceleration can now be further examined in some examples in Figure 2.23. For the case of Bader et al. (1988) we have $G_s \simeq 150\,\text{kg/m}^2\text{s}$, $z_b \simeq 1.5\,\text{m}$, $\bar{\epsilon}_{pd} \simeq \bar{\epsilon}_{pb} \simeq 0.2$, and $\rho_p \simeq 1700\,\text{kg/m}^3$. G_{sb}^* is then obtained as $1300\,\text{kg/m}^2\text{s}$ and $(G_s/G_{sb}^*)^2$ as 0.013. From Tanner et al. (1994) we have $G_s \simeq 110\,\text{kg/m}^2\text{s}$, $z_b \simeq 2.5\,\text{m}$, $\bar{\epsilon}_{pd} \simeq 0.04$, $\bar{\epsilon}_{pb} \simeq 0.08$, $\rho_p = 2500\,\text{kg/m}^3$. Then G_{sb}^* and $(G_s/G_{sb}^*)^2$ are $700\,\text{kg/m}^2\text{s}$ and 0.025, respectively. Accordingly, the acceleration effect is negligible in the bottom region of these examples.

If there exists a bubbling-bed-like bottom zone, solids are supposed to be splashed to the upper region. When the solid inventory is small the dilute region can dominate most of the riser space as in CFB combustors. The situation of Figure 2.21 then arises, where the bottom zone (bubbling bed) is connected to the splash zone and the splash zone to the dilute region. Furthermore, even when there exists a dense region with negligible acceleration effect, as in the Herb et al. (1989) and Bader et al. (1988) data in Figure 2.23, a transition zone is seen in the bottom, which is similar to the splash zone.

2.3.2 Formulation of core-annulus flow structure

To describe the solid upflow and downflow, let $w_{up} = w_C$ and $w_{down} = -w_A$ denote the solid upflow rate in the core and solid downflow rate in the annulus, respectively. At any height in a circulating fluidized bed riser, the sum of w_C and w_A is equal to the net flow rate w, i.e.

$$G_s A_t \equiv w = w_{up} - w_{down} = w_C + w_A \tag{2.30}$$

The time- and area-averaged solid velocity v_x ($x = C$ or A) can be defined in terms of w_x and time- and area-averaged particle volume fraction ϵ_{px} as:

$$v_x \equiv w_x/\rho_p A_x \epsilon_{px} \quad (x = C, A) \tag{2.31}$$

Then equation (2.30) can be rewritten as:

$$v_0 = \alpha^2 v_C \epsilon_{pC} + (1 - \alpha^2) v_A \epsilon_{pA} \tag{2.30}'$$

where v_0 is the superficial solid velocity.

The core-annulus flow has been explained first by radial gas velocity distribution and by phenomenological correlations for the distribution functions of particle concentration and velocity; second by turbulent diffusion of solids to the wall; and third by energy minimization with decreased wall friction by the solid downflow. The first approach was adopted by Shimizu et al. (1964), and recently for circulating fluidized beds by Berruti and Kalogerakis (1989), Rhodes (1990) and Pugsley et al. (1994). The attempt of Zenz (1994) to apply a 'flooding' type correlation to the present problem may also be classified into the first group. The second originates from the freeboard models of Horio et al. (1980) and Pemberton and

Davidson (1986). Further developments have been made for circulating fluidized beds by Bolton and Davidson (1988). Kunii and Levenspiel (1990, 1991, 1995) developed a similar three-phase model. The third phase originates from the Nakamura and Capes (1973) model for pneumatic conveying. Further developments for circulating fluidized beds have been made by Ishii *et al.* (1989), Li *et al.* (1988) and Bai *et al.* (1995).

Since the first approach has to deal with the gas velocity profile or suspension density, which is basically unknown, the second and the third approaches have often been adopted for the analysis of experimental data. However, the second explanation is only based on a material balance and cannot predict all the unknowns. The third explanation is based on time-averaged mass and momentum balances, but the gas–solid slip velocity versus drag force relationship must be known beforehand.

(a) Lateral solids transfer model and axial voidage profile (see Figure 2.26)
If it is assumed that the solids transfer flux from region x (e.g. core or annulus) to y (annulus or core) is proportional to the solid volume fraction in region x, then:

$$\frac{dw_{up}}{dz} = \frac{dw_{down}}{dz}$$
$$= \rho_p \pi D_C (k_{AC}\epsilon_{pA} - k_{CA}\epsilon_{pC})$$
$$= \rho_p \pi D_C k_{CA}(\epsilon_{pC,eq} - \epsilon_{pC}) \quad (2.32)$$

where k_{xy} is the lateral solid transfer coefficient from region x to region y. The core particle concentration $\epsilon_{pC,eq}$ in equilibrium with that in the annulus is

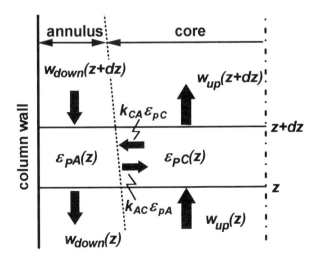

Figure 2.26 Lateral solids transfer model.

defined as:
$$\epsilon_{pC,eq} = K\epsilon_{pA} \quad (K \equiv k_{AC}/k_{CA} < 1) \tag{2.33}$$

For constant v_C and v_A the governing equations can be written as follows:

$$\frac{d\epsilon_{pC}}{dz} = \frac{4k_{CA}}{v_C D_t \alpha}(K\epsilon_{pA} - \epsilon_{pC}) - \frac{\epsilon_{pC}}{\alpha^2}\frac{d\alpha^2}{dz} \tag{2.32}'$$

and

$$\epsilon_{pC} = \frac{1-\alpha^2}{\alpha^2}\frac{(-v_A)}{v_C}\epsilon_{pA} + \frac{1}{\alpha^2}\frac{v_0}{v_C} \tag{2.30}''$$

ϵ_{pC} and ϵ_{pA} are linked to $\bar{\epsilon}_p$ by:

$$\bar{\epsilon}_p = \alpha^2 \epsilon_{pC} + (1-\alpha^2)\epsilon_{pA} \tag{2.34}$$

Dilute region: Difficulty of the constant α model and significance of the change in α Integrating equation (2.32)' subject to equation (2.34) under the constant α condition, we obtain:

$$\epsilon_{pC} = (\epsilon_{pC,d} - \epsilon_{pC\infty})\exp(-a(z - z_d)) + \epsilon_{pC\infty} \tag{2.35}$$

$$\bar{\epsilon}_p - \bar{\epsilon}_{p\infty} = (\bar{\epsilon}_{p,d} - \bar{\epsilon}_{p\infty})\exp(-a(z - z_d)) \tag{2.36}$$

where z_d is the height of the lower dense region and

$$a \equiv \frac{4k_{CA}}{v_C D_t \alpha}\left(1 - \frac{v_C}{(-v_A)}\frac{\alpha^2}{1-\alpha^2}K\right) \tag{2.37}$$

$$\epsilon_{pC\infty} \equiv \frac{-K}{(1-\alpha^2)}\frac{v_0}{(-v_A)}\frac{1}{1 - \frac{v_C}{(-v_A)}\frac{\alpha^2}{1-\alpha^2}K} \tag{2.38}$$

$$\bar{\epsilon}_{p\infty} = -\frac{1 + \frac{v_C}{(-v_A)}\frac{\alpha^2}{1-\alpha^2}K}{1 - \frac{v_C}{(-v_A)}\frac{\alpha^2}{1-\alpha^2}K}\left(\frac{v_0}{-v_A}\right) \tag{2.39}$$

Equation (2.36) is the same expression as the empirical correlation proposed by Zenz and Weil (1958). However, in the above model it is obvious that if v_0, v_C and $(-v_A)$ are all positive, either the decay factor a or the equilibrium particle concentration $\epsilon_{pC,\infty}$ must be negative. This is the fundamental difficulty of the constant parameter model discussed by Horio and Lei (1996). In Figure 2.27(a) the operating lines for the above two situations are demonstrated.

When α changes with height, the operating points in the ϵ_{pC} versus ϵ_{pA} plane form a curve instead of a straight line, as shown in Figure 2.27(b).

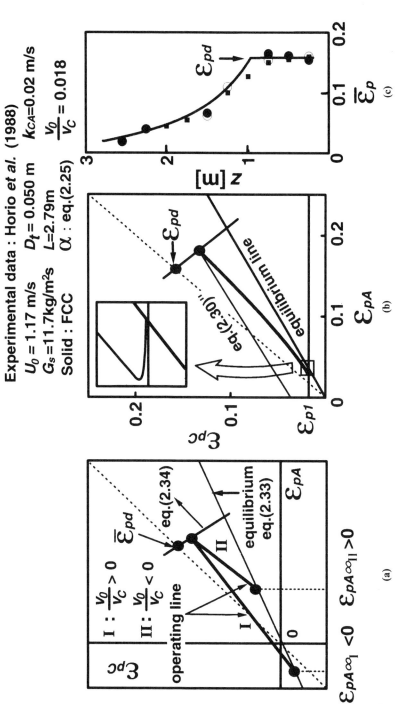

Figure 2.27. Two cases in the lateral solid transfer model (Horio and Lei, 1996). (a) Constant α and its difficulty. (b) Variable α case: operating line. (c) Variable α case: axial distribution of particle concentration

From equation (2.30)″ the operating line for variable α should satisfy:

$$\frac{d\epsilon_{pC}}{d\epsilon_{pA}} = \frac{1-\alpha^2}{\alpha^2}\frac{(-v_A)}{v_C}\frac{1}{1 - \frac{\epsilon_{pC} - \epsilon_{p1}}{\alpha^2(1-\alpha^2)}\left[\frac{d\alpha^2}{dz} \bigg/ \left(-\frac{d\epsilon_{pC}}{dz}\right)\right]} \quad (2.40)$$

For the integration of equations (2.32)′ and (2.40) the values of ϵ_{pA} and ϵ_{pd} at $z = z_d$ can be obtained from equations (2.30)″ and (2.34).

The boundary conditions at $z = L$ for equations (2.32)′ and (2.40) are derived from the assumption of an equilibrium and limit analysis on equations (2.30)′, (2.32)′, and (2.40) as

$$\epsilon_{pC}(L) \equiv \epsilon_{pC,1} = v_0/v_C \quad (2.41\text{-}1)$$

$$\epsilon_{pA}(L) \equiv \epsilon_{pA,1} = \epsilon_{pC,1}/K \quad (2.41\text{-}2)$$

$$d\epsilon_{pC}/d\epsilon_{pA} = 0 \quad (\text{at } z = L) \quad (2.41\text{-}3)$$

The operating line, accordingly, bends before it reaches the equilibrium line as shown in Figure 2.27(b). Figure 2.27(c) shows the comparison of the calculated and the observed axial distributions of $\bar{\epsilon}_p$.

Solution for the dense region For constant α, equation (2.35) also applies to the dense region. Substituting $z = -\infty$ and $\epsilon_{pC} = \epsilon_{pC,d}$ into equation (2.35), we find that only the following uniform distribution is possible for the lower dense region:

$$\epsilon_{pC} = \epsilon_{pC,d} = \frac{v_0 K_d}{(1-\alpha_d^2)v_{Ad} + \alpha_d^2 K_d v_{Cd}} \quad (2.42)$$

$$\bar{\epsilon}_p = \bar{\epsilon}_{p,d} = \frac{(1-\alpha_d^2 + \alpha_d^2 K_d)v_0}{(1-\alpha_d^2)v_{Ad} + \alpha_d^2 K_d v_{Cd}} \quad (2.43)$$

Empirical correlations As noted above, equation (2.36) is simply the Zenz and Weil (1958) correlation for $\bar{\epsilon}_p$, which has been applied by Kunii and Levenspiel (1991) for circulating fluidized bed conditions. For the decay factor a they proposed the following correlation which is similar to that of Lewis *et al.* (1962):

$$aU_0 = \begin{cases} 2\text{--}5 \text{ s}^{-1} & (d_p < 70\,\mu\text{m}) \\ 4\text{--}12 \text{ s}^{-1} & (d_p > 88\,\mu\text{m}) \end{cases} \quad (2.44)$$

Recently, Adánez *et al.* (1994) also proposed a correlation for a.

For CFB boilers Johnsson and Leckner (1995) presented the following expression to incorporate the splash zone into the previous dilute region

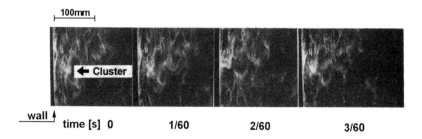

Figure 2.28 Diffusion of clusters toward the wall (FCC, $d_p = 61.3\,\mu\text{m}$, $\rho_p = 1780\,\text{kg/m}^3$, $U_0 = 0.58\,\text{m/s}$, $G_s = 0.22\,\text{kg/m}^2\text{s}$, $z = 1150\,\text{mm}$, shutter speed: 1/100 s, column: CFBA in Figure 2.35) (Horio and Kuroki, 1994).

model (see Figure 2.21):

$$\bar{\epsilon}_p = [\bar{\epsilon}_{p,b0} - \bar{\epsilon}_{p,exit}\exp(-a(L-z_{b0}))]\exp(-a_{splash}(z-z_{b0})) \\ + \bar{\epsilon}_{p,exit}\exp(a(L-z)) \quad (2.45)$$

where $a \simeq 0.23/(U_0 - u_T)$ and $a_{splash} \simeq 4u_T/U_0$.

To understand the decay process we need to look into the mechanism of lateral solids transport in the dilute region. For the freeboard of bubbling fluidized beds, Horio et al. (1980) and Pemberton and Davidson (1986) made it clear that the exponential decay voidage distribution results from lateral transport of solids to the solids downflow layer along the wall. The solids diffusion in the freeboard of bubbling fluidized beds is due to the special turbulence with swirls caused by bubble eruption dispersing particles and rising as 'ghost bubbles' (Pemberton and Davidson, 1986), lifted by buoyancy forces (Caram et al., 1984; Horio and Kuroki, 1994).

Solids diffusion in the dilute region of a circulating fluidized bed appears to be induced by well-developed turbulence. Figure 2.28 shows the lateral movement of clusters and downflow along the wall. This cluster diffusion from the core to the annulus causes cluster growth in the annulus, as already shown in Figure 2.8.

Bolton and Davidson (1988) confirmed that the following Pemberton and Davidson (1986) correlation

$$k_{ca} = 0.1\sqrt{\pi}u'/(1+S/12) \quad (2.46)$$

is applicable to circulating fluidized beds with:

$$u' = U_0(1 - 2.8Re^{-1/8}) \quad (2.47)$$

where S is the Stokes number and Re is the column Reynolds number.

For a particular column without solids reflux at its exit, i.e. $w_{down}(L) = 0$, equation (2.30) suggests that $w_{up}(L) = G_s A_t$. Therefore, from equations (2.31) and (2.36):

$$G_s A_t = \rho_p A_C v_C'[\bar{\epsilon}_{p\infty} + (\bar{\epsilon}_{pd} - \bar{\epsilon}_{p\infty})\exp(-a(L-z_d))] \quad (2.48)$$

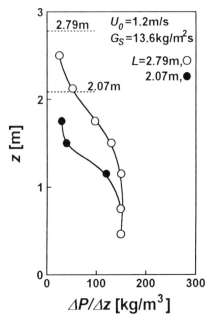

Figure 2.29 Effect of riser length on the axial distribution of pressure gradient ($d_p = 60\,\mu\text{m}$, $\rho_p = 1000\,\text{kg/m}^3$, column: see Figure 2.8) (Horio et al., 1988).

Hence for these conditions, once the gas velocity and G_s are given and $\bar{\epsilon}_{pd}$, $\bar{\epsilon}_{p\infty}$ and a are fixed, the dilute region height $(L - z_d)$ is determined. Therefore, when two columns of different heights are operated under exactly the same U_0 and G_s conditions, the length of the upper dilute region $(L - z_d)$ is maintained constant, but the dense region height takes different values. Figure 2.29 demonstrates this. However, this is valid only for sufficient solids hold-up in the downcomer. Accordingly, in CFB boilers it is not applicable, as discussed in section 2.3.3.

The effect of exit structure is expressed from the modeling viewpoint by the boundary condition at the exit to integrate equation (2.32). Kunii and Levenspiel (1995) have successfully reproduced the results of Bai et al. (1992) using their three-phase model.

(b) Momentum balance model with energy minimization principle (see Figure 2.30)

The above analysis is limited by the lack of a momentum balance. Although the core-annulus flow structure also appears in bubble columns and in bubbling fluidized beds (Miyauchi et al., 1981; Werther, 1977), the mechanism underlying the flow structure in circulating fluidized beds is different. In bubbly flows the upward flow of liquids or fluidized solids is induced by the wake-lifting effect of bubbles, as well as by the non-uniform drift of

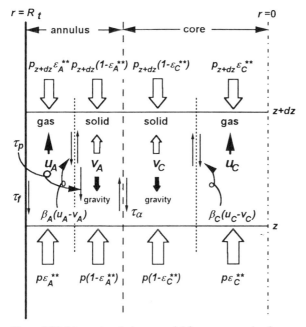

Figure 2.30 Momentum balance model for core–annulus flow.

the continuous phase during bubble passage. In tall fluidized beds of Geldart Group A powders, the buoyancy effect on the central bubble-rich region may also induce solids circulation. In the absence of bubbles in circulating fluidized beds, a different description must be found for the internal solids circulation. However, in much of the existing literature, little attention is paid to the clustering nature of suspensions, which may play an essential role in the formation of core/annulus flow.

For pneumatic transport, Nakamura and Capes (1973) developed a core-annulus model based on the Richardson and Zaki (1954) correlation for the gas–solid slip relationship. For clustering suspensions, Ishii et al. (1989) presented the clustering annular flow model based on a similar analysis, with an extended Richardson–Zaki correlation used earlier by Yerushalmi et al. (1978). Let us briefly discuss these two analyses (note that the expression for the gas–solid drag term is changed from that of Ishii et al. (1989) but the same expression for α can be obtained). The slip relationship for the region x, i.e. core or annulus, of both homogeneous and clustering suspensions can be summarized as:

$$u_{sl,x} \equiv u_x - v_x = u_{T,x}^{**}\epsilon_x^{**n}/\epsilon_x \qquad (x = C, A) \qquad (2.49)$$

where $u_{sl,x}$ is the gas–solid slip velocity and $u_{T,x}^{**}\epsilon_x^{**n}$ should be read as $u_{T,x}\epsilon_x^n$ for homogeneous suspensions and $u_{T,cl,x}\epsilon_x^{*n}$ for clustering suspensions, while u_T and $u_{T,cl}$ are mean terminal velocities of particles and clusters, respectively. $u_{T,cl}$ can be calculated by equation (2.17). The cluster-free space fraction ϵ^* is related to ϵ_x, the cross-section average voidage in the core or in the annulus, and ϵ_{cl}, the voidage in the cluster phase, by:

$$\epsilon_x^* = 1 - (1 - \epsilon_x)/(1 - \epsilon_{cl,x}) \qquad (x = C, A) \qquad (2.50)$$

For clustering suspension it can be confirmed that equation (2.49) can be written

$$u_x^* - v_x = u_{T,cl,x} \epsilon_x^{*n-1} \qquad (x = C, A) \qquad (2.49)'$$

where u_x^* is the gas velocity in the lean (i.e. cluster-free) phase. From equation (2.49) with equation (2.17), the drag force N_D acting on the cluster phase in a unit volume of bed can be written as:

$$N_D = \frac{\rho_{lean}}{3d_{cl}} \left[\frac{(u^* - v)}{\epsilon^{*n-1}} \right]^2 (1 - \epsilon^*) \qquad (2.51)$$

From the overall material balance for gas and solids, we may write:

$$U_0 = \alpha^2 U_{0C} + (1 - \alpha^2) U_{0A} \qquad (2.52)$$

$$v_0 = \alpha^2 v_{0C} + (1 - \alpha^2) v_{0A} \qquad (2.53)$$

The momentum balance equations with $\rho_g \ll \rho_p$ for steady state are given by:

gas in the core:

$$\epsilon_C^{**} \frac{dp}{dz} = -\beta_C(u_C - v_C) \qquad (2.54)$$

solids in the core:

$$(1 - \epsilon_C^{**}) \frac{dp}{dz} = \beta_C(u_C - v_C) - \frac{4\tau_\alpha}{\alpha D_t} - \rho_s^{**}(1 - \epsilon_C^{**})g \qquad (2.55)$$

gas in the annulus:

$$\epsilon_A^{**} \frac{dp}{dz} = -\beta_A(u_A - v_A) - \frac{4\tau_f}{(1-\alpha^2)D_t} \qquad (2.56)$$

solids in the annulus:

$$(1 - \epsilon_A^{**}) \frac{dp}{dz} = \beta_A(u_A - v_A) + \frac{4\alpha\tau_\alpha - 4\tau_p}{(1-\alpha^2)D_t} - \rho_s^{**}(1 - \epsilon_A^{**})g \qquad (2.57)$$

In the above expressions, ρ_s^{**} should be replaced by ρ_p for homogeneous suspensions and ρ_{cl}, i.e. cluster density, for clustering suspensions.

Solving the momentum equations for (dp/dz), we obtain the pressure gradient parameter $E \equiv -(1/\rho_p g)(dp/dz)$ as:

$$E = 1 - \epsilon + 4\tau_p/D_t \rho_p g$$

$$= \alpha^2(1 - \epsilon_C) + (1 - \alpha^2)(1 - \epsilon_A) + 4\tau_p/D_t \rho_p g \qquad (2.58)$$

where the wall–solids shear stress τ_p is written as:

$$\tau_p = (1/2)f_p(1-\epsilon)\rho_p|v_A|v_A \qquad (f_p = a|v_A|^b) \qquad (2.59)$$

Parameters a and b for the particle friction factor f_p are $(a,b) = (0.003, 0)$ from Stemerding (1962) or $(a,b) = (0.025\,\text{m/s}, -1)$ from Konno and Saito (1969).

Taking the derivative of E with respect to α and equating it to zero, we obtain the following expression for the area fraction of annulus $1-\alpha^2$, with which the pressure gradient is minimized. Using the Konno and Saito (1969) correlation, we can write

$$1-\alpha^2 = \frac{1}{\epsilon_C^{**}-\epsilon_A^{**}}\left[2\lambda_p(1-\epsilon_C^{**})(1-\epsilon_A^{**})\frac{\epsilon_C}{u_{T,C}^{**}}(u_{sl,C}-\bar{u}_{sl})\right]^{1/2} \qquad (2.60)$$

Here $\lambda_p \equiv au_{T,C}^{**2+b}/gD_t$, $u_{sl,C}$ is the slip velocity characteristic of the given suspension (see equation (2.49)); $\bar{u}_{sl}$ is the apparent overall slip velocity calculated from the operating conditions based on the voidage in the core, i.e.

$$\bar{u}_{sl} \equiv \frac{U_o}{\epsilon_C} - \frac{v_o}{1-\epsilon_C} \qquad (2.61)$$

This leads to the simple but significant conclusion that the core-annulus structure exists, i.e. $1-\alpha^2$ can have a real value only if the apparent overall slip velocity $\bar{u}_{sl}$ is smaller than the true slip velocity of the suspension in the core $u_{sl,C}$, which can only occur in clustering suspensions. As observed by Horio et al. (1992b) and Tanner et al. (1994), for ordinary circulating fluidized bed conditions $\bar{u}_{sl}$ can be 0.5–10 m/s for $U_0 = 1$–10 m/s, $G_s = 10$–100 kg/m^{2}s, $\rho_p \simeq 1000$ kg/m^3 and $\epsilon_C \simeq 0.8$–0.995. If we calculate $u_{sl,C}$ assuming that suspensions are homogeneous, its values for ordinary CFB particles, which are close to u_T ($\simeq$0.1–0.2 m/s), are usually much smaller than $\bar{u}_{sl}$ estimated above and cannot give real values for α.

By using the observed cluster size and voidage, Ishii et al. (1989) successfully explained the observed core diameter by equation (2.60) as shown in Figure 2.31, with d_{cl} predicted from equation (2.70).

To close the set of equations for the time-averaged macroscopic structure of core-annulus flow shown above, it is necessary to achieve energy minimization in the global sense (Horio and Takei, 1991). E must be minimized to the absolute minimum by adjusting all free variables in their plausible ranges, since the energy minimization operation that resulted in equation (2.60) only tentatively minimized the value of E with α for a given set of values of ϵ_C, ϵ_C^*, ϵ_A^* and λ_p, all of which are to be determined completely by the model. Based on a similar energy minimization principle, Li et al. (1991b) and Bai et al. (1995) developed a more detailed model for core-annulus flow.

To obtain a complete solution that is applicable to scale-up issues, it is necessary to establish a reliable slip velocity versus drag force relationship and a comprehensive formulation for clustering suspensions. The

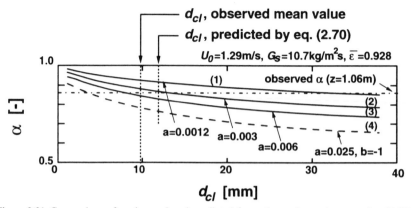

Figure 2.31 Comparison of α observed and predicted from cluster data using equation (2.60). ((1), (2), (3): τ_p from Stemerding correlation, (4): τ_p from Konno–Saito correlation) (Ishii et al., 1989).

one-dimensional formulation of Wirth (1991) for clustering suspensions is one such model, where it is assumed that clusters are rod-shaped vertical strands suspended both by the gas pressure gradient and by the impingement of lean phase particles and that the cluster phase is maintained at the minimum fluidization condition. The model consists of the following equations:

mass balance for gas:
$$U_0 = (1 - \epsilon^*)\epsilon_{mf} v_{cl} + U_{mf} + \epsilon^* u^* \quad (2.62)$$

mass balance for solids:
$$G_s = \rho_p(1 - \epsilon^*)(1 - \epsilon_{mf})v_{cl} \quad (2.63)$$

force balance for gas:
$$\left|\frac{dp}{dz}\right| = (\rho_p - \rho_f)(1 - \epsilon_{mf})(1 - \epsilon^*)g \quad (2.64)$$

force balance for clusters:
$$(1 - \epsilon^*)\left|\frac{dp}{dz}\right| + N_D = (\rho_p - \rho_f)(1 - \epsilon_{mf})(1 - \epsilon^*)g \quad (2.65)$$

For the drag force acting on clusters in a unit volume of the riser N_D, Wirth (1991) used the expression:

$$N_D = \lambda \rho_f u^*(u^* - v_{cl})/d_p \quad (2.66)$$

with an experimentally determined value of dimensionless parameter $\lambda = 0.00533$ and the lean phase particle velocity:

$$v^* = u^* - u_T \quad (2.67)$$

With this simple formulation, Wirth (1991) provided $|dp/dz|, G_s, U_0$ phase diagrams similar to those in Figure 2.15.

Instead of equations (2.64), (2.65) and (2.66) in the above analysis Horio and Ito (1996) used the following equations for $|dp/dz|$ and force balance for clusters, and equation (2.51) for N_D:

$$\left(\frac{dp}{dz}\right)_{cl} = -\frac{18\mu}{d_p^2} \frac{1-\epsilon_{cl}}{\epsilon_{cl}^{2(n-1)}} (u_{cl} - v)^2 \tag{2.68}$$

$$g(\rho_p - \rho_f)(1 - \epsilon) = (1 - \epsilon^*)\left(\frac{dp}{dz}\right)_{cl} + \epsilon^* \left(\frac{dp}{dz}\right)^* \tag{2.69}$$

Then we obtain

$$d_{cl} = \frac{1}{3(1-\epsilon_{cl})\epsilon^{*2(n-1)}} \frac{\rho_f}{g(\rho_p - \rho_f)} (u^* - v)^2 \tag{2.70}$$

From the assumption of bubble-like behavior of the lean phase flow the following relationship is derived:

$$\frac{d_{cl}}{D_b} = \left(\frac{3}{2}\phi_{cl}\right)^{1/3} \left(\frac{2}{\sqrt{3}}\left(\frac{\epsilon_{max}^*}{\epsilon^*}\right)^{1/3} - 1\right) \frac{1}{\gamma^2} [(1-\epsilon^*)(u^* - u_{cl}\epsilon_{cl})]^2 \tag{2.71}$$

where D_b is the mean diameter of bubble like voids formed between clusters, ϕ_{cl} is the aspect ratio of a cylindrical cluster, $\epsilon_{max}^* \simeq 0.74$ is the maximum possible value of ϵ^* and $\gamma \simeq 0.7$ is the bubble velocity coefficient. The cluster size and the voidage in the cluster phase thus determined agreed fairly well with observed values.

Some challenging experimental data necessary for future model development are now appearing, concerning, for instance, radial particle mass flux (Bolton and Davidson, 1988; Qi and Farag, 1993; Zhou et al., 1995), particle turbulence energy (Wang et al., 1993) and/or local gas–solid slip velocity (Horio et al., 1992b; Yang et al., 1992, 1993; Tanner et al., 1994).

2.3.3 Total pressure loop prediction

In Figure 2.29 it was shown that the dense region height z_d increases if the total column height is increased for the same U_0, G_s and other operating conditions. However, when the downcomer bed height z_D is small, part of the gas fed to the riser may travel backwards through the loop seal and up the downcomer. Such a gas flow direction change is bound to change the loop seal performance, particularly the solid recycle rate. Accordingly, for a fixed solid inventory there should be a critical column height above which the required G_s cannot be obtained. The issue of inventory and downcomer conditions has been raised by Weinstein et al. (1984) and Mori et al. (1989). As shown in Figure 2.32, different axial profiles can be obtained in a riser for the same U_0 and G_s conditions.

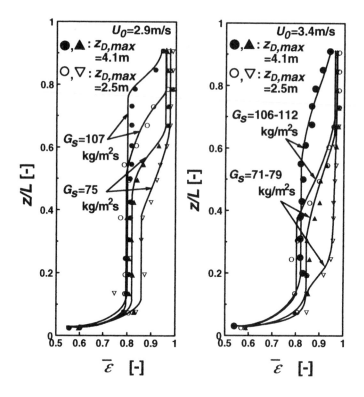

Figure 2.32 Effect of downcomer bed height on axial voidage distribution ($D_t = 0.152$ m, particles: FCC, $d_p = 49$ μm, $\rho_p = 1450$ kg/m³) (Weinstein et al., 1984).

To determine the dense region height for realistic situations, one must solve a set of equations for the whole pressure loop of a given circulating system. Since there are a variety of designs for the exit, the cyclone and the loop seal, the issue can be discussed here only from a rather conceptual point of view. The basic structure of a total pressure loop model should not differ greatly from models developed by Yang (1988) and Breault and Mathur (1989). Based on the terminology shown in Figure 2.33, material balance equations can be written as follows:

solids inventory: $\qquad W_t = W_R + W_D \qquad$ (2.72)

solids in riser: $\qquad W_R = \rho_p \bar{\epsilon}_{pd} A_t z_d + \rho_p A_t \int_{z_d}^{L} \bar{\epsilon}_p \, dz \qquad$ (2.73)

solids in downcomer: $\qquad W_D = \rho_p \epsilon_{pmf} \sum_i A_{Di} \Delta z_{Di} \qquad$ (2.74)

riser gas flow rate: $\qquad F_0 = U_0 A_t = F_R - F_{DR} \qquad$ (2.75)

HYDRODYNAMICS

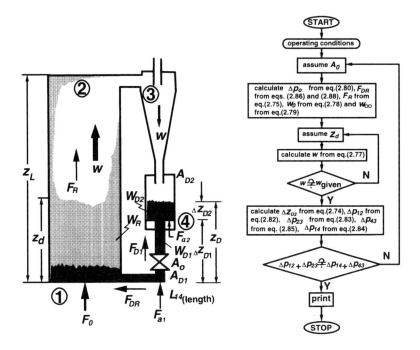

Figure 2.33 Simplified model for total pressure loop prediction.

downcomer gas flow rate: $\quad F_{D1} = U_{0D1}A_{D1} = F_{a1} - F_{DR} \quad$ (2.76)

solids circulation rate: $\quad w = w_\infty + (w_0 - w_\infty)\exp(-a(L - z_d)) \quad$ (2.77)

Assuming that all of the riser gas flows through the core region and applying equations (2.49) and (2.50), the upward solids flow rate in the dense region is obtained as:

$$w_0 = \rho_p A_t \left[U_{0R} - \alpha^2 u_{T,cl}\left(\frac{\bar{\epsilon}_d - \epsilon_{cl}}{1 - \epsilon_{cl}}\right)^n \right] \frac{1 - \bar{\epsilon}_d}{\bar{\epsilon}_d} \quad (2.78)$$

where $u_{T,cl}$ can be estimated using equations (2.17) and (2.70) once a value of ϵ_{cl} has been assumed. For w_∞, the Tanaka et al. (1972) correlation can be used for entrainment flux:

$$\frac{w_\infty}{A_t} = 0.046\rho_g(U_{0R} - u_T)Re_T^{0.3}\frac{(U_{0R} - u_T)}{\sqrt{gd_p}}\left(\frac{\rho_p - \rho_g}{\rho_g}\right)^{0.15} \quad (2.79)$$

For $\bar{\epsilon}_{pd}$, the Bai and Kato correlation (equation (2.14)) can be applied with U_0 replaced by U_{0R}. Parameters α and a can be determined from equations (2.25) and (2.44), respectively.

Solids flow rate through the loop seal is given by the following function of gas pressure drop across the constriction (or orifice) (Jones and Davidson, 1965):

$$w = C_{Do}(A_o/A_{D1})\sqrt{2\rho_p \epsilon_{mf} \Delta p_o} \qquad (2.80)$$

The pressure balance equation can be written as:

$$\Delta p_{12} + \Delta p_{23} = \Delta p_{14} + \Delta p_{43} \qquad (2.81)$$

where

riser pressure drop:
$$\Delta p_{12} = z_d \rho_p \bar{\epsilon}_{pd} g + \rho_p g \int_{z_d}^{L} \bar{\epsilon}_p \, dz \qquad (2.82)$$

cyclone pressure drop: $\quad \Delta p_{23} = 10 \gamma_{Cy}^2 \rho_f U_{0R}^2 \quad$ (see Yang, 1988) $\quad (2.83)$

downcomer pressure drop:

$$\Delta p_{14} = -\left[\frac{150\mu(1-\epsilon_{mf})}{\phi_s d_p \epsilon_{mf}} + 1.75\rho_f |u_{sl,D1}|\right] \frac{(1-\epsilon_{mf})}{\phi_s d_p \epsilon_{mf}} u_{sl,D1} L_{14} + \Delta p_o$$

(Ergun, 1952) $\quad (2.84)$

$$\Delta p_{43} = \rho_p \epsilon_{pmf} g \Delta z_{D2} \qquad (2.85)$$

orifice (or valve) pressure drop:

$$\Delta p_o = -\left[\frac{150\mu(1-\epsilon_{mf})}{\phi_s d_p \epsilon_{mf}} + \frac{1.75}{6}\rho_f |u_{sl,D1}| \frac{A_{D1}}{A_o}\right] \frac{(1-\epsilon_{mf})}{\phi_s d_p \epsilon_{mf}} \frac{A_{D1}}{4A_o} u_{sl,D1} D_o$$

(Leung and Jones, 1978) $\quad (2.86)$

The values of γ_{Cy} in equation (2.83) and slip velocity in the orifice or valve, $u_{sl,D1}$ in equation (2.86) are given by:

$$\gamma_{Cy} \equiv A_t/A_{Cy} \qquad (2.87)$$

$$u_{sl,D1} = F_{D1}/A_{D1}\epsilon_{mf} + w/\rho_p A_{D1} \epsilon_{pmf} \qquad (2.88)$$

The computation loop is illustrated in Figure 2.33. The hydrodynamic modeling of the riser allows prediction of w_0, the maximum available solids flow rate when there is no dilute region, and w_∞, the solids flow rate above the TDH.

The model predicts well the Weinstein *et al.* (1984) data as shown in Figure 2.34(a). When the downcomer bed height z_D is increased and a large flow resistance is allotted to the downcomer, the upper dilute region height becomes less sensitive to change in z_D for given values of U_0 and G_s. Furthermore, if the downcomer has a larger diameter, $(L - z_d)$ becomes less sensitive to change in total mass W_t. However, when the downcomer capacity is small, the dilute region height, as well as the real superficial gas velocity in the riser,

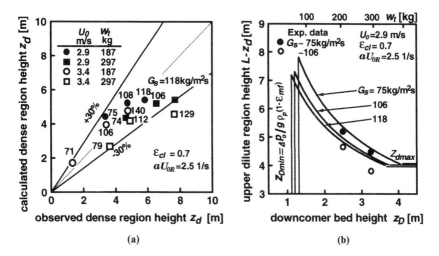

Figure 2.34 Prediction of dense region height (Lei and Horio, 1995). (a) Prediction from Weinstein *et al.* (1984) data. (b) Effect of downcomer bed height and diameter of downcomer on dense region height in riser.

do not remain constant for fixed G_s and U_0 due to the change in flow rate through the loop seal. This situation is encountered often in commercial plants where holding large amounts of solids in the downcomer is undesirable. Instead, for fixed G_s and U_{0R} the dilute region height is always constant. Note that there are situations where flow reversal could change the chemical atmosphere of the downcomer completely, a problem of possibly significant dimensions.

2.4 Scaling relationships

2.4.1 Scale-up and hydrodynamic scaling law

For successful scale-up of circulating fluidized bed processes the hydrodynamic scaling law is a powerful tool.

Three distinct phenomenological sub-processes (i.e. hydrodynamic gas–solid contact, heat and mass transfer and chemical reactions) take place in a circulating fluidized bed reactor. Since these sub-processes influence each other, 'scale-up' involves establishing a balance among hydrodynamics, heat and mass transfer and chemistry, such that the total output from the system satisfies a large-scale demand, maintaining the quality established in small-scale test plants. For scale-up to be successful we require both deductive and inductive approaches. In the deductive approach, hydrodynamics, heat and mass transfer, and chemistry are first examined separately as much as possible, and all information is then fed to a reactor model for prediction and reactor design. In the inductive approach, test plants, with design based on model predictions and empirical know-how, are operated to validate the reactor concept and to confirm the performances of equipment parts and material handling systems to see if they can withstand the scale change. A test program is conducted stepwise from the laboratory scale, through pilot plants to demonstration and commercial plants.

It should be kept in mind that the use of scale models based on the scaling law is, from a practical point of view, limited to hydrodynamics. The hydrodynamic time scale also changes as the scale of a system is varied. However, it is impossible to change the time scales of molecular processes, i.e. conduction, viscous energy dissipation and chemical reaction processes, without changing the balance with hydrodynamics.

As seen from Table 2.3, there have been three different approaches to the scaling law of fluidized beds. In a multiphase flow system, there is always a coexistence of flow structures of different length scales. In the case of circulating fluidized beds there are at least three scale ranks, i.e. the macroscopic issue of axial and radial time-average voidage and velocity distributions, the mesoscale issue of clusters and velocity fluctuations, and the microscopic issue of the flow field around each particle. To realize completely similar flow for all different scales is not only difficult but also unnecessary. In many practical situations similarity can be sacrificed in the meso- and microscale ranks. In approaches with (1) dimensionless analysis and (2) differential equations, it is often forgotten that the reference scales should be introduced not arbitrarily but through boundary conditions. Although fundamental value exists in these two approaches, (3) theoretical solutions and experimental correlations, when viable, appear to provide the most realistic approach.

2.4.2 Derivation of scaling law and experimental validations

The dimensionless expression of momentum equations (2.2) and (2.4) are written:

$$\left(\frac{\rho_f}{\rho_p}\right)\epsilon\left[\frac{\partial \hat{u}}{\partial \hat{t}} + (\hat{u}\cdot\hat{\nabla})\hat{u} - \frac{\vec{g}l}{U_0^2}\right] + \epsilon\hat{\nabla}\hat{p} + \frac{\beta l}{\rho_p U_0}\left(\hat{u} - \left(\frac{v_0}{U_0}\right)\hat{v}\right) = 0 \quad (2.89)$$

$$(1-\epsilon)\left[\left(\frac{U_0}{v_0}\right)\frac{\partial \hat{v}}{\partial \hat{t}} + (\hat{v}\cdot\hat{\nabla})\hat{v} + \left(\frac{U_0}{v_0}\right)^2\left(-\frac{\vec{g}l}{U_0^2} + \hat{\nabla}\hat{p}\right)\right]$$

$$-\frac{\beta l}{\rho_p U_0}\left(\frac{U_0}{v_0}\right)^2\left(\hat{u} - \left(\frac{v_0}{U_0}\right)\hat{v}\right) - \hat{\nabla}\cdot\hat{P}_s = 0 \quad (2.90)$$

Table 2.3 Comparison of criteria for hydrodynamic similarity of fluidized beds

	Authors	Theoretical background	Dimensionless groups other than $L_f/D_t, L_{ti}/D_t$	Necessary operating conditions ($m = D_t/D_t^\circ$)
Dimensional analysis	1 Fitzgerald et al. (1983, 1984)	Dimensional analysis	$d_p/D_t, gd_p/U_0^2$ $\rho_p^2 d_p^3 g/\mu^2, \rho_p/\rho_f$	$d_p = md_p^\circ; \rho_p = (\rho_f/\rho_f^\circ)\rho_p^\circ$ $U_0 = m^{1/2} U_0^\circ$ $m = (\rho_f^\circ \mu/\rho_f \mu^\circ)^{2/3}$: stiff
Differential equations	2-1 Glicksman (1984); Nicastro and Glicksman (1984)	Two-fluid model and Ergun correlation No consideration of boundary conditions d_p was taken as reference length scale	(a) Viscous regime ($\mathrm{Re}_p < 4$) $d_p/D_t, gd_p/U_0^2$ $\rho_p^2 d_p^3 g/\mu^2, \phi_s$ (b) Intermediate regime $d_p/D_t, gd_p/u_0^2$ $\rho_p^2 d_p^3 g/\mu^2, \rho_p/\rho_f$ (c) Inertial regime ($\mathrm{Re}_p > 1000$) $d_p/D_t, gd_p/U_0^2, \rho_p/\rho_f$	$d_p = md_p^\circ; \rho_p = m^{-3/2}(\mu/\mu^\circ)\rho_p^\circ$ $U_0 = m^{1/2} U_0^\circ, m$: free $d_p = md_p^\circ; \rho_p = (\rho_f/\rho_f^\circ)\rho_p^\circ$ $U_0 = m^{1/2} U_0^\circ$ $m = (\rho_f^\circ \mu/\rho_f \mu^\circ)$: stiff $d_p = md_p^\circ; \rho_p = (\rho_f/\rho_f^\circ)\rho_p^\circ$ $U_0 = m^{1/2} U_0^\circ, m$: free
	2-2 Zhang and Yang (1987)	Two-fluid model with no gas inertia term Gas–particle friction by Ergun equation No consideration on boundary conditions D_t is used as reference scale with no reasons given	(a) Viscous regime ($\mathrm{Re}_p < 4$) $g/D_t U_0^2$ $\rho_p^2 (d_p \phi_s)^4 g/\mu^2 D_t$ (b) Intermediate regime $g/D_t U_0^2, \rho_f D_t/\rho_p \phi_s d_p$ $\rho_p^2 (d_p \phi_s)^4 g/\mu^2 D_t$ (c) Inertial regime ($\mathrm{Re}_p > 400$) $g/D_t U_0^2, \rho_f D_t/\rho_p \phi_s d_p$	$\phi_s d_p = m^{1/4} (\mu \rho_p^\circ/\mu^\circ \rho_p)^{1/2} (\phi_s d_p)^\circ$ m: free $\phi_s d_p = m^2 (\rho_f/\rho_f^\circ)^3 (\mu/\mu^\circ)^2 (\phi_s d_p)^\circ$ $\rho_p = m^3 (\rho_f/\rho_f^\circ)^4 (\mu^\circ/\mu)^2 \rho_p^\circ$ m: free $\phi_s d_p = m(\rho_f/\rho_f^\circ)(\rho_p^\circ/\rho_p)$ $U_0 = m^{1/2} U_0^\circ, m$: free
	2-3 Glicksman (1988)	Two-fluid model (same as above) Boundary conditions considered (D_t enters from B.C. as reference scale)	Viscous regime ($\mathrm{Re}_p < 4$) $gD_t/U_0^2, U_0/U_{mf}$	$U_{mf} = m^{1/2} U_{mf}^\circ, U_0 = m^{1/2} U_0^\circ$ m: free

	2-4 Foscolo et al. (1990)	Two-fluid model with particle pressure gradient term	$gd_p^3\rho_f^2/\mu^2, \rho_f/\rho_p$ $U_0/u_T, L/d_p$	$d_p = md_p^\circ, \rho_p = (\rho_f/\rho_f^\circ)\rho_p^\circ$ $m = (\rho_f^\circ\mu/\rho_f\mu^\circ)^{2/3}$: stiff $U_0 = U_0^\circ(u_T/u_T^\circ)$
Solutions and correlations	3-1 Horio et al. (1986a, b)	Clift–Grace (1971) model for bubble coalescence Horio–Nonaka (1987) model for bubble splitting Davidson (1961) model for interstitial flow	(a) Group B, D particles $gD_t(U_0 - U_{mf})^{-2}$ for bubble distribution (b) Group A, B, D particles $gD_t(U_0 - U_{mf})^{-2}$ U_0/U_{mf}	$U_0 - U_{mf} = m^{1/2}(U_0 - U_{mf})^\circ$, m: free $U_0 - U_{mf} = m^{1/2}(U_0 - U_{mf})^\circ$ $U_{mf} = m^{1/2}U_{mf}^\circ, m$: free
	3-2 Horio et al. (1989)	Clustering annular flow model Modified Richardson–Zaki correlation for drag expression	gD_t/U_0^2 $v_0/U_0, u_T/U_0$ ρ_p/ρ_f (for $\epsilon \to 1$)	$U_0/U_0^\circ = v_0/v_0^\circ = u_T/u_T^\circ = m^{1/2}$ $(\rho_p/\rho_f) = (\rho_p/\rho_f)^\circ$ (for $\epsilon \to 1$) m: free

where $\hat{t} \equiv t/(l/U_0)$, $\hat{u} \equiv u/U_0$, $\hat{v} \equiv v/U_0$, $\hat{p} \equiv p/\rho_p U_0^2$, $\hat{P}_s \equiv P_s/\rho_p U_0^2$ and $\hat{\nabla} \equiv l\nabla$. When $\rho_f/\rho_p \ll 1$, equation (2.89) can be simplified to:

$$\epsilon\hat{\nabla}\hat{p} + \frac{\beta l}{\rho_p U_0}(\hat{u} - \hat{v}) = 0 \qquad (2.91)$$

From equation (2.6), the term $\beta l/\rho_p U_0$ in equation (2.91) may be expressed as:

$$\frac{\beta l}{\rho_p U_0} \cong \frac{gl}{U_0^2}\frac{U_0}{u_T}(1-\epsilon)\epsilon^{2-n} \qquad (\epsilon_{mf} \ll \epsilon \le 1) \qquad (2.92a)$$

$$\frac{\beta l}{\rho_p U_0} = \frac{gl}{U_0^2}\frac{U_0}{U_{mf}}(1-\epsilon_{mf})\epsilon_{mf}^2 \qquad (\epsilon \simeq \epsilon_{mf}) \qquad (2.92b)$$

The scaling law derived from equations (2.89), (2.90) and (2.91) is, therefore, stated as follows:

The flow field in a unit of length scale l, which is geometrically similar to a reference unit (denoted by superscript °), can be made similar, if the following four conditions are satisfied:

$$l/U_0^2 = l°/U_0^{°2} \qquad (2.93)$$

$$U_0/u_T = U_0°/u_T° \qquad (\epsilon_{mf} \ll \epsilon \le 1) \qquad (2.94a)$$

or

$$U_0/U_{mf} = U_0°/U_{mf}° \qquad (\epsilon_{mf} \le \epsilon \le 1) \qquad (2.94b)$$

$$v_0/U_0 = v_0°/U_0° \qquad (2.95)$$

$$\rho_f/\rho_p = \rho_f°/\rho_p° \qquad (2.96)$$

The last condition can be ignored when ρ_f/ρ_p is negligibly small.

Now, we have to choose the reference length scale l. Needless to say, particle size should not be taken as the reference scale, because our primary interest lies in the macroscopic flow structure of the plant scale. Thus, some characteristic length such as D_t is taken so that $m = l/l°$ represents the scale factor between geometrically similar columns. Accordingly, conditions (2.93) through (2.95) can be rewritten:

$$U_0/U_0° = v_0/v_0° = \sqrt{m} \qquad (2.97)$$

$$u_T = \sqrt{m}u_T° \qquad (\epsilon_{mf} \ll \epsilon \le 1) \qquad (2.98a)$$

$$U_{mf} = \sqrt{m}U_{mf}° \qquad (\epsilon_{mf} \le \epsilon \ll 1) \qquad (2.98b)$$

Conditions for particle properties are introduced only through the above relationship between u_T and $u_T°$ or U_{mf} and $U_{mf}°$. For the viscosity-dominant regime ($Ar \leq 104$), equations (2.98a) and (2.98b) both yield the same condition for d_p. That is:

$$\frac{d_p}{d_p^\circ} = m^{1/4} \left(\frac{\rho_p^\circ - \rho_f^\circ}{\rho_p - \rho_f} \cdot \frac{\mu}{\mu^\circ} \right)^{1/2} \quad (Ar \leq 104) \quad (2.99a)$$

For the inertia-dominant regime ($10^5 \leq Ar$):

$$\frac{d_p}{d_p^\circ} = m \frac{\rho_p^\circ - \rho_f^\circ}{\rho_p - \rho_f} \cdot \frac{\rho_f}{\rho_f^\circ} \quad (10^5 \leq Ar) \quad (2.99b)$$

As noted above, the judgment of the dominant mechanism is based on the Archimedes number Ar. The guideline of Glicksman (1988), i.e. $Re_p \equiv \rho_f U_0 d_p / \mu < 4$ for the viscosity-dominant regime, can be disregarded if the fluidizing gas velocity U_0 is considered as not being related to the criterion for particle size selection. In other words, equations (2.99) can be used regardless of the fluidizing gas velocity.

However, the argument based only on differential equations has the limitation of lacking information on macroscopic flow fields. If integrated solutions or their equivalents including semi-empirical correlations are utilized, the scaling law can be stated in a more realistic manner. For bubbling fluidized beds, for instance, the scaling law derived from the bubble coalescence/splitting model (Horio et al., 1986a) is stated as:

1. For Geldart Group B powders, the bubble fraction, bubble size distribution, solids circulation and mixing can be made similar among different scale models if the following condition is satisfied:

$$U_0 - U_{mf} = \sqrt{m}(U_0 - U_{mf})^\circ \quad (2.100)$$

2. For Group A powders, both bubble distribution and interstitial gas flow can be made similar if equation (2.98b) is satisfied, in addition to equation (2.100).

The above conditions have been experimentally validated for axial bubble size distribution, radial bubble flow distribution and solids mixing and segregation (Horio et al., 1986a, 1986b).

For the core-annulus flow in circulating fluidized beds, similar derivation of the scaling law from the integrated solution is possible (Horio et al., 1989). Let us briefly examine the derivation, which in turn gives us some idea how conditions (2.97) and (2.98) work.

The requirement for geometrically similar flow structure can be stated as:

1. same voidage distribution: $\epsilon(\zeta, \xi) = \epsilon^\circ(\zeta, \xi)$, $\epsilon_C^* = \epsilon_C^{*\circ}$ and $\epsilon_A^* = \epsilon_A^{*\circ}$
2. equal dimensionless core radius: $\alpha = \alpha^\circ$
3. similar division of gas flow to core and annulus: $u_A/u_C = u_A^\circ/u_C^\circ$

4. similar distribution of solids between core and annulus: $v_A/v_C = v_A^\circ/v_C^\circ$
5. same voidage in cluster: $\epsilon_{cl} = \epsilon_{cl}^\circ$

From equation (2.60) the annulus fraction $1 - \alpha^2$ is a function of the groups:

$$\frac{\epsilon_C}{1 - \epsilon_C} \frac{v_0}{u_{T,cl,C}} - \frac{u_0}{u_{T,cl,C}}, \quad \lambda_p, \quad \epsilon_C^{*n}, \quad \epsilon_A^{*n}$$

If all of the above four groups can be kept constant during the scale change, $1 - \alpha^2$ is unchanged. Therefore, with respect to λ_p, we can write:

$$\frac{\lambda_p}{\lambda_p^\circ} = \frac{D_t^\circ}{D_t} \left(\frac{u_{T,cl,C}}{u_{T,cl,C}^\circ}\right)^2 \equiv 1 \qquad \text{(for } \alpha = \alpha^\circ\text{)} \tag{2.101}$$

We then have:

$$u_{T,cl,C}/u_{T,cl,C}^\circ = \sqrt{m} \qquad \text{(for } \alpha = \alpha^\circ\text{)} \tag{2.102}$$

From conditions 1 through 4, equations (2.60) and (2.102) with equations (2.49) and (2.50), we reach the following necessary conditions for macroscopic flow structure:

$$U_0/U_0^\circ = v_0/v_0^\circ = u_{T,cl,C}/u_{T,cl,C}^\circ = \sqrt{m} \qquad \text{(for } \alpha = \alpha^\circ\text{)} \tag{2.103}$$

For the similarity of cluster sizes, substitution of equations (2.16) and (2.17) into equation (2.102) gives the following relationship for cluster size:

$$\frac{d_{cl}}{m d_{cl}^\circ} = \frac{\rho_{lean}(\rho_{cl} - \rho_{lean})^\circ}{\rho_{lean}^\circ(\rho_{cl} - \rho_{lean})}$$

$$= \frac{(1 - \epsilon_{lean})(\rho_p/\rho_f) + \epsilon_{lean}}{(1 - \epsilon_{lean}^\circ)(\rho_p/\rho_f)^\circ + \epsilon_{lean}^\circ} \cdot \frac{(\epsilon_{lean}^\circ - \epsilon_{cl}^\circ)((\rho_p/\rho_f)^\circ - 1)}{(\epsilon_{lean} - \epsilon_{cl})((\rho_p/\rho_f) - 1)} \tag{2.104}$$

Equation (2.104) implies that the cluster sizes become similar if the group on the right-hand side of equation (2.104) remains unity. Logically, two cases arise:

$$\frac{d_{cl}}{d_{cl}^\circ} = m \quad \text{if } \rho_f/\rho_p \ll 1 - \epsilon_{lean}, \qquad \rho_p = \rho_p^\circ \tag{2.105a}$$

$$\frac{d_{cl}}{d_{cl}^\circ} = m \quad \text{if } (\rho_p/\rho_f) = (\rho_p/\rho_f)^\circ \tag{2.105b}$$

Accordingly, there is a possibility of making cluster sizes similar between different scale models if the lean phase density is sufficiently high or if (ρ_p/ρ_f) is held constant.

With respect to requirement 5 ($\epsilon_{cl} = \epsilon_{cl}^\circ$) for the cluster phase, which should also be connected to the macroscopic flow field, the following conditions are required:

$$u_{sl,cl}/u_T = \epsilon_{cl}^{n-1} \quad \text{(Richardson–Zaki)} \quad (2.106)$$

$$u_{sl,cl}/U_0 = (u_{sl,cl}/U_0)° \quad (2.107)$$

Accordingly, the final condition reached is:

$$u_{sl,cl}/u_{sl,cl}° = u_T/u_T° = \sqrt{m} \quad (2.108)$$

The above set of equations (2.103), (2.104) or (2.105), and (2.108) is the same as the set of equations (2.97) and (2.98), but we now have more insight into the flow structure.

Scaling law validation has been successfully carried out by Ake and Glicksman (1988), Horio et al. (1989), Ishii and Murakami (1991), Chang and Louge (1992), Glicksman et al. (1993) and Glicksman et al. (1994). Figure 2.35 shows that the whole pressure loop, axial and radial profiles, pressure fluctuations and cluster size distributions can be made similar between two geometrically similar units.

In scaling experiments, attention must be paid also to particle shape and surface properties. Ake and Glicksman (1988) found that the shape of particles should not be too different in different scale models. By using coated particles, Chang and Louge (1992) found that the Coulomb friction of particles also affects pressure profiles, at least for dilute flow conditions.

2.5 Closing remarks

Complete description of circulating fluidized bed hydrodynamics is not yet possible after more than a decade of enthusiastic research. For practical purposes, the prediction of macroscopic flow structure can be performed based on the expressions introduced in this chapter. The scaling law can be applied to obtain more detailed information, with careful design of test units and particle properties and, of course, with sufficient instrumentation. Some significant issues still remaining are:

1. developing a general expression for the local gas–solid slip velocity versus drag force relationship for non-homogeneous clustering suspensions with corrections for wall effects;
2. developing quantitative description of the average length scale, shape and turbulent fluctuation of gas and solid motions; and
3. developing more precise description of boundary regions including the wall layer and the bottom zone; particular emphasis should be placed on the surface renewal of clusters and intermittent exchange of solids and gas between the wall layer and the core region.

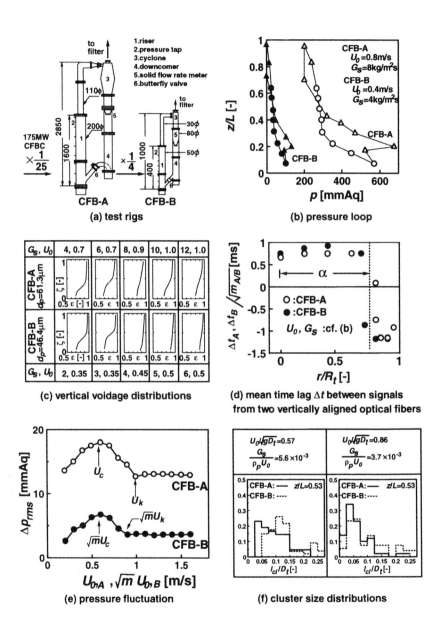

Figure 2.35 Validation of scaling low by Horio *et al.* (1989) ((a), (c), (d): Horio *et al.*, 1989; (e): Ishii and Murakami, 1991; (b), (f): Ishii, 1994) (CFB–A: FCC, $d_p = 61.3\,\mu\text{m}$, $\rho_p = 1780\,\text{kg/m}^3$; CFB–B: FCC, $d_p = 46.4\,\mu\text{m}$, $\rho_p = 1780\,\text{kg/m}^3$).

Stronger links between experimental, numerical and theoretical activities are now developing, which, sooner or later, will provide fruitful information for resolving these issues.

Acknowledgements

Many thanks to Professor Bo Leckner of Chalmers University of Technology who guided me to open my eyes to the flow structure of CFB boilers through his works and through valuable discussions. The author is indebted to Dr Elisa deCastro Boelman and her husband Mr Kees deJong for a critical reading of the manuscript. Many thanks are also due to my research associate Mayumi Tsukada and graduate students Mune Ito, Yuki Iwadate and Hongwei Lei for their great assistance.

Nomenclature

a	decay factor or parameter in equation (2.60), dimensionless or m/s
Ar	Archimedes number, $d_p^3 \rho_f(\rho_p - \rho_f)g/\mu^2$, dimensionless
A_t	cross-sectional area of riser, m^2
A_x	cross-sectional area of region x, m^2
b	parameter defined by equation (2.59), dimensionless
c_i	constant in equation (2.10), dimensionless
C_D	drag coefficient, dimensionless
d_{cl}	equivalent diameter of cluster, m
d_p	particle diameter, m
d_{p50}	median particle diameter, m
D_b	equivalent bubble diameter, m
D_C	core diameter, m
D_D	downcomer diameter, m
D_t	column diameter, m
E	pressure gradient parameter defined by equation (2.58), dimensionless
f_P	particle friction factor, dimensionless
$f(t)$	time-variation function in equation (2.7), dimensionless
F	gas flow rate, m^3/s
Fr_i	Froude number, U_i^2/gd_p, dimensionless
g	acceleration of gravity, m/s^2
$\vec{g}$	gravity acceleration vector, $(0, 0, -g)$, m/s^2
G_s	net solids mass flux, kg/m^{2}s
G_s'	local solids mass flux, kg/m^{2}s
G_{sb}^*	mass flux equivalent of gravity term defined by equation (2.29), kg/m^{2}s

Im	imaginary part
k_{xy}	lateral solids transfer coefficient from region x to y, m/s
k_1, k_2, k_3	x, y and z components of disturbance wave number vector, m^{-1}
K	equilibrium constant, k_{AC}/k_{CA}, dimensionless
l	length scale, m
l_{cl}	cluster length, m
L	riser height, m
L_f	fluidized bed height, m
L_i	length of part i, m
m	scale factor, dimensionless
M	virtual mass coefficient, dimensionless
n	constant in Richardson–Zaki correlation, dimensionless
n_i	constant in equation (2.10), dimensionless
N_D	drag force acting on clusters per unit riser volume, N/m^3
p	pressure, Pa
p_o	ambient pressure, Pa
P_s	particle–particle interaction stress tensor, Pa
r	radial coordinate, m
R	gas–solid interaction term defined by equation (2.5), Pa/m dimensionless
Re	Reynolds number, dimensionless
Re	real part
Re_i	particle Reynolds number, $d_p \rho_f U_i/\mu$ ($i = c, k, tr$), dimensionless
Re_p	particle Reynolds number, $d_p \rho_f U_0/\mu$, dimensionless
Re_T	particle Reynolds number, $d_p \rho_f u_T/\mu$, dimensionless
R_t	column radius, m
s_1, s_2	complex frequency introduced in equation (2.8), 1/s
S	Stokes number, $\rho_p d_p^2 u/18\mu D_t$, dimensionless
t	time, s
T	temperature, K
u	fluid velocity, m/s
u^*	gas velocity in lean phase, m/s
u_b	bubble rise velocity, m/s
u_{sl}	slip velocity, m/s
u_T	terminal velocity of particle, m/s
$u_{T,cl}$	terminal velocity of cluster, m/s
u_T^{**}	u_T or $u_{T,cl}$, m/s
u_x	U_{0x}/ϵ, m/s
U_c	onset velocity of bubbling-to-turbulent transition, m/s
U_i	U_c, U_k or U_{tr}, m/s
U_{i0}	U_i at ambient pressure, m/s
U_k	offset velocity of bubbling-to-turbulent transition, m/s
U_{mf}	minimum fluidization velocity, m/s
$U_{mf,cl}$	minimum fluidization velocity of cluster, m/s

U_{tr}	transport velocity, m/s
U_0	superficial gas velocity, m/s
U_{0x}	superficial gas velocity in region x, m/s
v	particle phase velocity, m/s
v'	local solid velocity, m/s
v_0	superficial solid velocity, G_s/ρ_p, m/s
v_{0x}	superficial solid velocity in region x, m/s
v_x	solid velocity in region x, $v_{0x}/(1-\epsilon)$, m/s
v^*	particle velocity in lean phase, m/s
$\dot{V}_b$	visible bubble flow rate per unit cross section, m/s
w	solid flow rate, kg/s
w_{down}	solid downflow rate, kg/s
w_{up}	solid upflow rate, kg/s
w_x	solid flow rate in region x (positive for upward flow), kg/s
W_D	solid hold-up in downcomer, kg
W_R	solid hold-up in riser, kg
W_t	total solid inventory, kg
x, y, z	Cartesian coordinates (z in upward vertical direction above distributor), m
z_b	bottom zone (including splash) height, m
z_{b0}	height of bottom bubbling zone, m
z_d	lower dense region height, m
z_D	downcomer bed height, m
α	dimensionless core diameter, D_C/D_t
β	gas–solid drag factor, kg/m³s
β'	parameter in equation (2.23b), dimensionless
γ	bubble velocity coefficient, dimensionless
γ_i	A_t/A_i, dimensionless
Δp_{ij}	pressure drop between points i and j, Pa
Δp_{rms}	root-mean-square pressure fluctuation, Pa
Δx	distance for disturbance amplitude to grow by a factor of $e = 2.718$, m
Δz_{Di}	height of section i in downcomer, m
ϵ	void fraction, dimensionless
ϵ_{cl}	void fraction of cluster phase, dimensionless
ϵ_b	bubble volume fraction, dimensionless
ϵ_p	local particle volume fraction, dimensionless
$\bar{\epsilon}_p$	cross-sectional average particle volume fraction, dimensionless
$\bar{\epsilon}_{pb}$	particle volume fraction in bottom zone, dimensionless
$\bar{\epsilon}_{pd}$	particle volume fraction in lower dense region, dimensionless
ϵ_{px}	average particle volume fraction in region x, dimensionless
$\epsilon_{px,d}$	ϵ_{px} at $z < z_d$, dimensionless
$\bar{\epsilon}_{p\infty}$	particle volume fraction in upper dilute region, dimensionless

ϵ_x	void fraction in region x, dimensionless
ϵ^*	void fraction of lean (i.e. cluster free) phase, dimensionless
ϵ^{**}	ϵ^* or ϵ_x, dimensionless
ϵ^*_{max}	maximum value of ϵ^*, dimensionless
ζ	z/L, dimensionless
λ	coefficient of momentum transfer between cluster and lean phases in equation (2.66), dimensionless
λ_p	$au_{T,C}^{**2+b}/gD_t$, dimensionless
μ	fluid viscosity, Pa·s
ξ	$2r/D_t$, dimensionless
ρ_{cl}	density of cluster phase, kg/m³
ρ_f	fluid density, kg/m³
ρ_g	gas density, kg/m³
ρ_{lean}	density of lean (i.e., cluster-free) phase, kg/m³
ρ_p	particle density, kg/m³
ρ_s^{**}	ρ_p or ρ_{cl}, kg/m³
τ_f	wall–fluid shear stress, Pa
τ_p	wall–solid shear stress, Pa
τ_α	core–annulus shear stress, Pa
ϕ	throughflow parameter, dimensionless
ϕ_{cl}	aspect ratio of cylindrical cluster, dimensionless
ϕ_s	shape factor, dimensionless
Φ	parameter in Table 2.1, dimensionless

Subscripts

A	annulus
a	aeration
b	bubble, bottom zone
b_0	bottom-bubbling bed
C	core
Cy	cyclone
cl	cluster, cluster phase
D	downcomer
DR	from downcomer to riser
d	lower dense region, downflow
dil	upper dilute region
eq	equilibrium
$exit$	riser exit
f	fluid
g	gas
$lean$	lean (i.e. cluster-free) phase
mf	minimum fluidization
o	orifice

p	particle
R	riser
s	solids
splash	splash zone
T	terminal
t	total
α	core/annulus boundary
∞	above TDH
0	superficial, ambient condition

Superscripts

$*$	cluster-free space (lean phase)
$-$	cross-sectional average
$\circ$	reference scale model
$\wedge$	dimensionless

References

Adánez, J., de Diego, L.F. and Gayán, P. (1993) Transport velocities of coal and sand particles. *Powder Technology*, 77, 61–68.

Adánez, J., Gayán, P., García-Labiano, F. and de Diego, L.F. (1994) Axial voidage profiles in fast fluidized beds. *Powder Technology*, 81, 259–268.

Ake, T.R. and Glicksman L.R. (1988) Scale model and full scale test results of a circulating fluidized bed combustor. *EPRI Seminar of Fluidized Bed Combustion Technology for Utility Applications*, Palo Alto, May 3–5.

Anderson, T.B. and Jackson, R. (1967) A fluid mechanical description of fluidized beds. *Industrial and Engineering Chemistry Fundamentals*, 6, 527–539.

Arena, U., Cammarota, A. and Pistone, L. (1986) High velocity fluidization behavior of solids in a laboratory scale circulating bed, in *Circulating Fluidized Bed Technology* (ed. P. Basu), Pergamon Press, Oxford, pp. 119–125.

Arena, U., Cammarota, A., Marzocchella, L. and Massimilla, L. (1989) Solid flow structures in a two-dimensional riser of a circulating fluidized bed. *Journal Chemical Engineering Japan*, 22, 236–241.

Avidan, A.A. and Yerushalmi, J. (1982) Bed expansion in high velocity fluidization. *Powder Technology*, 32, 223–232.

Bader, R., Findlay, J and Knowlton, T.M. (1988) Gas/solids flow patterns in a 30.5-cm-diameter circulating fluidized bed, in *Circulating Fluidized Bed Technology II* (eds P. Basu and J.F. Large), Pergamon Press, Oxford, pp. 123–137.

Bai, D.R. and Kato, K. (1994) Generalized correlations of solids holdups at dense and dilute regions of circulating fluidized beds. *Proceedings 7th SCEJ Symposium on CFB*, Society of Chemical Engineering Japan, Tokyo, 137–144.

Bai, D.R., Jin, Y. and Yu, Z.Q. (1991) Cluster observation in a two-dimensional fast fluidized bed, in *Fluidization '91: Science and Technology* (eds M. Kwauk and M. Hasatani), Science Press, Beijing, pp. 110–115.

Bai, D.R., Jin, Y., Yu, Z.Q. and Zhu, J.X. (1992) The axial distribution of the cross-sectionally averaged voidage in fast fluidized beds. *Powder Technology*, 71, 51–88.

Bai, D.R., Zhu, J.X., Jin, Y. and Yu, Z.Q. (1995) Internal recirculation flow structure in vertical upward flowing gas–solid suspensions, I. a core/annular model, *Powder Technology*, 85, 171–178.

Berruti, F. and Kalogerakis, N. (1989) Modeling of the internal flow structure of circulating fluidized beds. *Canadian Journal of Chemical Engineering*, **67**, 1010–1014.

Bi, H.T. and Fan, L.S. (1992) Existence of turbulent regime in gas–solid fluidization. *AIChE Journal*, **38**, 297–301.

Bi, H.T., Jiang, P.J. and Fan, L.S. (1991) Hydrodynamic behavior of the circulating fluidized bed with low density polymeric particles. *AIChE Meeting*, Paper 101d, Los Angeles.

Bi, H.T., Grace, J.R. and Zhu, J.X. (1993) On types of choking in vertical pneumatic systems. *International Journal of Multiphase Flow*, **19**, 1077–1092.

Bi, H.T., Zhou, J. and Grace, J.R. (1996) Annular wall layer thickness in circulating fluidized bed risers. *Canadian Journal of Chemical Engineering*, in press.

Bobkov, N.N. and Gupalo, Y.P. (1986) On the packet mechanism of mixing in a fluidized bed, in *Fluidization V* (eds K. Ostergaard and A. Sorensen), Engineering Foundation, New York, pp. 159–168.

Bolton, L.W. and Davidson, J.F. (1988) Recirculation of particles in fast fluidized risers, in *Circulating Fluidized Bed Technology II* (eds P. Basu and J.F. Large), Pergamon Press, Oxford, pp. 139–146.

Breault, R.W. and Mathur, V.K. (1989) High velocity fluidized bed hydrodynamic modeling. 1. Fundamental studies of pressure drop, 2. Circulating bed pressure drop modeling. *Industrial and Engineering Chemistry, Research*, **28**, 684–688 and 688–693.

Brereton, C.M.H. and Grace, J.R. (1994) End effects in circulating fluidized bed hydrodynamics, in *Circulating Fluidized Bed Technology IV* (ed. A.A. Avidan), AIChE, pp. 137–144.

Brereton, C.M.H., Grace, J.R. and Yu, J. (1988) Axial gas mixing in a circulating fluidized bed, in *Circulating Fluidized Bed Technology II* (eds. P. Basu and J.F. Large), Pergamon Press, Oxford, pp. 307–314.

Cai, P., Chen, S.P., Jin, Y., Yu, Z.Q. and Wang, Z.W. (1989) Effect of operating temperature and pressure on the transition from bubbling to turbulent fluidization. *AIChE Symposium Series*, **85**(270), 37–43.

Canada, G.S. and McLaughlin, M.H. (1978) Large particle fluidization and heat transfer at high pressures. *AIChE Symposium Series*, **74**(176), 27–37.

Canada, G.S., McLaughlin, M.H. and Staub, F.W. (1978) Flow regimes and void fraction distribution in gas fluidization of large particles in beds without tube banks. *AIChE Symposium Series*, **74**(176), 14–26.

Caram, H.S., Efes, Z. and Levy, E.K. (1984) Gas and particle motion induced by a bubble eruption at the surface of a gas fluidized bed. *AIChE Symposium Series*, **80**(234), 106–113.

Carotenuto, L., Crescitelli, S. and Donsi, G. (1974) High velocity behavior of fluidized beds: characteristic flow regimes. *Quaderni dell' Ingegnere Chemico Italiano*, **10**, 185–193.

Chang, H. and Louge, M. (1992) Fluid dynamic similarity of circulating fluidized beds. *Powder Technology*, **70**, 259–270.

Chen, B., Li, Y., Wang, F., Wang, S. and Kwauk, M. (1980) The formation and prediction of fast fluidized beds. *Chemical Metallurgy*, **4**, 20.

Clift, R. and Grace, J.R. (1971) Coalescence of bubbles in fluidized beds. *AIChE Symposium Series*, **67**(116), 23–33.

Davidson, J.F. (1961) Symposium on Fluidization – discussion. *Transactions Institution of Chemical Engineers*, **39**, 223–240.

Ergun, S. (1952) Fluid flow through packed columns. *Chemical Engineering Progress*, **48**, 89–94.

Fitzgerald, T., Bushnell, D.B., Crane, S. and Shieh, Y. (1983) Testing of cold scaled bed modeling for fluidized bed combustor, in *Proceedings 7th International Conference on Fluidized Bed Combustion, DOE/METEC*, vol. 2, pp. 766–780.

Fitzgerald, T., Bushnell, D.B., Crane, S. and Shieh, Y. (1984) Testing of cold scaled bed modeling for fluidized bed combustor. *Powder Technology*, **38**, 107–120.

Fortes, A.F., Joseph, D.D. and Lundgren, T.S. (1987) Nonlinear mechanics of fluidization of beds of spherical particles. *Journal Fluid Mechanics*, **177**, 467–483.

Foscolo, P.U., Felice, R.D., Gibilaro, L.G., Pistone, L. and Piccolo, V. (1990) Scaling relationships for fluidization: the generalized particle bed model. *Chemical Engineering Science*, **45**, 1647–1651.

Geldart, D. and Rhodes, M.J. (1986) From minimum fluidization to pneumatic transport – a critical review of the hydrodynamics, in *Circulating Fluidized Bed Technology* (ed. P. Basu), Pergamon Press, Oxford, pp. 21–31.

Gelperin, N.I. and Einstein, V.G. (1971) The analogy between fluidized beds and liquids, in *Fluidization* (eds J.F. Davidson and D. Harrison), Academic Press, London, pp. 541–568.

Glicksman, L.R. (1984) Scaling relationships for fluidized beds. *Chemical Engineering Science*, **39**, 1373–1379.

Glicksman, L.R. (1988) Scaling relationships for fluidized beds. *Chemical Engineering Science*, **43**, 1419–1421.

Glicksman, L.R., Hyre, M. and Woloshun, K. (1993) Simplified scaling relationships for fluidized beds. *Powder Technology*, **77**, 177–199.

Glicksman, L.R., Hyre, M.R. and Farrell, P.A. (1994) Dynamic similarity in fluidization. *International Journal Multiphase Flow*, **20**, 331–386.

Grace, J.R. and Tuot, J. (1979) A theory for cluster formation in vertically conveyed suspensions of intermediate density. *Transactions of the Institution of Chemical Engineers*, **57**, 49–54.

Han, G.Y., Lee, G.S. and Kim, S.D. (1985) Hydrodynamics of a circulating fluidized bed. *Korean Journal Chemical Engineering*, **2**, 141–147.

Hartge, E.U., Rensner, D. and Werther, J. (1988) Solid concentration and velocity patterns in circulating fluidized beds, in *Circulating Fluidized Bed Technology II* (eds P. Basu and J.F. Large), Pergamon Press, Oxford, pp. 165–180.

Hashimoto, O. (1989) Study of methanol synthesis reactor by turbulent fluidized beds, Doctoral Thesis, Nagoya Institute of Technology, Nagoya.

Hashimoto, O., Haruta, T., Mochizuki, K., Matsutani, W., Mori, S., Hiraoka, S., Yamada, I., Kojima, T. and Tsuji, K. (1988) Criteria for transition to turbulent fluidization in the high-velocity circulating fluidized bed. *Kagaku Kougaku Ronbunshu*, **14**, 309–315.

Hatano, H. and Kido, N. (1991) Visualization of solid particles flowing in circulating fluidized beds, in *Proceedings International Conference on Multiphase Flows, '91 – Tsukuba* (eds G. Matsui, A. Serizawa and Y. Tsuji), vol. 1, pp. 295–299.

Herb, B., Tuzla, K. and Chen, J.C. (1989) Distribution of solid concentrations in circulating fluidized bed, in *Fluidization VI* (eds J.R. Grace, L.W. Shemilt and M.A. Bergougnou), Engineering Foundation, New York, pp. 65–72.

Hirama, T., Takeuchi, H. and Chiba, T. (1992) Regime classification of macroscopic gas–solid flow in a circulating fluidized bed riser. *Powder Technology*, **70**, 215–222.

Horio, M. (1986) High velocity operation of fluidized beds. *Journal Powder Technology Japan*, **23**, 80–90.

Horio, M. (1991) Hydrodynamics of circulating fluidization – Present status and research needs, in *Circulating Fluidized Bed Technology III* (eds P. Basu, M. Horio and M. Hasatani), Pergamon Press, Oxford, pp. 3–14.

Horio, M. and Clift, R. (1992) A note on terminology: 'cluster' and 'agglomerates'. *Powder Technology*, **70**, 196.

Horio, M., Ishii, H., Kobukai, Y. and Yamanishi, N. (1989) A scaling law for circulating fluidized beds. *Journal of Chemical Engineering Japan*, **22**, 587–592.

Horio, M., Ishii, H. and Nisimuro, M. (1992a) On the nature of turbulent and fast fluidized beds. *Powder Technology*, **70**, 229–236.

Horio, M. and Ito, M. (1996) Prediction of cluster size in circulating fluidized beds. *Preprint DGS 22 for Fifth International Conference on Circulating Fluidized Beds*, Beijing.

Horio, M. and Kuroki, H. (1994) Three dimensional flow visualization of dilutely dispersed solids in bubbling and circulating fluidized beds. *Chemical Engineering Science*, **49**, 2413–2421.

Horio, M. and Lei, H. (1996) Significance of core diameter variation in the lateral solids transfer model for circulating fluidized beds, submitted to *AIChE Journal*.

Horio, M. and Morishita, K. (1988) Flow regimes of high velocity fluidization. *Japanese Journal Multiphase Flow*, **2**, 117–136.

Horio, M., Morishita, K., Tachibana, O. and Murata, N. (1988) Solid distribution and movement in circulating fluidized beds, in *Circulating Fluidized Bed Technology II* (eds P. Basu and J.F. Large), Pergamon Press, Oxford, pp. 147–154.

Horio, M., Mori, K., Takei, Y. and Ishii, H. (1992b) Simultaneous gas and solid velocity measurements in turbulent and fast fluidized beds, in *Fluidization VII* (eds O.E. Potter and D.J. Nicklin), Engineering Foundation, New York, pp. 757–762.

Horio, M. and Nonaka, A. (1986) A generalized bubble diameter correlation for gas–solid fluidized beds. *AIChE Journal*, **33**, 1865–1872.

Horio, M., Nonaka, A., Sawa, Y. and Muchi, I. (1986a) A new similarity rule for fluidized bed scale-up. *AIChE Journal*, **32**, 1466–1482.

Horio, M., Takada, M., Ishida, M. and Tanaka, N. (1986b) The similarity rule of fluidization and its application to solid mixing and circulation control, in *Fluidization V* (eds K. Ostergaard and A. Sorensen), Engineering Foundation, New York, pp. 151–158.

Horio, M. and Takei, Y. (1991) Macroscopic structure of recirculating flow of gas and solids in circulating fluidized beds, in *Circulating Fluidized Bed Technology III* (eds P. Basu, M. Horio and M. Hasatani), Pergamon Press, Oxford, pp. 207–212.

Horio, M., Taki, A., Hsieh, Y.S. and Muchi, I. (1980) Elutriation and particle transport through the freeboard of a gas–solid fluidized bed, in *Fluidization III* (eds J.R. Grace and J.M. Matsen), Plenum Press, New York and London, pp. 509–518.

Ishii, H. (1994) A study on flow structure and control of circulating fluidized beds, Doctoral Thesis, Tokyo University of Agriculture and Technology, Tokyo.

Ishii, H. and Horio, M. (1991) The flow structures of a circulating fluidized bed. *Advanced Powder Technology*, **2**, 25–36.

Ishii, H. and Murakami, M. (1991) Evaluation of the scaling law of circulating fluidized beds in regard to cluster behaviors, in *Circulating Fluidized Bed Technology III* (eds P. Basu, M. Horio and M. Hasatani), Pergamon Press, pp. 125–130.

Ishii, H., Nakajima, T. and Horio, M. (1989) The clustering annular flow model of circulating fluidized beds. *Journal of Chemical Engineering Japan*, **22**, 484–490.

Jackson, R. (1963) The mechanics of fluidized beds: I. The stability of the state of uniform fluidization. *Transactions of the Institution of Chemical Engineers*, **41**, 13–21.

Jiang, P.J. and Fan, L.S. (1991) Regime transition in gas–solids circulating fluidized beds. *AIChE Meeting*, paper 101e, Los Angeles.

Jin, Y., Yu, Z.Q., Qi, C. and Bai, D.R. (1988) The influence of exit structures on the axial distribution of voidage of fast fluidized bed, in *Fluidization '88: Science and Technology* (eds M. Kwauk and D. Kunii), Science Press, Beijing, pp. 165–173.

Johnsson, F. and Leckner, B. (1995). Vertical distribution of solids in a CFB-furnace, in *Proceedings 13th ASME International Conference on Fluidized Bed Combustion*, Orlando, pp. 671–679.

Johnsson, F., Svensson, A. and Leckner, B. (1992) Fluidization regimes in circulating fluidized bed boilers, in *Fluidization VII* (eds O.E. Potter, and D.J. Nicklin), Engineering Foundation, New York, pp. 471–478.

Jones, D.R. and Davidson, J.F. (1965) The flow of particles from a fluidized bed through an orifice. *Rheologica Acta*, **4**, 180–192.

Joseph, D.D., Singh, P. and Fortes, A. (1991) Nonlinear and finite size effects in fluidized suspensions, in *University of Minnesota Supercomputer Institute Research Report*, UMSI 91/232, Chapter 10.

Kehoe, P.W.K. and Davidson, J.F. (1971) Continuously slugging fluidized beds. *Institution of Chemical Engineers Symposium Series*, **33**(37), 97–116.

Konno, H. and Saito, S. (1969) Pneumatic conveying of solids through straight pipes. *Journal Chemical Engineering Japan*, **2**, 211–217.

Knowlton, T. (1995) Interaction of pressure and diameter on CFB pressure drop and holdup. Paper for *Workshop: 'Modeling and Control of Fluidized Bed Systems'*, Hamburg, May 22–23.

Kunii, D. and Levenspiel, O. (1990) Entrainment of solids from fluidized beds. *Powder Technology*, **61**, 193–206.

Kunii, D. and Levenspiel, O. (1991) Flow modeling of fast fluidized beds, in *Circulating Fluidized Bed Technology III* (eds P. Basu, M. Horio and M. Hasatani), Pergamon Press, Oxford, pp. 91–98.

Kunii, D. and Levenspiel, O. (1995). Effect of exit geometry on the vertical distribution of solids in circulating fluidized beds. *Powder Technology*, **84**, 83–90.

Kuroki, H. and Horio, M. (1994) The flow structure of a three-dimensional circulating fluidized bed observed by the laser sheet technique, in *Circulating Fluidized Bed Technology IV* (ed. A.A. Avidan), AIChE, New York, pp. 77–84.

Kwauk, M., Wang, N., Li, Y., Chen, B. and Shen, Z. (1986) Fast fluidization at ICM, in *Circulating Fluidized Bed Technology* (ed P. Basu), Pergamon Press, Oxford, pp. 33–62.

Lafanechere, L. and Jestin, L. (1995) Study of a circulating fluidized bed furnace behavior in order to scale it up to 600 MWe, in *Proceedings 13th ASME International Conference on Fluidized Bed Combustion*, Orlando, pp. 971–977.

Lanneau, K.P. (1960) Gas–solids contacting in fluidized beds. *Transactions Institution of Chemical Engineers*, **38**, 125–143.

Leckner, B. and Andersson, B. Å. (1992) Characteristic features of heat transfer in circulating fluidized bed boilers. *Powder Technology*, **70**, 303–314.

Leckner, B., Goriz, M.R., Zhang, W., Andersson, B.Å. and Jonsson, F. (1991) Boundary layers – first measurements in the 12 MW CFB research plant at Charmers University, *Proceedings of 11th International Conference on Fluidized Bed Combustion* (ed. E.J. Anthony), ASME, Montreal, pp. 771–776.

Lee, G.S. and Kim, S.D. (1982) The vertical pneumatic transport of cement raw meal. *Hwahak Konghak*, **20**, 207–216.

Lee, G.S. and Kim, S.D. (1988) Pressure fluctuations in turbulent fluidized beds. *Journal of Chemical Engineering Japan*, **21**, 515–521.

Lee, G.S. and Kim, S.D. (1990) Bed expansion characteristics and transition velocity in turbulent fluidized beds. *Powder Technology*, **62**, 207–215.

Lei, H. W. and Horio, M. (1996) Prediction of voidage distribution in circulating fluidized beds, *Proceedings of the 1st SCEJ Symposium on Fluidization*, Society of Chemical Engineers Japan, pp. 154–161.

Le Palud, T. and Zenz, F.A. (1989) Supercritical phase behavior of fluid-particle systems, in *Fluidization VI* (eds J.R. Grace, L.W. Shemilt and M.A. Bergougnou), Engineering Foundation, New York, pp. 121–128.

Leu, L.P., Lin, C.C. and Huang, J.W. (1990) Axial pressure distribution in turbulent fluidized beds, in *Proceedings of Asian Conference on Fluidized Beds and Three Phase Reactors* (eds K. Yoshida and S. Morooka), SCEJ/KICE/CICE, pp. 171–178.

Leung, L.S. and Jones, P.J. (1978) Coexistence of fluidized solids flow and packed bed flow in standpipes, in *Fluidization* (eds J.F. Davidson and D.L. Kearins), Cambridge University Press, Cambridge, pp. 116–121.

Lewis, W.K., Gilliland, E.R. and Bauer, W.C. (1949) Characteristics of fluidized particles. *Industrial and Engineering Chemistry*, **41**, 1104–1117.

Lewis, W.K., Gilliland, E.R. and Lang, P.M. (1962) Entrainment from fluidized beds. *Chemical Engineering Progress Symposium Series*, **58**(38), 65–78.

Li, H., Xia, Y., Tung, Y. and Kwauk, M. (1991a) Macro-visualization of cluster in a fast fluidized bed. *Powder Technology*, **66**, 231–235.

Li, J., Reh, L. and Kwauk, M. (1991b) Application of the principle of energy minimization to the fluid dynamics of circulating fluidized beds, in *Circulating Fluidized Bed Technology III* (eds P. Basu, M. Horio and M. Hasatani), Pergamon Press, Oxford, pp. 105–111.

Li, J., Tung, Y. and Kwauk, M. (1988) Energy transport and regime transition in particle-fluid two-phase flow, in *Circulating Fluidized Bed Technology II* (eds P. Basu and J.F. Large), Pergamon Press, Oxford, pp. 75–87.

Li, Y. and Kwauk, M. (1980) The dynamics of fast fluidization, in *Fluidization III* (eds J.R. Grace and J.M. Matsen), Plenum Press, New York, pp. 537–544.

Marzocchella, A. (1991) Fluidodinamica di sistemi solido–gas nelle 'colonne veloci' di letti fluidi circolanti, Ph.D. Thesis, University of Naples, Napoli.

Massimilla, L. (1973) Behavior of catalytic beds of fine particles at high gas velocities, *AIChE Symposium Series*, **69**(128), 11–13.

Matsen, J.M. (1982) Mechanism of choking and entrainment. *Powder Technology*, **31**, 21–34.

Miyauchi, T., Furusaki, S., Morooka, S. and Ikeda, Y. (1981) Transport phenomena and reaction in fluidized catalyst beds. *Advances in Chemical Engineering*, **11**, 275–448.

Mori, S., Hashimoto, O., Haruta, T., Yamada, I., Kuwa, M. and Saito, Y (1989) Fundamentals of turbulent fluidized catalytic reactor, in *Fluidization VI* (eds J.R. Grace, L.W. Shemilt and M.A. Bergougnou), Engineering Foundation, New York, pp. 49–56.

Moriya, S., Sakamoto, H., Kiya, M and Arie, M. (1983) Pressure fluctuation and fluid force acting on two cylinders. *Kikai Gakkai Ronbunsyu (B)*, **49**, 1364–1371.

Mueller, P. and Reh, L. (1994) Particle drag and pressure drop in accelerated gas–solid flow, in *Circulating Fluidized Bed Technology IV* (ed. A.A. Avidan) AIChE, New York, pp. 159–166.

Nakamura, K. and Capes, C.E. (1973) Vertical pneumatic conveying: a theoretical study of uniform and annular particle flow models. *Canadian Journal of Chemical Engineering*, **51**, 39–46.

Nicastro, M.T. and Glicksman, L.R. (1984) Experimental verification of scaling relationships. *Chemical Engineering Science*, **39**, 1381–1391.

Patience, G.S. and Chaouki, J. (1995) Solids hydrodynamics in the fully developed region of CFB risers, in *Preprints Fluidization VIII*, Tours, pp. 33–40.

Pemberton, S.T. and Davidson, J.F. (1986) Elutriation from fluidized beds – II. Disengagement of particles from gas in the freeboard. *Chemical Engineering Science*, **41**, 253–262.

Perales, J.F., Coll, T., Llop, M.F., Puigjaner, L., Arnaldos, J. and Casal, J. (1991) On the transition from bubbling to fast fluidization regimes, in *Circulating Fluidized Bed Technology III* (eds P. Basu, M. Horio and M. Hasatani), Pergamon Press, Oxford, p. 73–78.

Pritchett, J.W., Blake, T.R. and Garg, S.K. (1978) A numerical model of gas fluidized beds. *AIChE Symposium Series*, **176**(74), 134–148.

Pugsley, T.S., Berruti, F., Chaouki, J. and Patience, G.S. (1994) A predictive model for the gas–solid flow structure in circulating fluidized bed risers, in *Circulating Fluidized Bed Technology IV* (ed. A.A. Avidan), AIChE, New York, pp. 40–47.

Qi, C. and Farag, I. (1993) Lateral particle motion and its effect on particle concentration distribution in the riser of CFB. *AIChE Symposium Series*, **89**(296), 73–80.

Quadflieg, von H. (1977) Wirbelinduzielte Belastungen eines Zylinderpaares in inkompressibler Stroemungbei grosen Reynoldszahlen. *Forschung im Ingenieurwesen*, **43**, 9–18.

Reh, L. (1971) Fluidized bed processing. *Chemical Engineering Progress*, **67**, 58–63.

Rhodes, M.J. (1990) Modeling flow structure of upward-flowing gas–solid suspensions. *Powder Technology*, **60**, 27–38.

Rhodes, M.J., Hirama, T., Cerutti, G. and Geldart, D. (1989) Non-uniformities of solids flow in the risers of circulating fluidized beds, in *Fluidization VI* (eds J.R. Grace, L.W. Shemilt and M.A. Bergougnou), Engineering Foundation, New York, pp. 73–80.

Rhodes, M.J., Wang, X.S., Cheng, H. and Gibbs, B.M. (1992) Similar profiles of solids flux in circulating fluidized-bed risers. *Chemical Engineering Science*, **47**, 1635–1634.

Richardson, J.F. and Zaki, W.N. (1954) Sedimentation and fluidization. *Transactions Institution Chemical Engineers*, **32**, 35–53.

Satija, S. and Fan, L.S. (1985) Characteristics of slugging regime and transition to turbulent regime for fluidized beds of large coarse particles, *AIChE Journal*, **31**, 1554–1565.

Saxton, A.L. and Worley, A.C. (1970) Modern catalytic cracking design. *Oil and Gas Journal*, **68**, 82–99.

Shimizu, M., Mizuhata, Y. and Morita, N. (1964) Solid loading in dilute-phase fluidized beds. *Kagaku Kogaku*, **28**, 595–600.

Shin, B.C., Koh, Y.B. and Kim, S.D. (1984) Hydrodynamics of coal combustion characteristics of circulating fluidized bed. *Hwahak Konghak*, **22**, 253–258.

Son, J.E., Choi, J.H. and Lee, C.K. (1988) Hydrodynamics in a large circulating fluidized bed, in *Circulating Fluidized Bed Technology II* (eds P. Basu and J.F. Large), Pergamon Press, Oxford pp. 113–120.

Squires, A.M. (1982) Contribution toward a history of fluidization, in *Proceedings Joint Meeting of Chemical Industrial Engineering Society China and AIChE*, Beijing, pp. 322–353.

Squires, A., Kwauk, M. and Avidan, A.A. (1985) Fluid beds: at last, challenging two entrenched practices, *Science*, **230**, 1329–1337.

Stemerding, S. (1962) The pneumatic transport of cracking catalyst in vertical risers. *Chemical Engineering Science*, **17**, 599–608.

Takeuchi, H. and Hirama, T. (1991) Flow visualization in the riser of a circulating fluidized bed, in *Circulating Fluidized Bed Technology III* (eds P. Basu, M. Horio and M. Hasatani), Pergamon Press, Oxford, pp. 177–182.

Takeuchi, H., Hirama, T. and Leung, L.S. (1986a) On the regime of fast fluidization, in *Proceedings World Congress III Chemical Engineering*, vol. III, pp. 477–480.

Takeuchi, H., Hirama, T., Chiba, T., Baswas, J. and Leung, L.S. (1986b) A quantitative definition and flow regime diagram for fast fluidization. *Powder Technology*, **47**, 195–199.

Tanaka, I., Shinohara, H., Hirasue, H. and Tanaka, Y. (1972) Elutriation of fines from fluidized bed. *Journal Chemical Engineering Japan*, **5**, 51–57.

Tanaka, T., Yonemura, S., Kiribayashi, K. and Tsuji, Y. (1993) Cluster formation and particle-induced instability of gas–solid flows (numerical simulation of flow in vertical channel using the DSMC method). *Kikai Gakkai Ronbunshu (B)*, **59**, 2983–2989.

Tanner, H., Li, J. and Reh, L. (1994) Radial profiles of slip velocity between gas and solids in circulating fluidized beds. *AIChE Symposium Series*, **90**(301), 105–113.

Thiel, W.J. and Potter, O.E. (1977) Slugging in fluidized beds. *Industrial and Engineering Chemistry, Fundamentals*, **16**, 242–247.

Toomey, R.D. and Johnstone, H.F. (1952) Gaseous fluidization of solid particles. *Chemical Engineering Progress*, **48**, 220–226.

Tsukada, M. (1995) Fluidized bed hydrodynamics, heat transfer and high temperature process developments, Doctoral Thesis, Tokyo University of Agriculture and Technology, Tokyo.

Tsukada, M., Nakanishi, D. and Horio, M. (1993) The effect of pressure on the phase transition from bubbling to turbulent fluidization. *International Journal Multiphase Flow*, **19**, 27–34.

Tsukada, M., Nakanishi, D. and Horio, M. (1994) Effect of pressure on 'transport velocity' in a circulating fluidized bed, in *Circulating Fluidized Bed Technology IV* (ed. A.A. Avidan), AIChE, New York, pp. 209–215.

Tsuo, T.P. and Gidaspow, D. (1990) Computation of flow in circulating fluidized beds. *AIChE Journal*, **36**, 885–896.

van Breugel, J.W., Stein, J.J.M. and de Vries, R.J. (1969) Isokinetic sampling in a dense gas–solids stream. *Proceedings of the Institution of Mechanical Engineers*, **184**, 18–23.

Wang, T., Lin, Z.J., Zhu, C.M., Liu, D.C. and Saxena, S.C. (1993) Particle velocity measurements in a circulating fluidized bed, *AIChE Journal*, **39**, 1406–1410.

Wang, X.S. and Gibbs, B.M. (1991) Hydrodynamics of a circulating fluidized bed with secondary air injection, in *Circulating Fluidized Bed Technology III* (eds P. Basu, M. Horio and M. Hasatani), Pergamon Press, Oxford, pp. 225–230.

Weinstein, H., Graff, R.F., Meller, M. and Shao, M.J. (1984) The influence of the imposed pressure drop across a fast fluidized bed, in *Fluidization IV* (eds D. Kunii and R. Toei), Engineering Foundation, New York, pp. 299–306.

Werther, J. (1977) Strömungs mechanische grundlagen der Wirbelschichttechnik. *Chemie Ingenieuer Technik*, **49**, 193–202.

Werther, J. (1993) Fluid mechanics of large-scale CFB units, in *Circulating Fluidized Bed Technology IV* (ed. A.A. Avidan), AIChE, New York, pp. 1–4.

Werther, J. and Wein, J. (1994) Expansion behavior of gas fluidized beds in the turbulent regime. *AIChE Symposium Series*, **90**(301), 31–44.

Wirth, K.E. (1991) Fluid mechanics of circulating fluidized beds. *Chemical Engineering Technology*, **14**, 29–38.

Yang, C.V. and Chitester, D.C. (1988) Transition between bubbling and turbulent fluidization at elevated pressure. *AIChE Symposium Series*, **84**(262), 10–21.

Yang, W.C. (1976) A criterion for 'fast fluidization', in *Pneumatic Transport 3*, BHRA Fluid Engineering, paper E5-49-55.

Yang, W.C. (1988) A model for the dynamics of a circulating fluidized bed loop, in *Circulating Fluidized Bed Technology II* (eds P. Basu and J.F. Large), Pergamon Press, Oxford, pp. 181–191

Yang, Y.L., Jin, Y., Yu, Z.Q. and Wang, Z.W. (1992) Investigation on slip velocity distribution in the riser of dilute circulating fluidized bed. *Powder Technology*, **73**, 67–73.

Yang, Y.L., Jin, Y., Yu, Z.Q., Zhu, J.X. and Bi, H.T. (1993) Local slip behaviors in the circulating fluidized bed. *AIChE Symposium Series*, **89**(296), 81–90.

Yang, Y.R., Rong, S.X., Chen, G.T. and Chen, B.C. (1990) Flow regimes and regime transitions in turbulent fluidized beds. *Chemical Reaction Engineering and Technology*, **6**, 9.

Yerushalmi, J. and Cankurt, N.T. (1979) Further studies of the regimes of fluidization. *Powder Technology*, **24**, 187–205.

Yerushalmi, J. and Squires, A.M. (1977) The phenomenon of fast fluidization, *AIChE Symposium Series*, **73**(161), 44–50.

Yerushalmi, J., Turner, D.H. and Squires, A.M. (1976a) The fast fluidized bed. *Industrial and Engineering Chemistry, Process Design and Development*, **15**, 47–53.

Yerushalmi, J., Gluckman, M.J., Graff, R.A., Dobner, S. and Squires, A.M. (1976b) Production of gaseous fuels from coal in the fast fluidized bed, in *Fluidization Technology* (ed. D.L. Keairns), Hemisphere Publishing Corp., Washington, vol. II, pp. 437–469.

Yerushalmi, J., Cankurt, N.T., Geldart, D. and Liss, B. (1978) Flow regimes in vertical gas–solid contact systems, *AIChE Symposium Series*, **74**(176), 1–3.

Yonemura, S., Tanaka, T. and Tsuji, Y. (1995) Cluster formation in dispersed gas–solid flow (Effects of physical properties of particles), in *Proceedings Second International Conference on Multiphase Flows, '95 – Kyoto*, vol. 3, PT4-25-30.

Yousfi, Y. and Gau, G. (1974) Aérodynamique de l'écoulement vertical de suspensions concentrées gaz–solides, I. Régimes d'écoulement et stabilité aérodynamique. *Chemical Engineering Science*, **29**, 1939–1946.

Zenz, F.A (1994) Predicting the degree of particle refluxing in cocurrent upflow risers, in *Circulating Fluidized Bed Technology IV* (ed A.A. Avidan) AIChE, New York, pp. 594–599.

Zenz, F.A. and Weil, N.A. (1958) A theoretical–empirical approach to the mechanism of particle entrainment from fluidized beds. *AIChE Journal*, **4**, 472–479.

Zhang, M.C. and Yang, R.Y.K. (1987) On the scaling laws for bubbling gas–fluidized bed dynamics. *Powder Technology*, **51**, 159–165.

Zhang, W., Tung, Y. and Johnsson, F. (1991) Radial voidage profiles in fast fluidized beds of different diameters. *Chemical Engineering Science*, **46**, pp. 3045–3052.

Zheng, Q.Y. and Zhang, H. (1994) Experimental study of the effect of exit (end effect) geometric configuration on internal recycling of bed material in CFB combustor, in *Circulating Fluidized Bed Technology IV* (ed. A.A. Avidan), AIChE, New York, pp. 145–151.

Zhou, J., Grace, J.R., Qin, S., Brereton, C.M.H., Lin, C.J. and Zhu, J. (1994) Voidage profiles in a circulating fluidized bed of square cross-section, *Chemical Engineering Science*, **49** 3217–3226.

Zhou, J., Grace, J.R., Lim, G.J., Brereton, C.M.H., Qin, S. and Lim, K.S. (1995) Particle cross-flow, lateral momentum flux and lateral velocity in a circulating fluidized bed. *Canadian Journal of Chemical Engineering*, **73**, 612–619.

3 Gas mixing
UMBERTO ARENA

3.1 Introduction

The dispersion of gas phase in the riser of a circulating fluidized bed exerts great, sometimes even crucial, influence on the performance of a CFB gas–solids reactor. Incomplete mixing of secondary air streams into the gas–solids suspension rising from the bottom region of a CFB combustor may lead to hydrocarbon emissions from sub-stoichiometric zones as well as NO_x emissions from over-stoichiometric zones. Optimal axial and radial dispersion of the gas phase may be crucial in achieving high conversion and/or high selectivity in some gas-conversion processes such as catalytic cracking. An adequate understanding of the gas mixing process may shed light on other issues, such as the evolution of volatiles from fuel particles or the optimal distribution of feed points of a reactant. It is therefore essential for practical design as well as for successful modelling of reactor behaviour.

A distinction is usually made between macro-scale and micro-scale mixing. The former is related to gas and solids flow patterns inside the unit, which greatly affect vertical mixing phenomena, and, on a smaller scale, to the turbulent eddies produced by differences in vertical velocities, which exert a key role in lateral mixing (van Deemter, 1985). Micromixing is related to molecular transport phenomena and becomes of interest in fast chemical reactions, like combustion of coal volatiles (Stubington and Chan, 1990). The mixing of gas on the smallest scale will not be discussed further considering that it can generally be taken into account by means of mass transfer considerations (van Deemter, 1985).

The gas dispersion in a fluidized bed greatly depends on the prevailing flow regime. Moving from bed captive regimes typical of low velocity fluidized beds (LFBs) to transport regimes proper of CFBs, the mixing patterns necessarily become different. In a large-diameter bubbling bed, for instance, operated at a gas velocity of 1 m/s with sand or limestone particles of diameter at least 0.5 mm, gas mixing is qualitatively and quantitatively different from that occurring in a small-diameter high velocity riser, operated at a gas velocity of 15 m/s with FCC catalyst of mean diameter about 0.06 mm. As a consequence, investigations on mixing phenomena in captive (bubbling and slugging) fluidized beds have limited applicability to the more homogeneous high velocity fluidization regimes. Excellent surveys of experimental results and modelling studies on gas and solids mixing in fluidized beds have been published by Potter (1971) and van Deemter (1985). They could not adequately analyse gas mixing in CFB units since the majority of related studies date from the years following 1987. For the same reason, the paragraphs that Yerushalmi and Avidan (1985) and Yerushalmi (1986) dedicated to mixing phenomena in their reviews on high velocity fluidization regimes should be considered as the first approaches to the problem.

Starting from these fundamental works, an effort has been made to summarize the major gas mixing studies of the last decade. The objective is to explain existing discrepancies on the basis of the complex hydrodynamics of CFBs and of differences in apparatus, operating conditions, diagnostic techniques and models used to interpret the measurements. Analysis of the various investigations reported so far suggests that discrepancies result mainly from the different states of gas–solids flow in the riser. Due to the possible presence along the riser of various flow structures at the meso-scale (particle clusters and strands) and at the macro-scale (possible co-existence in the vertical direction of a bottom dense, a middle transition and a top dilute region, and, in the horizontal direction, of a dilute core and a dense annulus), gas mixing in the whole reactor is different from that occurring in a specific region (Li and Weinstein, 1989; Li and Wu, 1991). This also implies that in order to allow a correct analysis of each investigation, adequate information should be reported regarding the specific fluidization regime in the reactor, and/or in the particular zone of the reactor where the study has been carried out.

Another difficulty in comparing studies of this kind lies in the significant differences between the fluid mechanics of the two main groups of typical applications of CFB technology. Gas-conversion processes and solids-conversion processes have such different design and operating characteristics (Werther, 1993; Zhu and Bi, 1993) that gas mixing phenomenology as well as gas mixing requirements are necessarily different. In a gas phase reaction process, such as fluid catalytic cracking or Fischer–Tropsch synthesis, high gas velocity (6–28 m/s) and high solids circulation rate (300–1000 kg/m^{2}s) are essential. The output of the process is the gas phase so that gas backmixing is undesirable. In a gas–solids reaction process, such as solid fuel combustion or alumina calcination, the unit is operated at lower gas velocity (5–9 m/s) and at lower solids circulation rate (5–40 kg/m^{2}s). The final goal is the conversion of solids and the extent of gas backmixing may not be critical (Zhu and Bi, 1993). In other words, the very concept of gas (and solids) mixing may differ, depending on the particular process.

Most researchers so far have not sufficiently considered these aspects in the design of their laboratory or pilot-scale experiments, so that it is difficult, and sometimes impossible, to transfer the results to the design and operating conditions required by industrial applications.

3.2 Experimental studies

Studies reported by various investigators are summarized in Table 3.1, allowing comparison of the main aspects of experimental apparatus and procedures. Note that each group of researchers used different experimental equipment (in terms of geometry and size), different operating conditions (in terms of size and density of bed material, gas velocity, solids mass flux and regime of fluidization), different measurement techniques and different models to interpret the measurements. This variety of procedures highlights the difficulty of comparing the results and the need for common methods of investigation.

3.2.1 Experimental procedures

Gas mixing tests are usually carried out with the aid of an adsorbing or non-adsorbing gas tracer. Generally, non-adsorbing tracers have been used in mixing tests in order not to complicate the interpretation of data. On the other hand, the utilization of an adsorbing gas tracer, in combination with an adequate model for mixing and an approximate description of adsorption equilibria, often yielded more information than a non-adsorbing tracer (van Deemter, 1967, 1985; Krambeck et al., 1987). The crucial influence of gas adsorption on the results of tracer experiments in a fluidized bed has been clearly highlighted by Bohle and van Swaaij (1978). An exhaustive description of adsorbing versus non-adsorbing gas tracers, together with an accurate explanation regarding their appropriate utilization and kind of data that can be expected from each method, can be found in the publication by Nauman and Buffham (1983).

The techniques for tracer injection and detection, as well as the location of injecting and sampling probes along the riser axis and across its cross-section, may differ in gas mixing tests, depending on the scope of the investigation.

To obtain information about gas residence time distribution in the riser, a transient gas tracer test is used. Commonly, a given amount of the tracer gas is injected as a pulse at the riser bottom while continuous samples are withdrawn at downstream positions. Because of the short mean residence time in the riser, some authors believed that, particularly in a small-scale equipment, it is impractical to provide an inlet pulse of tracer without disturbing the flow and preferred to use a step change in inlet concentration (Brereton et al., 1988). Since an ideal pulse input is impossible to achieve experimentally, two measuring points are used to reduce or eliminate the error introduced by a non-ideal input; the distance between measurement points is also conveniently used in forming dimensionless quantities (Bischoff and Levenspiel, 1962). Different investigators have used different measurement volumes. Most (van Zoonen, 1962; Dry and White, 1989; Bai et al., 1992) preferred

Table 3.1 Major experimental studies on gas mixing in CFB risers

Investigator	Focus on[a]	Riser size	Bed solids size range (average size) density	U (m/s) G_s (kg/m²s)	Flow regime[b]	Tracer gas	Diagnostic technique	Notes	Assumed model
van Zoonen (1962)	A, R	0.05 m ID 10 m high	FCC catalyst 20–150 μm (65 μm) 1600 kg/m³	$U = 1.5$–12 $G_s = 100$–1000	Dilute transp. Dense transp.	H_2	Continuous tracer injection Samples downstream in radial direction (radial mixing) and at the riser exit (axial mixing) Thermal conductivity cell	Investigation on gas and solids, radial and axial, dispersion	Axial dispersed plug flow model
Cankurt and Yerushalmi (1978)	B	0.152 m ID 8.5 m high	FCC catalyst 0–150 μm (55 μm) 1074 kg/m³	$U = 0.2$–5.6 $G_s = 25$–145	Bubbling Turbulent Fast	CH_4	Continuous tracer injection Samples upstream along radial direction		
Yang et al. (1983)	B, R	0.115 m ID 8 m high	silica gel 160–315 μm (220 μm)	$U = 2.8$–5.5 $G_s = 32$–160	Fast transition Fast dilute	He	Continuous tracer injection Samples upstream and downstream along radial direction Thermal conductivity cell	Effect on D_{gr} of U, G_s and ϵ Radial variation of backmixing	Radial dispersed plug flow model
Adams (1988)	(A), R	0.3 × 0.4 m 4 m high	dolomite sand 200 μm 250 μm 1600 kg/m³ 3300 kg/m³	$U = 3.8$–4.5 $G_s = 30$–45	Fast dilute (?)	CH_4	Continuous tracer injection Total hydrocarbon analyser	Effect on D_{gr} of U, G_s and ϵ	Turbulent diffusion
Bader et al. (1988)	B, R	0.305 m ID 12.2 m high	FCC catalyst 76 μm 1714 kg/m³	$U = 3.7$–6.1 $G_s = 98$–177	Fast transition Fast dilute (?)	He	Steady state tracer injection Samples upstream and downstream along radial direction Thermal conductivity cell	Radial variation of backmixing Axial variation of radial mixing	Dispersed plug flow model (Klinkeberg solution)
Brereton et al. (1988)	A	0.152 m ID 9.3 m high	Sand 75–250 μm (148 μm) 2650 kg/m³	$U = 7.1$ $G_s = 0$–65	Fast (?) Dilute transp. (?)	He	Step change in tracer concentration Sample probe at the riser exit Thermal conductivity cell	Effect of riser exit geometry	Two-region dispersion model
Dry and White (1989)	A	0.09 m ID 7.2 m high	FCC catalyst 39–160 μm (71 μm) 1370 kg/m³	$U = 2$–8 $G_s = 36$–232	Turbulent Fast Dilute transp.	Ar	Pulse tracer injection Sample probe at the riser exit Mass spectrometer	D_g are reported by normalizing G_s at a value of 200 kg/m²s	One-dimensional dispersion model

Reference	Type[a]	Dimensions	Particles	Operating conditions	Flow regime	Tracer	Method	Purpose	Model
Li and Weinstein (1989); Weinstein et al. (1989)	B	0.152 m ID, 8 m high	FCC catalyst (59 μm), 1450 kg/m³	$U = 0.03$–4, $G_s = 0$–271	Bubbling, Turbulent, Fast, Dilute transp.	He	Continuous tracer injection. Samples at several upstream positions and along radial direction. Thermal conductivity analyser	Role of local fluidization regime	One-dimensional dispersion model
Li and Wu (1991)	A, B	0.09 m ID, 8 m high	FCC catalyst 20–164 μm (58 μm), 1575 kg/m³	$U = 1$–2, $G_s = 0$–30	Turbulent, Fast, Dilute transp.	H_2	Pulse tracer injection. Dual probe detection	Correlation for D_{ga} as a function of ϵ. Role of local fluidization regime	Axial dispersed plug flow model
Werther et al. (1991); Werther et al. (1992a)	B, R	0.4 m ID, 8.5 m high	sand 130 μm, 2600 kg/m³	$U = 3$–6.2, $G_s = 0$–70	Fast dilute	CO_2	Continuous trace injection. Sample probe moves across riser diameter. Infrared gas analyser	Investigation on radial mixing in the core zone	Dispersion model in the core zone
Bai et al. (1992)	A	0.140 m ID, 10 m high	silica gel 100 μm, 710 kg/m³	$U = 2$–10, $G_s = 10$–100	Turbulent, Fast dilute	organic substance	Pulse tracer injection. Sample probe at the riser exit. Chromatography	Gas and solids dispersion investigated simultaneously	One-dimensional dispersion model
Werther et al. (1992b)	B, R	0.4 m ID, 8.5 m high	sand 130 μm, 2600 kg/m³	$U = 3$, $G_s = 30$	Fast dilute	CO_2	Continuous tracer injection. Sample probe moves across riser diameter. Infrared gas analyser	Circumferential injection of tracer	Two-phase two-region model
Zheng et al. (1992)	R, S	0.102 m ID, 5.25 m high	resin 567, 701 μm, sand 364, 570 μm, 1392 kg/m³ 2560 kg/m³	$U = 2$–10, $G_s = 0$–30	Fast (?)	CO_2	Continuous tracer injection at riser axis. Sample probes at several downstream levels and moved across riser diameter. Infrared gas analyser	Effect on D_{gr} of U, G_s and $(1-\epsilon)$. Effect on D_{gr} of secondary air injection	One-dimensional dispersion model
Arena et al. (1993b); Marzocchella and Arena (1996)	S	0.120 m ID, 6.1 m high	glass ballottini 39 μm, 2540 kg/m³	$U = 3$ and 6, $G_s = 35$ and 55, $U = 15$, $G_s = 210$	Fast, Dilute transp.	CO_2	Continuous tracer injection in the lateral gas stream. Sample probe located at several downstream and upstream levels and moved across riser diameter. Infrared gas analyser	Effect of different lateral gas injection devices. Evaluation of mixing length	

[a] A, overall axial mixing measurements; B, local measurements of backmixing; R, local measurements of radial mixing; S, effect on gas mixing of a laterally injected gas stream. When not specifically indicated by the authors or derivable by published data, the flow regime is simply presumed and the symbol (?) indicates this lack of information.

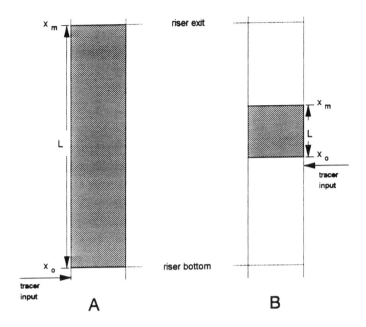

Figure 3.1 Possible test sections in gas RTD experiments in CFB risers. x_0 and x_m are the first and second measurement points, respectively.

to define the whole riser as the test section of their experiments (Figure 3.1A). Li and Wu (1991) identified as test section the part of the CFB riser between their dual probe detector (Figure 3.1B). This enabled them to check whether the measurement area was in the same flow regime, so that a specific mixing behaviour could be ascribed to a specific fluidization regime. This appears to be important since it partially reduces the influence of the complex CFB hydrodynamics on interpretation of experimental data. Measurements obtained from a test section covering the whole riser volume are affected by the mixing characteristics of various regions that are usually established along the riser. These experiments enable calculation of an effective axial mixing coefficient, evaluation of various degrees of mixing under different operating conditions and/or comparison of the mixing behaviour of gas and solids, but cannot generally be related to a specific fluidization condition.

To obtain information about gas backmixing, a steady-state mixing test is used. A continuous stream of a tracer gas is injected at a single point by a traversing probe, which can be located at several radial positions. The optimal values of the injection concentration must be determined carefully to ensure linearity and detectability for each flow regime (Li and Weinstein, 1989). Similarly, the injection rate should be as close as possible to the gas velocity in the riser to achieve a nearly isokinetic introduction (Bader *et al.*, 1988). Sample probes may be located at any radial position and at several elevations upstream of the tracer feeding point. Continuous tracer gas injection along the riser circumference may also be used to estimate the role of downflow near the wall (Werther et al., 1992b).

To obtain information about lateral gas mixing in the riser, a tracer gas is injected continuously on the axis of the riser. The tracer flow must be controlled to keep the injection gas velocity equal to or smaller than the local gas velocity (Werther et al., 1992a; Amos et al., 1993) because injection at a velocity greater than the maximum gas velocity on the riser centerline can give rise to a jet that increases radial dispersion (Werther et al., 1992a). Sampling probes located at several levels downstream of the injection point are traversed across the riser diameter to measure the radial variation in tracer gas concentration. No effect of sampling velocity has been observed at sampling rates smaller than the local gas velocity (Amos et al., 1993).

3.2.2 Mixing coefficients

Mixing processes in flow systems are often described with the aid of coefficients having the dimensions of a diffusion coefficient (Potter, 1971; van Deemter, 1985). In general, three mixing coefficients are needed to quantitatively describe gas mixing (Schügerl, 1967):

- The effective mixing coefficient D_g, which is a measure of the intensity of the overall gas dispersion in the direction of flow in the presence of a non-uniform velocity profile. D_g can be derived from the distribution of residence times.
- The axial (or backmixing) coefficient D_{ga}, which characterizes the contribution of dispersion in the flow direction. If there is no backflow, it is a measure of the intensity of diffusive mixing. D_{ga} can be evaluated from the tracer gas concentration upstream of the injection point.
- The lateral mixing coefficient D_{gr}, which characterizes the intensity of dispersion in the direction normal to flow. It can be evaluated from the tracer gas concentration downstream of and laterally displaced from the injection point.

Schügerl (1967) pointed out that the effective gas dispersion coefficient could be expressed by combining the contributions from axial and lateral dispersion using the relationship

$$D_g = D_{ga} + \beta v^2 D^2 / D_{gr} \qquad (3.1)$$

where v is the effective gas velocity, D the vessel diameter and β a dimensionless constant that characterizes the non-uniformity of the flow profile over the cross-section. β can be assumed to be 1/192 for a parabolic velocity distribution, about 0.5×10^{-3} for turbulent single-phase flow and 0 for a uniform velocity profile like that found by Schügerl for particles as small as 40 μm. In evaluating mixing coefficients from experimental data, the dimensionless Peclet number is obtained, defined as $Pe_g = UL/D_g$, where L is a characteristic dimension.

It should be noted that, unless detailed hydrodynamic conditions in the CFB system are known, true dispersion coefficients are difficult to measure. However, they are used here in order to simplify the analysis of existing literature.

3.2.3 Axial mixing

(a) Overall axial mixing measurements

Investigations on gas residence time distributions (RTD) in circulating fluidized bed risers have been carried out with different experimental units and under various operating conditions (Table 3.1). Almost all available data are summarized in Figures 3.2 and 3.3 in terms of D_g and $D_g/UL = 1/Pe_g$, respectively, as a function of flow parameters. It is likely that a more informative curve could be obtained by plotting the overall axial dispersion coefficients with respect to voidage. This is impossible, at this time, due to the almost total lack of information about the state of gas–solids flow in the measurement volumes.

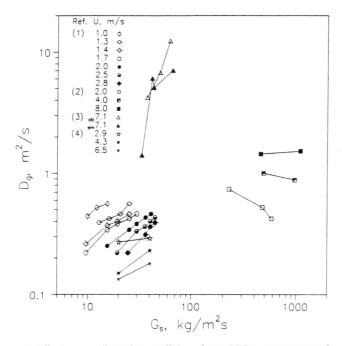

Figure 3.2 Effective gas dispersion coefficient from RTD measurements for different gas velocities and solids mass fluxes. Ref.: (1) Li and Wu, 1991; (2) van Zoonen, 1962; (3) Brereton et al., 1988 (ab = abrupt exit; sm = smooth exit); (4) Bai et al., 1992. See Table 3.1 for experimental details.

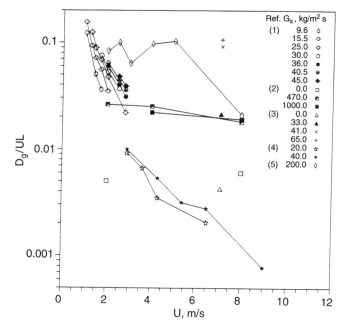

Figure 3.3 D_g/UL as a function of superficial gas velocity for several solids mass fluxes. Ref.: (1) Li and Wu, 1991; (2) van Zoonen, 1962; (3) Brereton et al., 1988 (sm = smooth exit); (4) Bai et al., 1992; (5) Dry and White, 1989. See Table 3.1 for experimental details.

Gas dispersion in an empty riser is much smaller than in the presence of circulating solids (see data by van Zoonen (1962) and Brereton et al. (1988) in Figure 3.3). The axial dispersion increases strongly as soon as solids particles enter the riser, i.e. for any $G_s > 0$. Li and Wu (1991) showed that when the gas velocity and solids mass flux are varied so that turbulent fluidization prevails in the riser, the spread in gas RTD curves becomes evident; as gas velocity increases further, at fixed solids mass flux, moving first into the fast fluidization and then into the dilute transport regimes, the spread clearly reduces, indicating a reduction in the amount of gas backmixing.

Experimental data reported in Figures 3.2 and 3.3 for $G_s < 100 \, kg/m^2 s$ confirm this result: the gas dispersion appears to increase as the solids mass flux increases (at fixed gas velocity) or as the gas velocity decreases (at fixed solids mass flux). In other words, gas mixing becomes evident as the suspension density increases, mainly due to more intense solids downflow along the riser walls (Brereton et al., 1988; Li and Wu, 1991; Bai et al., 1992). Li and Wu (1991) found for their own experiments that:

$$D_g = 0.1953 \epsilon^{-4.1197} \, m^2/s \tag{3.2}$$

This relationship should be valid for the entire fluidization regime spectrum from turbulent to pneumatic transport, but it has so far been tested only with the experimental data of the authors (Table 3.1).

It is noteworthy that data from RTD measurements carried out at very high values of G_s, i.e. under conditions close to those typical of gas-conversion processes, seem to follow an opposite trend (see data of van Zoonen (1962) and those of Dry and White (1989) at $U < 8$ m/s). This may be explained by noting that data showing this behaviour correspond to flow regimes close to dense transport, which differ substantially from the conditions corresponding to other data. Information so far reported on these high density CFBs is not sufficient to allow complete comprehension of the phenomena.

Comparison of the RTDs of gas and of solids shows that the axial dispersion of the gas is much lower than that of the particles (van Zoonen, 1962). The flow patterns of both gas and solids tend to be quite different from plug flow. The fluidization regime in the riser strongly affects the extent of gas and solids axial mixing (see also Chapter 4). An analysis of the RTD curves obtained by Bai *et al.* (1992), reproduced in Figure 3.4, confirms that the degree of axial dispersion of solids is larger, while the spread in gas residence times is considerable and should not be neglected. The difference between gas and solids RTD curves and the influence of solids circulation rates on axial dispersion of gas and solids both decrease as the gas velocity increases. The maximum spread in gas residence times and the most significant effect of the solids mass flux appear to occur at the lowest gas velocity and the highest gas–solids suspension density.

The above findings demonstrate that axial gas mixing in a CFB riser differs substantially from that in simple plug flow. Reported values of the reactor dispersion number D_g/UL (where L is the distance between measurement points) are generally between 0.002 and 0.2, indicating significant dispersion (Levenspiel, 1962).

(b) Local backmixing measurements

It has already been emphasized that the gas flow patterns in a fluidized bed depend a great deal on the fluidization regime and that, as a consequence, the performance of a CFB reactor is regime dependent. Local measurements of gas backmixing have been widely used to investigate this dependence, since they give the possibility of evaluating how chemical conversion and selectivity are affected by different gas flow patterns.

Many gas backmixing experiments have been reported for captive, low velocity fluidized beds, as extensively reviewed by Potter (1971) and van Deemter (1985). Relatively few gas backmixing experiments have been carried out in bed transport, high velocity regimes. The original experiments of Cankurt and Yerushalmi (1978), followed by those of Yang *et al.* (1983), Helmrich *et al.* (1986) and Bader *et al.* (1988), supported the idea that gas backmixing substantially decreases beyond the turbulent regime, so that plug flow of gas may generally be assumed in fast fluidized beds. This conclusion was refuted by Weinstein and his group (Weinstein et al., 1989; Li and Weinstein, 1989) who found considerable gas backmixing in fast beds. They illustrated how the extent of gas backmixing is governed by the flow structure in the test section between the injection and sample probes. Low velocity fluidized beds are characterized by a degree of heterogeneity or intermittency (alternation of dilute and dense phases), which may be assumed to be distributed in time rather than in space. As a consequence, a time-averaged measurement at a fixed point gives a result that appears to be averaged over this heterogeneity. High velocity fluidized beds are typically characterized by a space-distributed heterogeneity with a macroscopic flow structure showing distinct zones in both the vertical and horizontal directions. As a consequence, a time-averaged measurement at a fixed point does not average over this heterogeneity (Li and Weinstein, 1989).

There is evidence that in all fluidization regimes, gas backmixing is mainly determined by the downflow of solids particles. In bubbling fluidized beds, the motion of bubbles and circulation of gas through them, together with the motion of particles in and around the bubble wake and cloud, is the main cause of gas mixing. Solid particles moving down as a consequence of bubble movement give rise to a backmixing of gas, which increases as the superficial gas velocity increases. The degree of axial dispersion is high, except for the 'slow bubbles' in beds of coarse particles (see van Deemter, 1985, for differences between the 'slow' and 'fast' bubble regimes). Radial profiles of tracer concentration at several levels upstream of the tracer injection position are essentially flat (Figure 3.5), so that the radial positions of inlet points pro-duce no significant effect. The downflow of gas takes place throughout the whole bed cross-section (Cankurt and Yerushalmi, 1978; Li and Weinstein, 1989).

Turbulent fluidized beds are at the boundary between captive and transport regimes. The solids no longer form a continuous phase clearly separated by a discontinuous bubble phase, but are concentrated in clusters and strands of particles. Gas-solids contact is improved (Massimilla, 1973). All this corresponds to a more homogeneous flow structure or, better, to a heterogeneity that begins to be distributed in space rather than in time: it is possible to distinguish between a 'dense' and a 'transition' region of the turbulent bed. Radial profiles of tracer concentration (Figure 3.5) show increased concavity, suggesting that mixing patterns are dominated by downward flow of solids along the walls and upward flow of gas through a leaner core (Cankurt and Yerushalmi, 1978; Li and Weinstein, 1989; Li and Wu, 1991). The tracer gas detected in the dilute core is essentially due to mass exchange between the core and the wall annular region, where considerable circumferential mixing occurs so that the tracer gas can reach the side opposite to the injection point. This picture of gas and solids streamlines seems to be affected

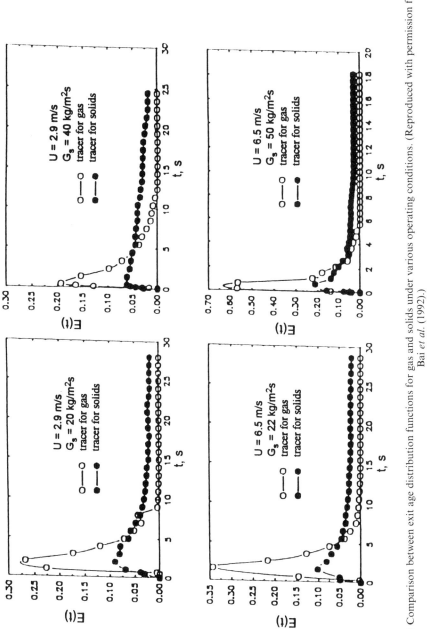

Figure 3.4 Comparison between exit age distribution functions for gas and solids under various operating conditions. (Reproduced with permission from Bai et al. (1992).)

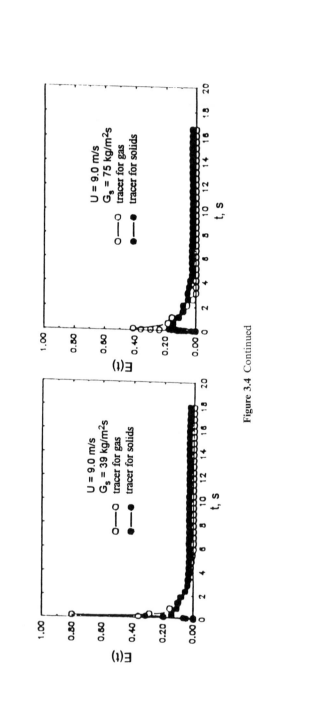

Figure 3.4 Continued

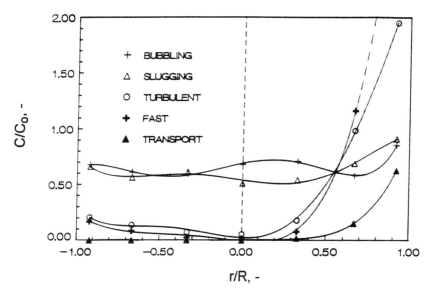

Figure 3.5 Comparison of backmixed tracer concentration in different fluidization regimes measured by means of a probe located 130 mm upstream of the injection level. Radial position of tracer inlet probe was $r/R = 0.89$. For turbulent and fast regimes the measurements were taken in the dense zone. (Reproduced with permission from Li and Weinstein (1989).)

by the small diameters (up to 152 mm) of the columns in which the investigations were carried out, but in any case the overall degree of gas backmixing is lower than in the captive regimes. Its extent in the transition region is slightly less than in the dense region (Figure 3.6), indicating that particle downflow is still extensive (Weinstein *et al.*, 1989).

Fast fluidized beds present the maximum degree of spatial heterogeneity. A distinction can usually be made in the vertical direction between a bottom dense, a middle transition, a top dilute and, depending on the design of the exit configuration, an outlet region. A two-phase structure may also be detected in the lateral direction, with a dilute suspension moving upward and a relatively dense phase falling downward. The probability of observing a specific phase at any particular instant greatly depends on the location: the lean phase generally dominates in the core, while the dense phase dominates in the wall region. Each of these regions exhibits quite different backmixing characteristics. Axial dispersion in the bottom relatively dense region is greater than in the transition and dilute zones (Li and Weinstein, 1989). Radial profiles of tracer gas concentrations show higher values near the walls and lower values across the core compared with those in a turbulent bed (Figure 3.5). An increase in superficial gas velocity reduces the levels of detected tracer concentrations, indicating that the gas velocity affects axial dispersion differently in captive and transport regimes. Backflow of gas in the central core of the bed

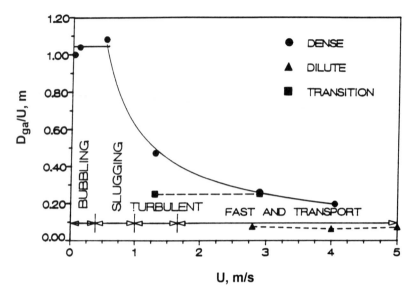

Figure 3.6 Ratio between length scale and Peclet number (a measure of backmixing depth) as a function of gas velocity over the regime spectrum. See Table 3.1 for experiment details. (Reproduced with permission from Li and Weinstein (1989).)

appears to be negligible, while the annular region is characterized by intense backmixing and still measurable circumferential mixing. This is confirmed by data of Werther *et al.* (1992b) obtained in a 400 mm ID riser: the tracer gas concentration was easily detected 1.58 m below the level of the circumferential injection and near the wall was even higher than 0.68 m above the injection plane. This marked backmixing in the annulus causes overall gas dispersion that cannot be neglected; the assumption of plug flow for gas in the fast fluidization regime is appropriate only for the dilute upward-moving core. The extent of backmixing can only be accounted for in a two-region model, with different hydrodynamic assumptions for core and annulus, so that different backmixing characteristics may be taken into account.

In the dilute transport regime the extent of gas backmixing is again related to downflow of solids particles. At very high gas velocities, when the particle trajectories are straight or nearly so, no gas backmixing is detected. At lower gas velocities, some of the particles appear to float near the wall while at still lower velocities particles begin to recirculate, i.e. they travel downward at the wall until they migrate back into the core and move upward again (Capes and Nakamura, 1973). Experiments carried out in this regime (Weinstein *et al.*, 1989) show that gas backmixing can still be considerable, depending on the intensity of solids downflow and, therefore, on solids mass flux at a given gas velocity.

3.2.4 Lateral mixing

Observations in CFB reactors of industrial dimensions show that lateral mixing is relatively poor. In particular, measurements by Couturier *et al.* (1991) and Zheng *et al.* (1991), substantially confirmed by data obtained from other commercial plants (Alliston, 1994), indicated that in large-scale CFB boilers oxygen-rich gas coexists in different parts of the riser with C-rich particles and CO-rich gas. O_2 and CO_2 concentrations measured by Couturier *et al.* (1991) in a 4×4 m cross-section boiler show that lateral gas profiles strongly depend upon initial release patterns and are essentially independent of height in the water-wall section, suggesting that little combustion takes place in that part of the combustor. The slow lateral mixing of solids coupled with the poor lateral gas mixing results in poor contact between oxygen and solid or gaseous fuels.

Table 3.1 summarizes information about radial gas dispersion studies in CFBs. Some authors measured tracer concentration profiles and evaluated a dispersion coefficient D_{gr} with no regard to the core-annulus structure (van Zoonen, 1962; Yang *et al.*, 1983; Zheng *et al.*, 1992). Others, like Werther and co-workers (1992a; 1992b), considered radial dispersion only in the dilute core and assumed that concentrations in the wall zone are constant.

It is generally accepted that lateral mixing in a single-phase flow is higher than that in a gas–solids suspension: experimental results by van Zoonen (1962), Adams (1988) and Zheng *et al.* (1992) show that as particles are added to the riser, the turbulence of the gas, and therefore its lateral dispersion, are reduced. For the range of voidage investigated by Adams ($\epsilon > 0.98$), it also appears that an increase in solids concentration further reduces radial dispersion. Measurements by Werther *et al.* (1992a) in the upper dilute core of a 400 mm ID riser provided values of $Pe_{gr,C} (= U_C D^* / D_{gr,C})$ independent of the gas velocity (and ≈ 465), confirming that, for $\epsilon > 0.98$, $D_{gr,C}$ decreases as gas velocity decreases, i.e. when, for fixed G_s, the suspension density increases. For $0.90 < \epsilon < 0.98$, Yang *et al.* (1983) showed an opposite effect of solids concentration on the degree of gas lateral mixing. This apparent discrepancy was clarified by radial gas dispersion experiments carried out by Zheng *et al.* (1992) using particles of different sizes and densities and covering the entire voidage range partially tested by previous authors. Results are reported in Figures 3.7a and 3.7b where D_{gr} is plotted as a function of solids mass flux (at several gas velocities) and as a function of $(1 - \epsilon)$, respectively. It appears that the effects of flow parameters are not monotonic: (i) the degree of dispersion is maximum in single-phase flow; (ii) addition of particles to establish a dilute transport regime in the riser leads to a damping of gas turbulence and to a consequent reduction of D_{gr}; (iii) further increase in gas–solids suspension density leads to an increase in the degree of turbulence and to enhanced gas mixing.

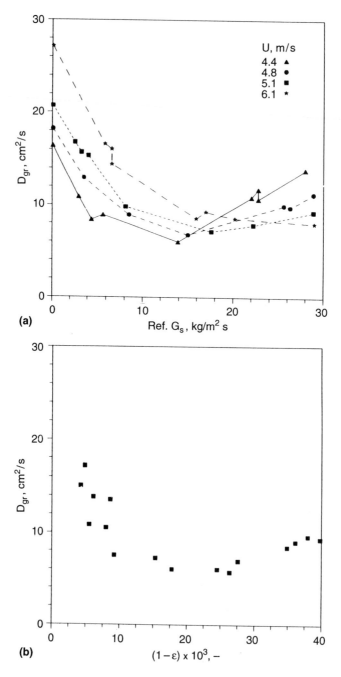

Figure 3.7 Radial gas dispersion coefficient as a function of (a) solids mass flux for different gas velocities and (b) solids volume fraction. Bed material is resin ($1392 \, kg/m^3$; 0.701 mm). (Redrawn with permission from Zheng et al. (1992).)

The above results indicate that solids particles can enhance or suppress turbulent intensity, thereby increasing or decreasing gas dispersion. The influence of solids particles on the turbulence structure must be taken into consideration. Hinze (1959) noted that the behaviour of a swarm of discrete particles in a turbulent fluid depends largely on the particle concentration and on the size of the particles with respect to the scale of turbulence of the fluid. Hetsroni (1989), also on the basis of results by Tsuji et al. (1984), found that the presence of particles with a low particle Reynolds number ($Re_p = v_{slip}d_p(\rho_p - \rho_g)/\mu_g$) tends to suppress the turbulence of the carrier fluid, while particles with $Re_p > 400$ tend to enhance turbulence. Gore and Crowe (1989) reorganized data by numerous authors as a function of the ratio of particle size d_p to the characteristic size of turbulent (large) eddies, l_e. The turbulence level was found to decrease for $d_p/l_e < 0.1$ and to increase for $d_p/l_e > 0.1$, under the assumption that the ratio l_e/D was equal to 0.1 everywhere across the pipe except near the wall (Hutchinson et al., 1971). The damping effect is related to the transformation of an eddy's energy into kinetic energy of the particle moved because of drag by the eddy. The opposite effect could be related to the turbulence that these larger particles create near the scale of the most energetic eddy (Gore and Crowe, 1989).

This latter consideration also suggests an explanation for the increased dispersion of gas when the solids concentration is further increased and clusters of particles continuously form: these clusters probably behave like single particles of larger size (i.e. size larger than $0.1 l_e$) contributing to increase the intensity of turbulence and hence the gas dispersion. Another explanation was suggested by Zheng et al. (1992): further increase in solids concentration leads to the formation of a solids layer falling downward along the walls with a certain degree of circumferential backmixing. This in turn enhances gas mixing in the horizontal direction so that D_{gr} increases at a certain transition point (indicated by the minimum in Figure 3.7b). It must be noted that most commercial CFB boilers operate with solids concentrations below this transition point, i.e. with a suspension density too low (usually less than $10\,\text{kg}/\text{m}^3$) for clusters or solids layers to form. Therefore, the poor lateral dispersion of gas is not improved.

3.2.5 Effect of design and operating variables

(a) Riser geometry

Riser diameter. Yerushalmi and Avidan (1985) suggested that the effect of the column diameter on gas dispersion coefficient is probably more than linear for small-diameter tubes, approximately linear for medium-size columns, and less than linear for large risers. Their observations appear to agree with the above cited results of change in the turbulent intensity as a function of d_p/l_e, under the assumption of l_e/D constant (Gore and Crowe, 1989).

Werther et al. (1992a) pointed out that large-scale units should be characterized by small values of d_p/l_e, which should lead to some reduction in turbulent intensity. They simulated the evolution of the degree of radial gas mixing in a CFB riser having an 8 m internal diameter and found that, if a reactant is injected at the riser centerline, its local concentration on the axis is still ten times higher than the mean concentration, even 20 m downstream of the injection level. This confirms that gas mixing in the upper core region of a large-scale CFB is relatively poor.

Riser wall geometry. Any roughness at the riser wall could significantly affect the uniformity of the flow structure, hence influencing the gas–solids contact efficiency and the conversion in the reactor: particles can be stripped off walls with an unobstructed smooth surface by the rising suspension more quickly than off walls with a vertical membrane surface (Wu et al., 1991). Jiang et al. (1991) studied this aspect in a catalytic CFB reactor but used a horizontal ring-type baffle. Experimental results showed that rings break up the solids layer in the wall region and increase the particle exchange between wall and core regions, leading to enhancement of lateral mixing of gas (and solids). For further discussion on different types of baffles and on major concerns with respect to their utilization in CFBs see Chapter 16.

Exit configuration. Recent experimental results (Senior and Brereton, 1992; Brereton and Grace, 1993; Zheng and Zhang, 1993) demonstrate that the geometry of the riser exit can greatly influence the performance of CFBs, by affecting pressure and voidage profiles, not only close to the roof, but also at a considerable distance down the column. Two main types of exit configurations are usually utilized in CFB risers (see also Chapter 16): 'abrupt exits', where the roof is higher than the top of the exit channel, and 'smooth exits', where short radius bends smoothly connect the top of the riser to the cyclone entrance. With an abrupt exit, a relatively dense suspension zone forms under the roof and many particles, especially heavier ones that cannot follow the gas streamlines into the exit ports, are reflected back down the column. This movement of solids probably contributes to axial gas dispersion in this area. With a smooth exit, the dense suspension zone disappears, the apparent suspension density continues to decrease over the top: relatively few particles are reflected back and a much greater proportion of solids leave directly through the exit.

Brereton et al. (1988) studied the influence of smooth and abrupt exits on the axial dispersion of gas at identical solids hold-ups. The base case, where no solids were present, gave similar results for both exit configurations, while operating the riser at an equivalent total solids hold-up the smooth exit showed a dramatic increase in axial gas dispersion compared to that of the abrupt exit.

(b) Cyclone

The cyclone is a highly turbulent space where gas mixing is much higher than in the riser. This plays an important role in CFB boilers where the use of a hot cyclone can have a positive influence not only on overall gas mixing but also on lowering CO emissions. Carbon monoxide is mainly generated by the partial oxidation of carbon and by the Bouduard reaction, i.e. it forms wherever char particles are present. In the cyclone, most of the particles are swept against the wall by the centrifugal force, so that only a few fine particles are present in the central gas vortex. The presence of this carbon-lean space, together with the intense gas mixing in the cyclone, results in enhanced combustion of CO, provided that temperature is about 850°C or higher.

(c) Particle density and size

The density and size of bed materials strongly affect the overall hydrodynamics and gas (and solids) flow patterns in CFB risers. Analysis of the data so far available suggests that an increase in both density and size of particulate material improve lateral gas dispersion in the riser.

Experimental evidence that an increase in particle density causes greater gas dispersion was found by Zheng (1994) in a 0.102 m ID riser, operated at the same values of U and G_s and alternatively fed with resin ($\rho_s = 1392\,\text{kg/m}^3$) and sand ($\rho_s = 2560\,\text{kg/m}^3$), both with an average diameter of 570 μm.

The above mentioned studies of Tsuji *et al.* (1984) and Gore and Crowe (1989) highlighted that particles of different sizes could enhance or suppress turbulent intensity and therefore increase or reduce radial gas dispersion. This finding has been confirmed by Zheng (1994), who found enhanced radial gas mixing when the average size of tested resin was increased from 567 to 701 μm.

3.2.6 Lateral injection of a gas stream

Most applications of circulating fluidized beds involve the injection of one or more streams of reactants along the riser. A secondary (sometimes even a tertiary) air stream is used in solid fuels combustion, where the practice of staged combustion is crucial for reducing NO_x emissions. Lateral feeding of reactants is widely adopted in gas-conversion processes such as thermal or catalytic cracking of heavy petroleum fractions, where fast mixing of different streams, together with the absence of gas and solids backmixing, would increase conversion and selectivity. In spite of this wide utilization of lateral injection of gas streams, almost all studies on various aspects of CFB hydrodynamics have been carried out by feeding the fluidizing gas at the bottom of the riser only. Recently a number of papers have focused attention on the effects that a laterally injected stream may have on the gas and solids patterns in the riser (Wang and Gibbs, 1991; Weinell et al., 1992; Zheng et al., 1992; Arena et al., 1993a; Brereton and Grace, 1993) and on the mixing phenomenology between the impinging streams (Arena et al., 1993b; Marzocchella and Arena, 1996).

(a) Effect of lateral injection on gas dispersion
As previously mentioned, 'secondary' air streams are particularly common in CFB boilers, where the total combustion air is split into a primary sub-stoichiometric stream, fed through the bottom distributor, and a secondary stream, injected at a higher riser level. This staged combustion is effective in limiting NO_x formation, but may also improve gas dispersion in the upper part of the riser. Experiments have been carried out at Tsinghua University by Zheng and his colleagues by using a secondary-to-primary air ratio of 0.7/0.3 and analysing gas samples taken from 0.6 to 2.0 m above the secondary air ports (Zheng et al., 1992). Some results are reported in Figure 3.8. The gas dispersion in the riser is enhanced by the secondary air jets, with the extent of this improvement depending on riser voidage. A larger solids concentration causes greater momentum of the rising suspension, which in turn leads to faster decay of the radial gas dispersion coefficient. The data in Figure 3.8 show that the reduction in the enhanced dispersion of gas may occur over very short distances. Thus, in a CFB boiler it could be useful to locate the secondary air jets near the carbon-rich region (Zheng et al., 1991).

It is necessary to know more about the influence of parameters related to secondary air stream injection (secondary-to-primary air ratio, lateral air injection device, level of secondary air injection, etc.) on the hydrodynamics of the gas–solids suspension and, therefore, on gas and solids flow patterns in the riser.

(b) Mixing between gas–solids suspension and lateral gas stream
The interaction between the rising gas–solids suspension and the laterally injected gas stream can substantially affect the performance of a CFB reactor. This appears to be particularly crucial for gas-conversion processes where rapid and uniform mixing of impinging streams is essential to ensuring fast vaporization of the feed and intimate contact between phases during their short residence time in the riser (Avidan and Shinnar, 1990; Zhu and Bi, 1993; Johnson et al., 1994). Two recent papers attempted to explore the characteristics of such interaction, focusing on gas and solids mixing phenomena downstream of the lateral injection. Operating conditions were close to those of gas-conversion processes ($U = 15\,\text{m/s}$; $G_s = 210\,\text{kg/m}^2\text{s}$) in one case (Arena et al., 1993b) and to those of CFB combustors ($U = 6\,\text{m/s}$; $G_s = 35\text{–}55\,\text{kg/m}^2\text{s}$) in the other (Marzocchella and Arena, 1996). A stream of carbon dioxide, as tracer, was mixed with the lateral air stream before injection into the riser. Two different feeding devices were

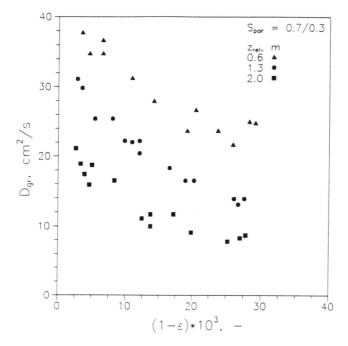

Figure 3.8 Effect of secondary air injection on gas dispersion in the riser as a function of solids fraction. S_{par} is the secondary-to-primary air ratio; z_{rel} is the distance from the secondary air inlet level. Bed material is sand (2560 kg/m^3; 0.570 mm). (Redrawn with permission from Zheng et al. (1992).)

used: in one, injection occurred through a slot over the whole circumference of the riser, while in the other, there were four equally spaced jets normally oriented with respect to the riser axis. The mechanism of interaction was found to change when the injection device was changed, resulting in different solids and gas distributions along the riser and in different values of the length necessary for complete mixing of gas and solids phases (Figure 3.9).

The crucial role of the feed-injection system was also highlighted by two papers from industry (Johnson et al., 1994; Miller et al., 1994) related to recent developments on FCC feed nozzles. The results indicate the necessity of combining the design of the riser with that of the feed-atomization device. For further discussion on FCC feed atomization and mixing see Chapter 13.

3.3 Mixing models

Several theoretical models have been proposed to account for gas dispersion in fluidized beds and utilized to interpret experimental measurements and

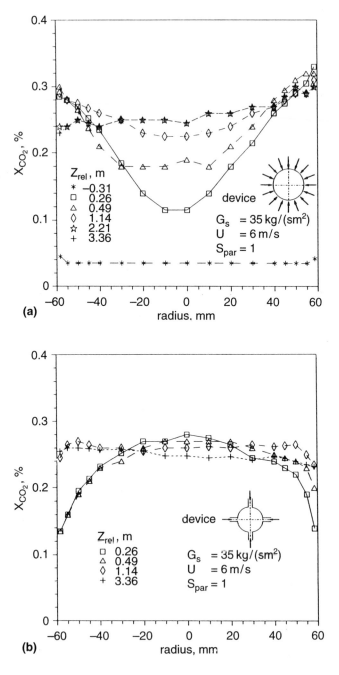

Figure 3.9 Radial concentration profiles of CO_2 injected, as tracer gas, in a lateral gas stream via two different feeding devices. Bed material is ballottini (2540 kg/m^3; 0.089 mm). (Redrawn with permission from Marzocchella and Arena (1996).)

provide parameters that can also be used in reactor and contacting models (Chapter 15). Understanding of the capabilities, assumptions and drawbacks of the main types of mixing models is of fundamental importance, particularly considering the discrepancies that sometimes arise from the adoption of different interpretative models.

Extensive and exhaustive studies on gas mixing modelling have been published in the last two decades by Potter and his colleagues (Potter, 1971; Fryer and Potter, 1976; Mao and Potter, 1986; White et al., 1992), which cover the entire fluidization regime spectrum from bubbling beds to dilute transport flows.

A summary of major contributions in the area of gas dispersion modelling in CFB risers is given in Table 3.2. No models exist that account for lateral stream injection. Gas mixing appears to be fairly well described by some of the available models as long as only primary gas is injected. It must also be noted that these models usually refer to the developed upper dilute zone of a CFB riser. The bottom region of the riser exhibits such intense gas and solids mixing that it is often considered a continuous stirred tank reactor (Weiss et al., 1987; Arena et al., 1991 and 1995). The same sometimes applies to the outlet region.

Two approaches to gas modelling in CFBs can be distinguished: the single-phase and the two-region approach. In this simple classification two-phase bubbling models (Grace, 1984) are not readily included due to the different type of heterogeneity (see section 3.2.2(b)) that characterizes the fluidization regimes usually establishing in CFBs. The utilization of this kind of model needs proper modifications, like those proposed in Chapter 15.

3.3.1 Single-phase approach

One of the first approaches to the modelling of gas mixing in high velocity fluidized beds was proposed by Wen (1979) who suggested that, in the presence of limited gas backmixing, a single-phase plug flow model could be assumed as an appropriate tool for mixing data interpretation. Yerushalmi and Avidan (1985) and Avidan and Edwards (1986) also suggested that for fine group A powders fluidized at high gas velocities, the gas–solids suspension can be treated as homogeneous and an axial dispersion model is appropriate to account for gas mixing.

In these models, generally known as dispersion models, there is a single phase flowing upward along the riser: the flow is turbulent, but not isotropically so that fluctuations in the vertical direction may differ from those in the lateral direction (van Deemter, 1985). These models use diffusion equations with modified diffusion coefficients (called 'dispersion coefficients') and represent the most popular methods to treat fluid mixing in reaction vessels. There are several dispersion models, from the most general to the most simplified, differing from each other for the increasing number of

Table 3.2 Major modelling studies on gas mixing in CFB risers

Author	Type of approach	Major assumptions	Validation with experiments	Major remarks
Yerushalmi (1986)	Two-phase	A lean phase moving upward and a dense phase falling downward are considered	Riser: 0.152 m ID, 8.5 m high FCC: 55 μm, 1074 kg/m^3 U = 0.21–1.52 m/s	The model predicts the diminishing percentage of gas backflow with U fairly well over the turbulent regime
Brereton et al. (1988)	Two-region	Gas in plug flow in the core Stagnant gas in the annulus Both regions well mixed radially R^*/R constant with height and = 0.85 K_g constant with height	Riser: 0.152 m ID, 9.3 m high Sand: 148 μm, 2650 kg/m^3 U = 7.1 m/s G_s = 0–65 kg/m^{2}s	The agreement with experimental RTD data is better when the continuity condition is relaxed
Kagawa et al. (1991)	Two-region	Gas in plug flow in the core Stagnant gas in the annulus Both regions well mixed radially R^*/R constant with height and = 0.85 $(1 - \epsilon_C) = 0.6(1 - \epsilon)$ and $(1 - \epsilon_A) = 2(1 - \epsilon)$	Riser: 0.1 m ID, 5 m high FCC: 46 μm, 2300 kg/m^3 U = 3 m/s G_s = 15, 80 kg/m^{2}s	Good fit to experimental data, particularly in the upper dilute zone
White et al. (1992)	Two-region	Gas in plug flow in the core Gas in the annulus alternatively considered stagnant, moving upwards or downwards Parameters are constant with height Slip velocity equal to the terminal velocity in the core and to zero in the dense zone	Riser: 0.109 m ID, 7.2 m high FCC: 71 μm, 1375 kg/m^3 U = 2–8 m/s G_s = 36–230 kg/m^{2}s Sand: 140 μm, 2650 kg/m^3 U = 4–6 m/s G_s = 65–260 kg/m^{2}s	The two-region model represents the data better than a single-phase plug flow model The annulus can be considered as stagnant rather than moving upwards or downwards K_g and $(R - R^*)$ increase with G_s and decrease with U Sand exhibits more plug-flow like behaviour than FCC at the same U and G_s

Reference	Model type	Description	Conditions	Remarks
Werther et al. (1992b)	Two-phase two-region	Dense phase only in the annulus Gas in plug flow in both the regions Gas radially mixed in the annulus while a gas radial dispersion is considered in the core R^*/R constant with height and $= 0.85$ Lean phase hold-up in the annulus $= 0.8$	Riser: 0.4 m ID, 8.5 m high Sand: 130 μm, 2600 kg/m^3 $U = 3$ m/s $G_s = 30$ kg/m^{2}s	It is necessary to include the core-annulus structure into the interpretation of the data, due to the quite different behaviour of the gas mixing in the two regions
Zethraeus et al. (1992)	Two-region	Gas velocity profile in the core is $U(r) = U_c(1 - r/R^*)^{1/2}$ Gas in the annulus is a slow upflow with a constant velocity of 0.5 m/s $\epsilon_C = 0.95$ and $\epsilon_A = 0.6$ Diameter of turbulent eddies is a linear function of radial position, being equal to 0.1D in the core and to 0.025D near the walls	Riser: 0.3 × 0.41 m, 4 m high Sand: 200 μm, 2800 kg/m^3 $U = 3.4$ and 5 m/s Mean bed density $= 10-54$ kg/m^3	Since the derivation of the equation for turbulent diffusion constant is only qualitative, it is necessary to introduce five model parameters to obtain a quantitative fitting of experimental data
Amos et al. (1993)	Two-region	Region 1 possesses the mixing behaviour of the core but covers all the riser cross-section Region 2 is a hypothetical extra air flow rate with the mixing properties of the annulus Gas in region 1 is considered in radially dispersed plug flow	Riser: 0.305 m ID, 6.6 m high Alumina: 71 μm, 2450 kg/m^3 $U = 2.6-5$ m/s $G_s = 20-120$ kg/m^{2}s	Results show that a systematic error is introduced by not accounting for the core-annulus structure. The apparent Peclet number $(=UD/D_{gr})$ is correlated with an apparent suspension Reynolds number $(=D(G_s + \rho_g U)/\mu_g)$ by an equation of the type $Pe = k_1 Re^{k_2}$
Zhang et al. (1993)	Single-phase	Radial velocity in the riser may be neglected	Commercial FCC unit radioactive tracer (Kr79)	One- and two-dimensional dispersion models have been developed using semi-infinite and convective boundary conditions
Kruse et al. (1995)	Two-dimensional two-phase	Lean phase and dense phase may appear at every axial and radial location Clusters have a constant solids concentration and descend with a velocity invariant with the radial position. A fluctuating motion of clusters in radial direction may exist Gas in the lean phase moves upward with local velocity depending on both the axial and radial position	Riser: 0.4 m ID, 9 m high Sand: 124 μm, 2600 kg/m^3 $U = 3-6.2$ m/s $G_s = 4-70$ kg/m^{2}s	The model requires some precalculated and measured data as input Results suggest that the model accounts for the main mixing phenomena in the riser upper dilute zone

assumptions and restrictions: a comprehensive description and comparison of them is given by Bischoff and Levenspiel (1962).

If one assumes that the fluid flows at the mean velocity (hence in plug flow), the dispersion model has two parameters, apart from the mean velocity in axial direction, due to the simultaneous presence of axial and transverse gas mixing. The differential equation describing the dispersed plug flow model is:

$$\frac{\partial C}{\partial t} = -\frac{U}{\epsilon}\frac{\partial C}{\partial z} + D_{ga}\frac{\partial^2 C}{\partial z^2} + \frac{D_{gr}}{r}\frac{\partial}{\partial r}\left(r\frac{\partial C}{\partial r}\right) + S + R_c \qquad (3.3)$$

where S and R_c are the source term and the rate of chemical reaction, respectively. If there is no variation in properties in the radial direction, i.e. only longitudinal gas mixing is observed, the equation defining the axial dispersed plug flow model is:

$$\frac{\partial C}{\partial t} = -\frac{U}{\epsilon}\frac{\partial C}{\partial z} + D_g\frac{\partial^2 C}{\partial z^2} + S + R_c \qquad (3.4)$$

where D_g is the only parameter of this model.

Results summarized in section 3.2.3 above demonstrate that flow patterns of gas inside a CFB riser can be substantially different from simple plug flow. As a consequence, the single-phase approach, though widely used due to its simplicity, is an oversimplification as it does not take into account the radial non-uniformities reported by many investigators. The inaccuracy increases as the suspension density increases, i.e. as the solids downflow along the walls becomes more intense. The single-phase approach may provide satisfactory agreement between mixing data and model predictions in some cases (Avidan and Edwards, 1986; Sun and Grace, 1992) but severe overpredictions have also been found in other cases, as reported by Jiang et al. (1991) and Ouyang et al. (1993) with respect to ozone conversion in laboratory-scale CFB risers.

3.3.2 Two-region approach

Most of the current models that treat gas mixing are based on the coreannulus structure that is usually established in the upper zone of CFB risers. The advantage is the ability to account for different gas mixing behaviour in two regions having quite different hydrodynamics. This approach was first utilized by Yerushalmi (1986) to predict gas backmixing in bubbling and turbulent regimes by means of a simple countercurrent model, while the first application of a two-region model to the fast fluidization regime was by Brereton et al. (1988). Table 3.2 gives basic information about the models so far developed which use this approach.

The modelling of radial dispersion of gas in the core region generally assumes that the suspension flows as a turbulent single phase inside a tube having the diameter of the core zone. The definition of this diameter is a debatable point. Some authors (see Table 3.2) assumed that the radius of the core region is independent of riser height for given values of U and G_s. They used $R^*/R = 0.85$ by measuring the solids flux along the riser cross-section: the point where the radial profile became negative was assumed as the boundary between core and annulus. Recent experiments (Zhou et al., 1995) assumed this boundary to be where the vertical time-mean particle velocity is 0, so obtaining different results. Amos et al. (1993) affirmed that the assumption of a fixed gas core boundary is unlikely to be correct and suggested that to overcome this drawback it could be useful to consider a region which covers the entire riser cross-section and has the mixing behaviour of the core region, and a hypothetical extra region that simulates the annulus behaviour but has no real space consistency.

Results obtained from the models of Werther et al. (1992b) and White et al. (1992) indicate that gas mixing is rather well described by dispersion in the core region and mass transfer between the lean and dense phases. The modelling of gas dispersion in the annulus is complicated by limited knowledge of the hydrodynamics along the wall. As a consequence, this region has been alternatively considered as a single phase, stagnant or moving in plug flow upwards or downwards, or even as a zone formed by two radially mixed phases, a lean phase moving upwards and a dense phase moving downwards. An assessment on the effect that these different assumptions can have on the performance of gas mixing models has been carried out by White et al. (1992). Their results confirm that a two-region model with radial segregation represents the RTD data more adequately than a single-phase plug flow model and indicate that the wall region can be considered as stagnant rather than moving either upwards or downwards. Deviations from plug flow appear to become more serious with an increase in solids flux, a decrease in gas velocity, or a reduction in particle size.

A further development of the two-region approach has been recently proposed by Werther's group (Kruse et al., 1995). The asumption of a simple core-annulus structure was substituted by continuous functions describing the variations of local flow parameters in the horizontal and vertical directions: at every axial and radial location within the upper dilute zone of the riser, an upflowing lean phase (the dilute suspension) and a downfalling dense phase (the clusters of solids) are considered. This 'two-dimensional/two-phase' model requires a series of precalculated or measured data (U, G_s, riser geometry, riser axial pressure profile, one radial profile of local G_s at one axial position) as input for the calculation procedure. The agreement between calculated and measured tracer gas profiles indicates that the model satisfactorily accounts for radial gas dispersion and gas backmixing in the upper dilute zone of a CFB riser.

3.4 Concluding remarks

Data summarized in this chapter highlight the great efforts that have been made in the last decade to reach a better understanding of gas mixing in circulating fluidized beds. In spite of these efforts, it is difficult at this time to reach more definitive conclusions based on the information, often strongly disconnected, that has been obtained so far. Two further obstacles must be overcome: an almost total lack of data from large-scale industrial CFB units and the significant differences in design and operating characteristics of gas-conversion and solids-conversion CFB processes. The former seems to be more crucial due to the difficulty of obtaining large-scale CFB units for research and the necessity of setting up appropriate diagnostic equipment for large risers. It would also be interesting to discuss and compare existing models when applied to units of this size. Such work should lead to the identification of requirements for gas mixing experiments and models resulting from different industrial applications of circulating fluidized beds, so that future research work may be correctly addressed.

Acknowledgements

The author is grateful to Dr Antonio Marzocchella for his help in the preparation of the manuscript and to Professor Roland Clift for his valuable suggestions.

Nomenclature

A	cross-sectional area, m^2
C	concentration or concentration of a tracer, kg/m^3
D	riser internal diameter or riser characteristic length, m
D^*	diameter of the core region, m
D_g	dispersion coefficient, m^2/s
d_p	particle size or particle Sauter mean diameter, m
$E(t)$	exit age distribution function, 1/s
g	acceleration due to gravity, m/s^2
G_s	solids mass flux, kg/m^{2}s
H	height, m
K_g	core-annulus mass exchange coefficient, m/s
l_e	length of turbulent eddies, m
L	characteristic dimension of the apparatus, m
Pe_g	Peclet number ($= UL/D_g$)
R	radius of the riser, m
R^*	radius of the core region, m

R_c	rate of chemical reaction in equations (3.3) and (3.4), kg/m^{3}s
r	radial coordinate, m
Re_m	$= D(G_s + \rho_g U)/\mu_g$, suspension Reynolds number
Re_p	$= v_{slip} d_p (\rho_p - \rho_g)/\mu_g$, particle Reynolds number
S	source term in equations (3.3) and (3.4), kg/m^{3}s
S_{par}	secondary-to-primary air ratio
t	time, s
U	superficial gas velocity, m/s
U_c	gas velocity in the vertical direction on the riser axis, m/s
U_s	superficial solids velocity, m/s
v	effective gas velocity, m/s
v_{slip}	slip velocity, m/s
v_T	particle terminal velocity, m/s
X_{CO_2}	volumetric percentage of CO_2 used as tracer gas
z	height coordinate, m
z_{rel}	distance from the secondary air inlet, m
β	constant in equation (3.1)
ΔP	pressure drop, Pa
ϵ	voidage averaged over riser cross-section
ϵ_{mf}	minimum fluidization voidage
ρ	density, kg/m^3

Subscripts

a	axial
A	annulus
C	core
g	gas
mf	minimum fluidization
o	at the injection level
r	radial

References

Adams, C.K. (1988) Gas mixing in fast fluidized beds, in *Circulating Fluidized Bed Technology II* (eds P. Basu and J.F. Large), Pergamon Press, Oxford, pp. 299–306.

Alliston, M. (1994) Private communication, Tampella Power Corp.

Amos, G., Rhodes, M.J. and Mineo, H. (1993) Gas mixing in gas–solids risers. *Chem. Eng. Sci.*, **48**, 943–949.

Arena, U., Malandrino, A. and Massimilla, L. (1991) Modelling of circulating fluidized bed combustion of a char. *Can. J. Chem. Eng.*, **69**, 860–868.

Arena, U., Cammarota, A., Marzocchella, A. and Massimilla, L. (1993a) Hydrodynamics of a circulating fluidized bed with secondary air injection, in *Proceedings of the 12th International Conference on FBC* (eds L. Rubow and G. Commonwealth), American Society of Mechanical Engrs., New York, pp. 899–906.

Arena, U., Marzocchella, A., Bruzzi, V. and Massimilla, L. (1993b) Mixing between a gas–solids suspension flowing in a riser and a lateral gas stream, in *Circulating Fluidized Bed Technology IV* (ed. A. Avidan), AIChE, New York, pp. 545–550.

Arena, U., Chirone, R., D'Amore, M., Miccio, M. and Salatino, P. (1995) Some issues in modeling bubbling and circulating fluidized bed coal combustors. *Powder Technol.*, **82**, 301–316.

Avidan, A. and Edwards, M. (1986) Modelling and scale-up of Mobil's fluid-bed MTG process, in *Fluidization V* (eds K. Ostergaard and A. Sorensen), Engineering Foundation, New York, pp. 457–464.

Avidan, A. and Shinnar, R. (1990) Development of catalytic cracking technology. A lesson in chemical reactor design. *Ind. Eng. Chem. Res.*, **29**, 931–942.

Bader, R., Findlay, J. and Knowlton, T.M. (1988) Gas/solids flow patterns in a 30.5 cm-diameter circulating fluidized bed, in *Circulating Fluidized Bed Technology II* (eds P. Basu and J.F. Large), Pergamon Press, Oxford, pp. 123–137.

Bai, D., Yi, J., Jin, Y. and Yu, Z. (1992) Residence time distributions of gas and solids in a circulating fluidized bed, in *Fluidization VII* (eds O.E. Potter and D.J. Nicklin), Engineering Foundation, New York, pp. 195–202.

Bischoff, K.B. and Levenspiel, O. (1962) Fluid dispersion-generalization and comparison of mathematical models. *Chem. Eng. Sci.*, **17**, 245–255 and 257–264.

Bohle, W. and van Swaaij, W.P.M. (1978) The influence of gas adsorption on mass transfer and gas mixing in a fluidized bed, in *Fluidization* (eds J.F. Davidson and D.L. Keairns), Cambridge University Press, Cambridge, pp. 167–172.

Brereton, C.M.H., Grace, J.R. and Yu, J. (1988) Axial gas mixing in a circulating fluidized bed, in *Circulating Fluidized Bed Technology II* (eds P. Basu and J.F. Large), Pergamon Press, Oxford, pp. 307–314.

Brereton, C.M.H. and Grace, J.R. (1993) End effects in circulating fluidized bed hydrodynamics, in *Circulating Fluidized Bed Technology IV* (ed. A. Avidan), AIChE, New York, pp. 137–144.

Cankurt, N. T. and Yerushalmi, J. (1978) Gas backmixing on high velocity fluidized beds, in *Fluidization* (eds J.F. Davidson and D.L. Keairns), Cambridge University Press, Cambridge, pp. 387–393.

Capes, C.E. and Nakamura, K. (1973) Vertical pneumatic conveying: An experimental study with particles in the intermediate and turbulent flow regimes. *Can. J. Chem. Eng.*, **51**, 31–38.

Couturier, M., Doucette, B., Stevens, D. and Poolpol, S. (1991) Temperature, gas concentration and solid mass flux profiles within a large circulating fluidized bed combustor, in *Proc. 11th Int. Conf. on FBC 'FBC Clean Energy for the World'* (ed. E.J. Anthony), American Society of Mechanical Engrs., New York, pp. 107–114.

Dry, R.J. and White, C.C. (1989) Gas residence time characteristics in high velocity circulating fluidized beds of FCC catalyst. *Powder Technol.*, **58**, 17–23.

Fryer, C. and Potter, O.E. (1976) Experimental investigation of models for fluidized bed catalytic reactors. *AIChE J.*, **22**, 38–47.

Gore, R.A. and Crowe, C.T. (1989) Effect of particle size on modulating turbulent intensity. *Int. J. Multiphase Flow.*, **15**, 279–285.

Grace, J.R. (1984) Generalized model for isothermal fluidized bed reactors, in *Recent Advances in Engineering Analysis of Chemically Reacting Systems*, chapter 13 (ed. L.K. Doraiswamy), Wiley Eastern, New Delhi.

Helmrich, H., Schugerl, K. and Janssen, K. (1986) Decomposition of $NaHCO_3$ in a laboratory and bench scale circulating fluidized bed reactor, in *Circulating Fluidized Bed Technology I* (ed. P. Basu), Pergamon Press, Oxford, pp. 161–166.

Hetsroni, G. (1989) Particles–turbulence interaction. *Int. J. Multiphase Flow.*, **15**, 735–746.

Hinze, J.O. (1959) *Turbulence*, 2nd edition, McGraw-Hill, New York, p. 460.

Hutchinson, P., Hewitt, G.F. and Dukler, A.E. (1971) Deposition of liquid or solid dispersion from turbulent gas streams: A stochastic model. *Chem. Eng. Sci.*, **26**, 419–439.

Jiang, P., Bi, H., Jean, R.-H. and Fan, L.-S. (1991) Baffle effects on performance of catalytic circulating fluidized bed reactor. *AIChE J.*, **37**, 1392–1400.

Johnson, D.L., Avidan, A.A., Schipper, P.H. and Miller, R.B. (1994) New nozzle improves FCC feed atomization, unit yield patterns. *Oil & Gas J.*, Oct. 24, 80–86.

Kagawa, H., Mineo, H., Yamazaki, R. and Yoshida, K. (1991) A gas–solid contacting model for fast fluidized bed, in *Circulating Fluidized Bed Technology III* (eds Basu, P., Horio, M. and Hasatani, M.), Pergamon Press, Oxford, pp. 551–556.

Krambeck, F.J., Avidan, A.A., Lee, C.K. and Lo, M.N. (1987) Predicting fluid-bed reactor efficiency using adsorbing gas tracers. *AIChE J.*, **33**, 1727–1734.

Kruse, M., Schoenfelder, H. and Werther, J. (1995) A two-dimensional model for gas mixing in the upper dilute zone of a circulating fluidized bed. *Can. J. Chem. Eng.*, **73**, 620–634.

Levenspiel, O. (1962) *Chemical Reaction Engineering*, John Wiley and Sons, Inc., New York, p. 264.

Li, J. and Weinstein, H. (1989) An experimental comparison of gas backmixing in fluidized beds across the regime spectrum. *Chem. Eng. Sci.*, **44**, 1697–1704.

Li, Y. and Wu, P. (1991) A study on axial gas mixing in a fast fluidized bed, in *Circulating Fluidized Bed Technology III* (eds P. Basu, M. Horio and M. Hasatani), Pergamon Press, Oxford, pp. 581–586.

Mao, Q.M. and Potter, O.E. (1986) Fluid-bed reaction with horizontal tubes as internals, in *Fluidization V* (eds K. Ostergaard and A. Sorensen), Engineering Foundation, New York, pp. 449–456.

Marzocchella, A. and Arena, U. (1996) Hydrodynamics of a circulating fluidized bed operated with different secondary air injection devices. *Powder Technol.*, **87**, 185–191.

Massimilla, L. (1973) Behavior of catalytic beds of fine particles at high gas velocities. *AIChE Symp. Series No. 128*, **69**, 11–13.

Miller, R.B., Niccum, P.K., Sestili, P.L., Johnson, D.L., Don, S. and Hansen, A.R. (1994) New developments in FCC feed injection and riser hydrodynamics, *AIChE 1994 Spring National Meeting*, April 12–21, Atlanta (Georgia).

Nauman, E.B. and Buffham, B.A. (1983) *Mixing in Continuous Flow Systems*, Wiley, New York.

Ouyang, S., Lin, J. and Potter, O.E. (1993) Ozone decomposition in a 0.254 m diameter circulating fluidized bed reactor. *Powder Technol.*, **74**, 73–78.

Potter, O.E. (1971) Mixing, in *Fluidization*, Chapter 7 (eds J.F. Davidson and D. Harrison), Academic Press, London, pp. 293–381.

Schugerl, K. (1967) Experimental comparison of mixing processes in two- and three-phase fluidized beds, in *Proc. International Symposium on Fluidization* (ed. A.A.H. Drinkeburg), Netherlands University Press, Amsterdam, pp. 782–794.

Senior, R.C. and Brereton, C.M.H. (1992) Modelling of circulating fluidized-bed solids flow and distribution. *Chem. Eng. Sci.*, **47**, 281–296.

Stubington, J.F. and Chan, S.W. (1990) On the phase location and rate of volatiles combustion in bubbling fluidized bed combustors. *Trans. IChemE*, **68**(A), 195–201.

Sun, G. and Grace, J.R. (1992) Effect of particle size distribution in different fluidization regimes. *AIChE J.*, **38**, 716–722.

Tsuji, Y., Morikawa, Y. and Shiomi, H. (1984) LDV measurements of an air–solid two-phase flow in a vertical pipe. *Int. J. Multiphase Flow.*, **139**, 417–435.

van Deemter, J.J. (1967) The counter-current flow model of a gas–solids fluidized bed, in *Proc. Int. Symp. on Fluidization* (ed. A.H.H. Drinkeburg), Netherlands University Press, Amsterdam, pp. 334–347.

van Deemter, J.J. (1985) Mixing, in *Fluidization 2nd edn.*, Chapter 9 (eds J.F. Davidson, R. Clift and D. Harrison), Academic Press, London, pp. 331–355.

van Zoonen, D. (1962) Measurements of diffusional phenomena and velocity profiles in a vertical riser, in *Proc. Symp. on The Interaction Between Fluids and Particles*, Instn. Chem. Engrs., London, pp. 64–71.

Wang, X.S. and Gibbs, B.M. (1991) Hydrodynamics of a circulating fluidized bed with secondary air injection, in *Circulating Fluidized Bed Technology III* (eds P. Basu, M. Horio and M. Hasatani), Pergamon Press, Oxford, pp. 225–230.

Weinell, C.E., Dam-Johansen, K., Johnsson, J.E., Gluchowski, Z. and Lade, K. (1992) Single particle behavior in circulating fluidized beds, in *Fluidization VII* (eds O.E. Potter and D.J. Nicklin), Engineering Foundation, New York, pp. 295–304.

Weinstein, H., Li, J., Bandlamudi, E., Feindt, H.J. and Graff, R.A. (1989) Gas backmixing of fluidized beds in different regimes and different regions, in *Fluidization VI* (eds J.R. Grace, L.W. Shemilt and M. Bergougnou), Engineering Foundation, New York, pp. 57–64.

Weiss, V., Fett, F.N., Helmrich, H. and Janssen, K. (1987) Mathematical modelling of circulating fluidized bed reactors by reference to a solid decomposition reaction and coal combustion. *Chem. Eng. Progress*, **22**, 79–90.

Wen, C.Y. (1979), in *Proc. N.S.F. Workshop on Fluidization and Fluid–Particles Systems VI* (ed. H. Littman), Renssealaer Polytechnic Institute, p. 317.

Werther, J. (1993) Fluid mechanics of large scale CFB units, in *Circulating Fluidized Bed Technology IV* (ed. A. Avidan), AIChE, New York, pp. 1–14.

Werther, J., Hartge, E.U., Kruse, M. and Nowak, W. (1991) Radial mixing of gas in the core zone of a pilot scale CFB, in *Circulating Fluidized Bed Technology III* (eds. P. Basu, M. Horio and M. Hasatani), Pergamon Press, Oxford, p. 593.

Werther, J., Hartge, E.U. and Kruse, M. (1992a) Radial gas mixing in the upper dilute core of a circulating fluidized bed. *Powder Technol.*, **70**, 293–301.

Werther, J., Hartge, E.U. and Kruse, M. (1992b) Gas mixing and interphase mass transfer in the circulating fluidized bed, in *Fluidization VII* (eds O.E. Potter and D.J. Nicklin), Engineering Foundation, New York, pp. 257–264.

White, C.C., Dry, R.J. and Potter, O.E. (1992) Modelling gas mixing in a 9 cm diameter circulating fluidized bed, in *Fluidization VII* (eds O.E. Potter and D.J. Nicklin), Engineering Foundation, New York, pp. 265–273.

Wu, R.L., Lim, C.J., Grace, J.R. and Brereton, C.M.H. (1991) Instantaneous local heat transfer and hydrodynamics in a circulating fluidized bed. *Int. J. Heat Mass Transfer*, **34**, 2019–2027.

Yang, G., Huang, Z. and Zhao, L. (1983) Radial gas dispersion in a fast fluidized bed, in *Fluidization* (eds D. Kunii and R. Toei), Cambridge University Press, Cambridge, pp. 145–152.

Yerushalmi, J. (1986) Gas mixing in high velocity fluidized beds, in *Gas Fluidization Technology*, Chapter 7 (ed. D. Geldart), Wiley, Chichester, UK, pp. 186–192.

Yerushalmi, J. and Avidan, A. (1985) Solid and gas mixing, in *Fluidization 2nd edn.*, Chapter 7 (eds J.F. Davidson, R. Clift and D. Harrison), Academic Press, London, pp. 278–287.

Zethraeus, B., Adams, C. and Berge, N. (1992) A simple model for turbulent gas mixing in CFB reactor. *Powder Technol.*, **69**, 101–105.

Zhang, Y.F., Arastoopour, H., Wegerer, D.A., Lomas, D.A. and Hemler, C.L. (1993) Experimental and theoretical analysis of gas and particle dispersion in large scale CFB, in *Circulating Fluidized Bed Technology IV* (ed. A. Avidan), AIChE, New York, pp. 473–478.

Zheng, Q.Y., Li, X., Ma, Z.W., Zhang, H., Wang, X.Y., Yu, L. and Xing, W. (1991) Design considerations for CFBC based on the results of fundamental research and operating experience of industrial CFB boilers, in *Proc. 11th Int. Conf. on FBC* (ed. E.J. Anthony), American Society of Mechanical Engrs., New York, pp. 1283–1288.

Zheng, Q.Y. (1994) Private communication, Tsinghua University, Beijing.

Zheng, Q.Y. and Zhang, H. (1993) Experimental study of the effect of exit (end effect) geometric configuration on internal recycling of bed material in CFB combustor, in *Circulating Fluidized Bed Technology IV* (ed. A. Avidan), AIChE, New York, pp. 145–151.

Zheng, Q.Y., Xing, W. and Lou, F. (1992) Experimental study on radial gas dispersion and its enhancement in circulating fluidized beds, in *Fluidization VII* (eds O.E. Potter and D.J. Nicklin), Engineering Foundation, New York, pp. 285–293.

Zhou, J., Grace, J.R., Lim, C.J. and Brereton, C.M.H. (1995) Particle velocity in a circulating fluidized bed riser of square cross section. *Chem. Eng. Sci.*, **50**, 237–244.

Zhu, J.-X. and Bi, H.-T. (1993) Development of CFB and high density circulating beds, in *Preprint Volume of Circulating Fluidized Bed Technology IV* (ed. A. Avidan), AIChE, New York, pp. 565–570.

4 Solids motion and mixing
JOACHIM WERTHER AND BERND HIRSCHBERG

4.1 Introduction

Circulating fluidized beds (CFB) exhibit very complex hydrodynamics, caused by interactions between the gas and solid phase. The motions of gas and solids are driven by many mechanisms that are difficult to identify and to describe.

The first section of the present chapter deals with investigations of particle motion in different regions of the riser. It is shown that different mechanisms of solids motion and mixing have to be distinguished. Bubbling fluidized beds are usually characterized by complete mixing of the solids. Whether this assumption is also justified in the CFB is examined in the following sections on axial and lateral solids mixing.

For the design of many physical and chemical processes it is important to understand the mechanisms and rates of solids mixing. For non-catalytic gas–solid reactions mean residence times and residence time distributions are of particular interest. They characterize the degree of mixing and give information about physical properties of solid particles in the riser. Insufficient solids mixing may cause hot spots, which should be avoided, especially for exothermic and fast reactions. A further aspect of solids mixing, which may be crucial for the performance of a cumbustor, is the lateral mixing of solids in the riser. This is important, for example, when the number and locations of solid feed points must be selected.

When solids of wide size distribution and/or different densities are fluidized, segregation of solids may occur. Usually, the solid particles that are easier to fluidize tend to be elutriated by the fluidization gas, while others remain at lower levels. As a result, a dynamic equilibrium is obtained between mixing and segregating tendencies.

The interrelation between mixing and segregation is of great interest, because, depending on the application, different mixing states may be desired. For combustors, for instance, perfect mixing of different components like coke, ash and limestone helps avoid hot spots and helps achieve uniform combustion. On the other hand, there are processes, e.g. agglomeration or prereduction of fine iron ores, that require segregation and separation of solids of different physical properties.

4.2 Particle motion and solids mixing mechanisms

Investigations of CFB hydrodynamics (see Chapter 2 as well as Hartge *et al.*, 1988; Zhang *et al.*, 1991; Wirth, 1990) have demonstrated that the riser can be divided into several regions with respect to their solids volume concentrations. As shown in Figure 4.1, four regions may be distinguished. The bottom zone, where acceleration of the solid particles takes place, is characterized by cross-sectional average solids volume concentrations, $\bar{C}$, of typically 0.1 to 0.2. After a transition zone lies a dilute zone, which occupies most of the riser height and is characterized by low solids volume concentrations ($>1\%$ for FCC risers, $<1\%$ for CFB combustors). At the riser top the exit geometry is responsible for the fluid dynamics in the exit zone. Solids mixing in these regions is discussed in detail in the following sections.

4.2.1 Particle motion in the bottom zone

Based on measurements of pressure fluctuations in the bottom zone of a CFB, Svensson *et al.* (1993) found that the bottom dense zone of a CFB shows hydrodynamic behavior similar to bubbling or turbulent fluidized beds, with fluidization gas passing through the bottom zone mostly in the form of voids. When these voids reach the surface of the bottom zone, they break and eject solids into the transition zone (Svensson *et al.*, 1993). The analogy to the movement of bubbles in bubbling fluidized beds has been supported by Werther and Wein (1994), who found the extrapolation of an existing bubble coalescence model to be sufficient to describe the size and velocity of voids.

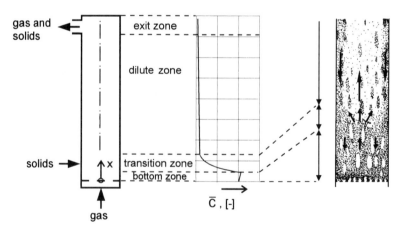

Figure 4.1 Zones of different solids volume fractions and solids motion in a CFB. (In sketch on right-hand side; arrows indicate main solids flow directions.)

Due to the lack of local investigations on solids mixing in the bottom zone of a CFB, it may initially be assumed that mixing mechanisms are similar to those in bubbling fluidized beds. According to Kunii and Levenspiel (1991), the transport in the wakes of rising voids is the essential mixing mechanism. In addition, the wakes of rising voids cause some lateral drift of the particles. Measurements of lateral mixing in a large fluidized bed by Bellgardt and Werther (1986) have shown that vertical mixing is at least one order of magnitude higher than lateral mixing.

4.2.2. Particle motion in the dilute zone

The dilute zone may be characterized by the existence of two 'phases', a lean and a dense phase. Studies of local hydrodynamics, e.g. by Hartge *et al.* (1988), have shown that the lean phase consists of an upward flowing dilute suspension, whereas the dense phase is formed by downward travelling particle clusters or strands. The dense phase is predominantly found in the region near the riser wall and shows solids concentrations at least an order of magnitude higher than in the lean phase. For simplicity the dense phase is often considered to be restricted to a layer close to the wall. The resulting core-annulus structure with a dense phase annulus and a lean phase core is basic to many current CFB models.

As a typical example of the solids motion in the dilute zone, radial profiles of local solids mass fluxes (Kruse and Werther, 1995), obtained by a suction probe (Rhodes, 1990), are given in Figure 4.2. The results are shown as

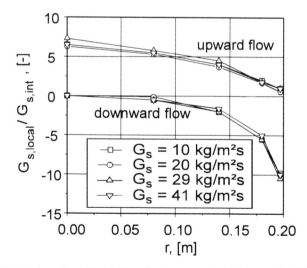

Figure 4.2 Radial profiles of local dimensionless upward and downward solids mass fluxes for $H = 15.6\,\text{m}$, $D = 0.4\,\text{m}$, ash/quartz sand ($200\,\mu\text{m}$), $U = 5\,\text{m/s}$, $x = 10.8\,\text{m}$; from Kruse and Werther (1995).

reduced solids fluxes (local solids mass flux divided by the net flux, $G_{s,int}$, obtained by averaging the local difference between upward and downward mass fluxes over the cross-sectional area), plotted against r/R. The upward solids mass fluxes reach their largest values at the riser center and decrease towards the wall, whereas the downward mass fluxes show a reverse tendency. At the wall, relatively large downward travelling mass fluxes have been found under these operating conditions. In agreement with findings by Rhodes et al. (1992a), these profiles were insensitive to changes in imposed solids circulation rates.

Similar flow patterns have been reported for large CFB boilers. For example, Figure 4.3 shows local solids flux measurements in a combustion chamber of a coal-fired boiler due to Werther (1993). The local solids mass fluxes are plotted against the distance from the membrane wall, y, measured from the fin of the membrane tube wall. Practically all solids downflow is seen to occur within 0.2 to 0.4 m from the wall.

This solids flow pattern is typical for CFB combustors, with external solids circulation rates, $G_s < 40 \, \text{kg/m}^2\text{s}$. FCC risers are typically operated at $G_s > 300 \, \text{kg/m}^2\text{s}$ (Werther, 1993). Under these conditions the horizontal profiles of the local solids net fluxes in a FCC riser exhibited a parabolic shape, but no net downflow near the wall (Azzi et al., 1990). At high gas and solids fluxes now used in industrial risers the solids may even travel up the wall (Avidan, 1995).

An additional characteristic of the dilute zone is the existence of a radial profile of local average solids velocities. By Laser Doppler Velocity (LDV)

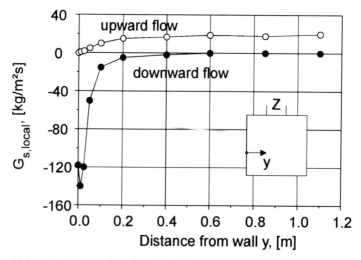

Figure 4.3 Local solids mass fluxes in a CFB boiler plotted against distance, y, from the wall for Flensburg power plant, $H = 28$ m, $U = 6.3$ m/s, $x = 17.3$ m, the y arrow indicates the position of traverse relative to the exit (Z) to the cyclone; from Werther (1993).

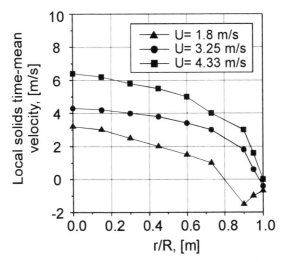

Figure 4.4 Radial profiles of solids velocities for $H = 11\,\text{m}$, $D = 0.14\,\text{m}$, FCC (54 μm), $G_S = 24.5\,\text{kg/m}^2\text{s}$, $x = 3.3\,\text{m}$; from Yang et al. (1992).

measurements Yang et al. (1992) found that the highest solids velocities occur at the riser center (Figure 4.4), where the mean solids velocities are about 1.5 to 2 times the superficial gas velocity. The velocity decreases towards the wall. Near the wall, negative values are obtained, indicating dominant downward flow of the solids. Note that the solids velocities reported for the wall region may be biased, since successful velocity measurements by LDV are more probable in the lean phase than in the dense phase. This is due to the lower risk of obstruction of the laser beam for smaller solids volume concentrations. A more detailed study of the motion of downward moving particles in regions near the wall has been conducted by Rhodes et al. (1992b) using a high-speed video camera. High density particle groups or swarms were observed to descend in contact with the wall at velocities in the range of -0.3 to $-0.4\,\text{m/s}$. A few millimeters from the wall, downflowing solids were found to fall with a velocity of $-1\,\text{m/s}$ as strands. Other authors (Hartge et al., 1988; Seiter, 1990; Golriz, 1992) report downflow velocities of solids near the wall in the range of -1 to $-2.4\,\text{m/s}$, most values lying between -1 and $-1.8\,\text{m/s}$.

Based on these hydrodynamic studies, it can be concluded that solids motion in the dilute zone is determined by the up- and downflowing phases. Due to countercurrent flow, intense solids mixing may be expected. In different mathematical model approaches, several authors have tried to describe the solids motion in this zone. All current models assume that solids mixing is assumed to be characterized by solids transfer between the core and the annulus regions. A distinction is made between two mechanisms, i.e. particle–particle collisions and particle turbulent diffusion.

Bolton and Davidson (1988) and Westphalen and Glicksman (1995) attribute the outward flux of core particles to gas-induced turbulent diffusion. The latter authors restrict their model to cases where solids volume concentrations are lower than 0.01. Such concentrations are usually encountered in CFB combustors (Werther, 1993). Under these conditions particle interactions may be assumed to be relatively unimportant. Gas turbulence then induces a corresponding fluctuation of the particles' velocities. The authors accordingly assume a radial solids flux from the core to the wall, proportional to the fluctuations of the radial velocity component of the particles. The solids flux is furthermore assumed to be proportional to the local solids volume concentration. The radial flux model is shown to give reasonable estimates of axial profiles of the cross-sectional averaged solids volume concentration, $\bar{C}_v(x)$, in small-diameter risers ($D < 0.4$ m).

The solids mixing models by Harris and Davidson (1993), Pugsley and Berruti (1995) and Senior and Brereton (1992) are based on interparticle collisions. Senior and Brereton (1992) postulated a flow structure with a transfer of particles between the annulus and core regions. As a possible mechanism by which upflowing core particles are deposited in the annulus region, they assumed that core particles gain lateral velocity by oblique particle–particle collisions. By analogy to the kinetic gas theory, the flow of core particles is therefore considered to be proportional to the core particle concentration. For the re-entrainment of particles from the annulus to the core phase, it is assumed that this is due to gas shear and core particles impinging on the annulus particle layer. From this, the authors derived solids mass transfer coefficients, which are related to additional model parameters, e.g. a wall layer disturbance factor and a viscous shear entrainment factor. They have to be determined by fitting the model to measured hydrodynamic data. Good prediction of axial suspension density profiles of a pilot-scale combustor was obtained.

Another model was proposed by Pugsley and Berruti (1995). Their model is related to the work of Wirth (1990), who presented a model for the momentum transfer arising from collisions between discrete particles and clusters. Since this momentum transfer is based on core particles striking the downward flowing particles and decelerating to the annular particle velocity, there is a resulting mass flux from the core to the annulus phase. The annulus-to-core flux is then calculated so that the solids material balance is satisfied. The model was successfully applied to the prediction of solids residence time distributions.

Harris and Davidson (1993) considered a particle transfer from the core to the annulus only. By analogy with reaction kinetics, their transfer rate is assumed to be proportional to the square of the core solids volume fraction. Based on momentum balances for the core and annulus, their predicted axial profiles of solids volume fractions in the core compare favorably with experimental data.

4.2.3 Particle motion in the transition zone

In the transition zone, a transition occurs from the dense bottom zone to the dilute zone with its low solids volume concentrations and its characteristic two-phase structure. In this region the frequency and size of clusters diminish with height (Senior and Brereton, 1992). Large amounts of solids are ejected by bursting voids from the bottom zone into the transition zone. Downfalling clusters carry solids from the dilute zone back into the transition zone. Hence, the transition zone is a region of high intensity solids mixing. However, solids mixing in the transition zone has not yet been investigated separately.

4.2.4 Particle motion in the exit zone

Two major types of exit geometries have been distinguished in the literature, i.e. an abrupt exit and a smooth exit. The latter is a smooth bend from the riser to the cyclone entrance, exhibiting negligible influence on the flow regime below. A more compact exit geometry is the abrupt exit, where the solids exit through a sharp 90° take-off just below the riser top. As shown in Figure 4.5, investigations using an abrupt exit (Brereton and Grace, 1993) have displayed increasing solids concentrations towards the top of the riser. The authors noted that this end effect is due to impacts of solids with the top of the riser. In particular, heavier particles that cannot follow the gas streamlines into the exit port, are reflected at the top of the riser, causing an accumulation of solids in this exit zone.

A comparison of solids mass flux profiles (Figure 4.6) for different exit geometries (Kruse and Werther, 1995) indicate that an abrupt exit leads in

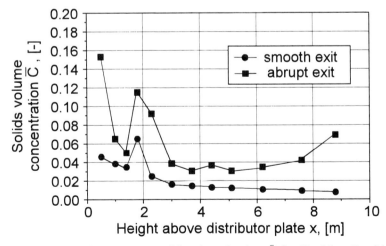

Figure 4.5 Effect of exit geometry on solids volume fraction, $\bar{C}$, for $H = 9.3$ m, $D = 0.15$ m, $U = 7.1$ m/s, $G_S = 73$ kg/m^{2}s, sand (148 μm); from Brereton and Grace (1993).

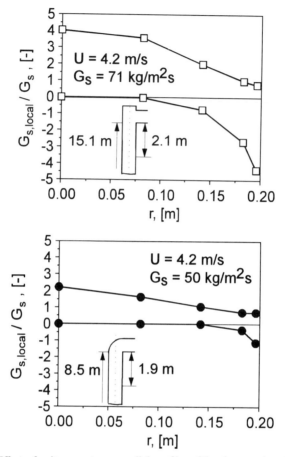

Figure 4.6 Effect of exit geometry on radial profiles of local upward and downward solid mass fluxes (upper figure: $H = 15.1$ m, $D = 0.4$ m, $x = 13$ m; sand (80 μm); lower figure: $H = 8.5$ m, $D = 0.4$ m, $x = 6.5$ m; sand (80 μm); from Kruse and Werther (1995)).

addition to more distinct radial profiles of solids mass fluxes near the riser top. Patience *et al.* (1990) therefore concluded that an abrupt exit causes intensified solids mixing.

4.3 Axial solids mixing

The preceding section has shown that the CFB exhibits different flow regions, each characterized by different solids flow patterns. It may therefore be expected that these different regions show different axial solids mixing characteristics.

4.3.1 Experimental techniques

Table 4.1 gives a compilation of the available literature. The experimental set-up consists in most cases of a pulse injection of tracer particles at some location into the riser and a detection device for measuring the pulse response at some distance downstream of the injection point. A typical experimental set-up is shown in Figure 4.7 together with a normalized response curve. Different tracer techniques have been used. The addition of ferromagnetic particles or NaCl particles has the disadvantage that repeated measurements successively increase the background tracer noise level. As a consequence the bed material must be replaced from time to time. The same is true for FCC particles impregnated with fluorescent dye, injected into an FCC riser by Kojima *et al.* (1989). The contamination of the bed material may be avoided by using irradiated tracer particles with a sufficiently short half-life. In the investigation by Wei *et al.* (1994) the tracer injection was replaced by introducing a light pulse by a flash to excite phosphor bed particles, avoiding contamination of the bed material. As solid tracer particles Bai *et al.* (1992) used bed particles that were marked with an organic substance whose desorption rate was detected by gas chromatography.

The design of solids mixing experiments presents a number of difficulties. The solids injection device should distribute the pulse of tracer solids into the riser in a way that is in accordance with the assumptions of the model subsequently used for the data evaluation. The same is valid for the detection device. If a locally measuring detector is used in connection with a one-dimensional modelling approach the question arises as to whether the measured tracer concentrations are representative of the whole cross-sectional area. Last but not least, the model used to interpret the response curve should consider the mixing mechanisms in all regions of the riser included in the section between injection and detection.

4.3.2 Axial solids dispersion model

A simple way to describe measured profiles of solids residence time distributions (RTD) is the dispersion model, represented by the differential equation

$$\frac{\partial C_T}{\partial t} = D_{ax,p} \frac{\partial^2 C_T}{\partial x^2} - \frac{\partial(v_p C_T)}{\partial x} \quad (4.1)$$

where C_T is the concentration of tagged particles. Equation 4.1 assumes convective transport of the solids with velocity, v_p, in the upward direction, superimposed by axial dispersion characterized by an axial dispersion coefficient, $D_{ax,p}$. This strongly simplified flow model is not in agreement

Table 4.1 Tracer techniques for investigating axial solids mixing

Author	Geometry	Tracer	Location of injection	Detection
Avidan (1980)	$D = 0.15$ m $H = 8.5$ m	Ferromagnetic particles	Dilute zone	Inductance bridge (cross-sectional average), at various locations in dilute zone
Bader et al. (1988)	$D = 0.305$ m $H = 12.2$ m	NaCl	Bottom zone	Sampling by suction probes on riser center-line, at various locations in dilute zone
Kojima et al. (1989)	$D = 0.05$ m $H = 3.6$ m	Injection of FCC labelled with fluorescent dye	Dilute zone ($x = 1.8$ m)	Fiber-optical probes at riser center-line, 0.2 m downstream of injection
Ambler et al. (1990)	$D = 0.05$ m $H = 3$ m	Irradiated sand	Bottom zone	Scintillation detector (cross-sectional average) at riser exit ($x = 3$ m)
Patience et al. (1990)	$D = 0.08$ m $H = 5$ m	Irradiated sand	Bottom zone	Scintillation detectors (cross-sectional average), at riser exit ($x = 4.72$ m)
Rhodes et al. (1991)	$D = 0.3$ m $H = 6.6$ m and $D = 0.152$ m $H = 6.2$ m	NaCl	Bottom zone	Sampling by suction probes located on riser center-line and wall, respectively, at $x = 4.4$ m and at riser exit ($x = 6.2$ m), respectively
Bai et al. (1992)	$D = 0.14$ m $H = 10$ m	Bed particles marked with an organic substance	Bottom zone	Detection of desorbed organic substance with gas chromatography at riser exit
Zheng et al. (1992)	$D = 0.09$ m $H = 10$ m	NaCl	Dilute zone ($x = 4.2$ m)	Sampling by suction probes at riser center-line, 0.24 m downstream of injection
Wei et al. (1994)	$D_{downcomer} = 0.14$ m $H_{downcomer} = 7.6$ m	Excitation of light emission of fine phosphor particles ($10\,\mu$m) in the bed by flash	Downcomer	Photomultiplier at various locations downstream of injection in the downcomer

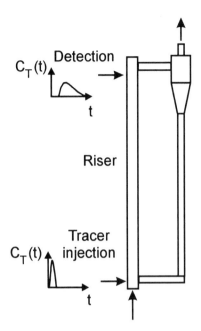
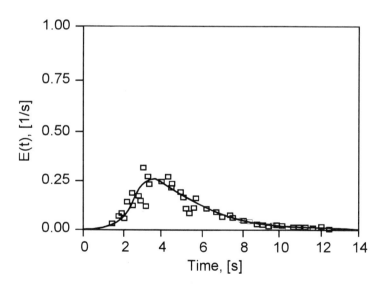

Figure 4.7 Experimental set-up used for measurements of solids RTD and normalized response curve ($H = 5$ m, $D = 0.83$ m, sand (275 μm), $U = 4.4$ m/s, $G_s = 28$ kg/m^{2}s, $L = 4.72$ m, tracer: irradiated sand, tracer injected just above distributor plate, detection at riser exit; from Patience et al. (1990)).

with the actual axial solids flow pattern in a CFB riser. However, numerical values of $D_{ax,p}$, obtained from an interpretation of measured solids RTDs with the simple axial solids dispersion model, give an indication of the extent of axial solids mixing in the CFB riser. Table 4.2 shows numerical values of axial solids dispersion coefficients, obtained by different authors. Some present their results in the form of axial solids Peclet numbers, defined by

$$Pe_{ax,p} = \frac{v_p L}{D_{ax,p}} \qquad (4.2)$$

where L is the distance between the injection and detection levels and v_p is a fictitious average solids velocity, defined by Patience *et al.* (1990) and Rhodes *et al.* (1991) as

$$v_p = \frac{L}{\bar{t}} \qquad (4.3)$$

The mean residence time $\bar{t}$, was determined by Rhodes *et al.* (1991) from the 50% value of the cumulative solids residence time distribution of the tracer, while Patience *et al.* (1990) proposed the first moment of the normalized RTD. Zheng *et al.* (1992) defined the solids velocity, v_p, on the basis of the external solids circulation rate as

$$v_p = \frac{G_s}{\rho_p \bar{C}_{tot}} \qquad (4.4)$$

The main differences between the numerical values of the axial solids dispersion coefficients and Peclet numbers, listed in Table 4.2, are probably due to the fact that the different investigations include different mixing regions inside the riser. For example, Patience *et al.* (1990) and Rhodes *et al.* (1991) injected the tracer into the bottom zone and detected it at the exit. Hence, the dispersion coefficients, $D^*_{ax,p}$, for both studies represent axial solids mixing averaged over the bottom zone, the transition zone and the dilute zone. The large values of $D^*_{ax,p}$ ranging from 0.1 to 20 m²/s, which is of the same order of magnitude as the dispersion coefficients of the gas phase discussed in the preceding chapter, are likely due to enhanced mixing in the bottom zone. Westphalen and Glicksman (1995) concluded therefore that the axial solids dispersion coefficient, $D_{ax,p,dz}$, in the dilute zone might be much less than the overall coefficient $D^*_{ax,p}$. The overall mixing coefficient is influenced by the proportion of the different regions included in the measurement. Since it is very difficult therefore to find unambiguous interpretations of $D^*_{ax,p}$, further discussion here focuses on investigations of axial solids mixing which have been restricted to the dilute zone.

Avidan (1980) and Kojima *et al.* (1989) reported dispersion coefficients in the range of 0.001 to 0.1 m²/s. Both authors injected the tracer into the dilute

Table 4.2 Range of experimental data on axial solids mixing ($D^*_{ax,p}$ and $Pe^*_{ax,p}$ indicate averaged solids mixing over bottom zone, transition zone and dilute zone; the index dz indicates that the corresponding mixing coefficient applies only to the dilute zone)

Author	Geometry	Bed material	Tracer	U (m/s)	G_s (kg/m²s)	$D_{ax,p,dz}$ (m²/s)	$D^*_{ax,p}$ (m²/s)	$Pe_{ax,p,dz}$ (—)	$Pe^*_{ax,p}$ (—)
Avidan (1980)	$D = 0.15$ m $H = 8.5$ m $L = 0.7$ m	FCC 50 μm	Ferromagnetic particles 50 μm	2.3–5.6	70–152	0.01–0.1			
Kojima et al. (1989)	$D = 0.05$ m $H = 3.6$ m $L = 0.1$ m and 0.2 m	FCC 60 μm	Fluorescent FCC 60 μm	1.5–2.15	?	0.0001–0.09 (80% of measurements between 0.001 and 0.03)			
Patience et al. (1990)	$D = 0.08$ m $H = 5$ m $L = 4.72$ m	Sand 277 μm	Irradiated sand 277 μm	4.1–6.3	45–161		0.2–1.7		3.5–9
Rhodes et al. (1991)	$D = 0.15$ m $H = 6.2$ m $L = 6.2$ m	Alumina 71 μm	NaCl 71 μm	2.8–5	5–65		7–19		
Rhodes et al. (1991)	$D = 0.3$ m $H = 6.6$ m $L = 6.6$ m	Alumina 71 μm	NaCl 71 μm	3–5	20–80		3–5		
Bai et al. (1992)	$D = 0.14$ m $H = 10$ m $L = 10$ m	Silica gel 100 μm	Silica gel marked with organic substance	2–10	10–100		only RTDs of gas and solids are given		
Zheng et al. (1992)	$D = 0.09$ m $H = 10$ m $L = 0.24$ m	FCC 47 μm	FCC saturated with NaCl	1.7–2.6	18–40			1.5–2.7	

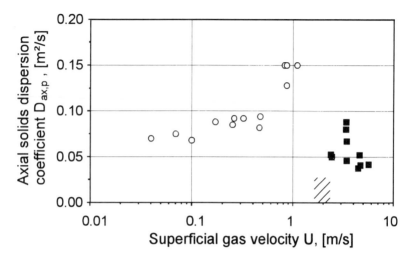

Figure 4.8 Axial dispersion of solids: A comparison of data from low velocity fluidized beds taken from Avidan and Yerushalmi (1985) with measurements of $D_{ax,p,dz}$ by Avidan (1980) and Kojima et al. (1989) in the dilute zone of a CFB riser. In all cases fine particles of FCC type have been used.

zone and detected it within 0.7 m (Avidan, 1980) and 0.1 m or 0.2 m (Kojima et al., 1989), respectively, above the injection point. Hence, the dispersion coefficients, $D_{ax,p,dz}$, in both studies truly represent axial solids mixing in the dilute zone. The numerical values are of the same order of magnitude as the dispersion coefficients presented by Bader et al. (1988) for axial gas mixing in the dilute zone (see Chapter 3).

Considering the limited database, it is not easy to generalize. Figure 4.8 shows the influence of the superficial gas velocity on the axial solids mixing. For comparison, data given previously by Avidan and Yerushalmi (1985) for low velocity fluidized beds have been added. Figure 4.8 indicates that in bubbling fluidized beds the increase of bubble frequency and size with increasing gas velocity are responsible for the distinct influence of U on solids mixing coefficients. Although the mixing coefficients in the CFB dilute zone are of the same order of magnitude as the mixing coefficients in the bubbling bed, it is obvious from Figure 4.8 that different mechanisms prevail in the two regimes. From Figure 4.8, solids mixing tends to increase with increasing gas velocity, U. It should be noted, however, that the underlying investigations

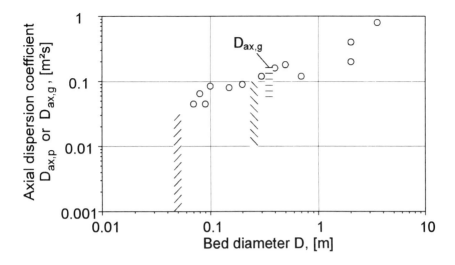

Figure 4.9 Solids mixing in bubbling fluidized beds of different diameters (compilation of data by Kunii and Levenspiel (1991)) and mixing in the dilute zone of CFB risers (range of data obtained by Avidan (1980) and Kojima *et al*. (1989), respectively, for solids mixing and by Bader *et al*. (1988) for gas mixing).

of Avidan (1980) and Kojima *et al*. (1989) were carried out in risers that differ in diameter by a factor of three. As is shown in Figure 4.9 the increase of the riser diameter has the effect of increasing the axial solids dispersion in the dilute zone. The tendency is similar to that in bubbling beds, as indicated by comparing with a compilation of bubbling bed data from Kunii and Levenspiel (1991).

In the case of the fine FCC particles used in the studies of Avidan (1980) and Kojima *et al*. (1989) it may be assumed that gas and solids motion in the dilute zone are largely governed by the same mechanisms, i.e. gas turbulence. It may therefore be concluded that the axial gas mixing coefficient in the dilute zone should at least give an estimate of the axial solids mixing coefficient. In Figure 4.9 the data of Bader *et al*. (1988) for gas mixing fit quite well into the general trend of increased mixing with increased bed diameter.

4.3.3 Core-annulus interchange model

Based on a core-annulus structure flow structure, Ambler *et al.* (1990) and Pugsley and Berruti (1995) presented mathematical models describing axial solids mixing in the dilute zone. In their models, solids mixing is considered to be due to particle transfer between the upward moving core and the downward falling annulus phase. Thereby, Pugsley and Berruti (1995) assumed that solids mixing is driven by interparticle collisions, as described above. In their approach, the transfer of solids between the two phases is described by two core-annulus interchange coefficients, $k_{ac}(x)$ and $k_{ca}(x)$, which are assumed to vary with height. Pugsley and Berruti (1995) successfully applied their model to describe a RTD measured by Patience *et al.* (1990). Note that this model is focused on the dilute zone. However, the experiments of Patience *et al.* (1990) include tracer injection into the bottom zone and detection at the top, i.e. the core-annulus model is applied to the whole riser including bottom, transition and dilute zones. The inclusion of the bottom zone may explain the strong dependence of k_{ca} on height close to the distributor, which has been observed by the authors.

4.4 Lateral solids mixing

4.4.1 Experimental techniques

The same measuring techniques used to determine axial solids mixing coefficients may generally be used to evaluate lateral mixing characteristics. Solids residence time distributions (RTD) have been used for this purpose by Wei *et al.* (1994) and Patience and Chaouki (1995).

Experimental methods that do not lead to contamination of the bed material and that are particularly suited for investigations of lateral mixing have been suggested by Westphalen and Glicksman (1995) and Koenigsdorff and Werther (1995). The method of Westphalen and Glicksman (1995), based on an earlier study by Valenzuela and Glicksman (1984), consists of quasi-steady state injection of heated solids combined with the detection of radial temperature profiles by thermistors downstream of the injection. Koenigsdorff and Werther (1995) use a vertical wire heater of 1 mm diameter to locally introduce heat into the riser, which is dispersed in the flow under the influence of convection and gas and particle dispersion. Temperature profiles are detected using Pt 100 resistance thermometers. This arrangement with the wire heater located on the riser axis was used to investigate solids mixing in the core zone. A similar set-up consisted of a circumferential wire heater mounted in three coils above each other near the riser wall. The resulting temperature profiles were used to investigate heat exchange between the lean and dense phases.

4.4.2 Evaluation of lateral solids mixing experiments

(a) Evaluation of solids residence time distributions (RTD)
Wei et al. (1994) interpreted their measurements on the basis of the two-dimensional dispersion model given by

$$\frac{\partial C_T}{\partial t} = D_{ax,p} \frac{\partial^2 C_T}{\partial x^2} + \frac{D_{r,p}}{r} \frac{\partial}{\partial r}\left(r \frac{\partial C_T}{\partial r}\right) - \frac{\partial(v_p C_T)}{\partial x} \quad (4.5)$$

to describe solids mixing in the downcomer ($D_{downcomer} = 0.14$ m) of their CFB. The results are given in the form of a solids radial Peclet number

$$Pe_{r,p} = \frac{U D_{downcomer}}{D_{r,p}} \quad (4.6)$$

which ranges from 70 to 300. This Peclet number range corresponds to lateral solids mixing dispersion coefficients, $D_{r,p}$, from 0.0026 to 0.0113 m²/s.

Patience and Chaouki (1995) applied equation (4.5) to describe their measurements of the solids RTD in the dilute zone. Axial solids dispersion was neglected because its contribution to solids mixing is small compared to axial solids convection. For tracer injection at $x = 1.75$ m and a cross-sectional averaged detection at $x = 4$ m, a solids RTD was obtained at $U = 8$ m/s and $G_s = 140$ kg/m²s, yielding a value of $D_{r,p}$ of 0.0025 m²/s.

(b) Evaluation of temperature profiles
The evaluation of the heating experiments suggested by Westphalen and Glicksman (1995) and Koenigsdorff and Werther (1995), respectively, requires consideration of the contributions of both the solid particles and the gas in the respective energy balances. Westphalen and Glicksman (1995) applied their heated solids injection techniques to the core region in the dilute zone of a 7 m tall, 0.2 m diameter CFB unit. Particles with $d_p = 180\,\mu m$ and a solids density of 2350 kg/m³ were used as bed material and for the injection at $x = 3$ m. An array of thermistor probes was located between 0 and 0.9 m above the injection location such that radial temperature profiles could be measured at different locations downstream of the injection. Models for the spread of tracer particles, transfer of heat from tracer particles to surrounding gas and particles and for heat transfer to the thermistors were used to estimate the radial particle diffusivity in the core region. Figure 4.10 compares measured and calculated temperature profiles at different distances from the solids injection. Unfortunately, the scatter of the resulting dispersion coefficients does not permit generalized tendencies to be deduced. The solids dispersion coefficient was of the order of 0.001 m²/s for the range of superficial gas velocities and solids concentrations tested. It should be noted that in all cases the solids dispersion coefficient turns out to be less than the gas dispersion coefficient calculated for single-phase turbulent flow.

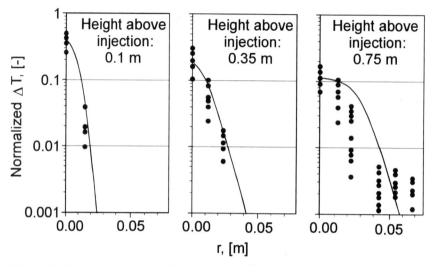

Figure 4.10 Measured and calculated temperature profiles obtained with the injection of heated solids (from Westphalen and Glicksman (1995); $D = 0.2$ m, $U = 3.5$ m/s, $G_s = 24$ kg/m^{2}s, average bed density 24 kg/m^3, $x = 3$ m, model calculation with $D_{r,p} = 0.00074$ m^2/s and $D_{r,g} = 0.0028$ m^2/s).

Koenigsdorff and Werther (1995) evaluate their heating experiments with a two-phase model that assumes a dense phase of downflowing particle strands dispersed in a continuous lean phase. The model is three-dimensional but assumes rotational symmetry. The extents of gas and particle dispersion in the lean phase are described by the dispersion coefficient, $D_{r,g}$ and $D_{r,p}$, respectively. A major contribution to solids mixing, particularly in the wall region, is due to gas and solids exchange between the two phases, which is accounted for by the interphase transfer coefficients, β_g and β_p, respectively. In the energy balances for the two phases an effective thermal conductivity of the lean phase in the radial direction, $\lambda_{eff,r}$, is used, defined by

$$\lambda_{eff,r} = (1 - C_l)(\lambda_g + D_{r,g}\rho_g c_{p,g}) + C_l D_{r,p}\rho_p c_{p,p} \tag{4.7}$$

The measurements were conducted in the dilute zone of a CFB riser of 0.2 m diameter and 3.5 m height. The bed material was SiC with a surface mean diameter of 60 μm. Measurements were first made with the heater on the riser axis located between $x = 1.95$ m and $x = 2.21$ m. Preliminary experiments in the empty column yielded radial temperature profiles shown in Figure 4.11. Fitting of the model to measurements yielded gas dispersion coefficients corresponding to

$$Pe_{r,g} = \frac{UD}{D_{r,g}} = 321 \tag{4.8}$$

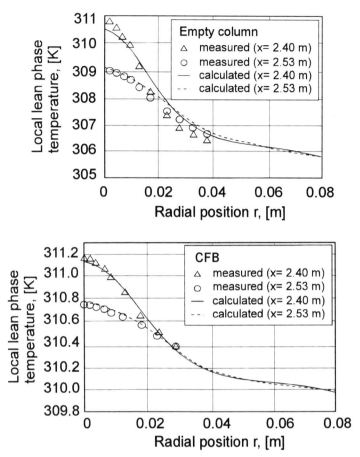

Figure 4.11 Horizontal temperature profiles obtained by electrical heating at $r = 0$ between $x = 1.95$ m and 2.21 m in the empty column (calculation with $D_{r,g} = 0.0025\,\text{m}^2\text{s}$) and under CFB conditions ($D = 0.2$ m, $U = 4$ m/s, $G_s = 20.2\,\text{kg/m}^2\text{s}$, and $C_l = 0.0019$, 45 W heating, calculation with $D_{r,g} = 0.0025\,\text{m}^2/\text{s}$, $D_{r,p} = 0.002\,\text{m}^2/\text{s}$; from Koenigsdorff and Werther (1995)).

The numerical value of 321 is within the range of data reported in the literature (Hinze, 1975; Martin *et al.*, 1992; Sherwood *et al.*, 1975; Werther *et al.*, 1992) for fully developed single-phase turbulent flow. As shown in Figure 4.11 the addition of solids to the riser leads to considerable flattening of the radial temperature profiles, which is due to the higher heat capacity of the particle-laden flow. In order to determine radial solids dispersion coefficients it was assumed that $D_{r,g}$ is unchanged when the particles were added. This simplifying assumption has no influence on the following calculations, since the temperature profile is governed by the presence and the motion of the solid particles due to their higher heat capacity. An evaluation of these measurements leads to numerical values of the Peclet number, $Pe_{r,p}$,

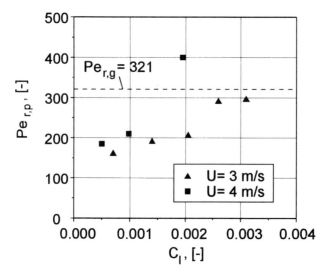

Figure 4.12 Influence of lean phase solids volume concentration, C_l, on the radial particle dispersion Peclet number, $Pe_{r,p}$. ($Pe_{r,g} = 321$ has been obtained for single-phase turbulent flow; from Koenigsdorff and Werther (1995)).

for radial solids dispersion:

$$Pe_{r,p} = \frac{UD}{D_{r,p}} \qquad (4.9)$$

$Pe_{r,p}$ turns out to be a function of the solids volume concentration in the lean phase, C_l (Figure 4.12). The fact that for the conditions investigated, i.e. $C_l < 0.003$, $Pe_{r,p}$ is generally smaller than $Pe_{r,g}$ means that particle dispersion is more pronounced than gas dispersion, contrary to the findings of Westphalen and Glicksman (1995). Together with the significant dependence of $Pe_{r,p}$ on C_l, this indicates a strong influence of particle–particle collisions on particle dispersion, since in the case of solely gas-induced particle mixing, particle dispersion coefficients would be less than gas dispersion coefficients. According to the results in Figure 4.12, particle dispersion is damped by increasing solids volume concentration in the lean phase.

The interphase mass transfer, i.e. the exchange of solids between the upflowing lean phase and the downflowing dense phase, yields a major contribution to solids mixing in the wall region. This effect was investigated with the circumferential wire heaters located at $x = 2.45$ m near the wall. Figure 4.13 shows the resulting radial temperature profiles measured 0.03 m downstream and 0.08 m upstream of the heating coils together with the respective model calculations. A satisfactory overall description for both the upstream and the downstream profiles is obtained with $\beta_g = \beta_p = 8.6\,\mathrm{s}^{-1}$. Averaging over all temperature profiles measured leads to $\beta_p = 5.4\,\mathrm{s}^{-1}$.

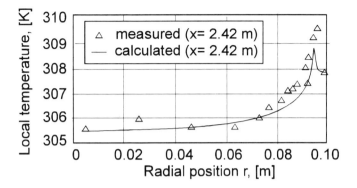
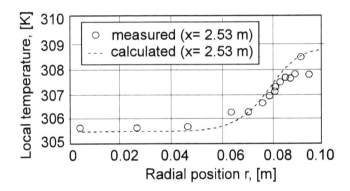

Figure 4.13 Temperature profiles in the dilute zone caused by electrical heating near the riser wall ($D = 0.2$ m, $U = 3$ m/s, $G_s = 6.8$ kg/m^{2}s and $\bar{C}_l = 0.0011$, 318 W heating at $x = 2.45$ m, calculation with $D_{r,g} = 0.0019$ m^2/s, $D_{r,p} = 0.0028$ m^2/s, $\beta_g = \beta_p = 8.6$ s^{-1}; from Koenigsdorff and Werther (1995)).

4.5 Fluidization of dissimilar particles/solids segregation

Closely related to solids mixing is the phenomenon of solids segregation, which occurs when dissimilar solids, i.e. solids of different sizes and/or densities, are fluidized. Solids segregation has received much attention in bubbling fluidized beds (e.g. Nienow and Chiba, 1985; Baeyens and Geldart, 1986; Kunii and Levenspiel, 1991). In recent years there has been growing interest in segregation effects in CFBs.

4.5.1 Experimental findings

A summary of relevant experimental studies is given in Table 4.3. Except for Chesonis *et al.* (1990), all authors have investigated segregation effects under steady state conditions. Chesonis *et al.*, (1990) injected a batch of petroleum

Table 4.3 Experimental investigations of fluidization of dissimilar particles in circulating fluidized beds

Author	Geometry	Particle mixture	Solids composition measurements	Results presented as
Chesonis et al. (1990)	$D = 0.1$ m $H = 6.4$ m	Bed: alumina (121 μm, 3500 kg/m^3); petroleum coke (168 μm, 2000 kg/m^3) injected at approx. $x = 0.5$ m	Suction sampling, coke contents from combustion	$\bar{C}(x)$, $\kappa(x)$
Nowak et al. (1990)	$D = 0.205$ m $H = 6.65$ m	FCC (46 μm, 2300 kg/m^3), silica sand (3 mm, 2300 kg/m^3)	Suction sampling, separation of solids by sieving	$\kappa(x)$
Bi et al. (1992)	$D = 0.1$ m $H = 6.3$ m	Fine particles: FCC (58 μm, 1153 kg/m^3) Coarse particles: polyethylene spheres (4.4 mm, 1010 kg/m^3) or glass beads (2 mm, 2500 kg/m^3)		$\bar{C}(x)$
Ijichi et al. (1992)	$D = 0.05$ m $H = 2.86$ m	Sand (222 μm, 2650 kg/m^3), Iron powder (238 μm, 7600 kg/m^3)	Suction sampling, magnetic separation of solids	$\bar{C}(x)$, $\kappa(x)$
Bai et al. (1994) and Nagakawa et al. (1994)	$D = 0.097$ m or 0.15 m $H = 3$ m	FCC (70 μm, 1700 kg/m^3), silica sand (320 μm, 2600 kg/m^3)	Suction sampling, separation of solids by sieving	$\bar{C}(x)$, $\bar{C}_{fine}(x)$, $\bar{C}_{coarse}(x)$ $\kappa(x)$
Jiang et al. (1994)	$D = 0.1$ m $H = 6.3$ m	Fine particles: FCC (89 μm, 1153 kg/m^3) or polyethylene resin (325 μm, 600 kg/m^3) Coarse particles: polyethylene spheres (4.4 mm, 1010 kg/m^3) or glass beads (2 mm, 2500 kg/m^3)		$\bar{C}(x)$
Hirschberg et al. (1995)	$D = 0.2$ m $H = 12$ m	Sand (220 μm, 2650 kg/m^3), Iron ore (230 μm, 4900 kg/m^3)	Suction sampling with subsequent measurements of mixture solids density	$\bar{C}(x)$, $\kappa(x)$
Na et al. (1996)	1.7 m × 1.4 m $H = 13.5$ m	Fine particles: sand (290 μm, 2600 kg/m^3) Coarse particles: stones (5.4 mm)	Suction sampling, separation of solids by sieving	$\kappa(x)$

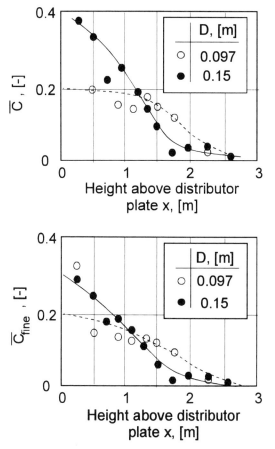

Figure 4.14 Comparison of axial distributions of the solids mixture volume concentrations, $\bar{C}$, and fines volume concentrations, $\bar{C}_{fine}$, in risers of different diameter ($U = 2\,\text{m/s}$, $G_s = 27\,\text{kg/m}^2\text{s}$, Bai et al. (1994)).

coke into their riser and followed the time-dependent coke concentration in a FCC bed at different locations inside the riser. They found that the coke mixed quickly in the bottom portion of the column before it was transported into the dilute zone.

Most authors have added coarse particles in a circulating fluidized bed of fine particles (Nowak et al., 1990; Bi et al., 1992; Bai et al., 1994; Jiang et al., 1994; Nagakawa et al., 1994; Na et al., 1996). As an example, Figure 4.14 shows measurements of Bai et al. (1994). In the upper figure the vertical solids volume concentrations calculated from the axial pressure profiles are shown, whereas in the lower figure the local composition of the solids mixture is shown as axial profiles of the local volume concentrations of the fine particles. Figure 4.14 compares the performance of risers of two different diameters under identical operating conditions. The enlargement of the column diameter leads to an increase of both the solids mixture and the fine solids hold-ups in the bottom section and a corresponding decrease in the upper part of the riser. Generally, the addition of coarse particles increase the hold-up and thus the residence time of solids in the riser. In the case of solids catalysed gas phase reaction, this may increase the conversion (Bi *et al.*, 1992).

Segregation effects are often undesirable. However, in circulating fluidized bed combustion segregation causes large coal feed particles to remain in the bottom part of the riser, leading to high combustion efficiencies. In the case of multi-solid fluidized beds (MSFB) coarse particles should always remain in the bottom part of the riser and should not be elutriated (Bai *et al.*, 1994; Na *et al.*, 1996).

Ijichi *et al.* (1992) and Hirschberg *et al.* (1995) investigated the behavior of mixtures of two components having nearly identical particle size distributions, but different solids densities in CFB risers of 0.05 m diameter and 0.2 m diameter, respectively. In both papers the influence of operating conditions on the axial solids composition profiles and on the approach to perfect mixing was investigated. Figure 4.15 shows axial composition profiles from the study by Hirschberg *et al.* (1995). For a given solids circulation rate the axial solids composition profiles become flatter with increasing superficial gas velocity.

4.5.2 Segregation mechanisms

The discussion of solids mixing inside the CFB riser above has shown that different mechanisms prevail in the different zones of the riser. The same will certainly be valid for segregation.

The bottom zone exhibits many similarities to a bubbling fluidized bed. It may therefore be expected that segregation in this zone is related to the size, frequency and motion of voids. As a consequence the segregation mechanisms identified by Nienow and co-workers (e.g. Nienow and Chiba, 1985; Rowe *et al.*, 1972) should be applicable. In the transition zone particles are ejected by erupting voids from the bottom zone. The transition zone is characterized by a decay of solids concentration from the high level in the bottom zone to the low level in the dilute zone. This decay indicates a strong segregation of the upflowing suspension into a downflow of solids returning into the bottom zone and a small upward flow continuing into the dilute zone. Since the transition zone acts like an elutriator, separation of particles may occur here according to their terminal settling velocities.

The mechanisms in the dilute zone are not as clear. In the upflowing lean phase, where the particles rise individually with comparatively large mean free paths between each other, segregation according to the particles' terminal velocities is probable. It is currently unclear whether segregation

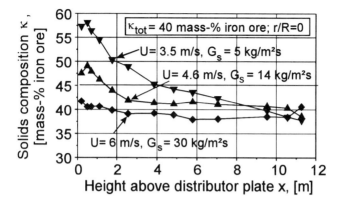

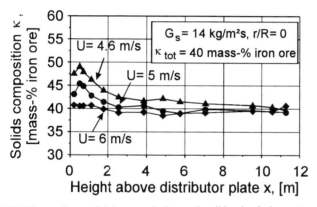

Figure 4.15 Effects of superficial gas velocity and solids circulation rate on axial solids composition profiles ($H = 12$ m, $D = 0.2$ m, sand (220 μm), iron ore (230 μm), from Hirschberg et al. (1995)).

occurs inside the dense phase clusters or strands and what contributions to the overall segregation behavior in this region are made by the interaction between the lean and dense phases. The exit region provides another space for segregation, due to impingement of coarse particles to the riser top, while fine particles tend to follow the exiting gas flow directly.

Figure 4.16 shows an axial solids volume concentration profile and the corresponding profile of the local solids composition from the investigations of Hirschberg et al. (1995). Throughout the dilute zone the solids composition is seen to be nearly constant, whereas the transition zone is characterized by a significant gradient of solids composition. It may therefore be concluded that at least in this case the transition zone is decisive for the segregation in the riser, leading to the observed enrichment of iron ore in the bottom zone. Indications that it is the particle's terminal velocity which is the physical property governing the segregation process may be taken from the plot of

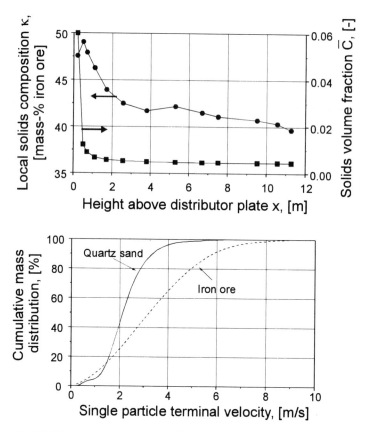

Figure 4.16 Solids volume concentration, $\bar{C}$, and solids composition profiles for a mixture of iron ore and sand ($H = 12$ m, $D = 0.2$ m, $U = 4.6$ m/s, $G_s = 13$ kg/m^{2}s, sand ($220\,\mu$m), iron ore ($230\,\mu$m), $\kappa_{tot} = 0.4$, from Hirschberg et al. (1995)) and their corresponding cumulative mass distributions of single particle terminal velocities.

the distributions of the single particle terminal velocities of the two components given in the same figure. The superficial gas velocity in the experiments was 4.6 m/s. At this velocity nearly 100 wt% of the quartz sand particles are entrainable, but only about 70 wt% of the iron ore particles. Thus, there is a tendency for iron ore particles to remain in the bottom section. In agreement with the terminal velocity distributions of the two components (Figure 4.16), negligible segregation was observed in further experiments at gas velocities exceeding 6 m/s, where over 90 wt% of the iron ore should be entrainable.

Nomenclature

C_T tracer concentration, dimensionless
c_p specific heat capacity, J/(kg K)

C	solids volume fraction, dimensionless
$\bar{C}$	cross-sectional averaged solids volume fraction, dimensionless
$\bar{C}_{coarse}$	cross-sectional averaged solids volume fraction of the coarse particles, dimensionless
$\bar{C}_{fine}$	cross-sectional averaged solids volume fraction of the fine particles, dimensionless
$\bar{C}_{tot}$	solids volume fraction averaged along riser height, dimensionless
d_p	mean diameter of particles, μm
D	inside diameter of CFB riser, m
D_{ax}	coefficient of axial dispersion, m^2/s
$D_{ax}{}^*$	coefficient of axial dispersion averaged over bottom zone, transition zone and dilute zone, m^2/s
$D_{downcomer}$	inside diameter of CFB downcomer, m
D_r	coefficient of radial dispersion, m^2/s
$E(t)$	normalized impulse response function, 1/s
G_s	solids circulation rate, kg/(m^{2}s)
$G_{s,local}$	local solids circulation rate, kg/(m^{2}s)
$G_{s,int}$	cross-sectional average solids circulation rate, kg/(m^{2}s)
H	riser height, m
L	axial distance between injection and detection point of tracer, m
k_{ac}	solids transfer coefficient for solids flow from annulus to core, m/s
k_{ca}	solids transfer coefficient for solids flow from core to annulus, m/s
Pe_{ax}	Peclet number for axial dispersion, dimensionless
$Pe_{ax}{}^*$	Peclet number for axial dispersion averaged over bottom zone, transition zone and dilute zone, dimensionless
Pe_r	Peclet number for radial dispersion, dimensionless
r	radial location, m
R	riser radius, m
t	time, s
$\bar{t}$	solids mean residence time, s
U	superficial gas velocity, m/s
u_t	terminal velocity, m/s
v_p	average solids velocity, m/s
x	height above distributor plate, m
y	distance from wall, m
β	coefficient of interphase transfer, 1/s
$\lambda_{eff,r}$	effective thermal conductivity of the lean phase in the radial direction, W/m^{2}K
λ_g	thermal conductivity of gas, W/m^{2}K
κ	solids mixture mass fraction of denser component, dimensionless
κ_{tot}	overall solids mixture composition in CFB loop, dimensionless
ρ	density, kg/m^3

Subscripts

g   gas
l   lean phase
p   particle
dz   dilute zone

References

Ambler, P., Milne, B., Berruti, F. and Scott, D. (1990) Residence time distribution of solids in a circulating fluidized bed: experimental and modelling studies. *Chemical Engineering Science*, **45**, 2179–2186.

Avidan, A. (1980) Bed expansion and solid mixing in high velocity fluidized beds. PhD dissertation, City College of New York.

Avidan, A. (1995), Mobil Research Center, Paulsboro, NJ, private communication.

Avidan, A. and Yerushalmi, J. (1985) Solids mixing in an expanded top fluid bed. *AIChE Journal*, **31**, 835–841.

Azzi, M., Turlier, P., Large, J. and Bernard, J. (1990) Use of a momentum probe and gammadensitometry to study local properties of fast fluidized beds, in *Circulating Fluidized Bed Technology III* (eds P. Basu, M. Horio and M. Hasatani), Pergamon Press, Oxford, pp. 189–194.

Bader, R., Findlay, J. and Knowlton, T. (1988) Gas/solids flow patterns in a 30.5 cm diameter circulating fluidized bed, in *Circulating Fluidized Bed Technology II* (eds P. Basu and J.F. Large), Pergamon Press, Oxford, pp. 123–137.

Baeyens, J. and Geldart, D. (1986) Solids mixing, in *Gas Fluidization Technology* (ed. D. Geldart), Wiley, Chichester, pp. 97–122.

Bai, D., Yi, J., Jin, Y. and Yu, Z. (1992) Residence time distribution of gas and solids in a circulating fluidized bed, in *Fluidization VII* (eds O.E. Potter and D.J. Nicklin), Engineering Foundation, New York, pp. 195–202.

Bai, D., Nakagawa, N., Shibuya, E., Kinoshita, H. and Kato, K. (1994) Axial distribution of solids holdups in binary solids circulating fluidized beds. *Journal of Chemical Engineering of Japan*, **27**, 271–275.

Bellgardt, D. and Werther, J. (1986) A novel method for the investigation of particle mixing in gas–solid systems. *Powder Technology*, **48**, 173–180.

Bi, H., Jiang, P., Jean, R. and Fan, L.-S. (1992) Coarse-particle effects in a multisolid circulating fluidized bed for catalytic reactions. *Chemical Engineering Science*, **47**, 3113–3124.

Bolton, L. and Davidson, J. (1988) Recirculation of particles in fast fluidized risers, in *Circulating Fluidized Bed Technology II* (eds P. Basu and J.F. Large), Pergamon Press, Oxford, pp. 139–146.

Brereton, C. and Grace, J. (1993) End effects in circulating fluidized bed hydrodynamics, in *Circulating Fluidized Bed Technology IV* (ed. A. Avidan), AIChE, New York, pp. 137–144.

Chesonis, D., Klinzing, G., Shah, T. and Dassori, C. (1990) Solids mixing in a recirculating fluidized bed, in *Circulating Fluidized Bed Technology III* (eds P. Basu, M. Horio and M. Hasatani), Pergamon Press, Oxford, pp. 587–592.

Golriz, M. (1992) Thermal and fluid-dynamic characteristics of circulating fluidized bed boilers. Thesis for the degree of licentiate of engineering, Chalmers University of Technology, Göteborg, Sweden.

Harris, B. and Davidson, J. (1993) A core/annulus deposition model for circulating fluidized beds, in *Circulating Fluidized Bed Technology IV* (ed. A. Avidan), AIChE, New York, pp. 32–39.

Hartge, E.U., Rensner, D. and Werther, J. (1988) Solids concentration and velocity patterns in circulating fluidized beds, in *Circulating Fluidized Bed Technology II* (eds P. Basu and J.F. Large), Pergamon Press, Oxford, pp. 165–180.

Hinze, J. (1975) *Turbulence*, 2nd edn, McGraw–Hill, New York.
Hirschberg, B., Werther, J., Delebarre, A. and Koniuta, A. (1995) Mixing and segregation of solids in a circulating fluidized bed, *Fluidization VIII Conference*, Tours, France, in press.
Ijichi, K., Tanaka, Y., Uemura, Y., Hatate, Y. and Yoshida, K. (1992) Solids holdup and concentration in a riser of circulating fluidized bed of two-component system. *Proc. of the 3rd Asian Conference on Fluidized Bed and Three-Phase Reactors* (eds H.S. Chun and S.D. Kim), Korea, pp. 433–441.
Jiang, P., Bi, H., Liang, S. and Fan, L.-S. (1994) Hydrodynamic behavior of circulating fluidized bed with polymeric particles. *AIChE Journal*, **40**(2), 193–206.
Koenigsdorff, R. and Werther, J. (1995) Gas and solids mixing in and flow structure modeling of the upper dilute zone of a circulating fluidized bed. *Powder Technology*, **82**, 317–329.
Kojima, T., Ishihara, K., Guilin, Y. and Furusawa, T. (1989) Measurement of solids behavior in a fast fluidized bed. *Journal of Chemical Engineering of Japan*, **22**(4), 341–346.
Kruse, M. and Werther, J. (1995) 2D gas and solids flow prediction in circulating fluidized beds based on suction probe and pressure profile measurements. *Chemical Engineering and Processing*, **34**, 185–203.
Kunii, D. and Levenspiel, O. (1991) *Fluidization Engineering*, 2nd edn, Butterworth–Heinemann, Boston.
Martin, M., Turlier, P., Bernard, J. and Wild, G. (1992) Gas and solid behavior in cracking circulating fluidized beds. *Powder Technology*, **70**, 249–258.
Na, Y., Yan, G., Sun, X., Cui, P., He, J., Karlsson, M. and Leckner, B. (1996) Large and small particles in CFB combustors. *Preprints of the 5th International Conference on Circulating Fluidized Beds*, Beijing.
Nagakawa, N., Bai, D., Shibuya, H., Kinoshita, H., Takarada, T. and Kato, K. (1994) Segregation of particles in binary solids circulating fluidized beds. *Journal of Chemical Engineering of Japan*, **27**, 194–198.
Nienow, A. and Chiba, T. (1985) Fluidization of dissimilar materials, in *Fluidization* 2nd edn (eds J.F. Davidson, R. Clift, D. Harrison), Academic Press, London, pp. 357–382.
Nowak, W., Mineo, H., Yamazaki, R. and Yoshida, K. (1990) Behavior of particles in a circulating fluidized bed of a mixture of two different sized particles, in *Circulating Fluidized Bed Technology III* (eds P. Basu, M. Horio and M. Hasatani), Pergamon Press, Oxford, pp. 219–224.
Patience, G. and Chaouki, J. (1995) Solids hydrodynamics in the fully developed region of CFB risers, *Fluidization VIII Conference*, Tours, France, in press.
Patience, G., Chaouki, J. and Kennedy, G. (1990) Solids residence time distribution in CFB reactors, in *Circulating Fluidized Bed Technology III* (eds P. Basu, M. Horio and M. Hasatani), Pergamon Press, Oxford, pp. 599–604.
Pugsley, T. and Berruti, F. (1995) A core-annulus solids interchange model for circulating fluidized beds and FCC risers, *Fluidization VIII Conference*, Tours, France, in press.
Rhodes, M. (1990) Modeling the flow structure of upward-flowing gas solids suspensions. *Powder Technology*, **60**, 27–38.
Rhodes, M.J., Zhou, S., Hirama, T. and Cheng, H. (1991) Effects of operating conditions on longitudinal solids mixing in a circulating fluidized bed riser. *AIChE Journal*, **37**, 1450–1458.
Rhodes, M., Wang, X., Cheng, H., Hirama, T. and Gibbs, B. (1992a) Similar profiles of solids flux in circulating fluidized-bed risers. *Chemical Engineering Science*, **47**, 1635–1643.
Rhodes, M., Mineo, H. and Hirama, T. (1992b) Particle motion at the wall of a circulating fluidized bed. *Powder Technology*, **70**, 207–214.
Rowe, P., Nienow, A. and Agbim, A. (1972) The mechanisms by which particles segregate in gas fluidized beds – binary systems of near-spherical particles. *Transactions of the Institution of Chemical Engineers*, **50**, 324–333.
Seiter, M. (1990) Radiale Feststoffkonzentration und axiale Feststoffgeschwindigkeit in wandnahen Bereichen zirkulierender Wirbelschichten. PhD dissertation, University Erlangen-Nürnberg, Germany.
Senior, R. and Brereton, C. (1992) Modeling of circulating fluidized bed solids flow and distribution. *Chemical Engineering Science*, **47**, 281–296.
Sherwood, T., Pigford, R. and Wilke, C. (1975) *Mass Transfer*, McGraw–Hill, New York.

Svensson, A., Johnsson, F. and Leckner, B. (1993) Fluid-dynamics of the bottom bed of circulating fluidized bed boilers. *Proc. XII International Conference on Fluidized Bed Combustion*, ASME, pp. 887–897.

Valenzuela, J. and Glicksman, L. (1984) An experimental study of solids mixing in a freely bubbling two-dimensional fluidized bed. *Powder Technology*, **38**, 63–72.

Wei, F., Wang, Z., Jin, Y., Yu, Z. and Chen, W. (1994) Dispersion of lateral and axial solids in a cocurrent downflow circulating fluidized bed. *Powder Technology*, **81**, 25–30.

Werther, J. (1993) Fluid mechanics of large-scale CFB units, in *Circulating Fluidized Bed Technology IV* (ed. A. Avidan), AIChE, New York, pp. 1–14.

Werther, J. and Wein, J. (1994) Expansion behavior of gas fluidized beds in the turbulent regime. *AIChE Symp. Series*, **90**(301), 31–44.

Werther, J., Hartge, E.-U. and Kruse, M. (1992) Radial gas mixing in the upper dilute core of a circulating fluidized bed. *Powder Technology*, **70**, 293–301.

Westphalen, D. and Glicksman, L. (1995) Lateral solids mixing measurements in circulating fluidized beds. *Powder Technology*, **82**, 153–168.

Wirth, K.-E. (1990) *Zirkulierende Wirbelschichten*, Springer–Verlag, Berlin.

Yang, Y., Jin, Y., Yu, Z. and Wang, Z. (1992) Investigations on slip velocity distribution in the riser of dilute circulating fluidized bed. *Powder Technology*, **73**, 67–73.

Zhang, W., Tung, Y. and Johnsson, F. (1991) Radial voidage profiles in fast fluidized beds of different diameters. *Chemical Engineering Science*, **46**, 3045–3052.

Zheng, C., Tung, Y., Li, H. and Kwauk, M. (1992) Characteristics of fast fluidized beds with internals, in *Fluidization VII* (eds O.E. Potter and D.J. Nicklin), Engineering Foundation, New York, 275–283.

5 Hydrodynamic modeling
JENNIFER L. SINCLAIR

5.1 Introduction

Experimental studies of risers have shown that these systems are often characterized by complex flow phenomena such as non-uniform spatial distribution of particles, large slip velocities between the phases, and the existence of several possible pressure gradients and solids holdups for specified values of gas and solid flow rates (see Chapter 2 – Hydrodynamics). The particle concentration profile influences the distribution of residence times of particles and may lead to recirculation of particles against the direction of their net motion. These effects are critically important in predicting the conversion in systems in which the particles react chemically with the gas or catalyse a gaseous reaction. Hence, an accurate understanding of the mechanism responsible for the cross-sectional distribution of solids is necessary to predict the performance of these systems. Empirical correlations have generally proven unsuccessful as they are typically limited to the database used to develop them and ignore the effect of radial non-uniformity of the basic variables. However, fundamentally based models have made some progress in this direction. These models can be used to predict how various parameters vary as system conditions change. This is very important, especially to determine the effects of scale-up, design and optimization. A review of these models is the subject of this chapter.

Numerical models describing the flow behavior of gas–solid mixtures can be classified as either Lagrangian or Eulerian according to the framework in which they are developed. In the Lagrangian approach, a separate equation of motion must be solved for each particle in the flow field. Because a large number of particle trajectories must be calculated in order to determine the average behavior of the system, they are generally restricted to very dilute flows as computational requirements are extremely high. An overview of current efforts in which the Lagrangian approach is used to describe the particle motion is given in section 5.9 below. In the Eulerian approach, the gas and solid phases are treated as separate interpenetrating continua. Although each phase is represented by only one equation of motion, these 'two-fluid' treatments contain a number of terms that must be chosen with great care. Unlike the Lagrangian models, however, the Eulerian models can be applied to flows of practical interest with a relatively small computational effort.

In order for a mathematical model to predict accurately all of the flow phenomena associated with gas–solid flows, each of the mechanical interactions in the system must be included in the model.

As outlined by Sinclair and Jackson (1989), these interactions, which depend on the mean and fluctuating components of the gas and solid velocity fields, include the following:

1. The interaction between the mean gas velocity and mean solid velocity that gives rise to, for example, the drag force between the two phases.
2. The interaction between the mean and fluctuating gas velocities that generates the gas-phase Reynolds stresses.
3. The interaction between the mean and fluctuating solid velocities that generates stresses in the solid assembly.
4. The interaction between the particles and the fluctuating gas velocity that results in an interfacial flux of the kinetic energy associated with random motion.

The earliest vertical gas–solid two-fluid flow models are characterized by steady, one-dimensional mass and momentum balances for each of the two phases. Representative works include those by Arastoopour and Gidaspow (1979a; 1979b) and Arastoopour *et al.* (1982). The balance laws allow for the developing flow of gas velocity, solid velocity and voidage. The corresponding radial variations and segregation patterns of particles, however, are ignored. Models that involve radial averaging must introduce empirical friction factors for both phases. These factors, or the effective force, must be adjusted in order to yield quantitative agreement with pressure drop measurements. Consequently, neither quantitative nor qualitative features of the overall behavior of denser suspensions can be represented correctly by one-dimensional models.

In order to predict radial as well as axial variations in vertical gas–solid flows, a description of shear and normal stresses associated with the solid phase is required. These stresses can either be accounted for in terms of an assumed value for the solid–phase viscosity (e.g. see Adewumi and Arastoopour, 1986) or, more recently, are based on a kinetic theory treatment. Recent models (e.g. Sinclair and Jackson, 1989; Ding and Gidaspow, 1990; Louge *et al.*, 1991; Pita and Sundaresan, 1993) have been able to predict qualitatively many of the flow phenomena associated with particle-laden flows by using the kinetic theory analogy and incorporating the first and third effects mentioned above. The quantitative accuracy of these models is uncertain, however, since they do not include descriptions of the interactions that involve gas-phase fluctuations or do not account for the random velocity associated with collections of particles. Other models have been proposed that do incorporate gas and particle velocity fluctuations at the level of single particles (Louge *et al.*, 1991; Bolio *et al.*, 1995), but these models are based on simplifying assumptions that limit their validity to systems of low particle concentrations. Dasgupta *et al.* (1994; 1996) and Hrenya and Sinclair (1996) have recently explored the role of large fluctuations in suspension density on the occurrence of segregation in dense suspensions. This chapter initially focuses on these and other two-fluid models.

5.2 Governing equations

In the two-fluid models, the particle phase is regarded as a continuum and the balance of momentum for the particle phase is obtained by some form of averaging. For example, in the set of governing equations proposed by Anderson and Jackson (1967), the equation of motion for the center of mass of a single particle is locally averaged over regions large enough to contain many particles but small with respect to the bounding container. The averaging procedure leads to terms that describe the variety of interactions associated with the velocity fields of the two phases. Accurate hydrodynamic predictions require correct descriptions for these various interactions. Many derivations of the governing equations for two-phase flows based on this type of averaging have appeared in the literature. Below we present a representative summary of the two-fluid governing equations, as well as the descriptions for the various interactions, which have appeared in several key papers in the fluidization field since 1985. (Note that if no description is given for a specific term in a model, the description follows that of the previous model listed.) These models have been applied to circulating and bubbling fluidized beds. As can be seen, even at the level of the governing equations differences exist between the models.

The two-fluid equations employed by Sinclair and Jackson (1989), Ocone *et al.* (1993), Bolio *et al.* (1995), Pita and Sundaresan (1991, 1993) are consistent with Anderson and Jackson (1967). For the papers by Pita and Sundaresan, an equivalent solid momentum balance (similar in form to that used by Sinclair and Jackson) can be obtained by multiplying the Pita and Sundaresan gas momentum balance by $-\epsilon/(1-\epsilon)$ and adding that balance to the solid momentum balance shown. Thus, the β appearing in Pita and Sundaresan's equations is equivalent to the voidage times the β appearing in the Sinclair and Jackson equations. The only other difference in this group of governing equations is that Sinclair and Jackson (1989) and Pita and Sundaresan make an approximation by bringing the effective gas viscosity outside the gradient in the gas momentum balance.

The extensive work of Gidaspow and co-workers in computer simulations of circulating and bubbling fluidized beds is outlined. Their direct numerical simulations of the full, time-dependent equations of motion are also based on the two-fluid model.

5.3 Summary of governing equations in various two-fluid models

Gas continuity
$$\nabla(\rho_g \epsilon v_g) = 0$$

Solids continuity
$$\nabla[\rho_p(1-\epsilon)v_p] = 0$$

Gas momentum (Sinclair and Jackson, 1989)
$$0 = -\nabla P + \beta(v_g - v_p) + 2\mu_{eg}\nabla S$$

where
$$\beta = \frac{\rho_p(1-\epsilon)g}{v_t \epsilon^{n-1}}$$

and
$$\mu_{eg} = \mu_g(1 + 2.5\nu + 7.6\nu^2)\left(1 - \frac{\nu}{\nu_0}\right)$$

Gas momentum (Ocone et al., 1993)
$$0 = -\nabla P + \beta(v_g - v_p) + 2\nabla(\mu_{eg}S)$$

Gas momentum (Bolio et al., 1995)
$$0 = -\nabla P + \beta(v_g - v_p) + 2\nabla[(\mu_{eg} + \mu_t)S]$$

with
$$\beta = \tfrac{3}{4}C_d \frac{(1-\epsilon)\rho_g |v_g - v_p|}{d_p \epsilon^{2.65}}$$

and μ_t given by a two-equation closure.

Gas momentum (Pita and Sundaresan, 1991)
$$0 = -\epsilon\nabla P + \epsilon\rho_g g + \beta(v_g - v_p) + 2\epsilon\mu_{eg}\nabla S$$

where
$$\beta = \tfrac{3}{4}C_d \frac{\epsilon(1-\epsilon)\rho_g |v_g - v_p|}{d_p \epsilon^{2.65}}$$

Gas momentum (Pita and Sundaresan, 1993)
$$\nabla[\rho_g \epsilon v_g v_g] = -\epsilon\nabla P + \epsilon\rho_g g + \beta(v_g - v_p)$$

Gas momentum (Louge et al., 1991)
$$0 = -\nabla(\epsilon P) + \beta(v_g - v_p) + 2\nabla[\epsilon(\mu_g + \mu_t)S]$$

where

$$\beta = \tfrac{3}{4}C_d \frac{(1-\epsilon)\rho_g |v_g - v_p|}{d_p}$$

and μ_t is given by a one-equation closure.

Gas momentum [Syamlal and Gidaspow (1985); Gidaspow (1986); Bouillard et al. (1989)]

$$\frac{\partial}{\partial t}[\rho_g \epsilon v_g] + \nabla[\rho_g \epsilon v_g v_g] = -\epsilon \nabla P + \epsilon \rho_g g + \beta(v_g - v_p)$$

with

$$\beta = \tfrac{3}{4}C_d \frac{\epsilon(1-\epsilon)\rho_g |v_g - v_p|}{d_p \epsilon^{2.65}}$$

except that

$$\beta = \tfrac{3}{4}C_d \frac{(1-\epsilon)\rho_g |v_g - v_p|}{d_p \epsilon^{2.65}}$$

in Model B of Bouillard et al. (1989).

Gas momentum [Gidaspow et al. (1989); Tsuo and Gidaspow (1990)]

$$\frac{\partial}{\partial t}[\rho_g \epsilon v_g] + \nabla[\rho_g \epsilon v_g v_g] = -\nabla P + \epsilon \rho_g g + \beta(v_g - v_p) + 2\nabla(\epsilon \mu_g S)$$

where

$$\beta = \tfrac{3}{4}C_d \frac{(1-\epsilon)\rho_g \rho_p |v_g - v_p|}{d_p \epsilon^{2.65}(\rho_p - \rho_g)}$$

Gas momentum [Ding and Gidaspow (1990); Ding et al. (1991)]

$$\frac{\partial}{\partial t}[\rho_g \epsilon v_g] + \nabla[\rho_g \epsilon v_g v_g] = -\epsilon \nabla P + \epsilon \rho_g g + \beta(v_g - v_p) + 2\nabla(\epsilon \mu_g S)$$

where

$$\beta = \tfrac{3}{4}C_d \frac{\epsilon(1-\epsilon)\rho_g |v_g - v_p|}{d_p \epsilon^{2.65}}$$

Gas momentum [Gidaspow et al. (1992); Gidaspow and Therdthianwong (1993)]

$$\frac{\partial}{\partial t}[\rho_g \epsilon v_g] + \nabla[\rho_g \epsilon v_g v_g] = -\epsilon \nabla P + \epsilon \rho_g g + \beta(v_g - v_p) + 2\nabla[\epsilon(\mu_g + \mu_t)S]$$

with μ_t given by a zero-equation closure.

Solids momentum (Sinclair and Jackson, 1989)

$$0 = (1-\epsilon)\rho_p g + \beta(v_g - v_p) - \nabla p_p + 2\nabla(\mu_p S_p)$$

where

$$p_p = \rho_p(1-\epsilon)T[1+2(1+e)(1-\epsilon)g_0] - \tfrac{4}{3}(1+e)\rho_p(1-\epsilon)^2 g_0 d_p \left(\frac{T}{\pi}\right)^{1/2} \nabla v_p$$

$$\mu_p = \tfrac{4}{5}(1+e)\rho_p(1-\epsilon)^2 g_0 d_p \left(\frac{T}{\pi}\right)^{1/2} + \tfrac{5}{96}\rho_p \pi d_p \left(\frac{T}{\pi}\right)^{1/2}$$

$$\times \frac{\{1+\tfrac{4}{5}(1+e)(1-\epsilon)g_0\}\{1+\tfrac{4}{5}(1+e)(1-\epsilon)g_0[\tfrac{3}{2}(1+e)-2]\}}{g_0\left\{\dfrac{(1+e)}{2}\left[2-\dfrac{(1+e)}{2}\right]\right\}}$$

and

$$g_0 = \frac{1}{1-(\nu/\nu_0)^{1/3}}$$

Solids momentum (Ocone et al., 1993)

$$0 = (1-\epsilon)\rho_p g + \beta(v_g - v_p) - \nabla p_p + 2\nabla(\mu_p S_p) - \nabla \sigma^f$$

with

$$|\sigma^f_{xy}| = \sigma^f_{yy} \sin\phi \quad (\phi = 28°)$$

where

$$\sigma^f_{yy} = 0.05 \frac{[(1-\epsilon)-0.5]^2}{[0.65-(1-\epsilon)]^3} \frac{N}{m^2} \quad \text{for } (1-\epsilon) > 0.5$$

Solids momentum (Bolio et al., 1995)

$$0 = (1-\epsilon)\rho_p g + \beta(v_g - v_p) - \nabla p_p + 2\nabla(\mu_p S_p)$$

Solids momentum (Pita and Sundaresan, 1991)

$$0 = -(1-\epsilon)\nabla P + (1-\epsilon)\rho_p g + \beta(v_g - v_p) - \nabla p_p + 2\nabla(\mu_p S_p)$$
$$+ 2(1-\epsilon)\mu_{eg} \nabla S$$

where

$$\mu_p = \tfrac{4}{5}(1+e)\rho_p(1-\epsilon)^2 g_0 d_p \left(\frac{T}{\pi}\right)^{1/2} + \tfrac{5}{96}\rho_p \pi d_p \left(\frac{T}{\pi}\right)^{1/2} \frac{\{1+\tfrac{8}{5}(1-\epsilon)g_0\}^2}{g_0}$$

Solids momentum (Pita and Sundaresan, 1993)

$$\nabla[\rho_p(1-\epsilon)v_p v_p] = -(1-\epsilon)\nabla P + (1-\epsilon)\rho_p g + \beta(v_g - v_p)$$
$$- \nabla p_p + 2\nabla(\mu_p S_p)$$

Solids momentum (Louge et al., 1991)

$$0 = (1-\epsilon)\rho_p g + \beta(v_g - v_p) - \nabla p_p + 2\nabla(\mu_p S_p)$$

where
$$p_p = \rho_p(1-\epsilon)T$$
and
$$\mu_p = \tfrac{5}{96}\rho_p \pi d_p \left(\frac{T}{\pi}\right)^{1/2}$$

Solids momentum [Syamlal and Gidaspow (1985); Gidaspow (1986); Bouillard et al. (1989) – Hydrodynamic Model A]

$$\frac{\partial}{\partial t}[\rho_p(1-\epsilon)v_p] + \nabla[\rho_p(1-\epsilon)v_p v_p] = -(1-\epsilon)\nabla P + (1-\epsilon)\rho_p g + \beta(v_g - v_p) + G(\epsilon)\nabla\epsilon$$

where
$$G(\epsilon) = 10^{(-8.76\epsilon + 5.43)} \; \frac{\text{N}}{\text{m}^2}$$

in Syamlal and Gidaspow (1985) and Gidaspow (1986) and
$$G(\epsilon) = \exp[-600(\epsilon - 0.376)] \; \frac{\text{N}}{\text{m}^2}$$

in Bouillard et al. (1989).

Solids momentum [Bouillard et al. (1989) – Hydrodynamic Model B]

$$\frac{\partial}{\partial t}[\rho_p(1-\epsilon)v_p] + \nabla[\rho_p(1-\epsilon)v_p v_p] = (1-\epsilon)\rho_p g + \beta(v_g - v_p) + G(\epsilon)\nabla\epsilon$$

Solids momentum [Gidaspow et al. (1989); Tsuo and Gidaspow (1990)]

$$\frac{\partial}{\partial t}[\rho_p(1-\epsilon)v_p] + \nabla[\rho_p(1-\epsilon)v_p v_p] = (1-\epsilon)\rho_p g + \beta(v_g - v_p) + G(\epsilon)\nabla\epsilon + 2\nabla[(1-\epsilon)\mu_p S]$$

with
$$G(\epsilon) = 10^{(-8.76\epsilon + 5.43)} \; \frac{\text{N}}{\text{m}^2}$$

and $\mu_p =$ specified constant.

Solids momentum [Ding and Gidaspow (1990); Ding et al. (1991)]

$$\frac{\partial}{\partial t}[\rho_p(1-\epsilon)v_p] + \nabla[\rho_p(1-\epsilon)v_p v_p] = -(1-\epsilon)\nabla P + (1-\epsilon)\rho_p g + \beta(v_g - v_p) - \nabla p_p + 2\nabla(\mu_p S_p)$$

where
$$p_p = \rho_p(1-\epsilon)T[1 + 2(1+e)(1-\epsilon)g_0] - \tfrac{4}{3}(1+e)\rho_p(1-\epsilon)^2 g_0 d_p \left(\frac{T}{\pi}\right)^{1/2} \nabla v_p$$

and

$$\mu_p = \tfrac{4}{5}(1+e)\rho_p(1-\epsilon)^2 g_0 d_p \left(\frac{T}{\pi}\right)^{1/2}$$

while

$$g_0 = \frac{0.6}{1-(\nu/\nu_0)^{1/3}}$$

in Ding and Gidaspow (1990) and

$$g_0 = \frac{1}{1-(\nu/\nu_0)^{2.5}\nu_0}$$

in Ding *et al.* (1991).

Solids momentum [Gidaspow *et al.* (1992); Gidaspow and Therdthianwong (1993)]

$$\frac{\partial}{\partial t}[\rho_p(1-\epsilon)v_p] + \nabla[\rho_p(1-\epsilon)v_p v_p] = (1-\epsilon)\rho_p g + \beta(v_g - v_p)$$
$$- \nabla p_p + 2\nabla(\mu_p S_p)$$

where

$$\mu_p = \tfrac{4}{5}(1+e)\rho_p(1-\epsilon)^2 g_0 d_p \left(\frac{T}{\pi}\right)^{1/2} + \tfrac{5}{96}\rho_p \pi d_p \left(\frac{T}{\pi}\right)^{1/2}$$
$$\times \frac{\{1+\tfrac{4}{5}(1+e)(1-\epsilon)g_0\}^2}{g_0\{(1+e)/2\}}$$

and

$$g_0 = \frac{0.6}{1-(\nu/\nu_0)^{1/3}}$$

Pseudo-thermal energy balance [Sinclair and Jackson (1989); Ocone *et al.* (1993)]

$$0 = \nabla(\kappa \nabla T) - p_p \nabla v_p + \mu_p 2 S_p \nabla v_p - \gamma$$

with

$$\kappa = 2\rho_p d_p g_0 (1+e)(1-\epsilon)^2 \left(\frac{T}{\pi}\right)^{1/2} + \tfrac{25}{128}\rho_p \pi d_p \left(\frac{T}{\pi}\right)^{1/2}$$
$$\times \left\{\frac{16}{(1+e)[41-\tfrac{33}{2}(1+e)]g_0}\right\}\{1+\tfrac{6}{5}(1+e)(1-\epsilon)g_0\}$$
$$\times \{1+\tfrac{3}{5}(1+e)^2[2(1+e)-3](1-\epsilon)g_0\}$$

and
$$\gamma = 3(1-e^2)(1-\epsilon)^2 \rho_p g_0 T \left[\frac{4}{d_p}\left(\frac{T}{\pi}\right)^{1/2}\right]$$

Pseudo-thermal energy balance (Bolio et al., 1995)
$$0 = \nabla(\kappa \nabla T) - p_p \nabla v_p + \mu_p 2 S_p \nabla v_p - \gamma - 3\beta T + \overline{\beta v'_{pi} v'_{gi}}$$

Pseudo-thermal energy balance [Pita and Sundaresan (1991); Pita and Sundaresan (1993)]
$$\tfrac{3}{2} \nabla[(1-\epsilon)\rho_p v_p T] = \nabla(\kappa \nabla T) - p_p \nabla v_p + \mu_p 2 S_p \nabla v_p - \gamma - 3\beta T$$

where
$$\kappa = 4\rho_p d_p g_0 (1-\epsilon)^2 \left(\frac{T}{\pi}\right)^{1/2} + \tfrac{25}{128}\rho_p \pi d_p \left(\frac{T}{\pi}\right)^{1/2} \left\{\frac{1}{g_0}\right\} \{1 + \tfrac{12}{5}(1-\epsilon)g_0\}^2$$

Pseudo-thermal energy balance (Louge et al., 1991)
$$0 = \nabla(\kappa \nabla T) - p_p \nabla v_p + \mu_p 2 S_p \nabla v_p - \gamma - 3\beta T + \overline{\beta v'_{pi} v'_{gi}}$$

with
$$\kappa = \tfrac{25}{128}\rho_p \pi d_p \left(\frac{T}{\pi}\right)^{1/2}$$

and
$$\gamma = 6(1-e^2)(1-\epsilon)^2 \rho_p T \left[\frac{4}{d_p}\left(\frac{T}{\pi}\right)^{1/2}\right]$$

Pseudo-thermal energy balance [Ding and Gidaspow (1990); Ding et al. (1991)]
$$\frac{3}{2}\frac{\partial}{\partial t}[(1-\epsilon)\rho_p T] + \tfrac{3}{2}\nabla[(1-\epsilon)\rho_p v_p T] = \nabla(\kappa \nabla T) - p_p \nabla v_p + \mu_p 2 S_p \nabla v_p - \gamma - 3\beta T$$

with
$$\kappa = 2\rho_p d_p g_0 (1+e)(1-\epsilon)^2 \left(\frac{T}{\pi}\right)^{1/2}$$

and
$$\gamma = 3(1-e^2)(1-\epsilon)^2 \rho_p g_0 T \left[\frac{4}{d_p}\left(\frac{T}{\pi}\right)^{1/2} - \nabla v_p\right]$$

Pseudo-thermal energy balance [Gidaspow et al. (1992); Therdthianwong (1993)]
$$\frac{3}{2}\frac{\partial}{\partial t}[(1-\epsilon)\rho_p T] + \tfrac{3}{2}\nabla[(1-\epsilon)\rho_p v_p T] = \nabla(\kappa \nabla T) - p_p \nabla v_p + \mu_p 2 S_p \nabla v_p - \gamma$$

where

$$\kappa = 2\rho_p d_p g_0 (1+e)(1-\epsilon)^2 \left(\frac{T}{\pi}\right)^{1/2}$$

$$+ \tfrac{25}{128}\rho_p \pi d_p \left(\frac{T}{\pi}\right)^{1/2} \left\{\frac{2}{(1+e)g_0}\right\} \{1 + \tfrac{6}{5}(1+e)(1-\epsilon)g_0\}^2$$

5.4 Earlier two-fluid models

The earlier two-fluid models of Gidaspow and co-workers [Syamlal and Gidaspow (1985); Gidaspow (1986)] dealt with the modeling of hydrodynamics and heat transfer in fluidized beds. In these models the presence of the gradient of gas-phase pressure in the solid momentum equation made the initial value problem ill-posed. To overcome this problem, a normal component of solids stress based on a solids stress modulus, $G(\epsilon)$, was added to the balances in order to stabilize the numerical solution. Various correlations and forms for $G(\epsilon)$ have been used by Gidaspow and co-workers in different papers, with a very wide range of values ascribed to $G(\epsilon)$ (Massoudi et al., 1992). In the paper by Bouillard et al. (1989), two different hydrodynamic equation sets are employed to predict porosity distributions in a fluidized bed with an immersed obstacle for erosion applications. Hydrodynamic Model A, which is similar in form to that given in Syamlal and Gidaspow (1985) and Gidaspow (1986), contains gas-phase pressure drops in both the gas and solid phases, while Hydrodynamic Model B retains the entire pressure drop in the fluid phase only. Hence, the drag coefficient for the two models differs by a factor of ϵ. For porosities greater than 0.8, the drag coefficient is given by an expression determined by Wen and Yu (1965); for porosities less than 0.8, the Ergun equation is applied. The solids stress modulus was physically interpreted as a Coulombic stress associated with particles in contact with each other at porosities below minimum fluidization.

Flow patterns in circulating fluidized beds were predicted using the models of Gidaspow et al. (1989) and Tsuo and Gidaspow (1990). These models were similar in form to the Hydrodynamic Model B of Bouillard et al. (1989) and were applied to bubbling beds with the addition of shear stresses in the gas and particle phases. The solids viscous term was added to account for the random collision of particles. Measured radial solids velocity distributions and solids momentum fluxes were used to determine the solids viscosity by applying an area-average mixture momentum equation (Gidaspow et al., 1989). Values of 5.09 poise for $520\,\mu\text{m}$ glass particles and 7.24 poise for $76\,\mu\text{m}$ FCC catalyst particles were computed using the data of Gidaspow et al. (1989) and Bader et al. (1988), respectively. The constant solids viscosity for the glass beads was then used in the simulations of Gidaspow et al. (1989) to predict radial profiles of solids concentration and solids velocity in a 7.62 cm ID riser at 4 s real time, which compared favorably to their experimental measurements. They were also able to predict the formation of wall clusters that descended and then disappeared at the bottom of the riser. The highest computed concentration in a cluster was 10% for the case considered (solid loading $m = 4$).

In the work of Tsuo and Gidaspow (1990), the same model was also used to simulate flow patterns in the riser section of a CFB with 520 μm glass particles for a total of 18 s real time. It took 4 s to fully fill the empty pipe with particles. The outlet mass flux of solids was found to oscillate with an average period of 5 s. The wall cluster descent speed was predicted to be approximately 1 m/s, consistent with the experimental observations of several researchers. Predictions for the gas and solid velocities and the solid volume fraction averaged from 10 to 15 s simulation time compared well with the experimental data of Luo (1987). Using the same system, the effects of superficial gas velocity, solids flow rate, and particle size on cluster formation were investigated by simulating 6 s of real fluidization time. The second set of experimental conditions chosen by Tsuo and Gidaspow (1990) to simulate were those of Bader *et al.* (1988), corresponding to the flow of 76 μm FCC catalyst particles in a 0.3 m ID riser. The simulation was carried out for 18 s of real time. Simulation results were then averaged over the 11 to 18 s period in order to obtain time-averaged values to compare with the measurements of Bader *et al.* (1988). The model predicted a core-annulus flow behavior. Quantitative comparisons of the simulation results with measurements of solid volume fraction and solid velocity were reasonably good. The predictions also showed a large discrepancy between the radially averaged porosity directly computed from the profile and the average porosity inferred from the pressure drop.

5.5 Kinetic theory

Sinclair and Jackson (1989) were the first to propose a fundamental description for solid-phase stresses in the context of a two-fluid model. The interaction of the particles and the gas was restricted to a mutual drag force, whose value depends on the concentration of the particles and the difference between the local average values of the velocities of the gas and particle phases. Since the gas does not slip freely at the wall of the pipe there is a gas velocity profile, and a corresponding profile of particle velocity was induced by the drag forces on the particles. As a result of this shearing in the particle phase, particles collided with each other, generating a random component of particle motion. This random particle motion is completely independent of the gas velocity fluctuations, and its magnitude can exceed that of the gas velocity fluctuations, providing that the particles are massive enough. The kinetic energy of this random motion is analogous to that of the thermal motion of molecules in a gas and, correspondingly, the random motion can be characterized by a 'particle temperature' proportional to the mean square of the random component of the particle velocity. The particle velocity fluctuations generate an effective pressure in the particle phase, together with an effective viscosity that resists shearing of the particle assembly. Both the effective pressure and the effective viscosity depend strongly on the 'particle temperature', so this must be found by solution of a separate differential equation representing a balance for the pseudo-thermal energy of random motion of the particles. Pseudo-thermal energy is generated by the working of the effective shear stresses in the particle phase, dissipated by inelastic particle–particle collisions and conducted according to gradients in the particle temperature.

The pseudo-thermal energy balance is obtained by subtracting the mechanical energy equation from the total energy balance, and introducing an energy transfer mechanism between the true thermal energy and pseudo-thermal energy, namely the inelasticity of particle–particle collisions. Constitutive expressions for the effective thermal conductivity of the particle phase and the effective pressure and viscosity, as well as the dissipation rate of pseudo-thermal energy, were developed by Lun *et al.* (1984) using an approach analogous to the methods employed in deriving the Chapman–Enskog dense-gas kinetic theory. Similar to the Chapman–Enskog theory, the particles are assumed to be spherical, frictionless, cohesionless and uniform in size in the theory of Lun *et al.* Unlike the kinetic theory, however, inelastic particle–particle collisions, in which pseudo-thermal energy is dissipated into real heat, are taken into account. Gidaspow (1994) presents the details of the derivation of the solid conservation equations for mass, momentum and granular energy starting with the Boltzmann equation.

An extensive review of the kinetic theory as applied to granular flow has been given by Campbell (1990). The kinetic theory has also been extended to account for non-uniform particle size by Jenkins and Mancini (1989) and Shen and Hopkins (1988), who studied a bimodal mixture of smooth, nearly elastic spheres in simple shear flow. These extensions are very important, as the profound influence of a non-uniform particle size distribution on flow properties in CFBs has been well documented by researchers such as Grace and Sun (1991). The effect of non-sphericity has been studied only to a limited extent in computer simulations of granular flow [e.g. Walton (1984) and Hopkins and Shen (1988)]. Lun (1991) has also developed a kinetic theory for the flow of slightly rough spheres. Arastoopour and co-workers (Seu-Kim and Arastoopour, 1993) are currently extending the kinetic theory to model cohesive particles and the resulting formation of particle agglomerates, in addition to carrying out experimental work (Arastoopour and Yang, 1992) with Group C particles using laser Doppler anemometry. Interstitial fluid effects of the constitutive theories have not been addressed rigorously to date.

5.6 Recent two-fluid models

The Sinclair and Jackson (1989) model revealed a remarkably rich variety of behavior in a vertical pipe over the whole range of possible flow conditions: co-current upflow, co-current downflow, and counter-current flow. Particle concentration was found to be high near the wall of the duct; depending on the elastic properties of particle–particle and particle–wall collisions, it could also be high near the axis of the tube. Particle recirculation was also observed in some situations, and choking and flooding phenomena were indicated. Building on the apparent success of the Sinclair and Jackson model, Pita and Sundaresan (1991) carried out a computational study of fully developed flow of gas–solid suspensions in vertical pipes to understand the predicted scale-up characteristics. They showed that the model could predict steady-state multiplicities wherein different pressure gradients and solids holdups could be obtained for the same gas and solid fluxes. The existence of multiplicities has been observed experimentally. The model predicted that when the tube diameter was small, the pressure gradient required to achieve desired solids and gas fluxes in the riser decreased as the tube diameter increased, a trend that has also been observed experimentally. They also predicted that operations in the counter-current mode at certain combinations of gas and solids fluxes were possible in large-diameter tubes, but not in small-diameter tubes. Finally, they pointed out that the model manifested an unsatisfactory degree of sensitivity to the inelasticity of the particle–particle collisions and the damping of particle-phase fluctuating motion by the gas. This sensitivity to the coefficient of restitution was also noted by Sinclair and Jackson (1989). Although the model ignored gas-phase turbulence and considered elastic particle–particle collisions only in the computational study, remarkable agreement with the experimental data of Bader *et al.* (1988) was obtained. Yasuna *et al.* (1995) performed extensive comparisons using the Sinclair and Jackson model and a large body of CFB data and, in general, found good agreement between the data and the predictions when elastic particle–particle collisions were considered.

A computational study of developing riser flow using the Sinclair and Jackson model has also been carried out by Pita and Sundaresan (1993). The only change in the model equations was that stresses in the gas phase were neglected, as their contribution is small in dense-phase flow. Various inlet and exit configurations were considered that permitted the assumption of cylindrical symmetry. They found that a core-annulus type of inlet configuration promoted internal recirculation of solids. They also found that circumferential gas injection caused the segregation of particles to the wall to occur more slowly and resulted in a drastic reduction in the recirculation of solids.

Ocone *et al.* (1993) extended the work of Sinclair and Jackson to include the case of ducts of arbitrary inclination. As a result of the particle compaction due to gravity, they included forces transmitted between particles due to sustained rolling or sliding contacts. An empirical frictional model represented this contribution to the particle-phase stress. The functional form for these frictional stresses followed earlier work of Johnson et al. (1990) who considered the case of dry granular flow in a chute. Ocone et al. (1993) predicted fully developed flows with the expected qualitative features, even in horizontal ducts. They explored the effects of the flow rates of the two phases, duct inclination angle and duct width on the resulting profiles.

It should be re-emphasized that while the above models produce qualitative and quantitative trends consistent with observed gas–solids flow and segregation patterns in real risers, these models all exhibited undue sensitivity to the coefficient of restitution. For perfectly elastic particle–particle collisions ($e = 1.0$) and inelastic particle-wall collisions, the models predict a high concentration of particles near the wall. However, if the coefficient of restitution of particle–particle collisions is merely decreased to 0.99, the segregation of particles appears at the core of the riser, contrary to the observed behavior in dense-phase flows. Evidently, not all mechanical interactions in the physical system are properly represented in the previous models.

For dilute-phase flow, the influence of gas-phase turbulence should have a profound effect on the resulting flow patterns. The dilute-phase flow model of Louge et al. (1991) was the first to include the effects of both gas-phase turbulence and particle collisions. Similar to the Sinclair and Jackson model, the solids stress was described using the kinetic theory analogy. Gas-phase turbulent stresses were modeled via a high Reynolds one-equation closure that required a pre-specified mixing length. Since this closure scheme was not valid in the viscous sublayer, a wall function was used to determine the velocity distribution in this region. Their model also contained a modified form of the expression proposed by Koch (1990) to approximate the correlation between the velocity fluctuations of the gas and those of the individual particles in dilute-phase flows. Model predictions for steady, fully developed, vertical pipe flow were compared with the laser Doppler velocimetry (LDV) non-invasive measurements of Tsuji et al. (1984). Significant flow features observed in the experimental data for dilute gas–solids flow in the case of larger particles were predicted. It was shown that inelastic particle–particle collisions can give rise to the observed large slip velocities, several times larger than the particle terminal velocity. The predicted solids velocity profiles were flatter than the corresponding gas velocity profiles, so that the relative velocity changed signs near the pipe wall, consistent with the experimental data. Although the predictions for both phases were in good agreement with the experimental data at low solids loadings, the predictive ability of the model worsened at higher loadings. More specifically, the model was unable to predict the flattening of the mean gas velocity profile with the addition of particles. Also, the model predictions for the fluctuating gas velocity showed large deviations from the experimental data. This inability of the model to give good predictions at higher loadings is not surprising, however, since the expressions used for the mixing length and wall functions are based on single-phase experimental data. Even for dilute two-phase flows, the universality of such functions begins to break down as the concentration of particles increases.

The model developed by Bolio et al. (1995) expanded on the efforts of Louge et al. (1991) by incorporating a low Reynolds k-ϵ closure to describe gas-phase stresses. This closure model, which was extended to account for the presence of a dilute particle phase, included the effects of both viscous and turbulent stresses, thereby eliminating the need for empirical wall functions. The model was able to predict a decrease in the slip velocity with increasing solids loading and decreasing particle size, as well as the flattening of the mean and fluctuating gas velocity profile with the addition of solids. Perhaps most significant were the comparisons of model predictions for the particle velocity fluctuations with the LDV data of Tsuji (1993). The predicted particle velocity fluctuations, generated by particle–particle and particle–wall interactions and described by kinetic theory, exceeded the gas fluctuations and were in excellent agreement with the data. The model predictions were also compared with the dilute-phase LDV data of Lee and Durst (1982) and Maeda et al. (1980).

The one flow variable not adequately captured in the model of Bolio et al. (1995) (and also in the model of Louge et al., 1991) is the turbulent kinetic energy of the gas phase since the rate of gas kinetic energy change is dominated by a dissipative mechanism where particles extract energy from the flow due to a drag effect. For all of the operating conditions investigated, the predicted turbulent intensity of the gas was significantly lower than the experimental measurements. However, for larger particles that do not follow the fluid motion, one might instead expect that generation due to the presence of wakes trailing each particle or vortex shedding would contribute to the gas velocity disturbance and would be the dominating factor in gas turbulence modulation in dilute-phase flows. Also, in the case of larger particles, where the particle relaxation time is very large relative to the particle–eddy interaction time, the particle velocity does not change appreciably during the time of interaction, and the contribution of the turbulence reduction to the total turbulence modulation should be small. A recent model by Bolio and Sinclair (1995) includes a theoretical description for turbulence generation taking into account the turbulent wakes behind larger particles, following a recent proposal by Yuan and Michaelides (1992). The model predicts that due to the significant flattening of the mean gas velocity profile with the addition of particles, and the corresponding decrease in turbulent kinetic energy production, a generation mechanism must be present in order to produce gas velocity fluctuations that are consistent with the experimental measurements, even in the case where the experimental results indicate net suppression of gas-phase turbulence.

For dense phase flow, recent studies have indicated that high-velocity gas–solids flows in vertical tubes are characterized by large, transient fluctuations in voidage (e.g. Plumpe *et al.*, 1993). In addition, time-dependent stresses have been observed in computer simulations of granular flow in the absence of an interstitial gas (Savage, 1992). For the solids loadings typically found in CFBs, the particles are the dominant contributors to both the weight of the suspension and inertial effects. For dense phase flow, all of this points to the formation of particle packets, similar to eddies in a single-phase turbulent fluid. Based on this analogy with single-phase turbulence, two regimes of two-phase flows can be identified. In the idealized 'laminar' regime, particles move individually [e.g. Louge *et al.* (1991), Bolio *et al.* (1995)]. In the idealized 'turbulent' regime, however, only the random motion of collections of particles occurs. A complete model for real dense phase flows and for flows of moderate loading must account for velocity fluctuations at both levels. The magnitude of the random velocity associated with collections of particles is defined by the turbulent kinetic energy of the particle phase (k_p), whereas the magnitude of the random velocity associated with individual particles is defined by the granular temperature. Based on limiting forms of the radial solids momentum balance, it can be shown that the solids volume fraction is inversely proportional to the granular temperature in the laminar regime and that the solids volume fraction is inversely proportional to the turbulent kinetic energy of the particle phase in the turbulent regime. Thus, there are two mechanisms responsible for particle segregation. As mentioned above, when only the laminar mechanism is considered, segregation is observed in the pipe center when the coefficient of restitution for particle–particle collisions is less than one. However, if the turbulent mechanism alone is considered, particle segregation always occurs at the wall since the particle-phase turbulent kinetic energy reaches a minimum at the wall. In the complete model, though, the segregation patterns depend on which of the two mechanisms dominates.

The first model application to dense-phase flows that incorporated this 'particle-phase turbulence' was given by Dasgupta *et al.* (1994). Although it was assumed that the local gas-particle slip was negligible, they showed that the non-uniform distribution of particles over the cross-section is a direct consequence of these unsteady particle fluctuations. Dasgupta *et al.* (1996) broadened the scope of their previous study by allowing for gas–particle slip, both at the scale of the mean and fluctuating velocities. Using this revised formulation, they calculated performance diagrams for fully-developed gas–particle flow in a vertical pipe over a range of operating conditions in the co-current upflow, counter-current flow and co-current downflow regimes. A comparison of model predictions with experimental data revealed good agreement in some cases and large departures in others. Although these two models did account for velocity fluctuations at both levels, the granular temperature was assumed to be constant such that the turbulent mechanism always dominated, thereby ensuring that the solids volume fraction always reached a maximum at the wall.

In the recent work of Hrenya and Sinclair (1995), both the granular temperature and the particle-phase turbulent kinetic energy are allowed to vary. In order to account for particle-phase turbulence, an approach similar to that used to describe single-phase turbulent flows is followed. Namely, the two-phase continuum equations derived by Anderson and Jackson (1967) are time-averaged. For steady, fully-developed flow in a vertical pipe, the equations are:

Equations of continuity: global, gas-phase, and solids-phase

$$\overline{\phi_g} + \overline{\phi_p} = 1; \quad \overline{\phi_g}\,\overline{V_{gr}} + \boxed{\overline{\phi'_g V'_{gr}}} = 0; \quad \overline{\phi_p}\,\overline{V_{pr}} + \boxed{\overline{\phi'_p V'_{pr}}} = 0$$

Gas-phase momentum balance (axial component)

$$\overline{V_{gr}}\left(\frac{d\overline{V_{gz}}}{dr}\right) = \frac{1}{r}\frac{d}{dr}\left[r\left(\nu_g \frac{d\overline{V_{gz}}}{dr} - \boxed{\overline{V'_{gr}V'_{gz}}}\right)\right] - \frac{1}{\rho_g}\frac{\partial \overline{P}}{\partial z} - \frac{\overline{\beta}}{\rho_g}(V_{gz} - V_{pz})$$

Solids-phase momentum balance (axial component)

$$0 = \frac{1}{r}\frac{d}{dr}\left[r\rho_p\left(G_1\nu_p \frac{d\overline{V_{pz}}}{dr} - \overline{\phi_p}\boxed{\overline{V'_{pr}V'_{pz}}}\right)\right] - \rho_p\overline{\phi_p}g_z + \overline{\beta(V_{gz} - V_{pz})}$$

Solids-phase momentum balance (radial component)

$$0 = -\frac{d}{dr}[\rho_p(G_2 T + \tfrac{2}{3}\overline{\phi_p}\boxed{k_p})] + \overline{\beta(V_{gr} - V_{pr})}$$

Fluctuation energy balance

$$0 = \frac{1}{r}\frac{d}{dr}\left[r\left(\overline{q_{PT,r}} + \tfrac{3}{2}\rho_p\overline{\phi_p}\boxed{\overline{V'_{pr}T'}}\right)\right] - \overline{\sigma_{rr}}\left(\frac{d\overline{V_{pr}}}{dr} + \frac{\overline{V_{pr}}}{r}\right) - \overline{\sigma_{rz}}\frac{d\overline{V_{pz}}}{dr}$$
$$- \gamma + \rho_p\overline{\phi_p}\boxed{\epsilon_p}$$

where an overbar indicates a time-averaged quantity and a prime refers to the fluctuating value. The boxes designate terms that arise from time-averaging and require further modeling. The final term in the fluctuation energy balance, which did not appear rigorously, was added based on an analogy with single-phase flows. In the same manner that turbulent kinetic energy of a single-phase fluid dissipates into true thermal heat, it is assumed that solids-phase turbulent kinetic energy is dissipated into pseudo-thermal heat (i.e. fluctuation energy).

The two competing mechanisms responsible for lateral segregation are represented mathematically by the two terms contained within the brackets in the radial component of the solids-phase momentum balance. The first term represents the laminar mechanism, while the second represents the turbulent mechanism. As mentioned above, if the turbulent mechanism dominates, the correct qualitative solids concentration profile will be obtained. However, it is interesting to note that even if the laminar term is much larger than the turbulent term, the sensitivity to the coefficient of restitution for particle–particle collisions may still be eliminated due to the presence of the two turbulent correlations present in the fluctuation energy balance (boxed terms). Turbulent conduction tends to flatten the granular temperature profile in the core of the pipe, resulting in a corresponding decrease in the solids concentration near the pipe center. Likewise, the extra source term has the same effect. Thus these two terms, depending on their magnitude, may remove the undue sensitivity to the coefficient of restitution, regardless of whether the laminar or turbulent mechanism dominates.

In order to form a closed set of governing equations, the turbulent correlations are modeled as follows. Although the gas-phase Reynolds stress does appear formally in the gas-phase momentum balance, its description is not crucial since the pressure drop and drag terms dominate in dense-phase flows, as also discussed by Pita and Sundaresan (1993). Thus, this correlation is described using a simple single-phase mixing length model. As a first and somewhat crude attempt at closure, the solid-phase Reynolds stress is also described via a mixing length model, except that both the eddy viscosity and Reynolds stress depend on the solids velocity field, rather than the gas velocity field. Dasgupta et al. (1994) employed a low Reynolds number k-ϵ model to represent the solids-phase Reynolds stress. It is acknowledged that single-phase turbulence models (both mixing length and k-ϵ models) do contain a certain amount of empiricism, and that this empiricism will not be the same for single-phase fluid turbulence as for particle-phase turbulence. However, the physical mechanisms governing the turbulence should be the same. The turbulent conduction term is modeled according to the Reynolds analogy, while the turbulent dispersion terms are described using the gradient assumption. Using this type of model, the flow predictions are much less sensitive to the coefficient of restitution, although the sensitivity has not been completely eliminated. An area of future work that needs to be explored is the effect of particle-phase turbulence on the kinetic theory closures.

The first transient model to incorporate the kinetic theory was given by Ding and Gidaspow (1990) and was used to compute porosities in two-dimensional fluidized beds. The predicted flow patterns were compared with the experimental results in Lin et al. (1985) who measured particle velocities in a cylindrical column using a radioactive particle tracking technique. The predicted porosity oscillations were in better agreement with the experimental data than those generated using the model of Bouillard et al. (1989), which erroneously predicted that the amplitude of the porosity oscillations was damped with time. The model was also able to predict the formation of bubbles. Ding et al. (1991) used the same model as Ding and Lyczkowski (1992) to compute erosion rates in a three-dimensional rectangular fluidized bed containing a single tube.

The most recent work of Gidaspow and co-workers has focused on simulations of the entire CFB loop. The models of Gidaspow et al. (1992) and Gidaspow and Therdthianwong (1993) incorporate a zero-equation closure for the gas-phase turbulence. In the former, species continuity equations were added to the simulation to predict the production of synthesis gas from char. In the latter, hydrodynamics and sorption of SO_2 by calcined limestone were studied in a CFB loop for conditions approximating those in a CFB combustor. Figures 5.1 and 5.2 present representative examples of the results of the transient simulations by Gidaspow; particle concentration distributions and SO_2 concentration distributions in an asymmetric and symmetric CFB loop are shown.

5.7 Boundary conditions

Complete formulation of these two-fluid models requires boundary conditions at the tube wall and centerline. At the axis of the pipe, symmetry clearly demands that the radial gradients of all the variables be zero. In general, it is not permissible to set the particle velocity or the pseudo-thermal or granular temperature equal to zero at a solid wall. Exceptions occur when the wall is sufficiently rough, minimizing particle slip, and when the bounding wall is sufficiently 'soft', creating highly inelastic particle–wall collisions. Johnson and Jackson (1987) derived a boundary condition for the particle velocity at the wall by equating the transfer rate of axial momentum from the particles to the wall to the limit of the tangential stress in the particle assembly on approaching the wall. This type of boundary condition has also been employed by Sinclair and Jackson (1989), Pita and Sundaresan (1991), Ocone et al. (1993), Pita and Sundaresan (1993), Bolio et al. (1995) and Bolio and Sinclair (1995). The transfer rate of axial momentum at the wall depends on the fraction of diffuse particle–wall collisions given by a specularity factor ϕ, which varies between zero for perfectly specular collisions and one when collisions are completely diffuse. Pita and Sundaresan (1991) have pointed out, however, that this boundary condition closely approximates the no-slip condition except in the case of very small values for the specularity coefficient.

Louge et al. (1991) employ a different type of solid velocity boundary condition at the wall. Recognizing that collisions in dilute-phase flow typically have low incident angles, they assume that these collisions involve sliding, and the normal and tangential components of the impulse are related by a coefficient of friction. Measurement of these particle–wall collision properties, the coefficient of friction and the specularity factor, have been made using stroboscopic photographs (Foerster et al., 1994) and particle image tracking velocimetry (Massah et al., 1994).

The earliest hydrodynamic models of Gidaspow and co-workers which employed the solids stress modulus [Syamlal and Gidaspow (1985), Gidaspow

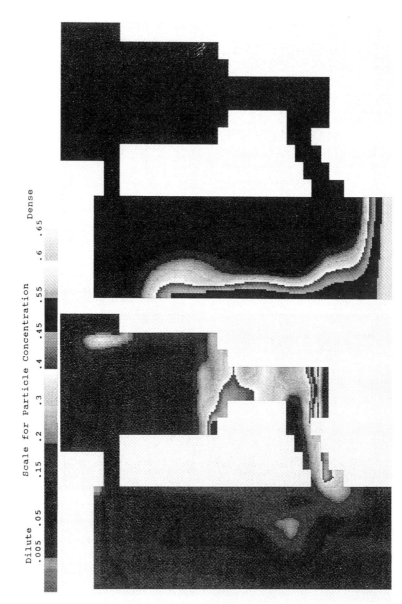

Figure 5.1 (a) Particle concentration distribution in a CFB for asymmetric solid feeding case at $V_{gin} = 5$ m/sec; shading denotes solid volume fraction at 20 sec. (b) SO_2 concentration distribution in CFB for asymmetric solid feeding case at $V_{gin} = 5$ m/sec; shading denotes SO_2 mole fraction at 20 sec.

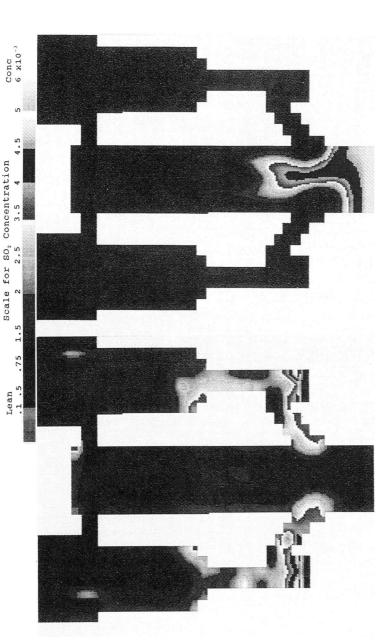

Figure 5.2 (a) Particle concentration distribution in a CFB for symmetric solid feeding case at $V_{gin} = 5$ m/sec; shading denotes solid volume fraction at 20 sec. (b) SO_2 concentration distribution in CFB for symmetric fold feeding case at $V_{gin} = 5$ m/sec; shading denotes SO_2 mole fraction at 20 sec.

(1986), Bouillard et al. (1989)] used no-slip boundary conditions (i.e. normal and tangential velocities for both phases were set equal to zero at the wall). The models of Gidaspow et al. (1989) and Tsuo and Gidaspow (1990) allowed for particle slip and applied a boundary condition given by Soo (1967) based on the interaction length of the particles with the fluid or the mean free path of the particles. All of the models of Gidaspow that employ the kinetic theory assume that the solid tangential velocity at the wall is proportional to its gradient at the wall and related by a slip parameter. [This is the same functional form as the boundary conditions of Sinclair and Jackson (1989) and Louge et al. (1991).] In the models of Gidaspow, the slip parameter is proportional to the particle diameter and inversely proportional to the solids volume fraction raised to the one-third power.

A boundary condition for the pseudo-thermal or granular temperature at the wall is found by equating the energy conducted to the wall by virtue of particle–particle collisions to the energy lost at the wall by particle–wall collisions and the energy generated by specular particle–wall collisions. A boundary condition of this sort has been applied by Sinclair and Jackson (1989), Pita and Sundaresan (1991), Louge et al. (1991), Ocone et al. (1993), Pita and Sundaresan (1993), Bolio et al. (1995) and Bolio and Sinclair (1995). In the models of Gidaspow that employ the kinetic theory, the energy flux at the wall is taken to be zero.

For the gas phase, the point value for the gas velocity is certainly zero at the wall. Hence, most models set the gas velocity equal to zero there. However, since the two-fluid governing equations are based on local average variables, the local averaged gas velocity will not vanish there. Any boundary condition for the local average gas velocity should degenerate into the no-slip condition at the limit of small particle concentration and should yield a flat gas velocity profile at high solids concentrations. A first attempt to formulate a boundary condition of this sort was made by Sinclair and Jackson (1989), based on a force balance on the fluid in a thin layer adjacent to the wall.

For models in which gas or particle-phase turbulence is described by a two-equation closure, the turbulent kinetic energy is set equal to zero at the wall. The dissipation rate at the wall is obtained by a reduction of the turbulent kinetic energy equation at the wall. In the one-equation turbulence closure model of Louge et al. (1991), the gradient of the turbulent kinetic energy is set equal to zero at the wall.

5.8 Other two-fluid models

Transient two-fluid models have also been employed at the Morgantown Energy Technology Center (METC) and at the Argonne National Laboratory. O'Brien and Syamlal (1993) at METC have developed a two-fluid model in which the commonly applied gas–particle drag correlation is corrected to account for the cluster formation associated with particles in the FCC size range. The correction was applied to the drag formulation used by O'Brien and Syamlal (1991) so that the model predictions agreed with the average pressure drop data of Bader et al. (1988). At Argonne, Ding and Lyczkowski (1992) studied erosion in three-dimensional fluidized beds using the model of Ding and Gidaspow (1990); Bouillard and Lyczkowski (1993) studied erosion and heat transfer in a CFB using the CLW Tech. Inc. computer code FLUCOMP/MOD1.

Another class of models that account for gas and solids-phase velocity fluctuations has been proposed primarily by researchers in the mechanical engineering community, namely Elghobashi and Abou-Arab (1983), Chen and Wood (1985), and Rizk and Elghobashi (1989). In all of these dilute-phase flow models, the effects of particle collisions are neglected and the particles are assumed to be small enough such that they follow the fluid motion very closely. Each of these three models incorporates a k-ϵ closure to describe gas-phase stresses, although only the model of Rizk and Elghobashi implements a low Reynolds form of this closure. Solids-phase stresses are described empirically as fractions of the corresponding gas-phase stresses; thus, these expressions are not valid for systems composed of larger particles in which the particle fluctuations often exceed those of the gas. Because the models are restricted to dilute flows involving small particles, the expressions proposed for the interaction between the fluctuating velocity components of the two phases only allow for the suppression of gas-phase turbulence.

Ahmadi and Ma (1990) used a phasic mass-weighted averaging technique to establish a thermomechanical formulation for turbulent multiphase flows. A closed system of field equations was obtained for determining the velocity, solids volume fraction and fluctuation kinetic energy of the two phases. Their model formulation included separate transport equations for the fluctuation kinetic energies of the particulate and fluid phases; it also included the particulate collisional stresses. Their model for the gas-phase has recently been upgraded to a two-equation low Reynolds number turbulence model (Cao and Ahmadi, 1995) and has been applied to dilute gas–particle turbulent flows in a vertical channel and compared with the experimental data of Tsuji et al. (1984). For the larger particles investigated by Cao and Ahmadi (1990), the model predictions did not, however, show that the particle velocity fluctuations exceeded the gas velocity fluctuations as would be expected. The model has not yet been applied to the case of dense-phase riser flow.

Johnson et al. (1991) developed a mathematical description within the context of mixture theory for a flowing mixture of solid particulates in a pipe. An assumed form of the stress tensor for the granular material is used. Their numerical results are not directly compared with experimental data, but are interpreted in terms of six dimensionless parameters depending upon the granular material properties (introduced in the assumed form for the solids stress tensor), the fluid density and viscosity, the drag coefficient, and the spin lift and slip-shear lift coefficients.

5.9 Semi-empirical models

One of the earliest attempts to model the core-annulus flow structure inherent in riser flow was proposed by Nakamura and Capes (1973). The particle flow was divided into a core region with particles carried upward and an annular region where particles travel downward. The gas velocity, particle velocity and solid volume fraction were different in the two regions. Two sets of model equations were written for the core and annulus regions. Many unknown parameters were introduced such as the size of the core and the shear stress across the core-annulus interface.

More recently, Rhodes and co-workers have proposed several models for CFB riser flow [Rhodes and Geldart (1987); Rhodes and Geldart (1989); Rhodes (1990)]. Rhodes et al. (1992) generated experimental measurements of solids mass flux profiles that indicated the presence of a 'similar profiles' regime. In this regime, radial profiles of reduced solids flux (local solids flux divided by the mean solids flux) were insensitive to changes in the mean solids flux at a given superficial gas velocity. Their experimental findings, together with those of other workers, were used to produce a semi-empirical model capable of predicting profiles of solids flux and solids velocity. Their model, containing two fitting parameters, was based on a given form for the reduced solids flux profile and an assumed parabolic solids volume fraction profile. Amos et al. (1993) developed a model for gas mixing in the riser of a CFB by considering radially dispersed plug flow in the riser core and a well-mixed side-stream in the riser annulus. In order to determine two fitting parameters in the model, experiments were conducted in a 0.3 m ID riser.

Berruti and co-workers have also developed several models based on the assumption of a core-annulus flow structure [Berruti and Kalogerakis (1989); Wong et al. (1992); Pugsley et al. (1993)]. In the work of Wong et al. (1992), for example, the riser was divided into three sections: an acceleration zone, a developed flow zone and a deceleration zone (for risers equipped with an abrupt exit configuration). The acceleration zone was modeled assuming the existence of a predominantly upward flowing gas–solids suspension. The top of the acceleration zone was taken as where the axial voidage profile becomes relatively constant. In the dense annular region of the developed flow zone, solids are assumed to move downward at the single particle terminal velocity with a voidage corresponding to that of a bubbling bed at minimum fluidization. The basis for the abrupt exit zone calculation was to express the net particle radial flux as a function of the solids core-to-annulus and annulus-to-core interchange coefficients. These coefficients, together with three other parameters that describe the core-annulus interface, were contained in the model.

The most comprehensive model in this semi-empirical class was proposed by Senior and Brereton (1992). A mechanistic approach was adopted, such that model parameters had physical significance. In their semi-empirical model, as with the others, dense particle sheets or streamers are assumed to fall at the walls while the core remains in dilute upflow. The novel aspect in their work, however, was the development of relationships for the interchange of solid particles between the streamers and the dilute suspension. They assumed that oblique particle–particle collisions account for lateral migration of core particles into the wall regions. Their predictions were tested against several sets of combustor data. Other semi-empirical models have also recently been proposed by Wirth (1991) and Harris and Davidson (1993).

5.10 Computer simulations

Recent advances in computer facilities and in numerical calculation techniques make it possible to analyse the flow of fluid–solids mixtures via computer simulation. These simulations have used the two-fluid model or involve discrete particle simulations. The work of Gidaspow and co-workers fits into the former category, while the works of Tsuji and co-workers and Sommerfeld and co-workers fall into the latter category. One advantage of the discrete particle simulations over two-fluid models is that the particle size and density distribution can be easily implemented by specifying the characteristics of the individual particles. The main limitation is the long computational time and extensive memory required to simulate real flows of large numbers of particles. In addition, models for particle–particle collisions, particle–wall collisions, and the interaction of the particles with the gas turbulence are also required in these discrete particle simulations.

In the earliest work by Tsuji and co-workers (Tsuji *et al.*, 1985), the Lagrangian method was used to calculate the trajectory of polyethylene particles ($50\,\mu m < d_p < 5000\,\mu m$) in pneumatic conveying ($m < 10$) in a horizontal pipe, assuming that the fluid phase was not affected locally by the particles except for the loss of pressure (one-way coupling). The motion of the solids phase in a prespecified fluid flow field (obeying the 1/7 power law) was obtained by integrating the equations of motion, which included lift due to particle rotation. The effects of particle–wall collisions were included by applying the impulse equations of classical mechanics and using the 'virtual wall model' or 'abnormal bouncing model', which involved three empirical parameters and enabled spherical particles to continue to bounce on the horizontal wall under the influence of gravity. In the work by Tsuji *et al.* (1987) the same particle–wall collision model was retained, but the effect of polystyrene particles ($d_p = 1000\,\mu\text{m}$; $m < 5$) on the fluid motion in the two-dimensional horizontal channel was taken into account (two-way coupling). Gas turbulence was described using a single-phase mixing length theory; the equation of particle motion included forces due to drag, gravity, and spin and shear lift. The effect of inter-particle collisions was first taken into account by Tanaka and Tsuji (1991) in a direct numerical simulation of gas–solids flow in a riser. They showed that the effects of particle–particle collisions are important at solids loading ratios less than one for $400\,\mu\text{m}$ to $1500\,\mu\text{m}$ particles. The collisional geometry during a time step was obtained from the relative motion of any pair of particles in the flow domain. This direct simulation was feasible by considering only a short segment of the riser and applying periodic boundary conditions.

More recently, Tsuji and co-workers used Cundall and Strack's (1979) Distinct Element Method (DEM) to describe contact forces between particles and between particles and the wall in two-dimensional fluidized beds with [Tsuji et al. (1993)] and without [Tanaka et al. (1993a)] partition walls. An inviscid gas was treated simultaneously with the motion of aluminum spheres ($d_p = 4000\,\mu\text{m}$). Drag and contact forces act on the particles, with the contact forces giving rise to changes in translational and rotational particle motion. The contact force model consisted of a spring, dash-pot and a friction slider, with specified values of the stiffness coefficient (from Hertzian contact theory), the coefficient of viscous dissipation (related to the coefficient of restitution), and the coefficient of friction. These simulations were successful in describing many of the salient features of fluidized beds.

Discrete particle simulations by Sommerfeld and co-workers have focused on the detailed modeling of particle–wall collisions. Wall roughness was simulated by assuming that particles collide with a virtual wall with randomly distributed inclination angles. Sommerfeld (1992) investigated three different particle–wall collision models, differing in the statistical treatment of the roughness, in which the change of the particle's translational and rotational velocities during the collision process was calculated from the impulse equations of classical mechanics, given values for the coefficient of restitution (which varied with impact angle) and the static and dynamic friction coefficients. Two types of collisions, sliding and non-sliding, were identified [as in the work of Tsuji et al. (1985)]. The type of collision was determined by the static friction coefficient, the coefficient of restitution, and the velocity of the particle surface relative to the contact point. By comparing the results of a numerical simulation of two-dimensional channel flow of $45\,\mu\text{m}$ ($m = 0.02$) and $108\,\mu\text{m}$ ($m = 0.17$) glass spheres in air, the best agreement with experimental results was obtained with a particle–wall collision model which assumed a Gaussian distribution for the roughness angle. This numerical simulation involved two-way coupling and incorporated particle size distribution effects. Gas turbulence was described using a k-ϵ model; the equation of particle motion included forces due to drag, gravity, and spin and shear lift. The effect of gas turbulence on the particle motion was modeled by a stochastic approach, the 'eddy lifetime concept', in which the instantaneous gas velocity was assumed to influence the particle motion during an interaction time. The interaction time of a particle was limited by two criteria, namely the turbulent eddy lifetime and the time for a particle to cross an eddy.

In a subsequent paper, Sommerfeld and Zivkovic (1992) included particle–particle interactions in the Sommerfeld (1992) model. The collision process for each particle traced through the flow field was calculated based on the average values of particle velocities and concentration in the control volume under consideration. This information yielded the probability of collisions for a particle during a time step. On the basis of this probability, it was decided in every time step, by generating a random number, whether or not a collision of a given particle would take place. If a collision did occur, a stochastic model governed the post-collision velocities of the particle. Using this model, simulations of dilute horizontal pipe flow with 40 μm glass spheres and two-way coupling were carried out. The importance of the different phenomena influencing the motion of the particles and the development of the particle mass flux distribution was also investigated in several two-dimensional simulations of horizontal channel flow with one-way coupling only. Solids loadings (m) from 6.5×10^{-6} to 3.25 were considered for 40 μm, 100 μm and 500 μm glass spheres. It was clearly demonstrated that particle–particle collisions are important, even for solids loadings less than one with the 40 μm particles. Recently, Sommerfeld and Qiu (1993) successfully applied a one-way coupling model which neglected particle–particle interactions to describe very dilute confined swirling flows with 45 μm glass beads. A distribution in particle size was also simulated by means of a log-normal distribution function.

One limitation in all of the above simulations is that they deal with dilute-phase flow only. This is because the number of particle trajectories that have to be computed simultaneously is limited by computer memory; this number is proportional to the solids loading ratio and to the pipe volume. Oesterle and Petitjean (1993) recently proposed a technique that makes it possible to simulate moderately dense suspension flows by computing a reasonable number of successive individual particle trajectories and by introducing artificial interparticle collisions during the trajectory calculation. The probability of these artificial collisions was predicted through an interactive process and depended on the local concentration and velocity of the solids phase according to well-defined rules. Sliding and non-sliding contacts and particle-wall impacts were considered, while wall roughness effects were neglected. The particle trajectories were predicted using linear and angular momentum conservation equations that included drag, gravity and spin lift. The influence of fluid turbulence was approximated by extending Tchen's theory (Hinze, 1975). The motion of 100 μm glass spheres in a horizontal pipe was simulated using this technique at solids loading ratios up to 20. However, the numerical predictions were obtained using one-way coupling, neglecting the influence of the particles on the gas velocity profile, and the gas velocity was assumed to obey the universal logarithmic law. Nevertheless, pressure drop predictions compared well with an empirical correlation of a large number of experimental data obtained at moderate loadings. This result is in contrast to previous dilute flow simulations that neglected interparticle collisions and revealed that collisions play an important role as soon as the loading ratio exceeds unity.

Nomenclature

C_D	single sphere drag coefficient
d_p	particle diameter
e	coefficient of restitution for particle–particle collisions
g	gravitational constant
g_0	radial distribution function
G_1	function that depends on e and g_0 (from kinetic theory)
G_2	function that depends on e and g_0 (from kinetic theory)
$G(\epsilon)$	solids stress modulus
k_p	particle phase turbulent kinetic energy
m	solids loading, i.e. ratio of solids mass flux to gas mass flux
n	coefficient in Richardson and Zaki drag expression
P	gas phase pressure
p_p	particle phase pressure
r	radial coordinate
$\mathbf{S}$	rate of deformation tensor
t	time
T	granular temperature
v_g	gas velocity
v_p	particle velocity
v_t	single particle terminal velocity
z	axial coordinate
β	drag coefficient for spherical particle in an assembly
ϵ	voidage
ϵ_p	particle phase dissipation of turbulent kinetic energy
ϕ_g	gas volume fraction
ϕ_p	solids volume fraction
γ	dissipation of granular energy
κ	particle phase thermal conductivity
ν	solids volume fraction
ν_0	solids volume fraction at closest packing

ν_g gas kinematic viscosity
ν_s particle phase kinematic viscosity
μ_g intrinsic gas viscosity
μ_{eg} effective gas viscosity
μ_t turbulent gas viscosity
μ_p particle phase viscosity
ρ_g gas density
ρ_p particle density
σ particle phase stress tensor

References

Adewumi, M.A. and Arastoopour, H. (1986) Two-dimensional steady state hydrodynamic analysis of gas–solids flow in vertical pneumatic conveying systems. *Powder Technology*, **48**, 67–74.

Ahmadi, G. and Ma, D. (1990) A thermodynamical formulation for dispersed multiphase turbulent flows—I. *Int. J. Multiphase Flow*, **16**, 323–340.

Amos, G., Rhodes, M.J. and Mineo, H. (1993) Gas mixing in gas–solid risers. *Chem. Eng. Sci.*, **48**, 943–949.

Anderson, T. and Jackson, R. (1967) Fluid mechanical description of fluidized beds. *Ind. Eng. Chem. Fundam.*, **6**, 527–539.

Arastoopour, H. and Gidaspow, D. (1979a) Analysis of IGT pneumatic conveying data and fast fluidization using a thermohydrodynamic model. *Powder Technology*, **22**, 77–87.

Arastoopour, H. and Gidaspow, D. (1979b) Vertical pneumatic conveying using four hydrodynamic models. *Ing. Eng. Chem. Fundam.*, **18**, 123–130.

Arastoopour, H., Lin, S. and Weil, S. (1982) Analysis of vertical pneumatic conveying of solids using multiphase flow models. *AIChE J.*, **28**, 467–473.

Arastoopour, H. and Yang, Y. (1992) Experimental studies on dilute gas and cohesive particles flow behavior using laser Doppler anemometer, in *Fluidization VII* (eds O.E. Potter and D.J. Nicklin), Engineering Foundation, New York, pp. 723–730.

Bader, R., Findlay, J., and Knowlton, T. (1988) Gas/solids flow patterns in a 30.5 cm diameter fluidized bed, in *Circulating Fluidized Bed Technology II* (ed. P. Basu and J.F. Large), Pergamon Press, New York, pp. 123–137.

Berruti, F. and Kalogerakis, N. (1989) Modelling the internal flow structure of circulating fluidized beds. *Can. J. Chem. Eng.*, **67**, 1010–1014.

Bolio, E.J. and Sinclair, J.L. (1995) Gas turbulence modulation in the pneumatic conveying of massive particles in vertical tubes. *Int. J. Multiphase Flow*, **21**, 985–1001.

Bolio, E.J., Yasuna, J.A. and Sinclair, J.L. (1995) Dilute turbulent gas–solid flow with particle–particle interactions. *AIChE J.*, **41**, 1375–1388.

Bouillard, J.X. and Lyczkowski, R.W. (1993) Hydrodynamics/heat transfer/erosion computer predictions for a cold small-scale CFBC, in *Circulating Fluidized Bed Technology IV* (ed. A.A. Avidan), AIChE, New York, 359–366.

Bouillard, J.X., Lyczkowski, R.W. and Gidaspow, D. (1989) Porosity distributions in a fluidized bed with an immersed obstacle. *AIChE. J.*, **35**, 978–922.

Campbell, C.S. (1990) Rapid granular flows. *Annu. Rev. Fluid Mech.*, **22**, 57–92.

Cao, J. and Ahmadi, G. (1995) Gas-particle two-phase turbulent flow in a vertical duct. *Int. J. Multiphase Flow*, **21**, 1203–1228.

Chen, C.P. and Wood, P.E. (1985) Turbulence closure model for dilute gas-particle flows. *Can. J. Chem. Eng.*, **63**, 349–360.

Cundall, P. and Strack, O. (1979) A discrete numerical model for granular assemblies. *Geotechnique*, **29**, 47–65.

Dasgupta, S., Jackson, R. and Sundaresan, S. (1994) Turbulent gas-particle flow in vertical risers. *AIChE J.*, **40**, 215–228.

Dasgupta, S., Jackson, R. and Sundaresan, S. (1996) Gas-particle flow in vertical pipes with high mass loading of particles. *Powder Technology*, in press.

Ding, J. and Gidaspow, D. (1990) A bubbling fluidization model using kinetic theory of granular flow. *AIChE J.*, **36**, 523–538.

Ding, J. and Lyczkowski, R.W. (1992) Three-dimensional kinetic theory modeling of hydrodynamics and erosion in fluidized beds. *Powder Technology*, **73**, 127–138.

Ding, J., Lyczkowski, R., Burge, S. and Gidaspow, D. (1991) Three-dimensional models of hydrodynamics and erosion in fluidized-bed combustors, presented at the 1991 AIChE Annual Meeting, Los Angeles, California

Elgobashi, S.E., and Abou-Arab, T.W. (1983) A two-equation turbulence model for two-phase flows. *Phys. Fluids.*, **26**, 931–938.

Foerster, S.F., Louge, M.Y., Chang, H. and Allia, K. (1994) Measurements of the collision properties of small spheres. *Phys. Fluids*, **6**, 1108–1115.

Gidaspow, D. (1986) Hydrodynamics of fluidization and heat transfer: supercomputer modeling. *Appl. Mech. Rev.*, **39**, 1–22.

Gidaspow, D. (1994) *Multiphase Flow and Fluidization: Continuum and Kinetic Theory Descriptions*, Academic Press, London.

Gidaspow, D., Rukmini, B. and Ding, J. (1992) Hydrodynamics of circulating fluidized beds: kinetic theory approach, in *Fluidization VII* (eds. O.E. Potter and D.J. Nicklin), Engineering Foundation, NY, p. 75.

Gidaspow, D. and Therdthianwong, A. (1993) Hydrodynamics and SO_2 sorption in a CFB loop, in *Circulating Fluidized Bed Technology IV* (ed A.A. Avidan), AIChE, New York, pp. 32–39.

Gidaspow, D., Tsuo, Y.P. and Luo, K.M. (1989) Computed and experimental cluster formation and velocity profiles in circulating fluidized beds, in *Fluidization IV* (eds. J.C. Grace, L.W. Schemilt and N.A. Bergougnou), Engineering Foundation, New York, pp. 81–88.

Grace, J.R. and Sun, G. (1991) Influence of particle size distribution on the performance of fluidized bed reactors. *Can. J. Chem. Eng.*, **69**, 1126–1134.

Harris, B.J. and Davidson, J.F. (1993) Modelling options for circulating fluidized beds; a core-annulus deposition model, in *Circulating Fluidized Beds Technology IV* (ed. A.A. Avidan), AIChE, New York, 32–39.

Hinze, J. (1975) *Turbulence*, 2nd edn, McGraw-Hill, New York.

Hopkins, M.A. and Shen, H.H. (1988) A Monte Carlo simulation of a rapid simple shear flow of granular materials, in *Micromechanics of Granular Materials* (eds M. Satake and J.T. Jenkins), Elsevier, London, pp. 349–364.

Hrenya, C.M. and Sinclair, J.L. (1996) Dense, turbulent gas–particle flows in vertical tubes. *AIChE J.*, submitted.

Jenkins, J.T. and Mancini, F. (1989) Kinetic theory for binary mixtures of smooth, nearly elastic spheres. *Phys. Fluids A.*, **1**, 2050–2057.

Johnson, P.C. and Jackson, R. (1987) Frictional–collisional constitutive relations for granular materials, with application to plane shearing. *J. Fluid Mech.*, **176**, 67–93.

Johnson, G., Massoudi, M. and Rajagopal, K.R. (1991) Flow of a fluid infused with solid particles through a pipe. *Int. J. Eng. Sci.* **29**, 649–661.

Johnson, P.C., Nott, P. and Jackson, R. (1990) Frictional–collisional equations of motion for particulate flows and their application to chutes. *J. Fluid Mech.*, **210**, 501–535.

Koch, D. (1990) Kinetic theory for a monodisperse gas-solid suspension. *Phys. Fluids A*, **2**, 1711–1723.

Lee, S.L. and Durst, F. (1982) On the motion of particles in turbulent duct flows. *Int. J. Multiphase Flow*, **8**, 125–146.

Lin, J.S., Chen, M.M. and Chao, B.T. (1985) A novel radioactive particle tracking facility for measurement of solids motion in gas-fluidized beds. *AIChE J.*, **31**, 465–473.

Louge, M., Mastorakos, E. and Jenkins, J.T. (1991) The role of particle collisions in pneumatic transport. *J. Fluid Mech.*, **231**, 345–359.

Lun, C.K. (1991) Kinetic theory for granular flow of dense, slightly inelastic, slightly rough spheres. *J. Fluid Mech.*, **233**, 539–559.

Lun, C.K., Savage, S., Jeffrey, D. and Chepurniy, N. (1984) Kinetic theories for granular flow; inelastic particles in Couette flow and slightly inelastic particles in a general flow field. *J. Fluid Mech.*, **140**, 223–256.

Luo, K.M. (1987) Experimental gas–solid vertical transport, PhD thesis, Illinois Institute of Technology, Chicago, Illinois.

Maeda, M., Hishida, K. and Furutani, T. (1980) Optical measurements of local gas and particle velocity in an upward flowing dilute gas–solids suspension. *Proc. Polyphase Flow and Transport Technology (Century 2-ETC, San Francisco)*, 211–216.

Massah, H., Shaffer, F.D., Sinclair, J.L. and Shahnam, M. (1994) Non-intrusive measurements of particle-wall collision properties. *Proceedings of the First International Particle Technology Forum*, Denver, CO, pp. 499–504.

Massoudi, M., Rajagopal, K.R., Ekmann, J.M. and Mathur, M.P. (1992) Remarks on the modeling of fluidized systems. *AIChE J.*, **38**, 471–472.

Nakamura, K. and Capes, C.E. (1973) Vertical pneumatic conveying: a theoretical study of uniform and annular particle flow models. *Can. J. Chem. Eng.*, **51**, 39–46.

O'Brien, T.J. and Syamlal, M. (1991) Fossil fuel circulating fluidized bed: simulation and experiment. *AIChE Symposium Series*, **87**, 127–136.

O'Brien, T.J. and Syamlal, M. (1993) Particle cluster effects in the numerical simulation of a circulating fluidized bed, in *Circulating Fluidized Bed Technology IV* (ed. A.A. Avidan), AIChE, New York, pp. 345–350.

Ocone, R., Sundaresan, S. and Jackson, R. (1993) Gas-particle flow in a duct of arbitrary inclination with particle-particle interactions. *AIChE J.*, **39**, 1261–1271.

Oesterle, B. and Petitjean, A. (1993) Simulation of particle-to-particle interactions in gas–solid flows. *Int. J. Multiphase Flow*, **29**, 199–211.

Pita, J.A. and Sundaresan, S. (1991) Gas–solid flow in vertical tubes. *AIChE J.*, **37**, 1009–1018.

Pita, J.A. and Sundaresan, S. (1993) Developing flow of a gas-particle mixture in a vertical riser. *AIChE J.*, **39**, 541–552.

Plumpe, J.G., Zhu, C. and Soo, S.L. (1993) Measurement of fluctuations in motion of particles in a dense gas–solid suspension in vertical pipe flow. *Powder Technology*, **77**, 209–214.

Pugsley, T.S., Berruti, F., Godfroy, L., Chaouki, J. and Patience, G.S. (1993) A predictive model for the gas–solid flow structure in circulating fluidized bed risers, in *Circulating Fluidized Bed Technology IV* (ed. A.A. Avidan), AIChE, New York, 40–47.

Rhodes, M.J. (1990) Modelling the flow structure of upward-flowing gas–solids suspensions. *Powder Technology*, **60**, 27–38.

Rhodes, M.J. and Geldart, D. (1987) A model for the circulating fluidized bed. *Powder Technology*, **53**, 155–162.

Rhodes, M.J. and Geldart, D. (1989) The upward flow of gas/solid suspensions. *Chem. Eng. Res. Des.*, **67**, 20–29.

Rhodes, M.J., Wang, X.S., Cheng, H. and Hirama, T. (1992) Similar profiles of solids flux in circulating fluidized-bed risers. *Chem. Eng. Sci.*, **47**, 1635–1643.

Rizk, M.A. and Elghobashi, S.E. (1989) A two-equation turbulence model for dispersed dilute confined two-phase flows. *Int. J. Multiphase Flow*, **15**, 119–133.

Savage, S.B. (1992) Instability of unbounded uniform granular shear flow. *J. Fluid Mech.*, **241**, 109–123.

Senior, R. C. and Brereton, C. (1992) Modelling of circulating fluidised-bed solids flow and distribution. *Chem. Eng. Sci.*, **47**, 281–296.

Seu-Kim, H. and Arastoopour, H. (1993) Analysis of gas/cohesive-particle in circulating fluidized bed using kinetic theory, presented at AIChE Annual Meeting, St. Louis, MO.

Shen, H.H. and Hopkins, M.A. (1988) Stresses in a rapid, simple shear flow of granular materials with multiple grain sizes. *Part. Sci. Technol.*, **6**, 1–15.

Sinclair, J.L. and Jackson, R. (1989) Gas-particle flow in a vertical pipe with particle–particle interactions. *AIChE J.*, **35**, 1473–1486.

Sommerfeld, M. (1991) On the particle velocity fluctuations in turbulent particulate two-phase flows, in *Turbulence Modification in Multiphase Flows*, ASME FED, **110**, 15–21.

Sommerfeld, M. (1992) Modelling of particle-wall collisions in confined gas–particle flows. *Int. J. Multiphase Flow*, **18**, 905–926.

Sommerfeld, M., Huber, N. and Wachter, P. (1993) Particle-wall collisions: experimental studies and numerical models, in *Gas-Solid Flows*, ASME FED, **166**, 183–191.

Sommerfeld, M. and Qiu, H.-H. (1993) Characterization of particle-laden, confined swirling flows by phase-Doppler anemometry and numerical calculation. *Int. J. Multiphase Flow*, **19**, 1093–1127.

Sommerfeld, M. and Zivkovic, G. (1992) Recent advances in the numerical simulation of pneumatic conveying through pipe systems, in *Computational Methods in Applied Sciences*, (eds C. Hirsch, J. Periaux and E. Onate), Elsevier, pp. 201–212.

Soo, S.L. (1967) *Fluid Dynamics of Multiphase Systems*. Blaisdell, Waltham, MA.

Syamlal, M. and Gidaspow, D. (1985) Hydrodynamics of fluidization: prediction of wall to bed heat transfer coefficients. *AIChE J.*, **31**, 127–135.

Tanaka, T. and Tsuji, Y. (1991) Numerical simulation of gas–solid two-phase flow in a vertical pipe, in *Gas–Solid Flows*, ASME FED, **121**, 123–128.

Tanaka, T., Kawaguchi, T., Nishi, S. and Tsuji, T. (1993a) Numerical simulation of two-dimensional fluidized bed: effect of partition walls, in *Gas–Solid Flows*, ASME FED, **166**, 17–22.

Tanaka, T., Yonemura, S., Kiribayashi, K. and Tsuji, Y. (1993b) Cluster formation and particle-induced instability of gas–solid flows (numerical simulation of flow in vertical channel using the DSMC method). *Transactions Japan Society of Mechanical Engineers, Part B*, **59**, 2982–2989.

Tsuji, Y. (1993) Osaka University, Personal communication.

Tsuji, Y., Kawaguchi, T. and Tanaka, T. (1993) Discrete particle simulation of two-dimensional fluidized bed. *Powder Technology*, **77**, 79–87.

Tsuji, Y., Morikawa, Y. and Shiomi, H. (1984) LDV measurements of an air–solid two-phase flow in a vertical pipe. *J. Fluid Mech.*, **139**, 417–434.

Tsuji, Y., Morikawa, Y., Tanaka, T., Nakatsukasa, N. and Nakatani, M. (1987) Numerical simulation of gas–solid two-phase flow in a two-dimensional horizontal channel. *Int. J. Multiphase Flow*, **13**, 671–684.

Tsuji, Y., Oshima, T. and Morikawa, Y. (1985) Numerical simulation of pneumatic conveying in horizontal pipe. *Kona*, **3**, 38–51.

Tsuo, Y.P. and Gidaspow, D. (1990) Computation of flow patterns in circulating fluidized beds. *AIChE J.*, **36**, 886–896.

Walton, O.R. (1984) Computer simulation of particulate flow. *Energy Technol. Rev.*, US-DOE Lawrence Livermore Lab, 24–44.

Wen, C. and Yu, Y. (1965) Mechanics of fluidization. *Chem. Eng. Prog. Symp. Ser.*, **62**, 100–106.

Wirth, K. (1991) Fluid mechanics of circulating fluidized beds. *Chem. Eng. Technol.*, **14**, 29–38.

Wong, R., Pugsley, T. and Berruti, F. (1992) Modelling the axial voidage profile and flow structure in risers of circulating fluidized beds. *Chem. Eng. Sci.*, **47**, 2301–2306.

Yasuna, J.A., Moyer, H.R., Elliott, S. and Sinclair, J.L. (1995) Quantitative predictions of gas-particle flow in a vertical pipe with particle–particle interactions. *Powder Technology*, **84**, 23–34.

Yuan, Z. and Michaelides, E.E. (1992) Turbulence modulation in particulate flows – a theoretical approach. *Int. J. Multiphase Flow*, **18**, 779–785.

6 Cyclones and other gas–solids separators
EDGAR MUSCHELKNAUTZ AND VOLKER GREIF

6.1 Introduction

A cyclone is a device that separates particulate solids from a fluid stream by a radial centrifugal force exerted on the particles. This force separates the solids from the gas by driving the solids to the cyclone wall, where they slide to the bottom outlet and are collected. Cyclones are widely used in conjunction with fluidized beds to remove solids from exit gas streams. Cyclones have no moving parts, are relatively inexpensive to construct, and maintenance costs are low.

Cyclones have been used to remove particulates from gas streams since the middle of the nineteenth century (Rietema and Verver, 1961). Early researchers (Rosin et al., 1932; van Tongeren, 1936; Alexander, 1949; Ter Linden, 1949; Lapple, 1951; Stairmand, 1951; Iinoya, 1953; Kaliski, 1958) recognized the importance of the cyclone, and conducted experiments designed to understand the operation of this mechanically simple but operationally complex device.

Many of these experiments (Ter Linden, 1949; Lapple, 1951; Stairmand, 1951, etc.) led to cyclone designs based on relative cyclone dimensions. More recently, Zenz (1975) developed an empirical cyclone design procedure which has achieved popular acceptance in the United States.

In Germany, Barth (1956) developed a theoretical model of cyclone operation, which was expanded by Dietz (1981). Muschelknautz (1970a; 1970b) further advanced the pioneering work of Barth, and developed a cyclone design technique that is in wide use in Europe. This chapter describes how the Muschelknautz design technique can be applied to cyclones, and especially cyclones operating in circulating fluidized bed systems.

In a coal-burning power plant utilizing a CFBC, cyclones are installed in order to separate solids as efficiently as possible from the exhaust gas and circulate them back to the combustor. In these plants, the solids in the circulating mass are about 90% fly ash of 5 to 500 μm particle size, about 5 to 10% partly burned coal of nearly the same size but lower density, and only about 1% very fine lime particles of 1 to 15 μm (Muschelknautz and Greif, 1994). The solids circulation rate is about 10 times the mass of the feed coal, and the unburnable part of the coal is about 3 to 10%. The solids circulate back to the combustor about 15 to 50 times or more.

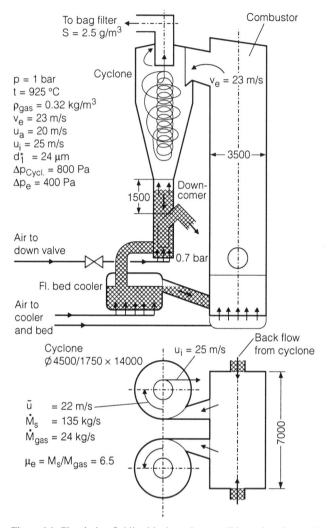

Figure 6.1 Circulating fluidized bed combustor. (Dimensions in mm.)

In petrochemical reactors, the necessary catalyst has to be recirculated to the fluidized bed or a riser. Particle size is usually about 10 to 200 µm, with a particle density of 1500 to 2500 kg/m³. The gas density depends on pressure and temperature, and is in a range of about 0.3 to 1.5 kg/m³. Its viscosity is about 30 to 60 × 10⁻⁶ Pa · s.

The loading of the gas flow

$$\mu_e = \frac{M_{solids}}{M_{gas}} \tag{6.1}$$

covers a range of about 1 to 10.

Figure 6.1 is a schematic drawing of a coal combustor in a power plant. The cross-section of the fluidized bed is rectangular. Two cyclones at the upper end of the bed are attached as closely as possible to the longer side of the cross-section. The solids are collected by the cyclones and circulated through downcomer tubes back to the combustion chamber. The downcomers end in a small fluidized bed called a J-valve or loop seal (see Chapter 7). These devices are essential in that they have to guarantee a constant backflow into the lower combustion zone. This has to occur against a typical pressure differential of 10 to 30 kPa at a low rate of aeration to prevent significant bypass of gas through the downcomer to the cyclone. A high bypass gas flow would carry a high rate of separated particles back into the cyclone. Around the cyclone axis there would be little centrifugal force and, therefore, there would be poor separation.

6.2 Particle size distribution

The widely used Rosin, Rammler, Sperling, and Bennet (RRSB) curves defined by the expression:

$$R(d) = \exp\left[-\left(\frac{d}{d'}\right)^n\right] \quad (6.2)$$

do not show the minimum diameter, d_{min}, nor the maximum diameter, d_{max}, of a particle size distribution. However, every size distribution has these limits. It is difficult to determine the limits from measured data. Often the limits have to be determined by a compromise, or by agreement between the builders and the users of cyclones and circulating fluidized beds. A much better description of the real particle size distribution, especially in the lower size range, is provided by the expressions

$$R(d) = \exp\left[-\left(\frac{d - d_{min}}{d' - d_{min}}\right)^n\right] \quad \text{for } d_{min} < d < d' \quad (6.3)$$

$$R(d) = \exp\left[-\left(\frac{d_{max} - d'}{d_{max} - d}\right)^{0.1} \cdot \left(\frac{d}{d'}\right)^n\right] \quad \text{for } d' < d < d_{max} \quad (6.4)$$

The equation for the higher particle size range, equation (6.4), is of lesser interest.

These equations are well accepted. They give good results for calculating separation efficiencies. Figure 6.2 shows residue curves of (a) particles fed into the cyclone (noted as 'Entrance' in Figure 6.2), (b) particles circulating in the cyclones with reduced loading and reduced maximum size, and (c) the finest size particles exiting the cyclone to the bag filters at the end of the process (noted as 'Fines' in Figure 6.2). The lower end of all the lines is at the same limit d_{min}. The other limits at d_{max} are different. All curves are

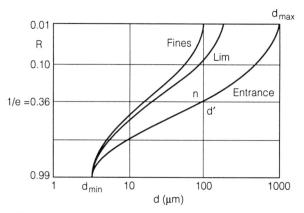

Figure 6.2 RRSB graphs of entry flow, limiting flow and fines.

forced to vertical lines at the limits. They result in very good agreement with measured data. It is especially helpful to define d_{min} in this way so that the calculated line shows the best agreement with the measured values in the range $R(d) > 0.9$. The calculation of cyclone separation efficiency was not very accurate until this improvement was made. The next important step was the development of the improved theory of limited loading in tangential flows. This theory states that the turning flow in cyclones is able to carry only a very low loading described by the relation

$$\mu_{lim} = \frac{M_{solid,lim}}{M_{gas}} \qquad (6.5)$$

A typical value of μ_{lim} is 0.025.

However, the loading of the entering flow is typically $\mu_e = 2.5$. Furthermore, the particle size distribution of the carried solids is much finer than at the entrance to the cyclone as shown in Figure 6.2. In order to calculate the limiting loading μ_{lim} of suspended particles and their size distribution, it is important to know accurately the flow rate and the centrifugal force in the separation chamber.

6.3 Cyclones with and without a vortex tube

The first cyclones for circulating fluidized beds had no vortex tube. These cyclones have a strong secondary flow of about 5 to 10% of the total flow traveling from below the top inward to the center (Figure 6.3). The main flow descends along the cylindrical/conical wall as a thin layer at an angle of about 15 to 20° to the horizontal plane. This flow reverses in the lower part of the separation chamber as a faster upflowing vortex. The diameter

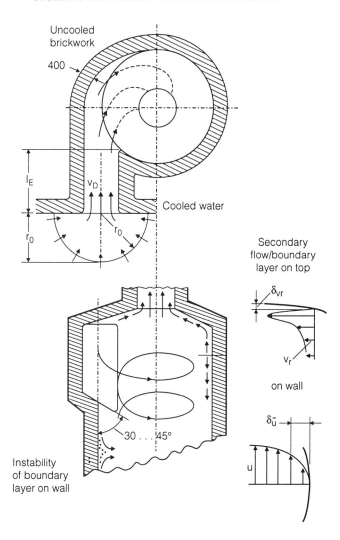

Figure 6.3 Gas flow in cyclones without an insert.

of this axially rotating flow is somewhat less than the outlet diameter, and it results in only slow moving gas which is not a significant portion of the gas flow-through.

The combined circumferential flows downward along the wall and upward in the inner vortex cause a considerable radial drop in static pressure in the separation chamber (Figure 6.23) according to the equation:

$$dp_{stat} = \rho_{gas} \frac{u(r)^2}{r} dr \qquad (6.6)$$

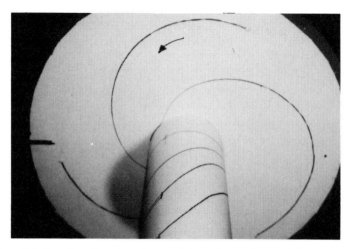

Figure 6.4 Secondary flow along the top of the cyclone and around the vortex tube.

In the slow-moving core there is quite a different distribution of velocity and static pressure, with

$$dp_{stat,core} = \frac{\rho_{gas}}{2} \cdot \frac{u_{core}^2}{r_{core}^2} \cdot r \cdot dr \quad . \tag{6.7}$$

The secondary flow boundary layer below the top-wall bridges at a distance of about 0.05 to 0.15 m to the wall where every velocity is zero. The centrifugal acceleration on the wall itself is zero. At a distance where the secondary flow is rotating with half the velocity of the main flow, its centrifugal acceleration is only one-fourth that of the main flow acceleration. The static pressure p_{stat} remains at a constant level along concentric circles in this layer, and is equal to that in the faster main flow. The radial pressure gradient dominates in the layer, resulting in a strong radial flow. Therefore, the secondary flow travels along steep spirals to the center (see Figure 6.4). This flow accelerates steadily, and results in little turbulence.

Cyclones with a vortex tube force the flow around the tube and down to the opening. On the way around the tube, the secondary flow rotates much slower than the surrounding main flow. This results in increased stability of those layers outside the tube and no turbulence, but a greater thickness of about 0.2 to 0.3 $(r_a - r_i)$. The secondary flow carries particles that need some time to be captured by the surrounding faster main flow. This is done at the same critical particle size as in the main flow below the insert in the separation chamber, within about three revolutions.

Trefz (1992a) shows measured as well as calculated data on the effect of vortex tube length on losses from the cyclone. Figure 6.5 shows a visual indication of the main results of this work, provided that the circumferential flow around the insert is steady. Because of the very limited particle loading

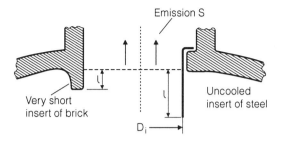

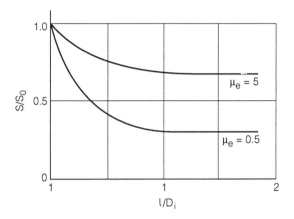

Figure 6.5 Influence of vortex tube length l on emission S from the cyclone.

in the inner cyclone separation chamber, most of the incoming solids are separated within the first three or four revolutions at the wall, Figure 6.6. At loadings of $\mu_e < 0.1$, some parallel single strands along the boundary layer increase flow stability because accumulations of particles induce a boundary layer 'suction' on the wall, which stabilizes the strands. Single strands of particles form on the wall while always moving along the same path.

6.4 Entrance duct and entrance velocities

The average particle size of the circulating solids in a CFBC covers a range of about 75 to 150 μm, while the gas velocity, v_{gas}, in the entrance duct varies from 15 to 25 m/s.

When particles at high loadings μ are conveyed to the top of the riser and into the horizontal entrance of the cyclone, their velocity, c_{solids}, is about 5 m/s. Because of their density (typically about 2500 kg/m³), and their size, they accelerate more slowly than the gas flow. An average Reynolds

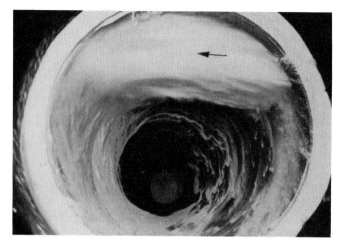

Figure 6.6 Strands on the wall in the separation compartment at $\mu_e = 7$. The photograph was taken though an open hole of 30 mm diameter in the cover sheet of the cyclone. The cyclone has an upper compartment of height $h = 0.2D_a$ (parts of it are the black colored edges on the photograph) superimposed on the lower compartment with the entrance slit. The white ring is the contact area between the two compartments with slightly different diameters.

number for the relative gas–solids flow is:

$$Re = \frac{(v_{gas} - c_{ein})d_{50}\rho_{gas}}{\eta_{gas}} \tag{6.8}$$

$$Re_{max} \approx \frac{(20-8)(125 \times 10^{-6})(0.32)}{45 \times 10^{-6}} \approx 10 \tag{6.9}$$

and with

$$w = v_{gas} - c_{ein} = 16 - 12 = 4\,\text{m/s} \qquad Re_{min} \approx 4 \tag{6.10}$$

In this range of Re, one can calculate the drag, W, on the particles from the settling velocity, w_s, and the weight of a particle cloud (Muschelknautz and Krambrock, 1993) as

$$W = M_s g \left(\frac{v_{gas} - c_{ein}}{w_s}\right)^{2-k} \tag{6.11}$$

For the range of Re shown above, a first estimate of the exponent k is

$$k \approx 0.75$$

An exponent of 0.75 means that 75% of the drag is caused by viscous forces, and only 25% by inertial (mass) forces. The settling velocity of a highly loaded flow is based on the settling velocity of single particles, w_{so}, i.e.

$$w_s = w_{so}(1 + 0.5\mu_e)^{0.25} \tag{6.12}$$

and is much greater than that for single particles because of the higher slip. The exponent k is a function of Re, and can be accurately calculated (Muschelknautz and Krambrock, 1993) from the relation:

$$k = \frac{A + (0.5)B\sqrt{Re}}{A + B\sqrt{Re} + C(Re)} \qquad (6.13)$$

and

$$w_{so} = \left(\frac{4/3(\Delta\rho_{solid,gas}g)d_{50}^{1+k}}{K\eta_{gas}^{k}\rho_{gas}^{1-k}}\right)^{1/(2-k)} \qquad (6.14)$$

where

$$K = \frac{A}{Re^{1-k}} + \frac{B}{Re^{0.5-k}} + C(Re^{k}) \qquad (6.15)$$

There are computer programs to calculate the particle speed as a function of length (Muschelknautz and Krambrock, 1993) along the entrance duct. Figure 6.7(a) shows the result of such calculations for a typical duct like that shown in Figure 6.1. For a limited length of the duct, the particles are accelerated to about 80 to 90% of the gas velocity at the end of the duct. Further acceleration can occur in longer ducts. Measurements in operating systems (Greif and Muschelknautz, 1994) confirm this. At the end of the duct, the inlet gas stream enters the free tangential flow field of the cyclone at constant static pressure. However, the drag on the particles is still present, and can still accelerate them. Only momentum exchange is possible between the solids and the gas flow, because the static pressure in the outer separation chamber has to be constant at each radial position, i.e.

$$M_{gas} \cdot \Delta v_{gas} = V_{gas} \cdot \Delta u_{gas} = -M_{solid} \cdot \Delta c_{solid} \qquad (6.16)$$

and thus

$$\Delta v_{gas} = \Delta u_{gas} = -\mu_e \cdot \Delta c_{solid} \qquad (6.17)$$

This exchange of momentum between the gas and solids occurs within a very short distance, Δl_c, and after the momentum exchange both velocities c_{solid} and v_{gas} with respect to u_{gas} are equal, i.e.

$$c_{ein} = v_{ein} = u_{ein} \qquad (6.18)$$

According to the curve in Figure 6.7(b), there is considerable additional pressure loss as the flow is accelerated by suction in the entrance area in front of the entrance duct and in the duct itself, as long as the duct is enclosed by walls. The total acceleration loss can be expressed as the sum of the loss due to the solids and the loss due to the gas, i.e.

$$dp = dp_{gas} + dp_{solid} \qquad (6.19)$$

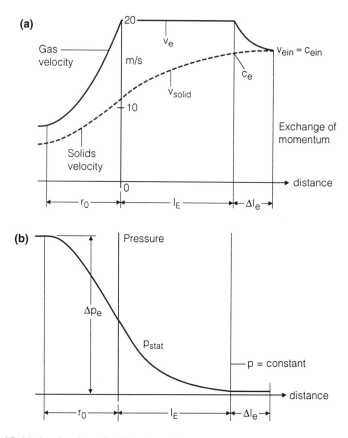

Figure 6.7 (a) Acceleration of solids along the entrance duct and exchange of momentum immediately after the inlet. (b) Change in static pressure along the entrance duct.

where

$$dp_{solid,acc} = -\rho_{gas} v_{gas}\, dv - \mu_e \cdot \rho_{gas} \cdot v_{gas} \cdot dc \qquad (6.20)$$

$$\Delta p_E = -\frac{\rho_{gas}}{2} v_{duct}^2 - \mu_e \rho_{gas} \bar{v}_{duct} \Delta c_{solid} \qquad (6.21)$$

$$\Delta p_E \approx -\frac{\rho_{gas}}{2} v_{duct}^2 \left[1 + \mu_e 2 \frac{\Delta c_{solid}}{\bar{v}_{duct}}\right] \qquad (6.22)$$

and where

$$\Delta c_{solid} \approx (0.8 \text{ to } 0.9) \cdot v_{duct} \qquad (6.23)$$

Friction on the walls caused by particle and gas flow may be neglected because it is small compared to the much higher acceleration effects. The total mass flow of gas and particles, $M_{gas}(1 + \mu_e)$, enters the cyclone separation zone in plug flow with nearly uniform velocity. The solids travel in

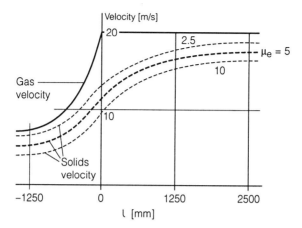

Figure 6.8 Influence of loading μ_e on acceleration of solids in the entrance duct of a particular cyclone, with $k = 0.75$.

straight lines, but the gas has to travel sideways at angles of about 20 to 40° through the dense cloud of particles. The gas flow, therefore, is decelerated by the particle drag. The best correlation for tangential velocity is given by the semi-empirical relation:

$$u_a = \frac{u_{ein}}{(1 + 0.5\mu_e)^{0.25}} \qquad (6.24)$$

with an accuracy of ±5%, if the average particle size is in the range of

$$d_{50} \approx 50 \text{ to } 200 \text{ } \mu m$$

The exponent 0.25 in equation (6.24) corresponds to the same k calculated for the higher loading, μ_e, according to equation (6.13).

Figure 6.8 shows that the acceleration in the entrance duct depends on the loading ratio, μ_e. With increasing μ_e, the velocity ratio c_{solid}/v_{gas} decreases. This means that the particle velocity at the exit of the duct decreases. Therefore, the pressure loss due to the acceleration of the particles (from equation (6.21)) increases linearly with increasing dust loading, μ_e.

Figure 6.9 shows a dimensionless plot of particle velocity for particles accelerating in the inlet duct. In this figure, the dust loading is taken into account in the definition of the abscissa.

6.5 Pressure drop and separation efficiency

6.5.1 General flow pattern

Generally, the three-dimensional flow field of a cyclone is characterized by a strong tangential velocity component, $u(r)$, and a weak radial component, v_r.

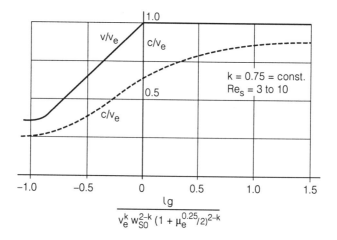

Figure 6.9 Dimensionless particle and gas velocities in the entrance duct.

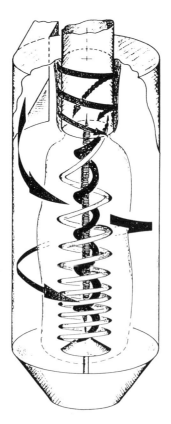

Figure 6.10 General flow patterns in a cyclone.

The axial component, v_{ax}, is directed downward between the outer radius of the cyclone body, r_a, and $0.7 \cdot r_a$ (Figure 6.10). For $r < 0.7 \cdot r_a$, v_{ax} is directed upward. Superimposed on the axial flow is the tangential velocity distribution, $u(r)$, described by the simple expression

$$u \cdot r^n = \text{constant} \quad (6.25)$$

The gas flow is directed downwards along a spiral, with a helical angle of about 15°. This flow field was first measured by Ter Linden (1949). Particles moving along the wall may possess larger helical angles, depending on the loading. The whirl exponent, n, in equation (6.25) is equal to 1 for ideal frictionless gas flow. For actual cyclones, n is less than 1, and decreases with increasing solids concentration in the inlet flow. For loadings of cyclones typical of circulating fluidized beds, n may approach 0. The value of n cannot be predetermined as a function of the cyclone shape and operating conditions.

At the top of the cyclone is a secondary flow through the boundary layer of about 10%. The particles carried by this secondary flow must be separated by the rotating flow around the vortex tube. Detailed investigations on this subject have been carried out by Trefz (1992a, b).

6.5.2 Separation efficiency of a cyclone according to the model of Barth/Muschelknautz

With the model of Barth (1956), equation (6.25) is no longer used. Friction is calculated as a function of a friction coefficient, λ_s, and the area offering resistance to friction inside the cyclone, A_R. The total friction coefficient consists of a gas component, λ_o, and a much larger component due to the particles. Muschelknautz and Krambrock (1970) measured λ_o, and found it was a function of the Reynolds number, the relative wall roughness, k_s/r_a, and the diameter of the cyclone (Figure 6.11). The variation of λ_s depends on the inlet loading ratio, μ_e, and is especially important for highly loaded cyclones. How λ_s varies was investigated by Muschelknautz and Brunner (1963), who found that

$$\lambda_s = \lambda_o + 0.25 \sqrt{\eta_{tot}\mu_e Fr_i \frac{\rho_{gas}}{\rho_{solid}(1-\epsilon_{strand})}} \left(\frac{r_a}{r_i}\right)^{-5/8} \quad (6.26)$$

where η_{tot} is the total cyclone separation efficiency, and Fr_i, the Froude number of the cyclone, is given by

$$Fr_i = \frac{v_i}{\sqrt{(2)(g)(r_i)}} \quad (6.27)$$

Knowing λ_s and the inner surface area, A_R, allows the velocity distribution of the tangential velocity component, $u(r)$ to be calculated by balancing the moments of momentum. Cyclone pressure drop and separation

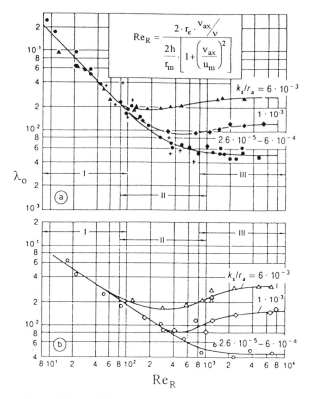

Figure 6.11 Cyclone friction coefficient for clean gas: (a) cylindrical cyclone body; (b) conical cyclone body.

efficiency are determined primarily by the tangential velocity of the particles and the gas in the rotational field inside the cyclone. High dust loadings cause significant friction on the cyclone wall and cause λ_s to increase and $u(r)$ to decrease (Figure 6.12). The tangential velocity can be calculated from:

$$u(r) = \frac{u_a(r_a/r)}{1 + \lambda_s \dfrac{A_R u_a}{2 V_{gas}} \sqrt{\dfrac{r_a}{r}}} \tag{6.28}$$

The outer tangential velocity, u_a, at radius r_a, depends on the type of cyclone inlet (i.e. tangential or spiral). The calculation of gas and particle behavior in the duct described above assumed u_a to be calculated for very highly loaded cyclones with a wide spiral inlet, as shown in Figure 6.13, according to equation (6.24), and when equation (6.18) applies, i.e.

$$u_a = v_{ein} \cdot (1 + 0.5 \cdot \mu_e)^{-0.25} \tag{6.29}$$

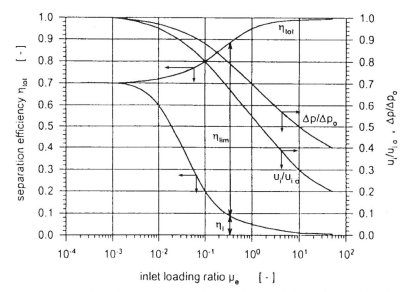

Figure 6.12. Separation efficiency, pressure drop, and tangential velocity as a function of entrance loading ratio.

The expansion of the stream after the inlet is characterized by the expansion/contraction coefficient α, expressed as a relationship between the tangential momentum based on the mean inlet radius, r_e, given by:

$$r_e = r_a - \frac{\Delta b}{2} \qquad (6.30)$$

and the momentum on r_2 (see Figure 6.15), i.e.:

$$\alpha = \frac{v_{ein} r_e}{u_2 r_2} \qquad (6.31)$$

For very low particle loadings (investigated by Barth and Leineweber, 1964) very long ducts, and fine particles, the particles are completely accelerated, and there is no expansion. The inlet jet contracts ($\alpha < 1$) due to the pressure field of the rotational flow, and v_e increases to u_2. For a tangential slit, Rentschler (1991) found that

$$\alpha = \frac{1 - \sqrt{1 + 4\left[\left(\frac{\beta}{2}\right)^2 - \frac{\beta}{2}\right]\sqrt{1 - \frac{1-\beta^2}{1+\mu_e}(2\beta - \beta^2)}}}{\beta} \qquad (6.32)$$

where β is the relative width of the slit, b_e/r_a. Knowing α, the pressure drop and separation efficiency for the cyclone can now be calculated.

The separation efficiency has to be divided into two parts due to different separation mechanisms. The restricted turbulence in the cyclone enables only

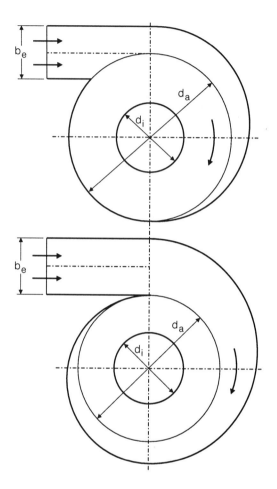

Figure 6.13 Inlet flow with half spiral (above) and full spiral (below) cyclones.

a certain loading of solids to be carried in the air stream. If the solids concentration in the gas from the riser exceeds this limited loading ratio, μ_{lim}, the excess mass is removed immediately after the cyclone inlet and forms strands or a continuous layer at the wall, Figure 6.6. Only a small fraction of the finer particle size distribution, Figure 6.14, remains in the gas flow, and undergoes centrifugal separation in the vortex of the cyclone. Both separation mechanisms are characterized by a cut size, d^*. The separation by exceeding the limited loading, μ_{lim}, is characterized by d_e^* (Trefz and Muschelknautz, 1993), where d_e^* is given as:

$$d_e^* = \sqrt{w_{s,50} \frac{18\eta_{gas}}{\Delta\rho_{gas,solid}\bar{z}_e}} \qquad (6.33)$$

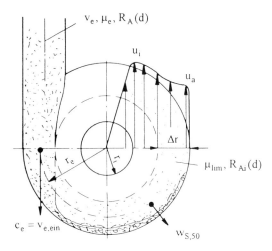

Figure 6.14 Solids separation after the cyclone inlet.

where it is assumed that Stokes' law applies, and $w_{s,50}$ is the settling velocity of a particle that is separated while moving along the clarification area, A_W, in the zone of downward flow under the mean centrifugal acceleration, $\bar{z}_e$. With a secondary flow of 10%, $w_{s,50}$ is given by

$$w_{s,50} = \frac{0.5 \times 0.9 V_{gas}}{A_W} \qquad (6.34)$$

The clarification area, A_W, includes the cylindrical part of the cyclone and the upper half of the cone because of the reversing vortex, Figure 6.15. The mean centrifugal acceleration, $\bar{z}_e$, along the streamline during the first turn in the cyclone, may be calculated from the tangential velocity, u_e, on the mean inlet radius, r_e, according to equation (6.31) and u_2 on radius r_2 that halves the cone of the cyclone body, i.e.

$$\bar{z}_e = \frac{u_e u_2}{\sqrt{r_e r_2}} \qquad (6.35)$$

with

$$u_e = \frac{u_a(r_a/r_e)}{1 + \frac{\lambda_s}{2}\frac{A_W}{0.9 V_{gas}} u_a \sqrt{\frac{r_a}{r_e}}} \qquad (6.36)$$

and

$$u_2 = \frac{u_a(r_a/r_2)}{1 + \frac{\lambda_s}{2}\frac{A_W}{0.9 V_{gas}} u_a \sqrt{\frac{r_a}{r_2}}} \qquad (6.37)$$

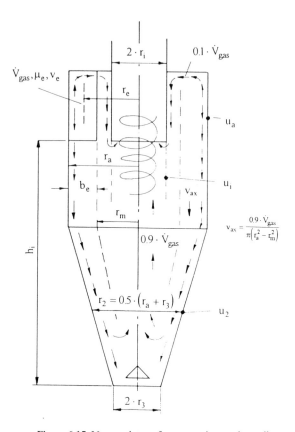

Figure 6.15 Nomenclature for separation at the wall.

The characteristic particle size, d_e^* determines the size distribution of the particles suspended inside the inner vortex (Figure 6.16). The complete residue curve of this 'inner feed material' may be approximated as an RRSB function as given in equation (6.2), i.e.

$$R_{Ai}(d) = \exp\left[-\left(\frac{d}{d_e^*/0.7^{1/n_{Ai}}}\right)^{n_{Ai}}\right] \quad (6.38)$$

The exponent, n_{Ai}, has been determined to be 1.2 for all feed distributions, $R_A(d)$, with $n_A < 1.2$. If the exponent n_A of the feed is greater than 1.2, it will equal n_{Ai} for the 'inner feed material'. Thus, the particle size distribution $R_{ai}(d)$ is only a function of the flow field in the cyclone and its geometry, Figure 6.14. For optimum results, the maximum and minimum particle sizes, d_{max} and d_{min}, have to be considered as indicated in equations (6.3) and (6.4).

Knowing $R_{Ai}(d)$, it is still required to calculate the amount of particles carried into the inner swirl. Generally, the limited loading ratio, μ_{lim}, depends

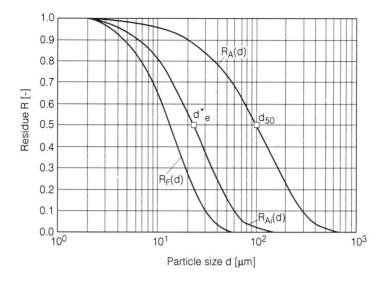

Figure 6.16. RRSB curves for feed $R_A(d)$, 'inner feed' $R_{Ai}(d)$, and fines $R_F(d)$.

on the ratio between the centrifugal force and the drag force (the Barth-number, Ba) acting on the dust cloud. It is proportional to the ratio d_e^*/d_{50}, where d_{50} is the mean particle size of the feed. Extensive and detailed investigations led to the determination of empirical constants, and found an influence of the feed loading, μ_e on μ_{lim} (Muschelknautz, 1970a, b; Trefz, 1992; Greif, 1996), i.e.

$$\mu_{lim} = 0.025 \frac{d_e^*}{d_{50}} (10\mu_e)^k \tag{6.39}$$

The exponent k is found from Figure 6.17. It is constant ($k = 0.15$) for inlet loading ratios $\mu_e > 0.1$, as is the case for most cyclones recirculating solids to fluidized beds. For small values of μ_e, k can increase to a maximum value of about 0.8.

For fine feed material, μ_{lim} is higher than for coarse feed materials. This means that separation efficiency decreases with finer feeds. The separation efficiency for this first separation mechanism is:

$$\eta_{lim} = 1 - \frac{\mu_{lim}}{\mu_e} \tag{6.40}$$

Generally, η_{lim} and μ_{lim} are strongly dependent on the inlet solid loading μ_e (Figure 6.12). With increasing μ_e the separation efficiency, η_{lim}, increases (Muschelknautz, 1970a, b). However, the inner tangential velocity decreases as μ_e increases, and the separation efficiency of the inner vortex also decreases.

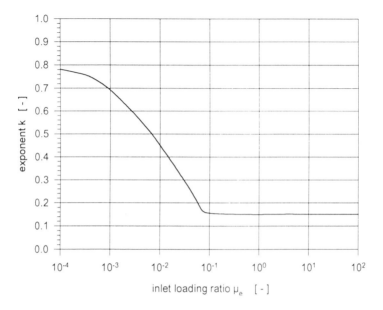

Figure 6.17 Exponent k for determination of limiting loading ratio μ_{lim}.

The second separation mechanism applies to the inner vortex, characterized by the cut size d_i^* in the main stream. It is given by the balance of the centrifugal force acting on the particles with a diameter of d_i^* and its drag, at the radius of the vortex tube, r_i. Assuming a secondary flow of 10%, d_i^* can be calculated from

$$d_i^* = \sqrt{\frac{18\eta_{gas}0.9V_{gas}}{(\rho_{solid} - \rho_{gas})u_i^2 2\pi h_i}} \qquad (6.41)$$

where u_i is the tangential velocity at radius r_i, calculated according to equation (6.28), and h_i is the height of the separation compartment below the gas outlet tube, Figure 6.15. Separation of the particles in the secondary flow along the cyclone top and around the vortex tube has to be carried out with the same cut size in order that the total separation efficiency of the cyclone does not get worse. This is accomplished by calculations with the velocity distribution of the secondary flow assumed to be symmetrical around the insert (Trefz, 1992a, b).

Because the inlet jet expands at high dust loadings, the velocity distribution around the vortex tube becomes asymmetrical when a tangential inlet is used, and the separation efficiency decreases. If a spiral (volute) inlet is used, the expanding jet is further from the vortex tube and thus will not disturb the secondary flow. Therefore, the velocity distribution remains symmetrical, and the separation efficiency of the secondary (vortex) flow remains high. This is the reason why a spiral inlet is better than a slotted inlet at high solid loadings.

The separation efficiency η_i due to the inner vortex is given by

$$\eta_i = \sum_{j=1}^{m} \eta_F(d_j) \Delta R_{Ai}(d_j) \qquad (6.42)$$

where η_F is a grade efficiency, as given in Figure 6.18a. In this equation, the residue curve for R_{Ai} is subdivided into m classes. The grade efficiency in the inner vortex, $\eta_F(d)$, is valid only if μ_e is less than μ_{lim}. If a cyclone operates at dust loadings such that μ_e is greater than μ_{lim}, which is normally the case, separation also occurs because the limited loading, μ_{lim}, has been exceeded. The consequence of this is a change in the total fractional efficiency as shown in Figure 6.18b (Greif, 1996).

With decreasing particle size the grade efficiency decreases, but not down to zero. Instead, it increases again, the amount depending on the inlet loading, μ_e. This means that very fine particles are carried by the coarse particles to the wall of the cyclone. This behavior has been observed in different commercial cyclones of 0.3 to 0.5 m diameter at temperatures between 20°C and 900°C, and at pressures up to 300 bar. This can also be determined by calculation using the theory presented.

A simpler way of obtaining η_i is to assume the cyclone separates as a sieve of mesh size, d_i^*. In this case, the inner vortex separation efficiency is:

$$\eta_i \approx R_{Ai}(d_i^*) = \exp\left[-\left(\frac{d_i^*}{d_e^*/0.7^{1/n_{Ai}}}\right)^{n_{Ai}}\right] \qquad (6.43)$$

In most cases, the accuracy of this method is sufficient, especially for calculating the high separation efficiency required for cyclones in circulating fluidized beds (Muschelknautz and Trefz, 1993; Muschelknautz et al., 1994).

The total separation efficiency then becomes

$$\eta_{tot} = \eta_{lim} + \eta_i = 1 - \frac{\mu_{lim}}{\mu_e} + \frac{\mu_{lim}}{\mu_e} R_{Ai}(d_i^*) \qquad (6.44)$$

If the loading in the feed, μ_e, is smaller than the critical loading ratio, μ_{lim}, then $\eta_{lim} = 0$ and the 'inner feed material' and the feed at the inlet have the same particle size distribution. The total separation efficiency in this case is:

$$\eta_{tot} \approx R_A(d_i^*). \qquad (6.45)$$

The amount of particle carryover is given by

$$S = (1 - \eta_{tot}) \cdot \mu_e \cdot \rho_{gas} \qquad (6.46)$$

The accuracy of S is approximately ±20% for a typical cyclone geometry.

The particle size distribution of the carryover, $R_F(d)$, can be determined in m size fractions by means of the total separation efficiency, the distribution of

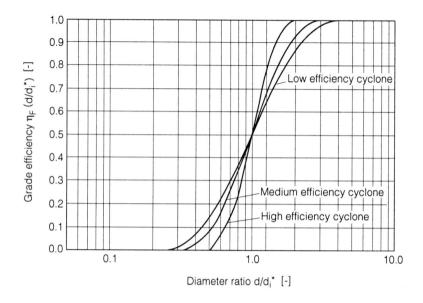

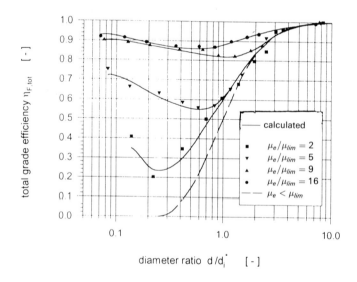

Figure 6.18 (a) Grade efficiency curve for separation in the inner vortex for three different cyclone geometries. (b) Total grade efficiency curves for separation due to exceeding the limited loading μ_{lim} for different inlet loadings μ_e, measured and calculated.

the 'inner feed', and the grade efficiency curve. Thus,

$$\Delta R_F(\bar{d}_j) = \frac{\mu_{lim}}{\mu_e} \frac{(1 - \eta_F(\bar{d}_j))}{1 - \eta_{tot}} \Delta R_{Ai}(\bar{d}_j) \tag{6.47}$$

The change in particle size distribution from the inlet to the outlet of a cyclone is shown in Figure 6.16.

6.5.3 Cyclone pressure drop

For changes in static dynamic pressure, all losses have to be calculated as total pressure drops. Following Barth (1956), the pressure drop of a cyclone is subdivided in two parts. There are (1) losses due to friction (gas at the wall and gas–solids at the wall), Δp_e, and (2) losses due to the flow field in the vortex tube, Δp_i. The total pressure loss of a cyclone, Δp_{tot}, is then:

$$\Delta p_{tot} = \Delta p_e + \Delta p_i \tag{6.48}$$

Δp_e depends on the wall friction coefficient, λ_s, which takes into account the mass of dust on the wall, the surface area of the separation chamber, A_R, and the mean tangential velocity at the wall. It is calculated analogously to the pressure losses of gas flowing over a flat plate, i.e.

$$\Delta p_e = \lambda_s \frac{A_R}{0.9 V_{gas}} \frac{\rho_{gas}}{2} (u_a u_i)^{3/2} \tag{6.49}$$

In general, the pressure drop in the vortex tube of a cyclone at low loadings is about 3 to 5 times higher than that caused by wall friction. In highly loaded cyclones, Δp_e and Δp_i are of the same order of magnitude. The primary reason for the high loss in the vortex tube is the very high velocity in the tube.

The tangential velocity in a potential vortex increases rapidly near the periphery. Therefore, the shear stress also increases enormously. This is why every vortex has an 'idling' core of radius r_{core}, Figure 6.19. In this core, there is no axial flow component. This is why the velocity in the remaining annular cross-section in the vortex tube, V_a, is about three times higher than the mean velocity of flow, v_i. With the superimposed tangential velocity component, the total velocity of the flow is:

$$w = \sqrt{u_i^2 + v_a^2} \tag{6.50}$$

This high kinetic energy is dissipated in the tube, and is the source of most of the Δp. The diameter of the 'idling' core is dependent on the velocity ratio, u_i/v_i, Figure 6.19. When this ratio <0.5, the core becomes unstable. With increasing u_i/v_i, the diameter of the core grows and with it the total velocity of the flow. This is why the pressure loss of the vortex tube depends

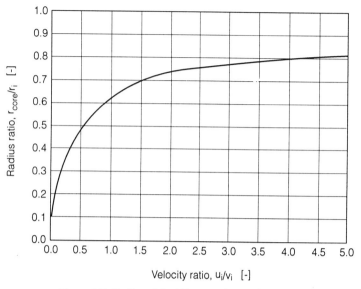

Figure 6.19 Radius of the idling core in rotational flow.

primarily on the velocity ratio. The pressure drop in the vortex tube, Δp_i, may be calculated from:

$$\Delta p_i = \left[2 + 3\left(\frac{u_i}{v_i}\right)^{4/3} + \left(\frac{u_i}{v_i}\right)^2\right]\frac{\rho_{gas}}{2}v_i^2 \quad (6.51)$$

where the mean velocity in the vortex tube is given by:

$$v_i = \frac{V_{gas}}{\pi r_i^2} \quad (6.52)$$

Pressure recovery can be achieved by decelerating the velocity by adding swirl vane inserts into the vortex tube to transform the kinetic energy into static pressure (Greif, 1996).

The pressure drop of a cyclone depends strongly on the inlet solids loading. A high inlet loading ratio leads to high friction, high λ_s, and rapid deceleration of the vortex. The inner tangential velocity, u_i, decreases compared to the inner tangential velocity without dust, u_{io}, Figure 6.12. The velocity ratio, u_i/v_i, diminishes and with it Δp_i. Therefore, for high dust loadings, the total pressure drop of the cyclone can be less than half of that for gas alone.

In addition to the actual pressure drop in the cyclone itself, the pressure drop due to the acceleration of gas and particles in the entrance duct, see Figure 6.8b, should also be included in the cyclone pressure drop.

6.6 Downcomer tube and fluidized bed seal with valve at its end

The ideal end of a downcomer tube would be a somewhat larger bed around the exit of the downcomer with the downcomer exit located nearly in its center, Figure 6.20. Outside the downcomer cross-section, air is used to fluidize the solids.

If only viscous forces act on the particles, the minimum fluidizing gas velocity, Figure 6.21, can be calculated from:

$$v_L = \frac{w_{so}}{40} \tag{6.53}$$

where w_{so} is the average settling velocity of a single particle according to Stokes law, i.e.

$$w_{so} = \frac{\rho_{solid} g d_{50}^2}{18 \eta_{gas}} \tag{6.54}$$

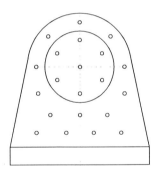

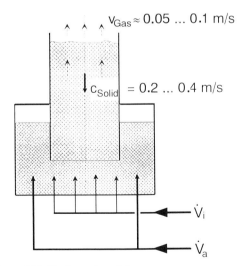

Figure 6.20 Ideal aeration of downcomer.

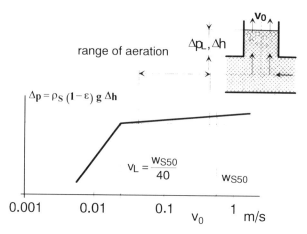

Figure 6.21 Aeration of downcomer duct and space below it.

Alternatively, any of the correlations for minimum fluidization velocity in the literature can also be used.

Beyond the minimum fluidization velocity, the mass of solids is fluidized. Beyond the settling velocity, w_{so}, the particles are pneumatically conveyed. According to Figure 6.21, there is a range of:

$$v_0 = \frac{1}{(5 \text{ to } 30)} w_{s50} \tag{6.55}$$

for sufficient fluidization of the solid mass to cause it to return to the riser through the loop seal or J-valve (see Chapter 7 for a discussion of these devices). In most cases no further fluidization is required in the downcomer itself, because part of the fluidizing gas may also enter the downcomer. If there is a large fluidized bed around the exit of the downcomer (Figure 6.20), and too few aeration nozzles are installed at the bottom of the fluidized bed (resulting in poor gas distribution), more fluidizing gas is required, i.e.

$$v_0 = 0.2 w_{s50} \tag{6.56}$$

In this case, a significant amount of gas can flow up the downcomer into the cyclone. This can result in a larger particle size distribution of the emitted dust. If the particles in the downcomer are larger than 3 to 5 times d_i^*, there is a short-circuit from the downcomer to the outlet of the cyclone. In this case, installing a 'Chinese hat' at the bottom of the cyclone can prevent this (see section 6.7).

The required distance to the end of the downcomer is:

$$a = (0.25 \text{ to } 0.5) \cdot D_{downcomer} \tag{6.57}$$

When starting a circulating fluidized bed, it is useful to add additional gas into the section below the downcomer duct. After start-up, this may be

CYCLONES AND OTHER GAS–SOLIDS SEPARATORS

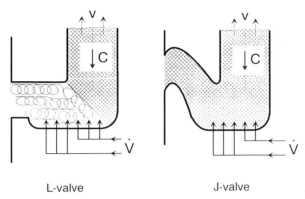

Figure 6.22 L- and J-valve at the end of downcomer duct.

reduced or shut down completely. There are other well-known simple valves of the L-type and J-type (Figure 6.22 and Chapter 7). Some of these may be aerated separately below and beside the downcomer cross-section.

6.7 Inserts in the separation zone

Usually there is no need for an insert such as a 'Chinese hat' in normal operating cyclones with tangential velocities, u_i, in a range of 25 to 35 m/s. If higher tangential velocities are required for better separation, higher velocities are necessary in the entrance duct as well as in the cyclone itself and additional pressure drop is required. If this is so, it can be helpful to separate the downcomer from the separation chamber in the cyclone by a Chinese hat installed in the center of the lower conical part, Figure 6.23. Its diameter should be approximately:

$$D_{hat} \leq (0.7 \text{ to } 0.8) d_i \qquad (6.58)$$

while the distance to the wall should be in the range of:

$$s \approx (0.1 \text{ to } 0.2) D_{hat} \qquad (6.59)$$

If this distance is kept as small as possible without blocking the mass flow of the solids from the cyclone, the space below the hat has essentially no circulation of gas and the static pressure is nearly the same as that at the edge of the insert, Figure 6.23. This leads to a very useful decrease in static pressure in the downcomer by approximately half. Very few solids short circuit to the cyclone, and its efficiency is increased to such a high level that, for CFB combustion, only unburnable fly ash is exhausted from the cyclone insert to the bag filters.

If there is an inner bypass of more than 3%, a Chinese hat still has a positive effect. It deflects the entrained particles in the upward-moving

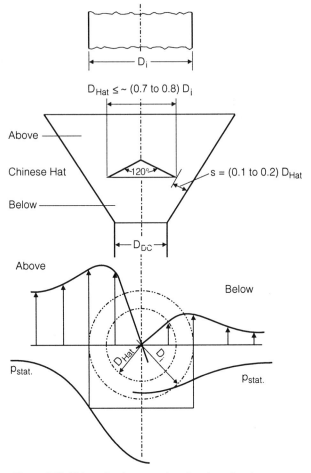

Figure 6.23 Chinese hat in separation chamber of cyclone.

bypass flow to a greater radius where their separation becomes easier. This is why the diameter of the Chinese hat has to be at least equal to the diameter of the core.

6.8 Other separators

For low gas flows in high pressure fluidized beds, lamella separators or simple gravity separation chambers are sometimes used. Figure 6.24 shows a schematic drawing of a lamella separator. Similar separators are used for droplet separation in chemical plants.

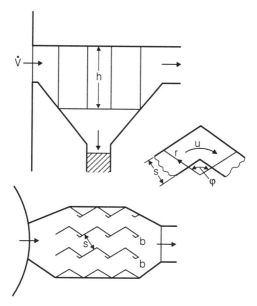

Figure 6.24 Lamella separator.

For every stage, the settling velocity w_s^* for a mean particle size, d_{50}, is calculated from:

$$w_s^* = \frac{\rho_{solid} z d_{50}^2}{18\eta_{gas}} \qquad (6.60)$$

This equation is analogous to equation (6.54) except that the acceleration due to gravity is replaced by the average centrifugal acceleration, z, in each lamella, i.e.

$$z = \frac{u^2}{r} \qquad (6.61)$$

w_s^* is also equal to:

$$w_s^* = \frac{\dot{V}_{gas}}{\varphi r h} \qquad (6.62)$$

where h is the height of each lamella perpendicular to the cross-section, Figure 6.24. The cut size of the separator is predictable if the gas flow rate is known.

The loading, μ_e, has a similar influence on the critical settling velocity and the limiting loading, μ_{lim}, of each lamella stage as it does in cyclones.

The pressure drop of one stage is

$$\Delta p = (0.5 \text{ to } 0.7)\frac{\rho_{gas}}{2} u^2 \qquad (6.63)$$

which is much lower than that of a cyclone at normal tangential velocities of 5 to 10 m/s.

These separators are relatively large because the distances between the lamellas themselves are relatively large, i.e.

$$s \approx 0.15 \text{ to } 0.25 \text{ m}$$

while the width, b, of the gutter collecting the particles should be ≈ 0.03 to 0.05 m.

Gravity separation chambers can be designed by the same equations with z replaced by the gravity, g. Because the separation efficiency is independent of the height of the separation chamber (as in settling tanks), improvement of the separation efficiency is possible by installing many settling zones on top of each other.

6.9 Closure

The cyclone theory described in this chapter has been shown to apply in several industrial applications such as:

1. cyclones in FCC units
2. CFBC cyclones
3. separation of solids after coffee and milk dryers
4. separation of fine metal powder

In all cases, cyclone pressure drop has been shown to be predicted to within ±10%. Cyclone losses were predicted to within ±20% of the measured losses. The diameters of the cyclones over which the theory has been applied have ranged from 0.6 to 8 meters. Operating conditions have ranged from:

Temperature: 20 to 950°C.
Dust loading: 0.5 to 15000 g/m^3
Pressure drop: 200 to 3000 Pa

Nomenclature

A, B, C	constants, dimensionless
A_R	total inner cyclone surface area, m^2
A_W	clarifying area, m^2
b_e	width of cyclone entrance, m
c	particle velocity, m/s
c_{ein}	solid velocity at the entrance, m/s
c_{solid}	velocity of solids, m/s
d	particle size, m
d_a	cyclone diameter, m

d_e^*	cut size for wall separation, m
d_i	diameter of vortex tube, m
d_i^*	cut size for separation in the vortex, m
d_{min}	minimum particle size, m
d_{max}	maximum particle size, m
d_{50}	mean particle size of the feed, m
d'	characteristic particle size of RRSB distribution, m
Fr_i	Froude number of cyclone, dimensionless
g	acceleration due to gravity, m/s²
h	height of cyclone or height of lamella separator (Figure 6.24), m
h_e	height of cyclone inlet, m
h_i	height of separation zone, m
k	exponent of limiting loading ratio μ_{lim}, dimensionless
k	exponent calculating drag, dimensionless
k_s	equivalent sand grain roughness of inner wall, m
l	lengthwise coordinate, equal to 0 at entrance to constricted duct immediately upstream of cyclone, m
l_E	length of cyclone entrance duct (Figure 6.3), m
M_{gas}	gas mass flow rate, kg/s
M_s	solid mass, kg
M_{solid}	solid mass flow rate, kg/s
n	exponent of tangential velocity distribution or of RRSB distribution, dimensionless
n_A	exponent in RRSB distribution of feed, dimensionless
n_{Ai}	exponent in RRSB distribution of 'inner feed', dimensionless
P	absolute pressure in cyclone, Pa
p_{stat}	static pressure, Pa
r_a	cyclone radius, m
r_c	$r_a - b_e/z$, m
r_e	mean inlet radius, equation 6.30, m
r_i	radius of gas outlet tube, m
r_m	mean cyclone radius $= \sqrt{r_a r_i}$, m
r_o	radius of region of acceleration just upstream of entrance duct (Figure 6.3), m
r_2	radius of the cone at half height, m
$R(d)$	residue, i.e. proportion of particles of diameter d still in suspension, dimensionless
$R_A(d)$	residue of the feed, dimensionless
$R_{Ai}(d)$	residue of the 'inner feed', dimensionless
$R_F(d)$	residue of the emitted dust, dimensionless
Re	Reynolds number, dimensionless
Re_R	Reynolds number defined in Figure 6.11, dimensionless
S	outlet solids concentration, kg/m³
S_o	solids concentration at exit with no outlet tube, kg/m³
s	lamella separator gap (Figure 6.24), m
t	temperature in cyclone, °C

u_a	tangential velocity at radius r_a, m/s
u_e	tangential velocity at radius r_e, m/s
u_{ein}	tangential velocity after the inlet, m/s
u_i	tangential velocity at radius r_i, m/s
u_{io}	tangential gas velocity with only gas flowing through cyclone, m/s
u_m	mean tangential velocity $= \sqrt{u_a u_i}$, m/s
u_2	tangential velocity at radius, $= r_2$, m/s
v	gas velocity, m/s
v_a	velocity in the annular cross-section of the vortex tube, m/s
v_{ax}	mean axial velocity in separation chamber, m/s
v_e	gas velocity in cyclone inlet duct, m/s
v_{ein}	gas velocity just inside cyclone barrel after exchange of momentum with solids, m/s
v_i	mean axial velocity in vortex tube, m/s
V_{gas}	volumetric flow rate, m^3/s
W	drag force, N
w_s	particle settling velocity, m/s
w_{so}	single particle settling velocity, m/s
z	centrifugal acceleration, m/s^2
α	expansion/contraction coefficient, dimensionless
β	relative width of cyclone inlet, dimensionless
Δb	enlargement due to expansion, m
Δc	difference of solid velocity, m/s
Δl_c	distance over which solids and gas exchange momentum after entering cyclone barrel (Figure 6.3), m
Δp	pressure loss, Pa
ΔP_{cycl}	pressure loss across cyclone, Pa
Δp_e	pressure loss in entrance duct, Pa
Δp_i	pressure loss of vortex tube, Pa
Δp_o	pressure drop for gas alone, Pa
Δp_{tot}	total pressure loss of cyclone, Pa
δ_u	boundary layer thickness at wall of cyclone, m
δ_{vr}	boundary layer thickness at root of cyclone, m
η_F	grade efficiency of inner separation, dimensionless
η_{gas}	gas viscosity, Pa·s
η_i	separation efficiency due to inner vortex, dimensionless
η_{lim}	separation efficiency due to wall, dimensionless
η_{tot}	total separation efficiency, dimensionless
λ_o	clean gas friction coefficient, dimensionless
λ_s	solids friction coefficient, dimensionless
μ_e	loading ratio in cyclone entrance, dimensionless
μ_{lim}	limited loading ratio, dimensionless
ν	gas kinematic viscosity, m^2/s
ρ_{gas}	gas density, kg/m^3
ρ_{solid}	solids density, kg/m^3
φ	angle in lamella separator (see Figure 6.24), radians

References

Alexander, R.M. (1949) *Fundamentals of cyclone design and operation*, Proceedings of Australian Institute of Mining and Metallurgy, Inc., **152, 153**, p. 203.
Barth, W. (1956) *Berechnung und Auslegung von Zyklonabscheidern aufgrund neuerer Untersuchungen*, BWK Band 8, Heft 1,S. 1–9.
Barth, W. and Leineweber, L. (1964) *Beurteilung und Auslegung von Zyklonabscheidern*, Staub, Bd. 24, Nr. 2, S. 41–53.
Dietz, P.W. (1981) Collection efficiency of cyclone separators. *AIChE J.* **27**, p. 888.
Greif, V. (1996) Reduzierung des Druckverlustes von Zyklonabscheidern durch Rückgewinnung der Drallenergie und Erweiterung der Grenzbeladungstheorie auf kleine und kleinste Staubbeladungen. Dissertation Universität Stuttgart.
Greif, V. and Muschelknautz, E. (1994) A new impeller probe for measurement of velocities at high temperatures and dust loadings in *Circulating Fluidized Bed Technology IV* (ed. A.A. Avidan), A.I.Ch.E., New York, p. 532–539.
Iinoya, K. (1953) *Memoirs of the Faculty of Engineering*, Nagoya University, **5**(2), September.
Kaliski, H. (1958) Vergleich von Fliehkraftentstaubern, *Freiberger Forschungsheft* A93, S. 78–100.
Lapple, C.E. (1951) *Dust and mist collection*, Air Pollution Abatement Manual, Manufacturing Chemists Association, Washington, DC.
Muschelknautz, E. (1970a) Auslegung von Zyklonabscheidern in der technischen Praxis, *Staub-Reinhaltung der Luft*, Bd. 30, Nr. 5, S. 187–195.
Muschelknautz, E. (1970b) Design of cyclone separators in the engineering practice, *Staub-Reinhaltung der Luft*, Vol 30, No 5, S. 187–195.
Muschelknautz, E. and Brunner, K. (1967) *Untersuchungen an Zyklonen*, CIT 39, Heft 9/10, S. 531–538.
Muschelknautz, E. and Greif, V. (1994) Fundamental and practical aspects of cyclones, in *Circulating Fluidized Bed Technology IV*, (ed. A.A. Avidan), A.I.Ch.E., New York, p. 20–27.
Muschelknautz, E. and Krambrock, W. (1970) Aerodynamische Beiwerte des Zyklonabscheiders aufgrund neuerer und verbesserter, *Messungen CIT 42*, Heft 5, S. 247–255.
Muschelknautz, E. and Krambrock, W. (1993) Pressure drop at pneumatic conveying, in *VDI Heat Atlas*, 6th edn, chap. Lh.
Muschelknautz, E. and Trefz, M. (1993) Pressure drop and separation efficiency in cyclones, *VDI Heat Atlas*, 6th edn, chap. Lj.
Muschelknautz, E., Greif, V. and Trefz, M. (1994) Zyklone zur Abscheidung von Feststoffen aus Gasen, *VDI Wärmeatlas*, 7. Auflage, Kap. Lja.
Rentschler, W. (1991) Abscheidung und Druckverlust des Gaszyklons in Abhängigkeit von der Staubbeladung, *VDI-Fortschrittbericht*, Reihe 3, Nr. 242.
Rietema, K. and Verver, C.G. (1961) *Cyclones in Industry*, American Elsevier Publishing Co., New York.
Rosin, P., Rammler, E. and Intelmann, W. (1932) Principles and limits of cyclone dust removal, *Zeitschrift Verein Deutscher Ingenieure*, **76**, p. 433.
Stairmand, C.J. (1951) The design and performance of cyclone separators. *Trans. Inst. Chem. Eng.*, **29**, p. 356–383.
Ter Linden, A.J. (1949) Investigations into cyclone dust collectors. *Proc. Inst. Mech. Engrs.* **160**, p. 233.
Trefz, M. (1992a) Boundary layer flows in cyclones and their influence on the solid separation, 2. *European Symposium on Separation of Particles from Gases*, Nurnberg/Germany.
Trefz, M. (1992b) Die verschiedenen Abscheidevorgänge im höher und hoch beladenen Zyklon unter besonderer Berücksichtigung der Sekundärströmung, VDI-Fortschrittbericht, Reihe 3, Nr. 295.
Trefz, M. and Muschelknautz, E. (1993) Extended cyclone theory for gas flow with high solids concentrations, *Chem. Eng. Technol.* **16**, p. 153.
van Tongeren, H. (1936) *Mechanical Engineering*, **58**, February, p. 127.
Zenz, F. A. (1975) Cyclone separators, Chapter 11, in *Manual on Disposal of Refinery Wastes*; Volume on Atmospheric Emissions, API Publications 931, Am. Pet. Inst. Refining., Washington, DC.

7 Standpipes and return systems
TED M. KNOWLTON

7.1 Introduction

Although chemistry is the initial driving force for the development of chemical processes, in many instances the key to successful process operation is how well the solids transport systems have been designed. This is especially true in circulating fluidized bed (CFB) processes, because these processes are dependent upon rapid and reliable circulation of solids.

Every CFB process contains a solids return system, but they can differ significantly in design. The two most common CFB processes, circulating fluidized bed combustion (CFBC) and fluid catalytic cracking (FCC), have completely different return systems. In a CFBC, the return systems are simple, consisting of only a cyclone, standpipe, and non-mechanical device to return the solids back to the bed (Figure 7.1). In FCC units, the return systems are more complex (Figure 7.2), because of the presence of a catalyst regenerator in the loop with the FCC riser (the circulating fluidized bed).

In all CFB recirculation systems, at least one standpipe is required to allow the solids to flow from the relatively low pressure region near the outlet of the circulating bed to a higher pressure region at the bottom. Explaining how standpipes and the other devices in a return system operate and interrelate is the purpose of this chapter.

7.2 Standpipes

A standpipe is essentially a length of pipe through which solids flow. Solids can flow through a standpipe in either dilute or dense phase flow. Standpipes can be vertical, angled, or a mixture of angled and vertical pipes.

The standpipe was invented by a research team from Jersey Standard in the 1940s (Campbell *et al.*, 1948) working on developing an FCC unit to produce gasoline for the war effort. The purpose of a standpipe is to transfer solids from a region of lower pressure to a region of higher pressure. This is schematically shown in Figure 7.3, where solids are being transferred from a fluidized bed at pressure P_1 to another fluidized bed operating at P_2, which is higher than P_1.

Solids can be transferred by gravity against an adverse pressure gradient if gas flows upward *relative* to the downward flowing solids. This relative

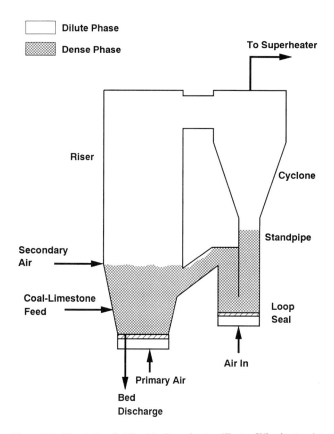

Figure 7.1 Circulating fluidized bed combustor (Foster Wheeler type).

gas–solids flow is then able to generate the required 'sealing' pressure drop. The direction of the actual gas flow relative to the standpipe wall can be either up or down and still have the relative gas-solids velocity, v_r, directed upwards. This is sometimes hard to understand, but it can be explained with the aid of Figure 7.4, and the definition of relative velocity. The relative gas–solids velocity, v_r, is defined as:

$$v_r = |v_s - v_g|$$

where v_s is the velocity of the solids, and $v_g = U/\epsilon$ is the interstitial gas velocity.

It is generally easier to visualize what is occurring in a solids transfer system by mentally traveling along with the solids. Therefore, the positive reference direction for determining v_r in this chapter will be the direction that the solids are flowing. For standpipe flow, this direction is downward.

In Figure 7.4 solids are being transferred downward in a standpipe from pressure P_1 to a higher pressure P_2. Solids velocities in Figure 7.4 are denoted

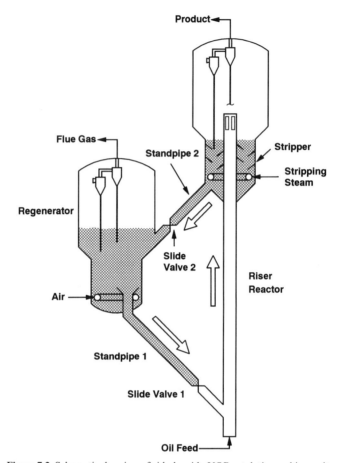

Figure 7.2 Schematic drawing of side-by-side UOP catalytic cracking unit.

by the length of the thick-lined arrows, gas velocities by the length of the dashed arrows, and the relative gas–solids velocity by the length of the thin-lined arrows.

In Case 1, solids are flowing downward, and gas is flowing upward relative to the standpipe wall. The relative velocity is directed upward, and is equal to the sum of the solids velocity and the gas velocity, i.e.

$$v_r = v_s - (-v_g) = v_s + v_g$$

For Case 2 solids are flowing down the standpipe. Gas is also flowing down the standpipe relative to the standpipe wall, but at a velocity less than that of the solids. For this case, the relative velocity is also directed upward, and is equal to the difference between the solids velocity and the gas velocity, i.e.

$$v_r = v_s - v_g$$

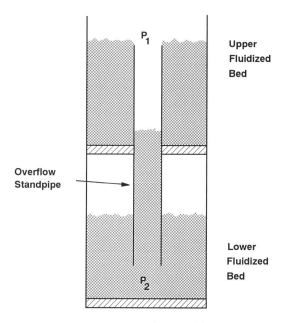

Figure 7.3 Overflow standpipe.

In both cases, if one were riding down the standpipe with the solids, the gas would appear to be moving upward.

The gas flowing upward relative to the solids generates a frictional pressure drop. The relationship between the pressure-drop-per-unit-length ($\Delta P/L$) and the relative velocity for a particular material is determined by the fluidization curve for that material. This curve is usually generated in a fluidization column where the solids are not flowing. However, the relationship also applies for solids flowing in a standpipe.

Nearly all circulating fluidized bed systems use either Geldart Group A or Geldart Group B solids. The fluidization curve for Geldart Group B solids differs from that for Group A solids. For both types of solids, as the relative gas velocity through the bed increases from zero, the $\Delta P/L$ through the bed increases linearly with v_r. This is the packed-bed region. At some v_r, the ΔP generated by the gas flowing through the solids is equal to the weight of the solids per unit area and the solids become fluidized. The relative velocity at this point is termed the interstitial minimum fluidization velocity, v_{mf}, or U/ϵ_{mf}. The $\Delta P/L$ at v_{mf} is designated as $(\Delta P/L)_{mf}$ and when divided by g is often referred to as the fluidized bed 'density' at minimum fluidization, because $\Delta P/(Lg)$ has the units of density.

Increases in v_r above v_{mf} do not lead to further increases in $\Delta P/L$. For Geldart Group B materials, nearly all of any gas flow in excess of that required at v_{mf} goes into the formation of bubbles. Therefore, as v_r increases

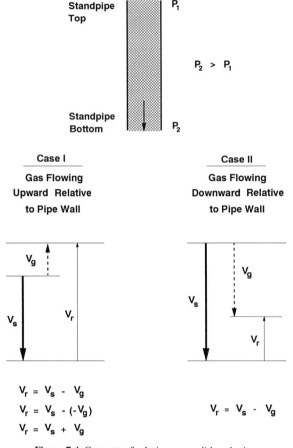

Figure 7.4 Concept of relative gas–solids velocity.

beyond v_{mf}, $\Delta P/L$ stays relatively constant, and then begins to decrease as the bubble volume in the bed increases.

For Geldart Group A materials, as v_r is increased above v_{mf} the solids expand without bubble generation for a certain velocity range. Because of this bubbleless expansion, $\Delta P/L$ decreases over this velocity range. The velocity where bubbles begin to form in Group A materials is called the minimum bubbling velocity, v_{mb}. Typical fluidization curves for Geldart Group A and Geldart Group B materials are shown in Figure 7.5.

Standpipes generally operate in three basic flow regimes – packed-bed flow, fluidized-bed flow, and a dilute-phase flow called streaming flow. The differences between these types of flow are discussed below:

1. *Packed-bed flow* In packed-bed flow v_r is less than v_{mf}, and the voidage in the standpipe is more or less constant. As v_r is increased,

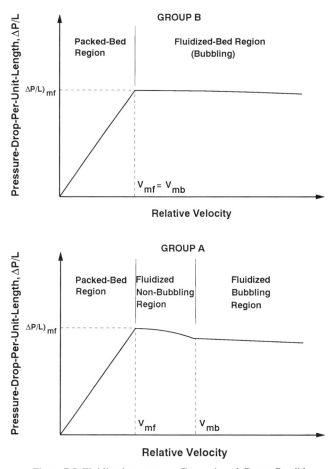

Figure 7.5 Fluidization curves – Group A and Group B solids.

$\Delta P/L$ increases in packed-bed flow. When a standpipe is operating in the moving packed-bed flow regime, a flow condition is sometimes reached which causes the standpipe to vibrate, and a loud 'chattering' of the solids occurs. This flow regime (often termed stick-slip flow) should be avoided, but no method presently exists to predict when it will occur.

2. *Fluidized-bed flow* In fluidized-bed flow, v_r is equal to or greater than v_{mf}. The voidage in the standpipe can (and generally does) change along the length of the standpipe, and $\Delta P/L$ does not change with increasing v_r. There are two kinds of fluidized-bed flow:

(a) bubbling fluidized-bed flow,
(b) non-bubbling, fluidized-bed flow.

When a Group B solid is fluidized, it always operates in the bubbling fluidized-bed mode since bubbles are formed at all relative velocities above v_{mf}. However, for Group A solids, there is an operating window corresponding to a relative velocity between v_{mf} and v_{mb}, where the solids are fluidized but no bubbles are formed in the standpipe. A standpipe operating with Group A solids and with a relative velocity above v_{mb} operates in the bubbling fluidized-bed mode.

Bubbles, especially large bubbles, are undesirable in a standpipe. If a standpipe is operating in the bubbling fluidized bed mode such that the solids velocity, v_s, is less than the bubble rise velocity, v_b, then bubbles will rise and grow by coalescence. The bubbles rising against the down-flowing solids hinder and limit the solids flow rate (Eleftheriades and Judd, 1978; Knowlton and Hirsan, 1978). The larger the bubbles, the greater the hindrance.

When the solids velocity in the standpipe is greater than the bubble-rise velocity, the bubbles travel down the standpipe relative to the standpipe wall. It is also possible for bubbles to coalesce and hinder solids flow when they are traveling downward. In this case, the small bubbles are carried downward faster than the larger bubbles. When they catch up to a larger bubble they coalesce, which results in even larger bubbles.

Bubbles also reduce the $\Delta P/Lg$ or 'density' of the solids in the standpipe. Thus, a standpipe operating in the bubbling regime will require a longer length to seal the same differential pressure than a standpipe in which the same solids are slightly above minimum fluidizing conditions.

Therefore, for optimum fluidized-bed standpipe operation:

(a) For Group B solids, v_r should be maintained just slightly above v_{mf}.
(b) For Group A solids, v_r should be maintained in a range below v_{mb} to just slightly above v_{mb}.

The relative velocity range, where it is best to operate standpipes with both Group A and Group B solids, is shown schematically in Figure 7.6. In both cases, operation in these areas will either reduce the formation of bubbles, or allow standpipe operation with only small bubbles.

3. *Streaming flow* Underflow standpipes (especially cyclone diplegs) sometimes operate in a dilute-phase streaming flow characterized by high voidages. A substantial amount of gas can be carried down the standpipe when operating in this mode (Geldart and Broodryk, 1991).

There are two basic types of standpipe configurations – the overflow standpipe (Figure 7.7a) and the underflow standpipe (Figure 7.7b). The overflow standpipe is so named because the solids 'overflow' from the top of the fluidized bed into the standpipe, and there is no bed of solids above the standpipe. In the underflow standpipe, the solids are introduced into

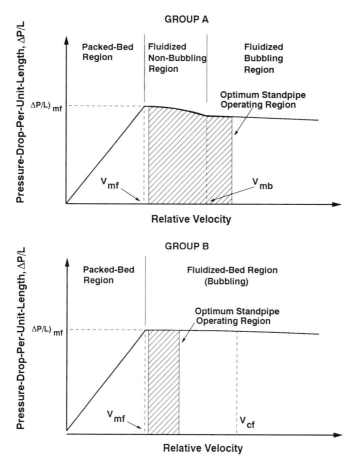

Figure 7.6 Optimum standpipe operating regions for Group A and B solids.

the standpipe from the underside, or bottom, of the bed or hopper, and a bed of solids is present above the standpipe. With this definition, a cyclone dipleg is classified as an overflow standpipe because there is no bed of solids above the entrance to the standpipe.

Each type of standpipe can be constrained at the top or the bottom. However, top-constrained standpipes are relatively rare, and nearly all standpipes used in circulating fluidized bed return systems are constrained at the bottom. The standpipes discussed in this chapter are all bottom-constrained standpipes.

In any gas–solids flow system, a pressure-drop loop can be defined such that the sum of the pressure drop components around the loop is zero. In many (but not all) pressure-drop loops, the standpipe is the *dependent* part of the loop. This means that the pressure drop across the standpipe

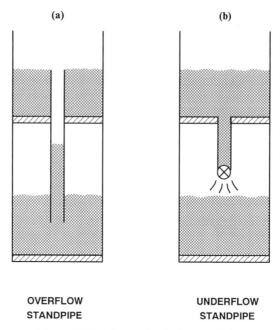

Figure 7.7 Overflow and underflow standpipes.

will automatically adjust to balance the pressure drop produced by the other *independent* components (which do not automatically adjust their pressure drops) in the loop. How the standpipe pressure drop adjustment is made is different for overflow and underflow standpipes.

Consider the overflow standpipe system shown in Figure 7.8a. Solids are being transferred at a low rate from the upper fluidized bed to the lower fluidized bed against the pressure differential $P_2 - P_1$. This pressure drop consists of the pressure drop in the lower fluidized bed from the standpipe exit to the top of the bed (ΔP_{lb}), the pressure drop across the distributor supporting the upper fluidized bed (ΔP_d), and the pressure drop across the upper fluidized bed (ΔP_{ub}), i.e.

$$P_2 - P_1 = \Delta P_{lb} + \Delta P_b + \Delta P_{ub}$$

This pressure drop must be balanced by the pressure drop generated in the overflow standpipe. If the standpipe is operating in the fluidized-bed mode at a $\Delta P/L$ equal to $(\Delta P/L)_{mf}$, the solids height in the standpipe, H_{sp}, will adjust so that the pressure buildup generated in the standpipe, ΔP_{sp}, will equal the product of $(\Delta P/L)_{mf}$ and H_{sp}, i.e.

$$\Delta P_{sp} = P_2 - P_1 = \Delta P_{lb} + \Delta P_d + \Delta P_{ub} = (\Delta P/L)_{mf}(H_{sp})$$

If the gas flow rate through the two beds is increased, ΔP_d increases while the pressure drops across the two fluidized beds essentially remain constant.

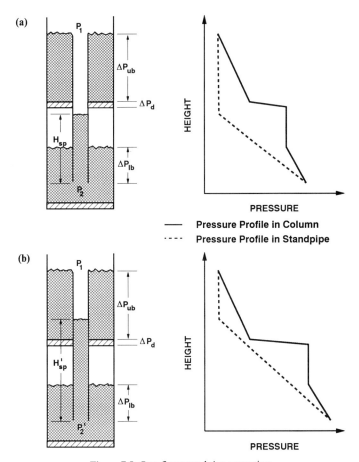

Figure 7.8 Overflow standpipe operation.

Therefore, $P_2 - P_1$ increases to $P'_2 - P_1$. The pressure drop across the overflow standpipe will also increase to $P'_2 - P_1$. The standpipe pressure drop increases because the height of fluidized solids in the standpipe increases from H_{sp} to H'_{sp}, i.e.

$$\Delta P_{sp} = P'_2 - P_1 = (\Delta P/L)_{mf}(H'_{sp})$$

This is shown schematically in Figure 7.8b, and the change is also reflected in the pressure diagram. If the increase in pressure drop is such that H_{sp} must increase to a value greater than the standpipe height available to seal the pressure differential, the standpipe will not operate.

As indicated above, the most common perception of an overflow standpipe is that it consists of a fluidized dense phase at the bottom (the height of which is proportional to the pressure drop across the dipleg), and a dilute phase in

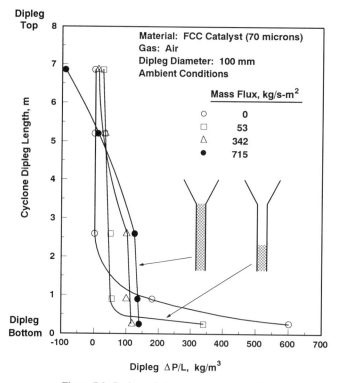

Figure 7.9 Cyclone dipleg pressure drop profiles.

the upper part (which does not contribute significantly to the sealing pressure drop in the standpipe). However, recent testing has shown that this picture is not necessarily correct for overflow standpipes operating at appreciable solids mass fluxes. Geldart and Broodryk (1991) reported that at high mass fluxes in an overflow standpipe, there is little or no difference between the densities in the top and bottom sections of the standpipe. This is illustrated in Figure 7.9, which shows recent data obtained at PSRI (1995). As seen in this figure, the typical dense phase at the bottom of the overflow standpipe (cyclone dipleg) only occurs at very low solids mass fluxes. At medium to high solids mass fluxes, the standpipe density was found to be relatively evenly distributed throughout the entire length of the standpipe. Although the exact mechanism of why and when this occurs is still being debated, it is known that substantial amounts of gas can be dragged down an overflow standpipe by the flowing solids. The amount of gas being transferred may be so great (as much as 1/3 of the gas entering the cyclone can be transferred down the dipleg under certain conditions; PSRI (1995)) that it prevents a dense phase from forming at the bottom, thus resulting in the types of profiles observed in Figure 7.9. In many cases, it is not desirable to have so much gas transferred down the standpipe with the solids. By increasing the pressure drop across the standpipe (increasing $\Delta P/L$ in the standpipe) or reducing solids mass flux through the standpipe, the amount of gas transferred down the standpipe can be substantially reduced. Practically, this means that the gas flow down the standpipe can be decreased by immersing the standpipe further in the fluidized bed (increasing $\Delta P/L$), increasing the diameter of the standpipe (reducing the mass flux), or reducing the length of the standpipe, if possible (increasing $\Delta P/L$).

Wirth (1995) conducted a study of a downcomer operating between a cyclone and a loop seal at the bottom of a downcomer. He also found that if the solids mass flux in the downcomer increased above a certain value, the downcomer lost its dense-phase seal at the bottom. This increased mass flux was accompanied by increasing amounts of gas flowing through the downcomer.

If the solids mass flux was increased further beyond a certain threshold value, an even greater flow of gas was observed in the downcomer with the result that even the upward-flowing part of the loop seal became dilute. Wirth found that this situation could be controlled by decreasing the solids flux through the downcomer. This could be accomplished in two ways: (1) by decreasing the rate of solids flowing around the unit, or (2) more practically, increasing the diameter of the downcomer. Increasing the diameter of the downcomer decreases the solids mass flux in the downcomer and, therefore, the amount of gas flowing through the downcomer.

This situation is also exacerbated by long standpipes. Therefore, to prevent this type of situation from occurring and causing poor loop seal operation in combustors, it is recommended that a large-diameter downcomer be used.

In Figure 7.10a, solids are transferred through an underflow standpipe (operating in the packed-bed mode) from the upper fluidized bed to the freeboard of the lower fluidized bed against the differential pressure $P_2 - P_1$. The differential pressure $P_2 - P_1$ consists of the pressure drop across the gas distributor of the upper fluidized bed ΔP_d. However, there is also a pressure drop across the solids flow control valve ΔP_v. Therefore, the standpipe pressure drop ΔP_{sp}, must equal the sum of ΔP_d and ΔP_v, i.e.

$$\Delta P_{sp} = (\Delta P/L)H_{sp} = \Delta P_d + \Delta P_v = P_2 - P_1 + \Delta P_v$$

Thus, for this packed-bed underflow standpipe case, the standpipe must generate a pressure drop greater than $P_2 - P_1$. This is shown as Case I in the pressure diagram of Figure 7.10b.

If the gas flow rate through the column is increased, ΔP_d will increase. If ΔP_v remains constant, then ΔP_{sp} must also increase to balance the pressure-drop loop. This is shown as Case II in the pressure diagram of Figure 7.10b. Unlike the overflow standpipe case, the solids level in the standpipe cannot rise to increase the pressure drop in the standpipe. However, the $\Delta P/L$ in

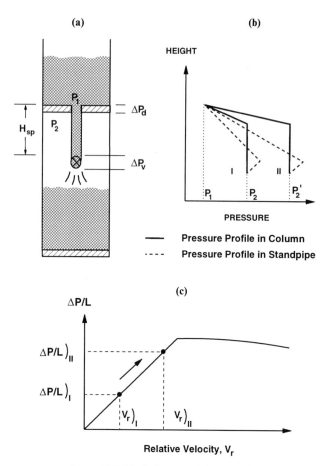

Figure 7.10 Underflow standpipe operation.

the standpipe must increase in order to balance the pressure drop around the loop. This occurs in a packed-bed standpipe because of an increase in v_r in the standpipe. This can be visualized with the aid of Figure 7.10c.

For Case I, the pressure drop in the loop was satisfied by having the standpipe operate at point I on the $\Delta P/L$ versus v_r curve, as shown in Figure 7.10c. When the pressure drop across the distributor increased, the v_r in the standpipe adjusted to generate a higher $\Delta P/L$, $(\Delta P/L)_{II}$, to balance the higher pressure drop.

If the pressure drop across the distributor is increased such that the product of $(\Delta P/L)_{mf}$ (the maximum $\Delta P/L$ possible in the standpipe) and the standpipe length, H_{sp}, is less than the sum of ΔP_v and ΔP_d, then the underflow standpipe will not seal.

As stated above, the standpipe is not always the component that adjusts to balance the change in pressure drop in a loop (i.e. it is not always the *dependent* part of the pressure-drop loop). In fluidized catalytic crackers, fluidized underflow standpipes are used that do not adjust for changes in pressure drop. This type of standpipe is one of the most widely used standpipes in industry. How does this type of fluidized underflow standpipe operate? Unlike the underflow standpipe in the previous example, the standpipe pressure drop will not adjust to balance pressure drop changes because the standpipe is operating in the fluidized bed mode. In the fluidized mode changes in the relative velocity will not cause changes in the pressure drop in the standpipe as in an underflow non-fluidized standpipe. Also, because it is an underflow standpipe and operating full of solids, changes in the bed height to balance the pressure drop are not possible. Therefore, this type of standpipe is designed so that its length is long enough to generate more pressure, or 'head', than required. The excess pressure generated by the standpipe is then 'burned up' across the slide valve in order to balance pressure drop changes in the other loop components.

With the fluidized underflow standpipe, aeration gas is added to the standpipe to maintain the solids in a fluidized state as they flow down the standpipe. As the solids flow down the fluidized underflow standpipe from a low pressure to a higher pressure, the gas in the standpipe is compressed, which causes the solids to move closer together. When the standpipe is operating at low pressures, the percentage change in gas density from the top of the standpipe to the bottom can be significant. If aeration is not added to the standpipe, the solids can defluidize near the bottom of the standpipe (Figure 7.11a). Defluidization of solids in the standpipe results in less pressure buildup in the standpipe and a reduction in the solids flow rate around the loop.

To maintain the solids in a fluidized underflow standpipe in a fluidized state, aeration gas is added to the standpipe. Generally, it is best to add the aeration uniformly along the standpipe instead of at one location (i.e. the bottom of the standpipe) because often a large bubble forms in the standpipe where all of the aeration gas is added. This is especially true if the solids flowing in the standpipe belong to Geldart Group A. Adding the correct amount of gas uniformly (every 2 to 3 meters) in a commercial fluidized underflow standpipe can prevent defluidization at the bottom of the standpipe (Figure 7.11b).

The amount of aeration required to maintain solids in a fluidized state throughout the standpipe was presented by Karri and Knowlton (1993) to be:

$$Q = 1000 \left[\frac{P_b}{P_t} \left(\frac{1}{\rho_{mf}} - \frac{1}{\rho_{sk}} \right) - \left(\frac{1}{\rho_t} - \frac{1}{\rho_{sk}} \right) \right]$$

In a commercial fluidized underflow standpipe, the amount of aeration theoretically required is added in equal increments via aeration taps located approximately 2 to 3 meters apart. Care should be taken not to overaerate

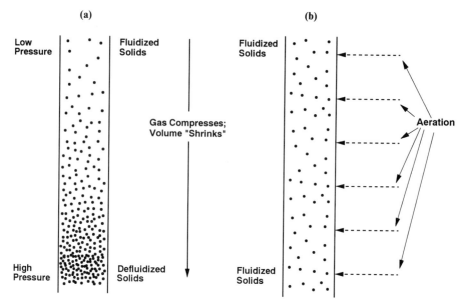

Figure 7.11 Adding aeration prevents defluidization.

the standpipe. If this occurs, large bubbles are generated in the standpipe, which hinder solids flowing down the standpipe. Thus, standpipes can be overaerated as well as underaerated.

7.3 Standpipes in CFB systems

The two most common processes using CFBs are circulating fluidized bed combustion and fluid catalytic cracking. There are substantial differences between the two processes in regard to the recirculation of solids.

One of the primary differences between CFBC and FCC units is the type of solids that are recirculated around the two types of units. The CFBC recirculates sand and ash with an average particle size of approximately 175 μm, while FCC units recirculate cracking catalyst with an average particle size of about 65 μm. The material recirculated in a CFBC is a Geldart Group B material, while the catalyst recirculated around the FCC unit is a Geldart Group A material. The mass fluxes in a CFBC range from about 20 to 80 kg/s m^2, while in an FCC unit the range is approximately 400 to 1000 kg/s m^2. As shown in Figures 7.1 and 7.2, the circulation system of an FCC unit is much more complex than that of the CFBC. The FCC unit utilizes two standpipes when recirculating solids, while there is only a single standpipe in the CFBC.

7.4 Standpipes in CFBCs

As indicated above, recirculation loops in CFBC systems are simpler than in FCC systems. CFBC systems have only one standpipe in their recirculation loop. In most combustors, the standpipe is an overflow standpipe as shown in Figure 7.1. However, other types of return systems are also used. Several of the different types of recirculation loops which are used by various manufacturers are shown in Figure 7.12. The different types of loops have evolved in response to different methods of controlling the temperature in the CFBC by removing heat from the exothermic combustion reaction. The two types of recirculation loops are categorized by whether the solids are automatically transferred back to the CFBC, or whether the solids flow into the CFBC is controlled by a valve.

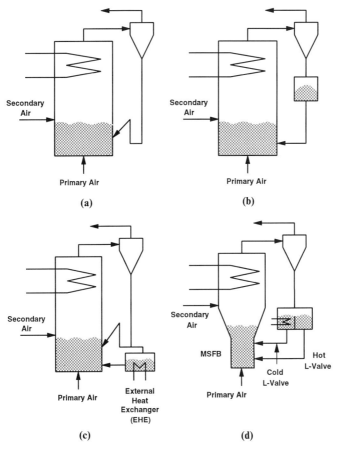

Figure 7.12 Four different industrial CFB configurations.

7.4.1 Automatic solids recirculation systems in CFBCs

The simplest type of return system utilized in CFBCs is shown in Figure 7.12a. In this type of recirculation system, the ash/limestone/char mixture from the CFBC is collected by a cyclone and flows into an **overflow** standpipe. The solids pass through the standpipe, into a loop seal (also sometimes called a J-valve, fluoseal, or siphon seal), and then back into the CFBC. In this type of design, solids that are collected by the cyclone are automatically transferred back into the bed via the loop seal. The loop seal is not a valve (it does not control the solids flow rate), but a device that transfers the solids back into the bed at the same rate that they enter the standpipe, i.e. automatically. This type of system, or a slightly modified version of it, is the most common type of recirculation system used in combustors. For example, it is the type generally used by Foster-Wheeler in CFB combustors. With this type of system, the heat produced by the exothermic combustion reaction is removed via heat exchange 'membranes' in the walls of the upper part of the riser.

A modified version of this type of recirculation system (developed by Lurgi) employs an external heat exchanger (EHE) in the recirculation system to aid in heat removal (Figure 7.12c). The EHE is an auxiliary fluidized bed containing cooling coils. Some of the solids from the standpipe are routed to the EHE, cooled, and then returned to the bed. The reaction temperature is then moderated by the cooler solids.

In this type of return system, the high pressure point in the system is at the bottom of the standpipe. The pressure drops across the part of the loop seal where solids move upward (ΔP_{ls}), across the CFB (ΔP_{CFB}), and across the cyclone (ΔP_{cy}) are balanced by the pressure drop across the standpipe (ΔP_{sp}):

$$\Delta P_{ls} + \Delta P_{CFB} + \Delta P_{cy} = \Delta P_{sp}$$

As shown above, when the pressure drop in the independent part of the loop increases or decreases, the pressure drop in the overflow standpipe adjusts to these changes by a raising or lowering of the level of the fluidized solids in the standpipe, respectively.

7.4.2 Controlled solids recirculation systems in CFBCs

A different type of recirculation system in which the solids flow rate back to the CFBC is not automatic but is controlled in some fashion, is shown in Figure 7.12b. In this system, the solids from the cyclone (sometimes an inertial separator is used in place of a cyclone to minimize pressure drop) pass directly into a surge vessel and then through an **underflow** standpipe and non-mechanical L-valve before being returned to the CFBC. This type of recirculation loop differs from those of Figures 7.11a and 7.11c in that:

(a) the standpipe is an underflow standpipe, not an overflow standpipe,
(b) the standpipe operates in packed-bed flow, not fluidized-bed flow, and
(c) the solids in the recirculation loop are controlled by the L-valve, and are not automatically returned to the CFB as with a loop seal.

With this type of recirculation loop, the temperature in the CFBC is controlled by the amount of solids returned to the fluidized bed via the L-valve. For this type of system to operate satisfactorily, a surge vessel must be placed in the system above the underflow standpipe to absorb imbalances or surges in the solids flow rate. A typical system of this type was developed by Studsvik (Kobro and Brereton, 1985).

For this system, the high-pressure point in the recirculation loop is at the aeration point of the L-valve, and

$$\Delta P_{L\text{-}valve} + \Delta P_{CFB} + \Delta P_{cy} = \Delta P_{sp} + \Delta P_{surge}$$

The pressure drop across the surge vessel is so low that essentially the pressure balance becomes:

$$\Delta P_{L\text{-}valve} + \Delta P_{CFB} + \Delta P_{cy} = \Delta P_{sp}$$

In this recirculation system, changes in return leg pressure drop occur because of changes in the relative velocity in the underflow standpipe. If the $\Delta P/L$ in the standpipe ever reaches that at minimum fluidization ($\Delta P/L_{mf}$), the standpipe cannot 'absorb' any more pressure drop, and the solids flow limit of the system is reached at that point.

A third type of recirculation system (developed by Batelle) is shown in Figure 7.12d. In this system, the solids collected by the cyclone flow into a partitioned fluidized bed. One side of the partitioned bed is cooled, while the other side of the bed remains hot. An underflow standpipe and L-valve combination are used to control the flow rate of both the hot and cold solids into the CFBC. This type of system is similar to that of Figure 7.12b in that the standpipe used in the recirculation loop is an underflow standpipe. However, the pressure drop balance is somewhat different because the solids in the 'surge' vessel in this system are fluidized. In the recirculation loop of Figure 7.12c they are not fluidized. Therefore, the surge vessel pressure drop in this system is not negligible and:

$$\Delta P_{L\text{-}valve} + \Delta P_{CFB} + \Delta P_{cy} = \Delta P_{sp} + \Delta P_{surge}$$

Care must be taken with return systems employing a fluidized surge not to make the fluidized bed level in the surge vessel too high. If the fluidized solids in the surge vessel develop more pressure than is dissipated by the L-valve, CFB, and cyclone, then the standpipe pressure drop will be negative. This means that gas will pass down the standpipe faster than the solids. If the gas passing down the standpipe is enough to cause the solids to flow around the restricting bend, the L-valve will not shut off. In this case it is not the L-valve's 'fault' that the system will not shut off, but bad design that has caused the problem.

7.5 Standpipes in FCC units

The recirculating loops of FCC units are significantly different than those of CFBCs. FCC catalyst is a Geldart Group A material, while the recirculating material in a CFBC is a Geldart Group B material. The primary standpipes in FCC units are underflow fluidized-bed standpipes, not underflow packed-bed standpipes or overflow fluidized-bed standpipes. In addition, standpipes in FCC units can be either completely vertical, completely angled, or a combination of vertical and angled sections. Nearly all FCC units incorporate two standpipes in their loop systems. Solids flow around the system is controlled by a slide valve (or a plug valve in some cases) in each standpipe.

A typical FCC riser cracker loop system developed by UOP (called a side-by-side unit) is shown in Figure 7.2. A pressure balance around this unit gives:

$$\Delta P_{regen} + \Delta P_{sp1} - \Delta P_{sv1} - \Delta P_{riser} + \Delta P_{stripper} + \Delta P_{sp2} - \Delta P_{sv2} = 0$$

In this unit, hot catalyst is introduced into a dilute-phase riser where it is contacted with crude oil. The hot solids vaporize the oil and catalytically crack it into lower-molecular weight hydrocarbons in the riser. The catalyst reactivity is significantly reduced by carbon deposition during this step. Therefore, the catalyst is transferred to a regenerator in which the carbon is burned off the catalyst. This heats up the catalyst and restores its reactivity. However, before the catalyst is returned to the regenerator via an angled standpipe/slide valve combination, hydrocarbon vapors are removed from the catalyst in a steam stripper. From the regenerator, the catalyst is transferred down another angled standpipe/slide valve combination and is injected into the bottom of the riser.

In most FCC units, the solids flow rate around the system is controlled by a slide valve. FCC units designed by the M.W. Kellogg Company use another type of valve called a cone, or plug valve, to control the solids flow rate (Wrench et al., 1985).

Solids flow in standpipes in FCC units such as those shown in Figure 7.2 are generally controlled in the following manner. In Standpipe 1, Slide Valve 1 controls the temperature at the outlet of the riser by varying the flow of regenerated solids to the riser. The slide valve in standpipe 2 controls the fluidized solids level in the stripper (King, 1992). Generally, slide valves in FCC standpipes are operated with a pressure drop between about 15 kPa (the minimum required for good control of the catalyst) and 100 kPa. Higher pressure drops result in excessive valve wear. Good valve design will result in the valve operating between 25% and 75% open, with the valve port area about 25% to 50% of the standpipe open area. A typical

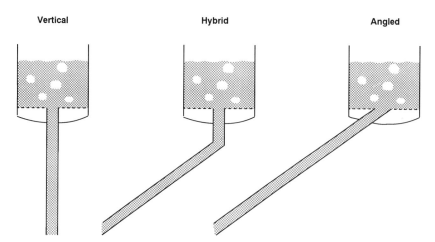

Figure 7.13 Vertical, angled and hybrid fluidized underflow standpipes.

design is for the valve to operate in the middle of the ranges shown above (i.e. at about 50 kPa pressure drop and 50% open) as reported by King (1992).

Underflow fluidized standpipes in FCC units are operated in a vertical configuration, completely angled configuration, or a hybrid configuration in which both vertical and angled sections are present (Figure 7.13). Angling a standpipe is a very convenient way to transfer solids between two points which are separated horizontally as well as vertically. However, it has been found (Karri and Knowlton, 1993; Yaslik, 1993) that angled underflow fluidized standpipes do not perform as well as vertical standpipes.

Sauer *et al.* (1984) and Karri and Knowlton (1993) studied hybrid angled standpipe operation using transparent standpipes to allow visual observation of the flow. Both found that the gas and solids separated in the standpipe, with the gas bubbles flowing up along the upper portion of the standpipe while the solids flowed down along the bottom portion of the standpipe (Figure 7.14). The pressure buildup in the hybrid standpipe was lower than that in the vertical standpipe, and Karri and Knowlton (1993) reported that the maximum solids mass flux possible in a hybrid angled underflow fluidized standpipe was less than that attainable in a vertical underflow fluidized standpipe (Figure 7.15). The principal reason for this is that the rising bubbles in the angled section of the standpipe become relatively large at a low solids flow rate. At a certain solids mass flux, the bubbles become large enough to bridge across the vertical section at the top of the standpipe, hindering the solids flow. When this occurs, the maximum solids flow rate in the hybrid angled standpipe has been achieved.

Karri *et al.* (1995) showed that the solids flow rate through a hybrid angled standpipe can be increased if a bypass line (Figure 7.16) is added between the top of the angled section of the standpipe and the freeboard of the bed above

234 CIRCULATING FLUIDIZED BEDS

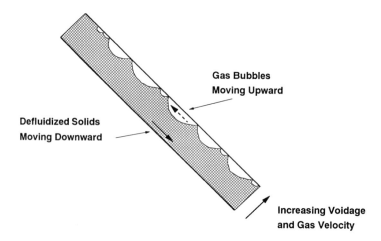

Figure 7.14 Gas and solids flow in an angled standpipe.

it. The effect of adding the bypass line at the top of the standpipe can be seen in Figure 7.15. The bypass line allows the bubble gas from the angled section to bypass the vertical section of the pipe so that large bubbles are not formed there. Thus, the solids flow rate can be increased. Karri *et al.* (1995) reported that if the bypass was used, the solids flow rate could be increased to such a value that the solids velocity in the hybrid standpipe became greater than the

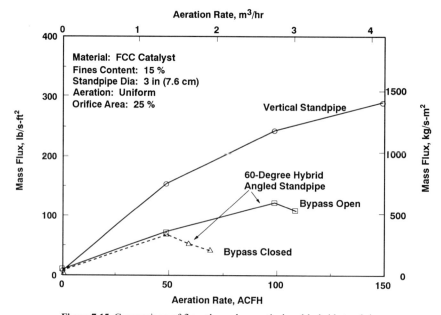

Figure 7.15 Comparison of flow through a vertical and hybrid standpipe.

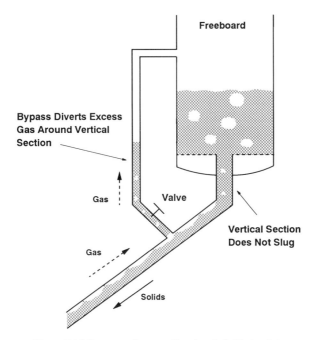

Figure 7.16 Bypass to increase flow in a hybrid standpipe.

bubble rise velocity, and the bubbles were carried down the standpipe with the solids. When the bubbles were being carried down the standpipe by the solids, the bypass line could then be closed and the standpipe would operate without slugging in the vertical section.

Even though vertical standpipes can transfer solids more efficiently than hybrid angled standpipes, true angled standpipes (those containing no vertical section) are operated satisfactorily in large FCC units with Geldart Group A catalyst. However, these standpipes are relatively short, and are designed so that the mass flux through them is not too high so that they can be operated satisfactorily. Yaslik (1993) found that a long angled standpipe had a limited solids circulation rate relative to vertical standpipes. Thus, when operating a hybrid angled standpipe or a true angled standpipe it is essential to: (1) utilize a bypass or vent line to dissipate large bubbles at the top of the line, and (2) keep the line as short as possible so that the large slugs will not have as great a length in which to form.

7.6 Laboratory CFB systems

Laboratory CFB systems differ from industrial systems in several respects, depending on whether CFBC or FCC unit systems are being simulated.

Nearly all laboratory units recycle a fixed inventory of solids during operation, and do not have simultaneous feed and discharge. Also, most small and large laboratory systems utilize only one standpipe in a single recirculation loop for simplicity. This simulates the single-standpipe-loop CFBC well, but not the double-standpipe-loop FCC units. The various types of loop systems can be classified as (1) non-valved (NV), (2) valved with no solids reservoir (VNSR), and (3) valved with solids reservoir (VWSR).

The simplest type of laboratory recirculation system is the batch NV system, which does not have a valve in the return leg to control the solids flow rate. This type of system uses a fluidized standpipe in conjunction with an automatic solids return device (generally a loop seal or automatic L-valve) to seal against the pressure at the bottom of the CFB (Figure 7.17). Because the solids circulation flow rate in the batch NV system cannot be controlled easily, this type of system is generally not practical for the controlled laboratory study of CFBs or return systems. The NV type of return system is commonly used in industry because of its simplicity (Goidich *et al.*, 1991; Lundqvist *et al.* 1991; Mii *et al.*, 1991; Plass *et al.*, 1991). However, it is used in conjunction with continuous feed and discharge of solids, and not as a batch NV system.

The solids flow rate (and the axial density profile in the riser) in a batch NV system depends upon the gas velocity in the riser and/or the height of solids in

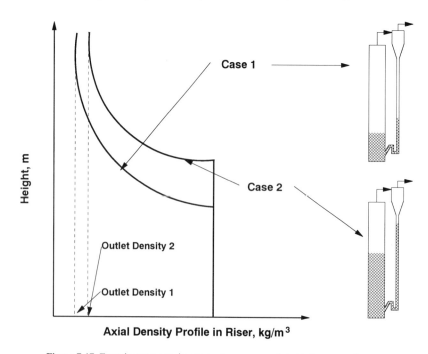

Figure 7.17 Entrainment rate increases to accommodate higher mass fluxes.

the standpipe (inventory). At high gas velocities in the riser and low fluidized bed heights in the standpipe, the limit of the solids mass flux in the riser is set by the height of solids in the standpipe. This is because the low height of solids in the standpipe cannot develop enough head to 'push' the solids into the riser at a rate high enough to equal the maximum solids-carrying capability of the gas flowing through the riser. Thus, the solids flow rate is less than that at choking for that particular gas velocity in the riser.

At lower gas velocities in the riser and/or higher levels of fluidized solids in the standpipe, the limiting mass flux in the riser (as well as the axial solids density in the riser) depends primarily on the solids inventory in the riser. For this situation, the height of solids in the standpipe produces a bottom standpipe pressure sufficient to be able to 'push' solids into the riser at a rate equal to or greater than the choking solids flow rate at that gas velocity. When this condition is achieved, the axial density profile in the system depends upon the inventory in the standpipe.

Consider Case 1 in Figure 7.17 in which the height of solids in the standpipe develops a pressure at the bottom of the standpipe great enough so that the maximum solids-carrying capacity of the gas flowing through the riser is exceeded. Even when the solids-carrying capacity of the riser gas flow rate is exceeded, the solids flow rate in the CFB is not limited. Instead, the axial density profile in the riser changes to adjust to the increasing solids flow rate. What happens in this situation is that a dense phase forms at the bottom of the riser, and rises to a height such that the entrainment rate from the dense phase equals the solids flow rate into the bottom of the riser. This aspect of riser flow was explained by Rhodes and Geldart (1985). When the inventory in the system increases (Case 2 in Figure 7.17), the solids flow rate into the riser increases. In order to balance the increased solids flow rate, the height of the dense phase in the riser increases, so that the entrainment rate from the dense phase equals the new solids flow rate.

A typical valved return system with no solids reservoir (VNSR) is shown in Figure 7.18. This type of loop system is a much better laboratory system than the batch NV system. Because of the valve in the standpipe, the solids flow rate in the system can be much more accurately controlled than in the NV system. However, even though there is a valve in the VNSR system, operation is still sensitive to solids inventory.

Consider the VNSR system shown in Figure 7.18a in which the fluidized solids in the standpipe are at a relatively low height. The solids flow rate in the system (and therefore the solids mass flux in the riser) depends on the pressure developed at the bottom of the standpipe (which is dependent upon the height of the solids in the standpipe) as well as the pressure drop across the controlling valve. For the same valve opening and the same gas flow in the riser, if the height of solids in the standpipe is increased (Figure 7.18b), the pressure at the bottom of the standpipe is also increased, which leads to an increase in the solids flow rate. Therefore, in this system, the

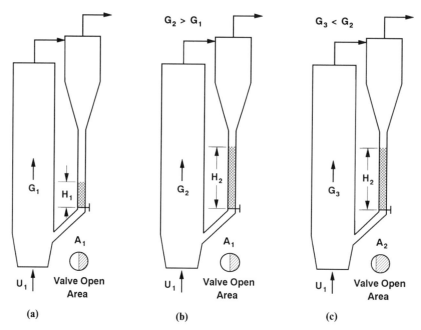

Figure 7.18 Valved return systems with no solids reservoir.

effect of increasing solids inventory in a VNSR system is to increase the **maximum** solids flow rate attainable in the system. Also, the higher the standpipe and riser combination, the greater the solids flow rate that is possible in the system. However, because of the valve in the system, various fluidized solids seal height/solid flow rate combinations are possible. In Figure 7.18c, the inventory in the system is the same as that in Figure 7.18b, but the solids flow rate is smaller. This is because the valve opening is less. This results in a higher pressure drop across the valve and a lower solids flow rate around the recirculation loop. Thus, this type of system is much more flexible than the NV system discussed above. However, the maximum solids flow rate in the system is still dependent upon the inventory (i.e. height) of solids in the standpipe.

With both the NV and VNSR systems, if the solids flow rate around the system is increased at a constant gas velocity, the inventory in the riser increases. This reduces the amount of solids available in the return standpipe, which in turn reduces the height of solids in the standpipe and, therefore, the maximum pressure that the standpipe can produce.

A valved system with a solids reservoir (VWSR) is even more flexible than the VNSR system, and also can be constructed so that the solids flow rate in the system is essentially independent of the solids inventory in the system. There are two basic types of VWSR systems: those with a fluidized standpipe and reservoir, and those with a non-fluidized standpipe and reservoir.

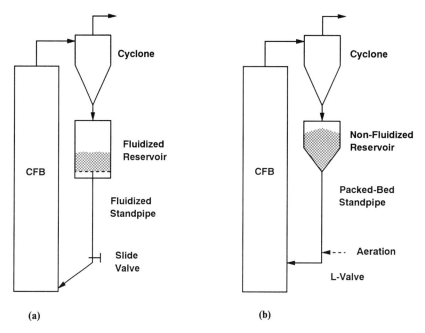

Figure 7.19 Valved return systems with solids reservoir.

Consider the 'fluidized' VWSR system shown in Figure 7.19a. In this system, a large fluidized reservoir is placed in the return line. The maximum solids flow rate in the system is proportional to the total fluidized height of the solids in the standpipe and fluidized reservoir.

When it is desired to increase the solids flow rate in the system, the valve at the bottom of the reservoir/standpipe return system is opened further. It can continue to be opened until the maximum solids flow rate in the system is achieved, or the limiting mass flux in the riser is attained. In this system, if more solids are added to the system and the inventory of the reservoir is large relative to the inventory in the riser and standpipe, it will take a large amount of solids to achieve an increase in the maximum solids flow rate attainable (because of the large-reservoir capacity). Thus, practically, the maximum solids flow rate possible in the system is not a function of the solids inventory. Obviously, increasing the height of the solids in the fluidized bed would enable higher solids flow rates in the system at a particular valve opening, but this would entail the use of a relatively large quantity of solids. Even if the maximum solids inventory possible in the riser is achieved, the level of the solids in the fluidized bed is not significantly changed because the volume of the reservoir is much larger than the volume of the riser.

Now consider the non-fluidized VWSR system shown in Figure 7.19b, in which the valve in the system is a non-mechanical L-valve. With this system, it is required that the standpipe operate in the packed-bed mode. The

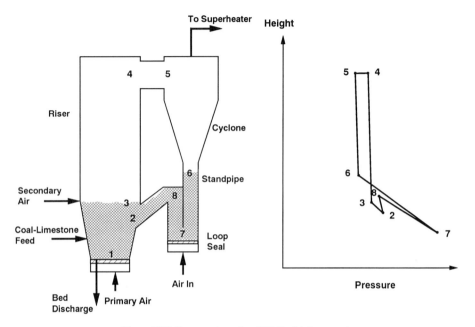

Figure 7.20 Pressure loop for CFBC with loop seal.

maximum solids flow rate in the system is dependent primarily upon the height of the standpipe in the system, because very little pressure drop is generated by the large-diameter packed-bed reservoir. The maximum solids flow rate in the system is attained when the standpipe is incipiently fluidized. If more aeration is added to the L-valve after this point is reached, bubbles form in the standpipe and reduce the solids flow rate. Therefore, what sets the maximum solids flow rate in the system is the height of the standpipe, not the inventory of solids in the system. A higher **maximum** solids flow rate in the non-fluidized standpipe return system can be achieved if the solids reservoir is fluidized. However, the maximum solids flow rate is still achieved when the packed-bed standpipe becomes fluidized.

As shown in Figure 7.12, commercial CFBC systems are generally simple NV systems (Figure 7.12a) or VWSR systems (Figures 7.12b and 7.12d). Static pressure versus height diagrams for the simple NV recirculation system and for non-fluidized and fluidized VWSR systems are shown in Figure 7.20, and Figures 7.21a and 7.21b, respectively.

7.7 Non-mechanical solids flow devices

A non-mechanical solids flow device is one that uses only aeration gas in conjunction with its geometric shape to cause particulate solids to flow

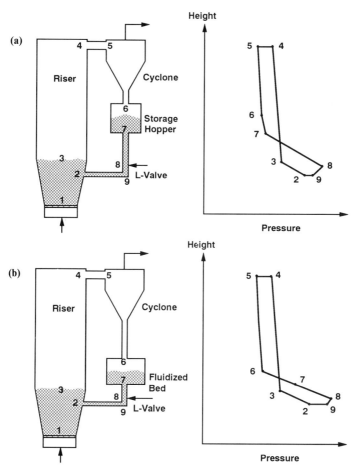

Figure 7.21 Pressure loops for CFBs with reservoirs.

through it. Non-mechanical solids flow control devices have several advantages over mechanical solids flow devices:

1. They have no moving mechanical parts which are subject to wear and/or seizure. This feature is especially beneficial when operating at elevated temperatures and pressures.
2. They are inexpensive because they are constructed of ordinary pipe and fittings.
3. They can be fabricated 'in-house' which avoids long delivery times, often associated with the purchase or replacement of large mechanical valves.

Non-mechanical devices can be operated in two different modes:

1. in the valve mode to control the flow rate of particulate solids,
2. in an 'automatic' solids flow-through mode.

242 CIRCULATING FLUIDIZED BEDS

There is often confusion as to how these modes differ, and what kind of non-mechanical devices should be used for a particular application. Each mode of operation is discussed below.

7.7.1 Non-mechanical valve mode

In the valve mode of operation, the solids flow rate through the non-mechanical device is controlled by the amount of aeration gas added to it. The most common types of non-mechanical valves are the L-valve and the J-valve. These devices are shown schematically in Figure 7.22. The primary differences between these devices are their shape and the direction in which they discharge solids. Both devices operate on the same principle. It is harder to fabricate a smooth 180-degree bend for a typical J-valve. Therefore, the J-valve can be approximated and configured more simply by the geometry shown in Figure 7.22c.

The most common non-mechanical valve is the L-valve (so named because it is shaped like the letter 'L') because it is easiest to construct, and also because it is slightly more efficient than the J-valve (Knowlton et al., 1981). Because the principle of operation of non-mechanical valves is the same, non-mechanical valve operation is presented here primarily through a discussion of the characteristics of the L-valve.

Solids flow through a non-mechanical valve because of drag forces on the particles produced by the aeration gas. When aeration gas is added to a non-mechanical valve, gas flows downward through the particles and around the constricting bend. This relative gas–solids flow produces a frictional drag force on the particles in the direction of flow. When this drag force exceeds the force required to overcome the resistance to solids flow around the bend, the solids flow through the valve.

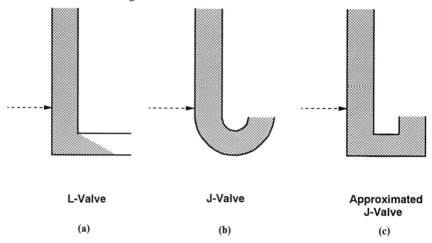

Figure 7.22 The most common non-mechanical valves.

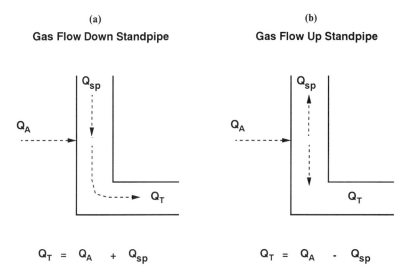

Figure 7.23 Gas flow around L-valve bend.

The actual gas flow that causes the solids to flow around the L-valve is not just the amount of aeration gas added to the valve. If gas is traveling down the moving packed-bed standpipe (which occurs in most cases) with the solids, the amount of gas that flows around the L-valve bend, Q_T, is the sum of the standpipe gas flow, Q_{sp}, plus the aeration gas flow, Q_A, as shown in Figure 7.23a. If the gas is flowing up the standpipe (which is the case when the standpipe is operating with low solids flow rates and/or large solids), then the amount of gas flowing around the bed, Q_T, is the difference between the aeration gas flow and the gas flowing up the standpipe as shown in Figure 7.23b.

When aeration is added to a non-mechanical valve, solids do not begin to flow immediately. The initial aeration gas added is not enough to produce the frictional force required to start solids flow. Above the threshold amount of gas required to initiate solids flow, additional aeration gas added to the valve causes the solids flow rate to increase, and reducing the amount of aeration to the valve causes the solids flow rate to decrease. In general, there is little hysteresis in the aeration-versus-solids flow rate curve for a non-mechanical valve.

Non-mechanical valves work best with materials having average particle sizes between 100 and 5000 microns. These materials are in Geldart Groups B and D. Materials with average particle sizes greater than about 2000 microns require substantial amounts of gas to generate the drag forces required to make the solids flow around the constricting bend. This is because larger solids have less surface area available for the generation of the drag forces required to produce flow through a non-mechanical valve.

These larger materials work best in non-mechanical valves if there are smaller particles mixed in with the larger ones. The smaller particles fill the void spaces between the larger particles and decrease the voidage of the solids mixture, thereby increasing the drag on the entire mass of solids when aeration is added, and causing the solids to flow through the constricting bend.

In general, Geldart Group A materials (with average particles sizes from approximately 30 to 100 microns) do not work in L-valves. Group A materials retain air in their interstices and remain fluidized for a substantial period of time when they are added to the standpipe attached to the non-mechanical valve. Because they remain fluidized, they flow through the constricting bend like water, and the L-valve cannot control the solids. Although most Group A materials cannot be used with non-mechanical devices operating in the control mode, materials at the upper end of the Group A classification which contain few fines (particles smaller than 44 microns) can and have been controlled in non-mechanical valves.

Geldart Group C materials (with average particle sizes less than about 30 microns) have interparticle forces that are large relative to body forces and are very cohesive (flour is a typical example). These materials do not flow well in any type of pipe, and do not flow well in L-valves.

Non-mechanical valves are used extensively in CFBC systems where Geldart Group B solids are used. They are not used in FCC circulating systems where Geldart Group A solids are used.

A non-mechanical device operating in the valve mode is always located at the bottom of an underflow standpipe operating in moving packed-bed flow. The standpipe is usually fed by a hopper, which can either be fluidized or non-fluidized. Knowlton and Hirsan (1978) and Knowlton et al. (1978) have shown that the operation of a non-mechanical valve is dependent upon the pressure balance and the geometry of the system.

Consider the riser/L-valve return system shown in Figure 7.19b. The high-pressure point in such a recycle loop is at the L-valve aeration point. The low-pressure common point is at the bottom of the cyclone. The pressure balance around the recycle loop is such that:

$$\Delta P_{L\text{-}valve} + \Delta P_{CFB} + \Delta P_{cy} = \Delta P_{sp} + \Delta P_{surge\ hopper}$$

The pressure drop across a non-fluidized surge hopper is generally negligible, so that the above equation may be written:

$$\Delta P_{sp} \cong \Delta P_{L\text{-}valve} + \Delta P_{CFB} + \Delta P_{cy}$$

The moving packed-bed standpipe is the dependent part of the pressure-drop loop in that its pressure drop adjusts to exactly balance the pressure drop produced by the sum of the pressure drops on the independent side of the loop. However, there is a maximum pressure drop per unit length ($\Delta P/L$) that the moving packed-bed standpipe can develop. This maximum value is the fluidized-bed pressure drop per unit length, $(\Delta P/L)_{mf}$, for the material.

The independent pressure drop can be increased by increasing any or all of the pressure drops across the riser, the cyclone, or the L-valve. For a constant gas velocity in the riser, as the solids flow rate into the bed is increased the independent part of the pressure-drop loop increases. The moving packed-bed standpipe pressure drop then increases to balance this increase. It does this by automatically increasing the relative gas–solids velocity in the standpipe. Further increases in the solids flow rate can occur until the $\Delta P/L$ in the moving packed-bed standpipe reaches the limiting value of $(\Delta P/L)_{mf}$. Because of its reduced capacity to absorb pressure drop, a short standpipe reaches its maximum $\Delta P/L$ at a lower solids flow rate than a longer standpipe. Thus, the maximum solids flow rate through a non-mechanical valve depends upon the length of standpipe above it.

As indicated above during the discussion of how an underflow, packed-bed standpipe operates, the pressure drop in such a standpipe is generated by the relative velocity, v_r, between the gas and solids. When v_r reaches the value necessary for minimum fluidization of the solids, a transition from packed-bed to fluidized-bed flow occurs. Any further increase in v_r results in the formation of bubbles in the standpipe. These bubbles hinder the flow of solids through the standpipe and cause a decrease in the solids flow rate.

To determine the minimum standpipe length required for a particular solids flow rate, it is necessary to estimate the pressure drop on the independent side of the pressure drop loop at the solid flow rate required. The minimum length of standpipe necessary, L_{min}, is:

$$L_{min} = (\Delta P_{independent})/(\Delta P/L)_{mf}$$

The actual length of standpipe selected for an L-valve design should be greater than L_{min} to allow for the possibility of future increases in the solids flow requirements and to act as a safety factor. Standpipe lengths are typically designed to be 1.2 to 2 times L_{min}, depending on the length of the standpipe.

To determine the diameter of an L-valve, it is necessary to select the linear solids velocity desired in the standpipe. Nearly all L-valves operate over a linear particle velocity range of between 0 and 0.3 m/s in the standpipe. Velocities greater than about 0.3 m/s can result in stick-slip flow in the standpipe. Although an L-valve may theoretically be designed for any linear solids velocity in the standpipe above it, a value near 0.15 m/s is usually selected to allow for substantial increases or decreases in the solids flow rate.

It is desirable to add aeration to an L-valve as low in the standpipe as possible. This results in maximizing standpipe length and minimizing L-valve pressure drop – both of which increase the maximum solids flow rate through the L-valve. However, if aeration is added too low in the standpipe

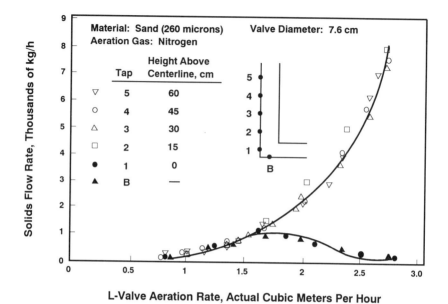

Figure 7.24 Effect of aeration location on L-valve operation.

of the L-valve, gas bypassing results and solids flow control can be insensitive and/or ineffective.

Knowlton and Hirsan (1978) found that aeration was most effective if it was added at a length-to-diameter (L/D) ratio of 1.5 or more above the centerline of the horizontal section of the L-valve. Aeration added at the centerline of the horizontal section or at the bottom of the centerline of the standpipe was found to bypass directly to the top of the horizontal section. Thus, it was not being efficiently utilized to drag the solids through the constricting bend. The aeration tap locations and the solids flow rate versus aeration rate curves for each aeration tap location investigated in their study are shown in Figure 7.24. In summary, L-valve aeration should be added above the constricting bend. To assure good operation (i.e. to prevent bypassing), it should be added at an L/D of greater than 1.5 above the centerline of the horizontal section of the L-valve.

In order for an L-valve to operate properly, the L-valve horizontal section length must be kept between a minimum and maximum length. The minimum length must be greater than the horizontal length, H_{min}, to which the solids flow due to their angle of repose. For design purposes, the minimum horizontal length should be between 1.5 and 2 times H_{min}. In general, the shorter the L-valve, the better the operation of the device.

If the L-valve is operating properly, solids flow through it in small pulses at a relatively high frequency, and for all practical purposes, the solids flow is steady. If the L-valve horizontal section is too long, solids flow can become intermittent (slugging, stopping, and then surging again). In L-valves with long horizontal sections, a slug of solids builds up to such a size that it blocks the pipe and solids flow stops momentarily. Gas pressure then builds up behind the slug until it becomes so great that the slug collapses and the solids surge momentarily through the L-valve. Another slug then builds up after the gas pressure is released, and the pattern repeats. This cycle does not generally occur if the L-valve horizontal length is less than an L/D ratio of about 8 to 10. Adding additional gas to the horizontal section prevents slug formation in long L-valves and results in smoother solids flow. The extra gas does not affect control of the valve.

The extra gas for the horizontal section can be added in several ways. Adding aeration taps to the bottom of the L-valve works well except that the taps tend to become plugged if the gas to the taps is shut off. Two other methods have been adopted by various L-valve operators. In one, gas is added to the horizontal section via taps installed along the top of the L-valve section. This prevents plugging of the taps when gas to them is shut off. However, this gas is not as effective as gas added to the bottom because some of it bypasses along the top of the L-valve. Another method is to insert a tube with holes drilled along its bottom into the lower part of the horizontal section. This tube can be inserted and withdrawn as required. For hot L-valves, this tube may warp and move to the top of the pipe. Therefore, it should be prevented from moving to the top of the section by means of a loose clamp that can allow for longitudinal expansion.

Knowlton and Hirsan (1978) have determined the effect of varying geometrical and particle parameters on the operation of the L-valve. The effects of these parameters are summarized below.

L-valve aeration requirements increase with:

1. increasing vertical section diameter
2. increasing particle size
3. increasing particle density

L-valve pressure drop increases with:

1. increasing solids flow rate
2. increasing particle density
3. decreasing horizontal section diameter

The aeration rate, solids flow rate, and pressure drop relationships in an L-valve have been estimated by Geldart and Jones (1991) and Yang and Knowlton (1993). The Yang and Knowlton method is somewhat involved. It relates L-valve flow to a variable solids area (A_0). This technique first requires estimating or measuring the bulk density at minimum fluidization, the terminal velocity of the average particle size of solids flowing through the L-valve, and the L-valve pressure drop. From these values and the L-valve diameter and horizontal length, the relationship between Q_T, Q_{sp}, and A_0, and W_s (the solids flow rate through the L-valve) is determined by solving the following four equations by trial and error:

1. The Jones and Davidson (1965) equation relating valve pressure drop to solids flow rate and valve open area, i.e.

$$\Delta P_{L\text{-}valve} = \frac{1}{2\rho_p(1-\epsilon_{mf})}\left[\frac{W_s}{C_D A_0}\right]^2$$

2. The relationship between the amount of aeration added to the L-valve (Q_A), the amount of gas flowing down the standpipe (Q_{sp}), and the amount actually flowing around the L-valve bend (Q_T):

$$Q_T = Q_a + Q_{sp}$$

3. An equation estimating the amount of gas flowing down the standpipe (this equation assumes no slip between the solids and gas in the standpipe):

$$Q_T = \frac{W_s \epsilon_{mf}}{\rho_p(1-\epsilon_{mf})}$$

4. The empirical equation relating Q_T and the variable L-valve area (A_0):

$$\frac{Q_T}{(\pi/4)D_v^2 L_v} = 1.9 + 7.64 \frac{U_t A_0}{(\pi/4)D_v^2 L_v}$$

This technique relates L-valve aeration and solids flow rate to within about ±40%. The procedure was developed from data on L-valves ranging in size from 50 to 150 mm in diameter.

If possible, it is recommended that the basic pressure drop as a function of aeration relationship be obtained from a small L-valve test unit. Alternatively, these parameters may be extrapolated from the data presented by Knowlton and Hirsan (1978).

People are often surprised that L-valve aeration gas requirements at high temperatures are small when compared to aeration requirements at low temperature. The reason for this is that the viscosity of a gas increases at high temperatures. Therefore, the drag on the particles for the same external aeration rate increases. This results in an increase in the solids flow rate through the L-valve. Because the viscosity of a gas can increase by a factor of 2 to 2.5 as the temperature is increased, the amount of gas required to achieve a certain solids flow rate at high temperature can be as little as half of that at low temperatures for the same aeration rate.

In addition, if particles flowing through the L-valve are Geldart Group B particles that lie near the AB boundary, increasing system temperature can cause the particles to cross to the A side of the boundary (Grace, 1986). If this occurs, the L-valve may then experience flow problems at high temperatures not observed at low temperature.

7.7.2 Automatic solids flow devices

Non-mechanical devices can also be used to automatically pass solids through them. In the automatic mode, they serve as a simple flowthrough device without controlling the solids. If the solids flow rate to the device is changed, the device automatically adjusts to accommodate the changed flow rate. These devices are primarily used to assist in sealing pressure (in conjunction with a standpipe) and to reroute the solids where desired. There are several types of these non-mechanical devices. The most popular are the straight cyclone dipleg, a loop seal (also often called a J-valve, siphon seal, or fluoseal), and an L-valve (in this case the 'valve' title is a misnomer since in this mode it is not controlling the solids flow).

In circulating fluidized bed systems, the most frequent application of automatic non-mechanical devices is to recycle collected cyclone 'fines' back to the CFB. The cyclone discharge is at a lower pressure than the desired return point in the CFB, so the solids must be transferred against a pressure gradient. In the past, lockhoppers or rotary valves were sometimes used to transfer these solids – generally with less than satisfactory results. The best and simplest way to return the fines is to use a non-mechanical device. The simplest non-mechanical device used to recycle cyclone fines back to a fluidized bed is a straight cyclone dipleg. However, the dipleg must be immersed in a fluidized, dense-phase bed (or fitted with a trickle valve or another control device at its discharge end if it is not immersed in the bed) in order to perform properly. Since cyclones in CFBs are generally external to the bed and located some distance from the CFB, a vertical cyclone dipleg cannot be used. Angled diplegs tend to slug and perform poorly for essentially the same reason as angled standpipes (see above). Therefore, loop seals, seal pots, and L-valves are employed to return the solids to the bed. A V-valve, another type of solids return device, can also be used in this manner. A discussion of the various types of automatic non-mechanical devices used is given below.

(a) Seal pot. A seal pot is essentially an external fluidized bed into which the cyclone fines discharge via a straight dipleg (Figure 7.25a). The solids and gas from the cyclone and the fluidizing gas for the seal pot are discharged via a downwardly angled overflow line to the desired return point. Seal pots have been used for many years and are reliable devices. However, one must be sure that the fluidizing gas velocity is high enough that the lateral solids transfer rate in the fluidized bed is greater than the solids flow rate through the dipleg for good operation. This is generally accomplished by increasing the fluidizing gas velocity in the seal pot until satisfactory operation is achieved.

With a seal pot, the solids in the dipleg rise to a height necessary to seal the pressure drop around the recycle solids transfer loop. In general, seal pot transfer problems are associated more with the standpipe than the seal pot

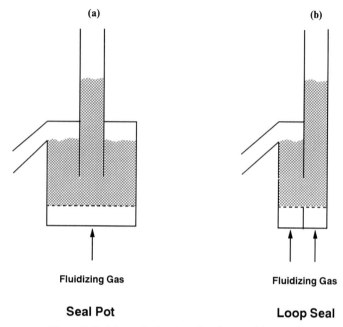

Figure 7.25 Schematic drawing of seal pot and loop seal.

itself. These problems depend to a large extent on the particle size of the solids. Geldart Groups A and B particles can be transferred with relative ease. Cohesive Group C particles are difficult to transfer unless auxiliary means (vibration, pulsing) are used to assist the non-mechanical device. For low solids flow rates, it is recommended that a straight horizontal opening be used at the dipleg exit. Experience has shown that diplegs with mitered bends require a relatively high solids mass flux through them to operate properly.

Aerating standpipes is sometimes necessary in order to achieve optimum solids flow through them. However, too much aeration is just as detrimental as too little aeration. Too much aeration leads to bubble formation and slugging. Often seal pots operate satisfactorily with no additional aeration added to the standpipe, especially with Group B solids. However, it is usually wise to add aeration taps into the standpipe discharging into the sealpot. They will be there if needed, and do not have to be used if they are not. An aeration point at the bottom of the standpipe and aeration points approximately every 2 to 3 m of solid seal height should be sufficient. It is also best to use a separate aeration control rotameter for each aeration point, especially in research units. In a commercial unit, restriction orifices are generally sufficient to ensure equal flows to each aeration location.

(b) Loop seal. The loop seal (Figure 7.25b) is essentially a variation of the seal pot. Like the seal pot, it is composed of a standpipe and a fluidized-bed section. However, the solids from the standpipe enter the fluidized bed from the side. This allows the fluidized bed portion of the device to be smaller in diameter, resulting in a smaller transfer device, lower fluidization gas requirements, and more efficient operation.

The height of the vertical flow portion of the loop seal can be increased to 'insulate' the operation of the loop seal from the pressure fluctuations in the bed to which it is discharging. Increasing the height of the vertical portion of the loop seal also helps prevent 'blowing' the seal leg of the loop seal if a pressure upset occurs. However, a large pressure upset will still blow the seal and cause other operational problems as well.

It is essential to fluidize the upflow section of the loop seal in order for Geldart Group B solids to flow smoothly through the loop seal. For Geldart Group A solids, little or no fluidization gas may be required to be added to the upflow section, because they may not be defluidized when flowing around the bed. With both types of solids, the minimum amount of aeration gas necessary to produce smooth, steady flow is what should be used. Too much gas results in slugging and unsteady flow. It is also recommended that aeration at the base of the dipleg and aeration added to the upflow section be separated so that the amount of aeration added below the bottom of the dipleg and the amount added for the upflow section can be varied independently of each other.

As noted above, Wirth (1995) found that using a cyclone dipleg too small in diameter caused an excessive amount of gas to be carried down the dipleg. At high mass fluxes, the amount of gas flow was so great that it caused the upward-flowing part of the loop seal to become dilute. This problem was alleviated by increasing the dipleg diameter to decrease the mass flux in the dipleg. Because the amount of gas carried down the dipleg is proportional to the solids mass flux in the downcomer, increasing the dipleg diameter decreased the solids mass flux for the same solids flow rate, and also decreased the amount of gas carried down the dipleg.

(c) N-valve. A novel type of non-mechanical device called the N-valve was used by Hirama *et al.* (1986). However, this type of device is not recommended because gas channeling back from the angled section can cause large bubbles in the standpipe just as with the hybrid standpipes discussed earlier.

(d) V-valve. The V-valve (Figure 7.26) is a close relative of the loop seal, and also acts as an automatic flow device at the base of standpipes operating in the overflow mode (Li *et al.*, 1982). The V-valve consists of an angled diverging section connected to a standpipe. A circular aperture in the standpipe allows the solids to flow from the standpipe into the diverging section.

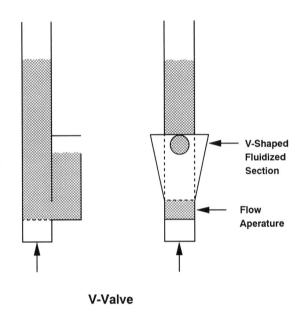

Figure 7.26 Schematic drawing of V-valve.

The included angle of the diverging section is usually small – between 5 and 10 degrees. Greater angles cause uneven distribution of the gas and solids.

As with the loop seal, the V-valve will not operate if the solids in the upflow (diverging) section are not fluidized. When the solids in the diverging section are fluidized, solids flow from the standpipe, through the aperture and diverging section into the fluidized bed. As with the loop seal, the upflow section of the V-valve prevents pressure surges from causing the dipleg to 'blow'.

(e) L-valve The L-valve can also operate automatically at the bottom of an overflow standpipe. Its operation is somewhat different than that of the other non-mechanical devices. Chan *et al.* (1988) showed how the L-valve operates in the automatic mode. Consider the L-valve shown in Figure 7.27a in which solids are flowing through the L-valve at rate W_1, while aeration is being added to the L-valve at a constant rate Q_A. The solids above the aeration point are flowing in the fluidized-bed mode, and are at an equilibrium height, H_1. For Geldart Group B solids the particles below the aeration point are flowing in the moving packed-bed mode. The fraction of aeration gas required to produce solids mass flow rate W_1 flow around the elbow with the solids, while the remaining portion of the aerating gas flows up the standpipe.

If the solids flow rate into the standpipe above the L-valve increases from W_1 to W_2 (Figure 7.27b), the solids level in the standpipe initially rises because solids are being fed to the L-valve faster than they are being

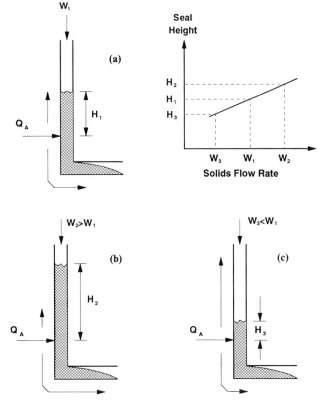

Figure 7.27 Depiction of automatic L-valve operation.

discharged. The increased height of solids in the standpipe causes the pressure at the aeration point to increase relative to the bed. If a constant aeration flow to the aeration point is maintained by a critical orifice or a control valve, a greater fraction of the aeration gas flows around the L-valve elbow causing an increase in solids flow through the L-valve. If enough aeration gas is added to the L-valve, the system reaches equilibrium at a point where a larger fraction of the aeration gas is flowing around the L-valve bend to cause solids to flow at rate W_2. The height of solids in the standpipe above the L-valve reaches equilibrium at increased height, H_2, and the system is again in balance. If the solids flow rate decreases from W_1 to W_3 (Figure 7.27c), the solids level in the standpipe above the L-valve falls, decreasing the pressure at the aeration point relative to the bed. This results in less gas flowing around the L-valve bend. The reduced flow of gas through the L-valve results in a lower solids flow rate. Balance is reached when solids are flowing through the L-valve at rate W_3, and the solids in the standpipe are at equilibrium height H_3.

In the description above, gas in the standpipe was assumed to be traveling upward relative to the standpipe wall. It is also possible for gas in the standpipe to travel downward relative to the standpipe wall. In fact, gas travels downward with the solids for most standpipe operation.

The L-valve may not work in the automatic mode if it is used to discharge Geldart Group A solids into a dilute-phase environment. This is because Group A solids may not deaerate and will flush through the L-valve and not form a pressure seal. However, Geldart Group A solids will work in an automatic L-valve if it is discharging into a dense-bed. An automatic L-valve works well with Geldart Group B solids when discharging into both dense-phase and dilute-phase media.

The L-valve will not operate automatically over an infinite range of solid flow rates. At some increase in the solids flow rate in the example given above, not enough gas is available to fluidize the solids in the standpipe and they will defluidize. This causes the standpipe to fill with defluidized solids to a level that balances the pressure-drop loop. The automatic L-valve can also work if the solids in the standpipe are in the packed-bed mode. However, because the $\Delta P/L$ in a packed bed can vary over a wide range, the height of solids required for pressure sealing in the standpipe can also vary. This makes it difficult to control the height of solids in the standpipe, and they can easily back up into the cyclone, resulting in significant loss of material. To prevent this from happening, more aeration must be added to the L-valve.

Although the automatic L-valve is geometrically simpler and less expensive to install than the loop seal, for the reason given above, the automatic L-valve can be more difficult to operate when using Group B materials. Loop seal operation requires fewer adjustments, and is often preferred over the L-valve for automatic operation with Group B materials. For Group A materials, either automatic device can be used.

As noted above, it is generally best to design a non-mechanical device by obtaining the required data in a cold-flow test unit using the actual solids to be transferred. If this is not possible, solids of similar size and density can be used. Non-mechanical valve or automatic non-mechanical device design has not yet reached a stage where it can be done analytically with great confidence. Cold modeling in a unit with 76- to 100-mm-diameter transfer lines is relatively inexpensive and minimizes operational problems when incorporating the device into a pilot plant or commercial system.

7.8 Cyclone diplegs and trickle valves

As noted above, cyclone diplegs are really overflow standpipes. It is important that they be designed correctly. Poor operation of cyclone diplegs usually results in poor collection efficiency for the cyclone to which it is attached.

Diplegs attached to first-stage cyclones generally give few operational problems because the solids flux rate down them is high. Typical design fluxes for a first-stage cyclone dipleg in FCC units are 350 to 750 kg/m²s. Second-stage cyclone diplegs are the ones that tend to have flow problems because the solids in second-stage diplegs are finer (sometimes approaching cohesive Group C size), and because the solids mass flux through the dipleg is low. Both factors result in more sluggish solids flow, which can sometimes lead to blockage. The minimum dipleg diameter recommended for all commercial cyclone diplegs is about 100 mm. This diameter is large enough to prevent most bridging. Obviously, smaller research units will have smaller diameter diplegs, with the associated plugging problems.

Cyclone diplegs are often designed with a mitered bend at its end (Figure 7.28a). This type of bend generally offers no problem for a first-stage dipleg

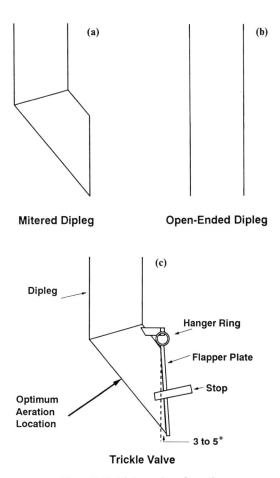

Figure 7.28 Dipleg end configurations.

because of the high mass fluxes through them, but can cause problems in a second-stage dipleg. When a mitered bend is used, the bend causes a restriction in the flow path. If the solids mass flux is not high enough, the dipleg can plug. It is much better to use a straight end dipleg (Figure 7.28b) instead of a mitered bend if the mass flux through the dipleg is low (less than $100\,kg/m^2 s$).

During start-up, fluidizing gas preferentially flows up the dipleg of a cyclone until a bed is established to seal it. This occurs because gas does not have to undergo the pressure losses of the cyclone entrance and cyclone barrel when passing up the dipleg. Many processes operate with high gas flows in the unit during start-up. First-stage diplegs generally have enough solids flow through them to allow a seal to be established in spite of the upward gas flow in the dipleg. However, the solids flow through a second-stage dipleg is generally too small (and the gas flow even greater than through the first-stage dipleg) to establish a seal unless a device to prevent this gas flow is attached to the end of the dipleg. A device called a trickle valve (Figure 7.28c) is generally used to prevent this bypass gas flow. A trickle valve is essentially a loosely hung plate that hangs vertically in front of the dipleg discharge opening. Trickle valves allow the establishment of a solids seal upon start-up, but can cause problems during operation. Typical problems are plugging (if the trickle valve is not designed correctly) or erratic pulsating flow. It is preferable to avoid trickle valves and to modify the start-up procedure to allow a lower gas flow rate through the unit until the second-stage cyclone dipleg is sealed. However, this is not always possible, and trickle valves are often a necessary 'evil'.

In order for the trickle valve to function properly and not bind during operation, it is attached to the dipleg by loose hanger rings. The dipleg opening is generally inclined about 3 to 5 degrees from the vertical so that the flapper plate exerts a positive closing force equal to the moment of its weight about the pivot point of the valve.

Geldart and Kerdoncuff (1992) and Bristow and Shingles (1989) studied trickle valve operation in cold flow models with 100 mm diameter diplegs. Four types of solids flow through the trickle valve were found to occur: (1) constant trickling, (2) dumping, (3) trickling–dumping, and (4) flooding. Constant trickling is the most desirable type of flow because it is smoother and because dipleg blockage is least likely to occur. Constant trickling occurs when solids discharge continually through the valve maintaining a constant height of solids in the dipleg. Flow in this regime was found to occur at high solids velocities in the dipleg (between about 2 and 25 cm/s), high aeration rates equivalent to about 3 cm/s in the dipleg and at the location shown in Figure 7.28c (other aeration locations were not nearly as effective), high cyclone differential pressures, and high trickle valve opening torques. All of these parameters tend to increase the amount of gas aeration or leakage rate into the dipleg. Bristow and Shingles (1989) also found that gas leakage through the trickle valve and/or aeration was absolutely necessary for satisfactory trickle valve operation.

The dumping mode of trickle valve operation occurs at low solids velocities, low aeration and/or leakage rates, low cyclone pressure drops, and low valve torques. These are the opposite of what causes the more desirable constant trickling flow. This mode of flow causes more frequent blockage of solids and may cause excessive wear in the upper part of the dipleg discharge.

Trickling–dumping is the regime intermediate between constant trickling and dumping. Material continuously trickles out of the valve while the solids periodically dump when the level in the dipleg reaches the critical height to cause the valve to open.

Flooding occurs when the solids flow into the dipleg is greater than the solids discharge through the valve. This causes solids to back up into the

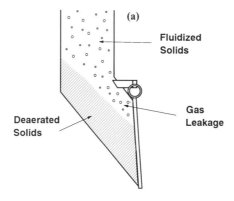

Gas Leakage Through Trickle Valve

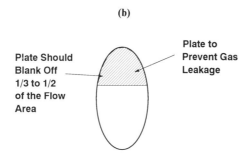

Plate Location for Erosion Prevention

Figure 7.29 Trickle valve leakage and erosion prevention.

cyclone. Flooding is generally caused by extremely high solids flow rates or excessive leakage or aeration.

Trickle valve diplegs in commercial units often exhibit erosion in the upper part of the flapper. This is due to the fact that gas leakage is preferentially through the upper part of the trickle valve (Figure 7.29a). The relatively high leakage gas velocities (up to 60 cm/s in the dipleg, but much higher in the space between the flapper and the dipleg) carry solids with them and eventually erode the dipleg and flapper (a 'horseshoe' pattern occurs on the flapper) in this area. This can be prevented by adding a plate across the upper part of the dipleg (Figure 7.29b). This reduces the leakage rate and prevents erosion. Bristow and Shingles (1989) found that a plate covering at least 1/3 of the dipleg area and up to 1/2 of the area was effective in reducing the gas leakage rate into the dipleg. However, covering less of the dipleg opening than this was ineffective, while covering too much of the opening resulted in blockage of the solids. Although covering the upper part of the dipleg opening can prevent erosion, erosion probably occurs because the trickle valve is operating in the dumping mode. It is much better to modify the dipleg/trickle valve operation to cause constant trickling flow to occur than to install the plate restrictions.

Nomenclature

A_0	L-valve variable solids flow area, m^2
C_D	orifice coefficient
D	pipe diameter, m
D_v	L-valve diameter, m
g	gravitational constant, 9.81 m/s^2
H_{sp}	standpipe height, m
H'_{sp}	changed standpipe height, m
L	length, m
L_{min}	minimum standpipe length required to seal, m
L_{sp}	standpipe length, m
P	pressure, Pa
P'	changed pressure, Pa
P_b	pressure at bottom of standpipe, Pa
P_t	pressure at top of standpipe, Pa
ΔP	pressure drop, Pa
ΔP_{II}	higher packed-bed pressure drop, Pa
ΔP_{CFB}	CFB pressure drop, Pa
ΔP_{CFBC}	circulating fluidized bed combustor pressure drop, Pa
ΔP_{cy}	cyclone pressure drop, Pa
ΔP_d	distributor pressure drop, Pa
$\Delta P_{independent}$	independent part of loop pressure drop, Pa

ΔP_{lb}	lower fluidized bed pressure drop, Pa
ΔP_{ls}	loop seal pressure drop, Pa
$\Delta P_{L\text{-}valve}$	L-valve pressure drop, Pa
ΔP_{regen}	regenerator pressure drop, Pa
ΔP_{riser}	riser pressure drop, Pa
ΔP_{sp}	standpipe pressure drop, Pa
ΔP_{sp1}	first standpipe pressure drop, Pa
ΔP_{sp2}	second standpipe pressure drop, Pa
$\Delta P_{stripper}$	stripper pressure drop, Pa
ΔP_{surge}	surge vessel pressure drop, Pa
ΔP_{sv1}	first slide valve pressure drop, Pa
ΔP_{sv2}	second slide valve pressure drop, Pa
ΔP_{ub}	upper fluidized bed pressure drop, Pa
ΔP_v	valve pressure drop, Pa
$\Delta P/L$	pressure drop per unit length, Pa/m
$(\Delta P/L)_{mf}$	pressure drop per unit length at minimum fluidization, Pa/m
Q	aeration required to prevent defluidization, m^3/s
U	superficial gas velocity, m/s
U_t	terminal velocity, m/s
v_g	interstitial gas velocity, U_g/ϵ, m/s
v_{mb}	minimum bubbling velocity, m/s
v_{mf}	interstitial minimum fluidization velocity, U_{mf}/ϵ, m/s
v_r	relative gas/solid velocity, m/s
v_s	solids velocity, m/s
w_s	solids flow rate, kg/s
ϵ	voidage
ϵ_{mf}	voidage at minimum fluidization
ρ_{ak}	skeletal density, kg/m^3
ρ_{mf}	fluidized-bed density at minimum fluidization, kg/m^3
ρ_p	particle density, kg/m^3
ρ_t	fluidized-bed density at top of standpipe, kg/m^3

References

Bristow, T.C. and Shingles, T. (1989) Cyclone dipleg and trickle valve operation, in *Fluidization VI* (eds. J.R. Grace, L.W. Shemilt and M.A. Bergougnou), Engineering Foundation, New York, pp. 161–168.

Campbell, D.L., Martin, H.Z. and Tyson, C.W. (1948) US Patent 2,451,803.

Chan, I., Findlay, J. and Knowlton, T.M. (1988) Operation of a nonmechanical L-valve in the automatic mode, Paper presented at the Fine Particle Society Meeting, Santa Clara, CA.

Eleftheriades, C.M. and Judd, M.R. (1978) The design of downcomers joining gas-fluidized beds in multistage systems. *Powder Technology*, **21**, 217.

Geldart, D. and Broodryk, N. (1991) Studies on the behaviour of cyclone diplegs, Presented at the Annual Meeting of the AIChE, Los Angeles, CA. November 18–21.

Geldart, D. and Jones, P. (1991) The behaviour of L-valves with granular powders. *Powder Technology*, **67**, 163.

Geldart, D. and Kerdoncuff, A. (1992) The behaviour of secondary and tertiary cyclone diplegs. Presented at the AIChE Annual Meeting, Miami Beach.

Goidich, S., McGee, A.R. and Richardson, K. (1991) The NISCO cogeneration project 100-MWe circulating fluidized bed reheat steam generator, in *Proceedings 11th International Fluidized Bed Combustion Conference* (ed. E.J. Anthony), ASME, New York, p. 57.

Grace, J.R. (1986) Contacting modes and behaviour classification of gas–solid and other two-phase suspensions. *Can. J. Chem. Eng.*, **64**, 353–363.

Hirama, T., Takeuchi, H. and Horio, M. (1986) Nitric oxide emission from circulating fluidized bed coal combustion, in *Proceedings Ninth International Fluidized Bed Combustion Conference*, ASME, New York, p. 898.

Jones, D.R.M. and Davidson, J.F. (1965) *Rheol. Acta*, **4**, 180.

Karri, S.B.R. and Knowlton, T.M. (1993) Comparison of group A solids flow in hybrid angled and vertical standpipes, in *Circulating Fluidized Bed Technology IV* (ed. A. Avidan), AIChE, New York, pp. 253–259.

Karri, S.B.R., Knowlton, T.M. and Litchfield, J. (1995) Increasing solids flow rates through a hybrid angled standpipe using a bypass line, in *Preprints Eighth International Conference on Fluidization* (eds. C. Laguérie and J.F. Large), PROGEP, France, p. 1075.

King, D. (1992) Fluidized catalytic crackers, an engineering review, in *Fluidization VII* (eds. O.E. Potter and D.J. Nicklin), Engineering Foundation, New York, pp. 15–26.

Knowlton, T.M. and Aquino, M.R.Y. (1981) A comparison of several lift-line feeders. Paper presented at the Second Congress of Chemical Engineering, Montreal, Canada.

Knowlton, T.M. and Hirsan, I. (1978) L-valves characterized for solids flow. *Hydrocarbon Processing*, **57**, 149.

Knowlton, T.M., Hirsan, I. and Leung, L.S. (1978) The effect of aeration tap location on the performance of a J-valve, in *Fluidization II* (eds. J.F. Davidson and D.L. Keairns), Cambridge University Press, Cambridge, pp. 128–133.

Kobro, H. and Brereton, C. (1985) Control and fuel flexibility of circulating fluidised bed. Studsvik Arbetsrapport Technical Note.

Li, X., Liu, D. and Kwauk, M. (1982) Pneumatically controlled multistage fluidized beds – II, in *Proceedings of the Joint Meeting of Chemical Engineers, SIESC and AIChE*, Chem. Industry Press, Beijing, p. 382.

Lundqvist, R., Basak, A.K., Smedley, J. and Boyd, T.J. (1991) An evaluation of process performance and scale-up effects for Ahlstrom Pyroflow circulating fluidized bed boilers using results from the 110 MWe NUCLA boiler and a 0.6 MWth pilot plant, in *Proceedings 11th International Fluidized Bed Combustion Conference* (ed. E.J. Anthony), ASME, New York, p. 131.

Mii, T., Fuji, E., Takebayashi, S., Murata, A., Fukui, J., Shiraishi, E. and Koyanagi, T. (1991) Two year operational experience of commercial multi-solid fluidized bed boilers in Japan, in *Proceedings 11th International Fluidized Bed Combustion Conference* (ed. E.J. Anthony), ASME, New York, p. 373.

Plass, L., Beißwenger, H. and Anders, R. (1991) Large size power plants working according to the atmospheric and pressurized CFB technology, in *Proceedings 11th International Fluidized Bed Combustion Conference* (ed. E.J. Anthony), ASME, New York, p. 175.

PSRI Technical Report RR-181 (1995) An investigation of the effect of solids mass flux on the axial density profile and the amount of gas entrained in a submerged cyclone dipleg, Chicago, IL.

Rhodes, M.J. and Geldart, D. (1985) The hydrodynamics of recirculating fluidized beds, in *Circulating Fluidized Bed Technology* (ed. P. Basu), Pergamon Press, Toronto, p. 193.

Sauer, R.A., Chan, I.H. and Knowlton, T.M. (1984) The effects of system and geometrical parameters on the flow of Class-B solids in overflow standpipes. *AIChE Symposium Series* **234**(80), p. 1.

Wirth, K.E. (1995) Fluid mechanics of the downcomer in circulating fluidized beds, in *Preprints Eighth International Conference on Fluidization* (eds. C. Laguérie and J.F. Large), PROGEP, France, p. 105.

Wrench, R.E., Wilson, J.E. and Guglietta, G. (1985) Design features for improved cat cracker operations. Presented at the first South American Ketjen Catalyst Seminar, Rio de Janeiro, Brazil.

Yang, W.-C. and Knowlton, T.M. (1993) L-valve equations. *Powder Technology*, **77**, 49.

Yaslik, A.D. (1993) Circulation difficulties in long angled standpipes, in *Circulating Fluidized Bed Technology IV* (ed. A. Avidan), AIChE, New York, pp. 484–489.

8 Heat transfer in circulating fluidized beds
LEON R. GLICKSMAN

8.1 Introduction

Circulating fluidized beds involving combustion or exothermic reaction commonly require heat transfer to the bed walls. The heat transfer helps to control the bed temperature and serves as a primary means to generate steam or hot water from the bed.

Circulating beds are characterized by a substantial variation in average cross-sectional solids density from the distributor to the outlet; there may be corresponding variations in local bed-to-wall heat transfer with bed height.

It is important to design the bed for the proper rate of heat transfer. Overestimating the rate of heat transfer to the walls results in the need for additional heat transfer surface or operation at elevated bed density, which presents a penalty in fan operating power.

There is no guarantee that heat transfer results obtained in a laboratory bed or pilot plant are directly transferable to a much larger commercial design.

8.1.1 General observations

Heat transfer results are reported in terms of the heat transfer coefficient, h, which relates the rate of heat transfer from the bed to the surface to the overall temperature difference and the surface area:

$$q = hA(T_{bed} - T_{wall}) \qquad (8.1)$$

Although the gas or particles near the wall may undergo a temperature change it is more convenient to write h in terms of the difference between the mean bed temperature at the cross-section in question and the wall temperature. q and h may be defined for a small local area or may be quantities averaged over a large area of the surface.

Observations of heat transfer in both small-scale laboratory columns and larger commercial beds indicate that h increases with the cross-sectional average solids density. h also increases with elevated temperatures. When the superficial gas velocity is varied, if the solids recycle rate is also adjusted to keep the cross-section average solids density constant, h varies little, if at all. In some instances h is found to increase when the mean particle diameter is decreased. The vertical length of the active heat transfer surface influences h: longer surfaces result in lower values of h as well as a decreased influence of particle size. Finally, h tends to increase with bed diameter at a fixed cross-sectional-averaged particle concentration. Preliminary evidence indicates that h is also a function of surface roughness; even small amplitude roughness may lead to noticeable changes in h.

Although the interest in circulating fluidized beds has increased substantially over the past decade, our understanding of their hydrodynamics and heat transfer is still far short of the state of knowledge for bubbling fluidized beds. At present, there is not a definitive set of experimental data or predictive models that allow designers to predict confidently the heat transfer rate for a new circulating bed design. There is a dearth of information available for large commercial units.

The goal of the present chapter is to develop a physical understanding of the heat transfer process in circulating beds. We will show that heat transfer is intimately tied to hydrodynamics, especially to particle and gas behavior close to the heat transfer surface. The hydrodynamic behavior near the wall, as best we understand it, will help us gain an understanding of general trends in observed heat transfer behavior and lead to simple predictive models.

A number of different models have been proposed. Each requires several key parameters which are known, at best, approximately. Given the uncertainty in the required parameters, models that are unnecessarily detailed or complex are unwarranted and do not lead to better predictive capability. Emphasis is given to models that are straightforward and incorporate the relevant physics.

The great majority of available heat transfer data were obtained from small laboratory-scale beds. These are reviewed in light of the physical models with an eye toward validity, general correlation, and extrapolation of the results to commercial-scale systems. Trends of the data, in many cases, can be explained by proper understanding of the physical process.

This chapter is arranged so that physics of bed hydrodynamics and heat transfer is dealt with (sections 8.2 and 8.3) before complete heat transfer models are discussed (section 8.4). Section 8.5 deals with advanced considerations of bed hydrodynamics and can be safely omitted by designers primarily interested in working relationships. Use of small-scale experiments to model larger beds is the subject of section 8.6. This is followed by a brief consideration of heat transfer measurement techniques and uncertainty. Experimental results for laboratory-scale beds and large commercial beds are discussed in sections 8.8 and 8.9, respectively. Regrettably the latter is too brief. Order of magnitude estimates and working relationships are developed in section 8.10, followed by a consideration of heat transfer augmentation (section 8.11).

8.2 Hydrodynamics

Our primary focus will be on heat transfer between a circulating bed and the bed walls. The heat transfer process is controlled by the hydrodynamics of the solid and gas mixture in the vicinity of the wall. Although the wall hydrodynamics is important to heat transfer, relevant information has only recently come to light. As described in previous chapters, the overall structure of a circulating or fast bed includes a core with clusters of particles and individual particles moving upward in the gas stream. The particles actively circulate in the core where the temperature is near uniform. In an annular region near the walls where the gas velocity is reduced, the particles tend to fall downward. The width of this zone, whose boundary is commonly defined as the point where the net solid flux (upward minus downward) is zero, tends to be a modest portion of the bed diameter. Clusters or individual particles at the mean bed temperature enter the annular region from the core. There will be a lateral temperature gradient near a cooled wall.

The radial deposition of solids from the core to the wall has been likened to a radial diffusion process (Bolton and Davidson, 1988). Glicksman and Westphalen (1994) argued that in the upper dilute region of the bed, the radial flux is due primarily to radial motion of dilute collections of particles rather than radial motion of concentrated clusters. (The particles may form more concentrated clusters or strands within the annular region or at the wall inhibiting their motion by diffusion back into the core.) In the lower portion of the bed, particle motion to the wall may be largely due to particles ejected from the dense region near the base of the bed with a radial component of velocity.

The downflowing layer near the wall exhibits considerable short-time excursions in local concentration and layer thickness as clusters appear and are replaced by dilute gas–solids mixture (Louge *et al.*, 1990; Leckner *et al.*, 1991; Brereton and Grace, 1993; Soong *et al.*, 1994). Thus the view of this annular layer as a homogeneous continuum with a time invariant concentration is misleading.

High speed video observation of a CFB wall shows clusters or streamers of particles. Rhodes *et al.* (1992) describe these as arch-shaped swarms. The clusters form and flow down the wall for some appreciable distance, of the order of the bed width or longer. For smooth surfaces, there is no evidence that particles approach the surface, contact it briefly and quickly rebound from it.

Figure 8.1 shows successive high speed images at the wall under moderately dilute conditions. A local area of the wall is alternately covered by a dense cluster or streamer, followed by a dilute mixture of gas and particles (Katoh *et al.*, 1990; Li *et al.*, 1991; Lints and Glicksman, 1993). Figure 8.2 shows a typical time-varying output from a phonographic stylus that extended one particle diameter into the bed from the wall. Each pulse is an

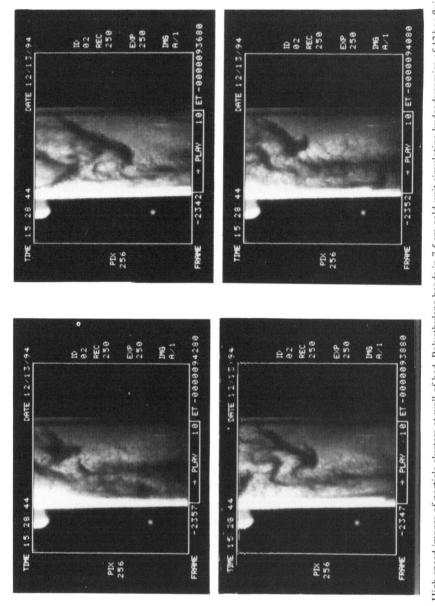

Figure 8.1 High speed images of particle clusters at wall of bed. Polyethylene beads in 7.5 cm cold unit simulating hydrodynamics of 12 bar fluidized bed combustor with 45 cm diameter. Time interval between frames is 1/50 s.

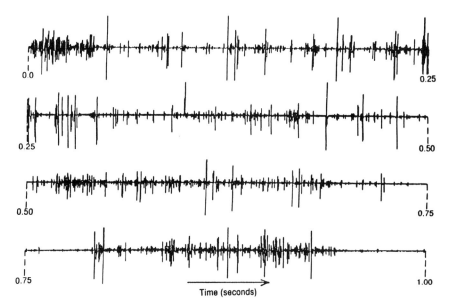

Figure 8.2 One-second signal output from a particle impact probe extending approximately 1.5 particle diameter from the wall. Each pulse is caused by a single particle strike on the probe (Lints, 1993).

individual particle or a closely spaced group of particles impacting the probe. There are irregular time periods of low concentration interspersed with periods of high activity involving clusters whose concentrations cover a wide range. At some point the clusters or swarms become unstable and leave the wall. Very small roughness elements, e.g. forward facing steps with a height of one particle diameter, are seen to dislodge the clusters from the wall (Katoh et al., 1990).

The fraction of the wall covered by dense swarms or clusters is set by the balance between cluster formation due to radial deposition at the wall and cluster shedding from the wall. Conditions in the core control the former, while wall conditions such as surface geometry and possibly gas velocity or shear stress control shedding.

Average wall coverage can be obtained from the results of a particle impact probe (Figure 8.2) and from capacitance probe measurement at or near the wall (Wu, 1989; Dou, 1990; Louge et al., 1990). The wall coverage varies with the cross-section average solids fraction as shown in Figure 8.3 (Lints, 1992). Also included are video data obtained by Katoh et al. (1990), Li et al. (1991) and Rhodes et al. (1991). As the average solids fraction in the cross-section increases, a larger portion of the wall is covered by clusters. The wall of a small laboratory riser has a continuous coverage when the cross-section average solid fraction exceeds several percent.

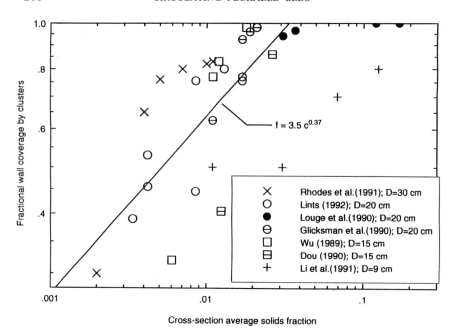

Figure 8.3 Fraction of wall covered by clusters as a function of cross-sectional average solid fraction (Lints and Glicksman, 1993).

Note that there is an influence of column diameter in Figure 8.3 with larger risers achieving higher coverage at the same cross-sectional average concentration. This is to be expected since the ratio of total vertical solids flux, proportional to cross-section area, to total radial flux, proportional to circumference, should remain constant at a given cross-sectional density. From this it follows that the radial solids deposition rate per unit area and wall coverage increases with bed diameter. All of the data in Figure 8.3 are for modest-sized risers, but the trend may well continue for large commercial beds.

Lints (1992) compiled data for the average cluster solid concentration at the wall from different investigators. For most cases the cluster solid volumetric concentration is in the range of 0.1 to 0.2, much lower than the emulsion phase voidage of a bubbling bed.

Visual studies suggest that clusters are not in direct contact with smooth walls. Using an impact probe that could be positioned at various distances from the wall, Lints and Glicksman (1993) measured the gas layer thickness between the wall and the nearest clusters for 182 μm sand. The gas layer thickness was found to be less than one particle diameter, and it decreased with increasing cross-sectional average solids concentration (see Figure 8.4). Wirth *et al.* (1991) used gamma rays to measure the solid-to-wall spacing and reported values of 1 mm, much larger than those found by direct contact measurements.

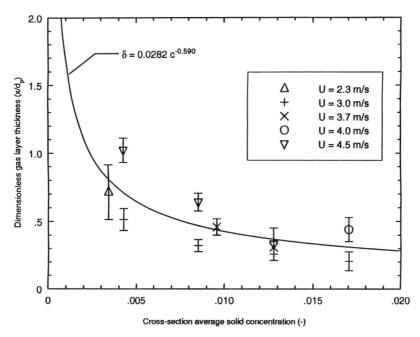

Figure 8.4 Thickness of gas layer between wall and clusters (Lints and Glicksman, 1993).

Several measurements of cluster velocity near the wall have been made (see Table 8.1). Most have been based on visual observations. The falling velocity of solid particles close to the walls appears to be minimally influenced by particle size or density. This would suggest that the solids are acting together as a cluster or wave at the wall.

Most observations give a value of approximately 1 m/s for the cluster falling velocity. Rhodes *et al.* (1991) reported a swarm velocity of 0.3–0.4 m/s with a region of higher velocity, 1 m/s, behind the swarms at the wall. Wu *et al.* (1991) found a velocity of 1.62 m/s from cross-correlations of instantaneous heat transfer results at two vertical locations and a velocity of 1.26 m/s from high speed cinematography. Measurements of individual clusters falling at the wall of an air-filled bed also indicated that the falling velocity was independent of the superficial gas velocity (Glicksman, 1988). This suggests that the velocity is set by a balance between gravity and a cluster–wall interaction that must be hydrodynamic if the cluster is not in direct contact with the wall.

The appropriate model for heat transfer between the particle and the wall will depend on the relative time of cluster–wall contact and the thermal time constant of a particle. If the maximum cluster velocity, v_{max}, is set by a balance of laminar shear stress and gravity, then the shear stress can be represented as Mgv/v_{max} where v is the time-varying velocity of a cluster

Table 8.1 Measurement of cluster velocity

Researcher(s)	Particles	Superficial gas velocity (m/s)	Column diameter (cm)	Velocity measurement technique	Cluster velocity at wall (m/s)
Bader et al. (1988)	FCC, 76 μm	3.7	30.5	Pitot tube	0.3 → 0.9
Gidaspow et al. (1989)	Glass, 520 μm	5	7.6	External visual	1.1
Glicksman (1988)	80 μm Sand 600 μm Aluminum	0–6.5	10	Visual	1.2 → 2
Golriz and Leckner (1992)	Silica sand, 260 μm	3.5–6.5	170 × 170 membrane wall	Cross-correlation of double thermocouple	1.4 → 2.2
Hartge et al. (1988)	FCC, 85 μm CFB ash, 120 μm	2.9–3.7	40	Intrusive light reflection probe	0.2 → 2.8
Horio et al. (1988)	FCC, 60 μm	1.17–1.29	5.0	Intrusive light reflection probe	0.45 → 0.65
Nowak et al. (1991)	FCC, 46 μm	4.0	20.5	Intrusive light reflection probe	0.6 → 1.0
Rhodes et al. (1991)	Alumina, 70 μm	3–4	30.5	External visual	0.1 → 2.2
Wirth and Seiter (1991)	Phosphorescent ZnS, 50 μm	0.6–1.9	16.8	External visual	1.0 → 3.0
Wu et al. (1991)	Sand, 171 μm	7	15.2	External visual	0.64 → 2.77
Wu et al. (1991)	Sand, 171 μm	7	15.2	Cross-correlation of double thermocouple	1.62

Table 8.2 Contact time for clusters at the wall for a given vertical distance

Vertical distance (cm)	Time (eq. (8.3)) $v_{max} = 1\,\text{m/s}$ (s)	Time (eq. (8.3)) $v_{max} = 0.5\,\text{m/s}$ (s)	Time $v=$ constant $= v_{max}$ (s)	Time free fall, no resistance, no upper value on v_{max} (s)
0.01	0.049	0.053	0.01	0.045
0.1	0.19	0.25	0.10	0.14
1	1.1	2.1	1.1	0.45
10	10.1	30.1	10	1.4

of mass M. The equation of motion can be written as

$$M\frac{d^2 x}{dt^2} = Mg - \frac{Mgv}{v_{max}} \qquad (8.2)$$

yielding with zero initial velocity:

$$x = \frac{v_{max}^2}{g}\left[e^{-(gt)/v_{max}} - 1\right] + v_{max} t \qquad (8.3)$$

The contact times for clusters at the wall can be expressed in terms of the vertical distance the cluster traverses before leaving the wall (Table 8.2). For distances between 10 and 100 cm, the contact time is roughly 0.1 s to 1 s.

8.3 Heat transfer fundamentals

Heat can be transferred from the core of the bed to the wall by several different mechanisms. Heated particles at the core temperature move to the wall and transfer their energy while in contact with the wall. This is termed particle convection. Since the particles seldom touch the wall, most of the heat transfer must take place through the gas layer separating the particle and the wall. Nevertheless, the motion of the particle from core to wall is the primary mechanism for energy transfer, and the overall process is properly termed particle convection. Particle convection takes place at areas of the wall covered by comparatively dense clusters, groups or strands.

For the remainder of the wall area, uncovered by clusters, the wall is contacted by gas or a very dilute particle–gas mixture. The sparse collection of particles may aid in the heat transfer, but the primary mechanism is contact of the wall by gas which is at or near the bulk temperature while gas motion is the primary means of transferring energy from the core or inner portion of the annulus to the wall. In some instances one might suspect that the gas also transfers heat to particles or clusters in the annulus near the wall. This mechanism for heat transfer to the uncovered wall is termed gas convection. The gas motion at the wall is tied to turbulent fluctuations in the overall gas flow and may be augmented by the motion of the neighboring particle clusters. With gas convection the gas motion acts as the primary means to transport energy across the annulus to the wall, as well as being the agent for heat transfer at the wall. In contrast, with particle convection the particle motion brings the energy (in the form of hot particles) to the proximity of the wall where heat is transferred to the wall by solid contact and/or through the intervening gas layer.

At elevated temperature, radiation serves to augment the heat transfer both to the uncovered surface as well as the surface covered by clusters. To obtain the overall average heat transfer to the wall some investigators have advocated simple superposition of particle convection, gas convection and radiation, each calculated in the absence of the other two mechanisms. In general, this approximation will cause errors. If clusters cover a substantial portion of the heat transfer surface, then adding particle convection and gas convection, each calculated for the entire surface, overestimates the total. Radiation heat transfer can also be substantially overestimated by use of superposition.

8.3.1 Particle convection

In all but the most dilute cases, particle convection is a primary heat transfer mechanism. This has been confirmed by Ebert et al. (1993), where bed-to-wall heat and mass transfer experiments were carried out in the same circulating bed. With single-phase gas flow the two experiments gave identical results when they were compared using the heat-to-mass transfer analogy. In fast fluidization, the particles enhanced the rate of mass transfer by roughly 50 to 100% over results for single-phase gas flow, while the particles enhanced the heat transfer by an order of magnitude over single-phase gas flow results for the same flow conditions. Thus, enhanced energy transfer set up by particle motion and heat capacity is much more important than the particle influence on the hydrodynamics.

Particle convection is due to heat transfer from a cluster or stream of particles, initially at the bulk bed temperature, which forms at the surface. Wu et al. (1991), Louge et al. (1990), and Dou et al. (1992) simultaneously measured the instantaneous heat transfer to a small probe together with the local particle concentration. They found a close correlation between peaks in time-resolved heat transfer coefficient and maxima in the particle concentration, indicating the role of high-density clusters in augmenting the surface heat transfer.

Werdermann and Werther (1993) and Leckner et al. (1991) observed a thermal boundary layer of thickness roughly equivalent to the annular region in larger circulating beds. If particles moving from the core to the wall can traverse this region without significant chance of collision, then

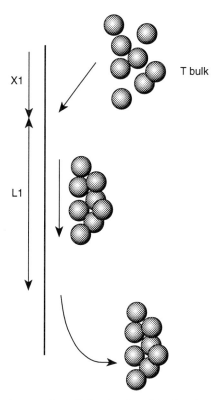

Figure 8.5 Single cluster at the wall showing deposition locations and contact length.

the temperature of particles arriving at the wall should be close to the bulk temperature. If, however, there is a high probability of multiple collisions when traversing the boundary layer region, the temperature of the particles arriving at the wall will be altered.

The schematic of particle convection is illustrated in Figure 8.5. A cluster appears at the wall at location x_1. It falls along the surface transferring heat to the wall. After falling a distance L_1, it separates from the surface and mixes with the core or the annular flow region. The average heat transfer coefficient is made up of the contributions due to the whole array of clusters that arrive at different positions, x_1, x_2 etc. and fall different distances before they depart. Moreover, one must distinguish results from a typical commercial design where the active heat transfer surface covers a large fraction of the riser height and experiments in which the active heat transfer probe only covers a short vertical distance along the wall.

When the cluster first reaches the wall the heat transfer takes place between the wall and the particles immediately adjacent to the wall. These particles are initially at the bulk riser temperature and heat transfer occurs by steady-state conduction through the gas layer separating the particles and the wall. (For a 100 μm thick air layer the time to reach steady state conduction is one millisecond.) At high temperature this is augmented by radiation. For a single particle within the cluster, the energy balance in the absence of radiation is

$$\frac{\rho_p c_p \pi d_p^3}{6} \frac{dT}{dt} = \frac{4k_g}{d_p} \frac{\pi d_p^2}{4} (T_W - T) \qquad (8.4)$$

where we have used the gas layer thickness from Figure 8.4 corresponding to a cross-section average concentration of 1.5 to 2%. The particle thermal time constant becomes

$$\tau_p = \frac{\rho_p c_p d_p^2}{6k_g} \qquad (8.5)$$

For spherical particles with density and specific heat similar to sand or limestone fluidized with air, Table 8.3 gives the thermal time constant for different diameters and mean air temperatures. If the contact time is much less than the thermal time constant, the particle temperature can be taken as constant at T_B and the heat transfer is confined to conduction across the gas gap. For contact time periods of the order of one-tenth to one-third the thermal time constant, equation (8.4), applies for the entire contact time. For contact times greater than $\tau_p/3$, the particle temperature changes appreciably and heat transfer takes place between particles one or more rows removed from the wall. This is shown schematically in Figure 8.6. Note that as contact time increases relative to τ_p the heat transfer rate decreases since the particles closest to the wall approach the wall temperature. Comparing the thermal time constant with the calculated contact time of Table 8.2, it can be seen that for 100 μm particles in contact with short, 10 or 20 mm long active heat transfer surfaces, the particle temperature remains essentially unchanged and the heat transfer rate should be high. If the cluster remains on an active heat transfer surface over roughly a 1 m length, then the surface layer will have cooled considerably and the heat

Table 8.3 Thermal time constant for particles adjacent to wall, equation 8.5

Particle diameter (μm)	Mean air temperature (C°)	Thermal time contact (s)
50	100	0.03
100	100	0.12
250	100	0.84
100	425	0.07*
250	425	0.48*

* Note, in these cases, radiation augments gas conduction reducing the time constant.

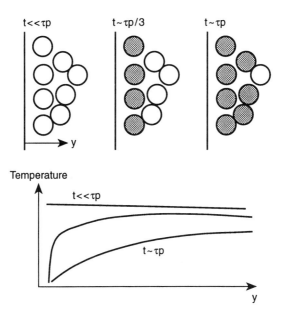

Figure 8.6 Transient cooling of clusters. Contact times greater than or less than the thermal time constant of a particle yield different thermal penetration depths into cluster.

transfer rate at the end of the path will be considerably reduced from that at the beginning. Furthermore, consideration of heat transfer between particles several layers removed from the wall must be included.

Mickley and Fairbanks (1955) suggested that a cluster of particles at the wall of a bubbling bed could be modeled as a homogeneous material with a constant effective conductivity and heat capacity. If the cluster is thicker than the thermal penetration depth, less than 1 mm for a contact time of 1 s, the transient heat transfer solution for a semi-infinite body can be used. The instantaneous heat transfer coefficient and the time-mean heat transfer coefficients are, respectively:

$$h_t = \frac{q(t)}{T_B - T_W} = \sqrt{\frac{k_e \rho_p c_p \epsilon_c}{\pi t}} \tag{8.6}$$

$$h_H = 2\sqrt{\frac{k_e \rho_p c_p \epsilon_c}{\pi \tau_c}} \tag{8.7}$$

where t and τ_c are the instantaneous and overall contact times, respectively, and k_e is the effective conductivity of the particle–gas mixture in the emulsion and ϵ_c is the cluster solid fraction by volume, $1 - \epsilon$. Gelperin and Einstein (1971) give an empirical relationship for the effective conductivity which fits most data:

$$\frac{k_e}{k_g} = 1 + \frac{(1-\epsilon)[1-(k_g/k_p)]}{(k_g/k_p) + 0.28\epsilon^{0.63(k_p/k_g)^{0.18}}} \qquad (8.8)$$

For sand and gas the calculated value of k_e generally is between 2 and 4 times the gas conductivity. Equation (8.7) tends to overpredict the heat transfer since contact resistance at the wall is neglected.

Equation (8.7) gives the mean value of the heat transfer coefficient and mean heat transfer from the cluster to the wall averaged over its time of contact with the wall. Observations of the wall indicate that the falling distance or time of contact varies with each cluster.

The mean falling time of N clusters is,

$$\tau = \frac{1}{N}\sum_{i=1}^{N} t_i \qquad (8.9)$$

and h_H based on the mean falling time is

$$h_H(\tau) = 2\sqrt{\frac{k_e \rho_p c_p (\epsilon_c)}{\pi \tau}} \qquad (8.10)$$

while the true mean value of h_H should be obtained by averaging individual values of h_H evaluated for each cluster with its own distinct falling time:

$$\overline{h_H} = \frac{1}{N}\sum_{i=1}^{N} h_H(t_i) \qquad (8.11)$$

Assuming that the falling times for clusters are equally probable between 0.25τ and 1.75τ, the mean heat transfer calculated using equation (8.10) or (8.11) differ by only 10%. If the falling time distribution follows a normal distribution, when the standard deviation is equal to or less than the mean value, $h_H(\tau)$ and $\overline{h_H}$ agree within 14%. Similarly, variations in the cluster density are closely represented by use of the mean cluster density.

The Mickley renewal type model does not give accurate results for short contact times since it does not account for the gas gap between the cluster and the wall. The effective heat transfer at the wall gap can be represented as

$$h_W = \frac{k_g}{\delta d_p} \qquad (8.12)$$

where δ is the dimensionless gap thickness, given for example by the data of Figure 8.4. A good approximation is to consider the wall layer and the Mickley renewal model setting up thermal resistances that act in series. This gives an overall cluster-to-wall heat transfer coefficient of

$$h_c = \left[\frac{1}{\overline{h_H}} + \frac{1}{h_w}\right]^{-1} \qquad (8.13)$$

Gelperin and Einstein (1971) among others have shown that equation (8.13) is a good approximation to the exact solution. Basu and Nag (1987) proposed a similar model for circulating beds.

When the region of temperature change within a cluster is only several particle diameters thick the homogenous model is questionable. Nevertheless, equation (8.13) does agree with measured values even in the limit of very low contact times (Gloski et al., 1984).

The proper values of the parameters to use with equations (8.7), (8.12), and (8.13) are discussed in section 8.4. h_c applies to the portion of the surface covered by clusters. The overall bed-to-wall heat transfer can be represented as

$$h_{overall} = fh_c + (1-f)h_g \tag{8.14a}$$

where f is the fraction of the surface covered by clusters and h_g is the gas convective heat transfer coefficient to the portion of the surface left uncovered.

8.3.2 Gas convection

Relatively scant attention has been paid to gas convection between the bed and the uncovered portion of the heat transfer surface. Some investigators use the single-phase correlation for flow of gas alone (Wu, 1989; Kunii and Levenspiel, 1991). Others implicitly assume the entire surface is covered by clusters (Sekthira et al., 1988; Dou, 1990; Mahalingam and Kolar, 1991), ignoring h_g. In very dilute flows where only a small fraction of the heat transfer surface is covered by clusters, gas convection will be important.

Lints (1992) surveyed existing CFB heat transfer measurements at cross-section averaged densities below 50 kg/m^3 and obtained an estimate for h_g by linearly extrapolating measurements to zero density. He found that the zero density extrapolations are greater than corresponding values obtained from existing single-phase gas flow correlations. This is in agreement with the trends found by Ebert et al. (1993) in a circulating bed using mass transfer measurements. The overall mass transfer coefficient increased with the presence of the particles but was insensitive to the particle concentration. These results suggest that the presence of even a modest fraction of particle clusters tends to enhance the gas convection on the measured surface. At low values of f the clusters act as roughness elements or turbulence promoters. This can be approximated by assuming the gas heat transfer to the uncovered surface acts similar to gas heat transfer to a series of short, 100 mm long heat transfer surfaces. The presence of individual particles increases the effective conductivity of the dilute phase by about 10% when the solids volume fraction is 3%.

8.3.3 Radiation heat transfer

At elevated bed temperatures the rate of heat transfer to the wall increases due to the increase of gas conductivity and the contribution of radiation heat transfer. Radiation acts in parallel with gas convection at the uncovered wall areas and it increases the heat transfer from the cluster to the wall.

When a cluster initially contacts the wall, it has a uniform temperature. If we take the cluster as having an effective emissivity ϵ_{eff} and assume that the wall and cluster act as gray bodies, the radiant transfer can be written as

$$q_{rad,clusters} = \frac{\sigma(T_c^4 - T_W^4)}{(1/\epsilon_W) + (1/\epsilon_{\mathit{eff}}) - 1} \qquad (8.14b)$$

If the clusters are isothermal, the effective emissivity will be higher than the particle surface emissivity because of the re-entrant geometry of the particle array. Note that the radiation is dependent on the cluster temperature. Two different conditions are possible depending on the solids concentration of the cluster. If the cluster has a solids concentration near 0.5, as in a bubbling bed, the radiant transfer takes place between the wall and first two rows of the cluster. In that case, as the cluster moves down the surface and its temperature decreases, the radiant transfer will quickly diminish due to the nonlinear dependence on the cluster temperature. For example, with an initial cluster temperature of 800°C and a wall temperature of 100°C, when the surface temperature of clusters adjacent to the wall falls to 600°C, conduction across the gas gap is decreased by 29% while radiation is reduced by 56%. For long transit lengths approaching 1 m, h_H is much smaller than h_W. That is, the surface temperature of the cluster approaches the wall temperature and the radiation contribution has a minimal impact, as shown by numerical calculations by Fang et al. (1995b). It is wrong to assume that bed-to-wall radiation based on the bulk and wall temperatures can simply be added to the particle convective component for the fraction of surface covered by clusters. In general, the radiation contribution will be much more modest when the cluster solids concentration is near 0.5.

Measurements of cluster concentration in laboratory sized risers indicate that the average solids fraction in a cluster at the wall is in the range of 0.1–0.2 for cross-section averaged concentrations of 0.04 or smaller. For these wall clusters the average inter-particle spacing has grown so that only a modest percentage of radiation to the wall is emitted by the particles in the first row adjacent to the wall. Most of the radiation incident on the wall is emitted by particles in the interior of the cluster. At a cluster solids fraction of 0.2, the radiation mean free path is approximately 5 times larger than the average inter-particle spacing.

As a lower concentration cluster moves down the wall, the particles directly adjacent to the wall cool, and radiation from these particles is rapidly reduced. The particles several rows from the wall are still near their initial temperature and they continue to radiate to the wall. After one second of contact, which corresponds to a cluster displacement of 1 meter, the total radiation from the cluster to the wall is still approximately half its initial value. The radiation level is maintained because the mean free path for radiation is larger than the thermal penetration distance for conduction into the cluster.

For the portion of the surface uncovered by clusters, radiation from particles closer to the center of the riser is incident on the wall. If the material is well mixed so that all of the particles are at uniform temperature, the effective emissivity of the particle cloud exchanging radiation with the wall can be determined. Since the solid particles approximate gray bodies with continuous surface emissivities over all wavelengths, the emissivity of the cloud can be closely estimated by the mean beam length, given by Hottel and Sarofim (1967):

$$\epsilon_{eff} = 1 - \exp(-KL_m) \qquad (8.15)$$

For a long cylindrical riser L_m is about 88% of the bed diameter. For other shapes L_m can be approximated as 3.5 times the ratio of bed volume to surface area. K is the mean extinction coefficient. For black body particles whose diameter is much larger than the wavelength of appreciable radiation, the extinction coefficient can be given as

$$K = n \frac{\pi d_p^2}{4} = \frac{3}{2} \frac{(1-\delta)}{d_p} \qquad (8.16)$$

where n is the number of particles per unit volume and δ is the volume fraction of gas.

Alternatively,

$$K = \frac{3}{2} \epsilon_p \frac{(1-\delta)}{d_p} \qquad (8.17)$$

where ϵ_p, the particle emissivity, is now included. Han et al. (1992) measured the radiative component of flux to the wall of a 4.9 m diameter riser. The effective emissivity using equations (8.15) and (8.17) with ϵ_p equal to 0.5 is compared to the measured value in Figure 8.7. This approximation does a reasonable job of fitting the data. Even for such a modest riser diameter, the effective emissivity rapidly approaches unity. When a riser of 100 cm diameter contains 100 μm particles, with a particle surface emissivity of 0.5, the effective emissivity reaches 0.95 at a solids fraction of 4×10^{-3} or a density of 10 kg/m^3. For larger commercial units the emissivity approaches unity at much smaller particle loadings.

For large units, if the particles are well mixed at the mean bed temperature, T_B, radiation heat transfer to the portion of the surface not covered by

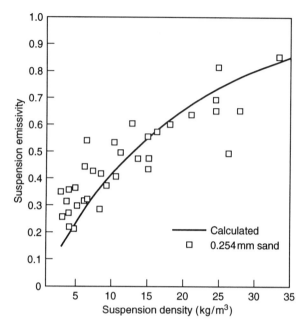

Figure 8.7 Predicted and measured suspension emissivity versus suspension density for particles in gas flow, 200 to 600°C, 49 mm diameter column (Han et al., 1992).

particles can be estimated as

$$q_r = \epsilon_W \sigma (T_B^4 - T_W^4) \tag{8.18}$$

where ϵ_W is the wall emissivity.

In limited experiments in a pilot plant with a 1.7 m × 1.7 m cross-section (Leckner et al., 1991), and in large commercial systems (Werdermann and Werther, 1993), the existence of a thermal boundary layer with particle temperatures below the mean bed temperature was found in a region roughly comparable to the downflowing annular width. In this case, radiation from the core to the wall is reduced if the emissivity of the boundary layer is appreciable. When the emissivity of the layer approaches one, most of the radiation to the wall comes from the cooled layer. Wedermann and Werther carried out approximate calculations to indicate that the cooled layer reduced the radiative flux to the wall by almost one-half.

The presence of the cooled layer is probably due to radiative cooling of particles in the annular region where the mixing with the core is not intense enough to maintain a uniform temperature. The estimated radiative mean free path is roughly half the thermal boundary layer thickness, which is consistent with this explanation. In smaller units such an effect will probably not be seen because the annular region is thinner and relatively transparent so that more of the particles in the well-mixed core section take part in direct radiative transfer with the wall.

8.4 Heat transfer models

An understanding of the hydrodynamic and heat transfer phenomena near the wall of a circulating bed must be combined to produce a quantitative prediction of the heat transfer. Visual observations and high-speed video images taken at the wall of the bed together with order of magnitude considerations have led most investigators to propose predictive models for particle convection based on renewal models similar to Mickley's.

Subbarao and Basu (1986) proposed a renewal model for circulating beds. Some early applications of the renewal model implicitly assumed that the entire wall was covered by dense-phase clusters; this should only be valid at high bed densities. Later embodiments of renewal model for circulating fluidized beds explicitly include length effects (Glicksman, 1988). Wu et al. (1990) assumed that the heated solids falling along the surface mixed with fresh material at the bulk bed temperature with all strands or clusters leaving in phase at explicit renewal distances. The general trend of the resulting model is correct, however the in-phase assumption gave a series of heat transfer maxima and minima with distance and required several additional fitted parameters. In practice, the clusters or strands form and detach at random locations with a range of contact times leading to a more continuous behavior of the heat transfer coefficient with height (Fang et al., 1995a).

8.4.1 Parameter values

The renewal models give reasonable trends that agree with the observed heat transfer results. The difficulty in using them for quantitative predictions is the lack of accurate values for the parameters needed in the models. The greatest uncertainty concerns the hydrodynamic parameters. Considering the renewal model at elevated temperature, equations (8.7), (8.12), (8.13), (8.14), and (8.17), the unknown parameters are:

- f: the fraction of surface covered by clusters
- δ: the dimensionless gas layer thickness between the surface and the falling clusters
- ϵ_c: the average cluster solids concentration
- τ_c: the average contact time between clusters and wall
- h_g: dilute phase gas convective heat transfer coefficient
- ϵ_p: particle emissivity
- ϵ_W: wall emissivity

8.4.2 Wall resistance

For small isolated heat transfer surfaces the wall resistance should be the predominant resistance since the particles do not have time to substantially change temperatures as they pass over the surface. Wu et al. (1990) found the value of gas layer thickness between the wall and the particles that best fits the short surface heat transfer data of Basu and Nag (1987) and Kobro and Brereton (1986) as well as their own data. This included particles with diameters 87 to 250 μm and solid density of about 10 to 80 kg/m^3. Reasonable agreement was found with δ, the ratio of gas thickness to particle diameter, $= 0.4$. The direct measurement of the cluster to surface distance by Lints, Figure 8.4, resulted in comparable values between one-half and one-third of a particle diameter. However, Lints found that the gap decreased as the cross-section averaged solids density increased while Wu's comparisons suggested a modest increase of gap width with solids density. Interestingly this value for the gas layer thickness is not far from that measured by Decker and Glicksman (1981) and Glicksman et al. (1994b) for surfaces immersed in bubbling beds; they found an average value of 1/6 for δ.

8.4.3 Cluster wall coverage

Data by several investigators for the fraction of wall covered by clusters are shown on Figure 8.3. In this case only measurements made immediately at the wall are included for wall coverage. Note that coverage increases with cross-sectional averaged solids fraction.

Data for large risers are missing; the trend in Figure 8.3 suggests that at the same cross-sectional average solids density, larger columns have a higher percentage wall coverage. Thus the data represent a lower limit to cluster coverage at the wall of a large commercial bed.

8.4.4 Cluster solids concentration

The average cluster solids concentration at the wall is found to be a function of cross-section average concentration. Lints (1992) found the average cluster solids concentration could be closely represented by

$$\epsilon_c = 1.23(1 - \epsilon_{\text{cross section}})^{0.54} \tag{8.19}$$

A correlation for wall concentration by Zhang et al. (1991) agrees closely with the smaller bed results at higher cross-section averaged solids concentration.

8.4.5 Contact time

If the clusters remain at the wall for a substantial time, the transient conduction in the cluster is the dominant resistance for heat transfer. In the semi-infinite model, the parameter which is most difficult to predict is the contact time of the cluster at the wall. In the previous section it was demonstrated that the distribution of contact times is a minor factor if the mean contact time can be established (see also Fang et al., 1995a).

Investigators have more or less agreed on the measured values of the falling velocity of the clusters at the wall (see Table 8.1). The contact time can be estimated by taking a constant falling velocity

$$t_c = \frac{L_c}{u_c} \qquad (8.20)$$

The key unknown factor becomes the mean distance a cluster falls along the wall before it is displaced. If the contact distance is long, e.g. 1 m or more, then the precise value of L_c is not crucial. Furthermore, as shown in Table 8.3, for long contact distances the acceleration of the cluster from its initial velocity to u_c has a modest influence on t_c. On the other hand, for shorter contact distances, e.g. 100 mm, acceleration effects are important.

Overriding these considerations is the limited knowledge available on the magnitude of the mean contact distance. Visual and video observations at the smooth wall of a 0.2 m diameter circulating bed at the author's laboratory showed some clusters that could be viewed falling along the wall for a distance of at least 0.5 m. The identifiable clusters may not be representative of the average.

Wu et al. (1991) used two fast heat transfer probes with different vertical spacings. From the maximum cross-correlation at differing separation distances they estimated a mean residence length of strands; they caution that at longer separation distances the degree of cross-correlation decreases rapidly. For smooth walls they found that the mean residence length could be expressed as

$$L_c = 0.0178 \bar{\rho}^{0.596} \qquad (8.21)$$

with L_c in m and $\bar{\rho}$ in kg/m^3. Visual observations of a vertical membrane wall indicated that the clusters remained at the wall for a longer falling distance. Wu et al. (1989) found that the local heat transfer coefficient continually decreased with distance over a 1 m long membrane surface. This suggests that L_c is of the order of 1 m for a membrane wall.

Burki et al. (1993) measured the local heat transfer versus height for a bed with smooth round walls. For 0.245 mm sand they found a length of approximately 0.3 m where h decreased with distance from the top. For a 97 μm FCC catalyst, the length over which h decreased was 0.45 m, suggesting L_c is larger than the values given by Wu for smooth surfaces. Interestingly, Burki et al. (1993) also found in the circulating riser a thermal entrance region at the lower end of the heat transfer surface (the heater section started 11 m above the distributor) similar to their results for single-phase gas flow.

Golriz and Leckner (1992) inserted a horizontal obstacle at the membrane wall of a circulating bed combustor. The obstacle deflected downward moving strands away from a section of the wall. They found that the thermal boundary layer at the wall (roughly 0.2 m thick) had a higher temperature for a length of 1.5 m below the obstacle. Based on these observations they stated that the residence length of strands in the zone near the surface is greater than 1.5 m.

At present, it can be concluded that vertical membrane walls result in longer residence lengths for clusters at the wall than flat smooth surfaces. The specific values to use for L_c are still uncertain. For commercial beds, results are even more uncertain since small roughness elements and obstacles disrupt the clusters moving them away from the wall and reducing L_c.

8.4.6 Parametric trends

Using the correlations for δ, f, and ϵ_c shown in Figures 8.4 and 8.3 and equation (8.19), respectively, and with $u_c = 1$ m/s, the relative magnitude of the heat transfer resistances can be evaluated and the influence of bed parameters investigated. Figure 8.8 illustrates the relative magnitude of the wall heat transfer coefficient, h_W, given by equation (8.12), to the time-averaged heat transfer coefficient for the homogeneous cluster, equation (8.7). As the mean distance the cluster falls along the wall, L_c, increases,

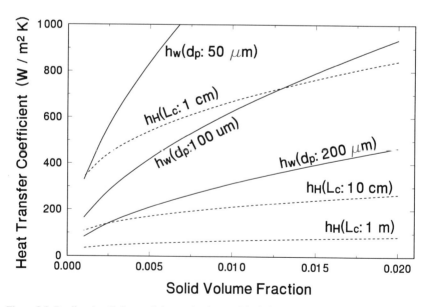

Figure 8.8 Predicted wall, h_W, and cluster, h_H (or emulsion), heat transfer coefficient for different particle diameters, d_p, and mean contact distances at the wall, L_c, at 20°C.

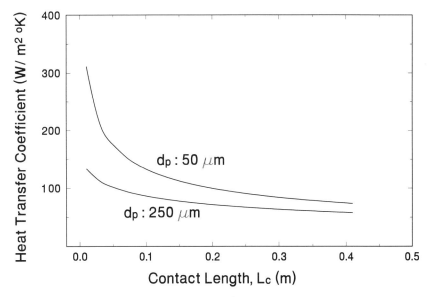

Figure 8.9 Predicted overall heat transfer coefficient for two different particle sizes versus mean contact distance at the wall. Solid volume fraction: 0.01; f, 0.7; U_g, 7 m/s; T, 20°C.

heat transfer through the cluster becomes the dominant resistance. When L_c has a value of 10 mm and the particle diameter is 0.1 mm, the two resistances to heat transfer are roughly equivalent.

The average heat transfer coefficient decreases as the mean distance traveled by the cluster along the wall increases (Figure 8.9). These results were obtained with a wall coverage of 70%. For longer mean distances, predictions for different particle diameters merge since the wall resistance becomes negligible.

Using these relationships for a small unit at room temperature and at a volumetric concentration of 10^{-3}, the predicted particle and gas convection are roughly equivalent. At a volumetric concentration of 10^{-2}, particle convection is dominant.

Figure 8.10 shows a comparison of heat transfer mechanisms at elevated temperatures where radiation is important. At the low volumetric concentration of solids, radiation to the surface uncovered by clusters is the largest mode of heat transfer. The surface is assumed to radiate to particles at the mean temperature. Shading of the wall by cooler particles in the annulus is not considered so this is an upper limit for radiation. As the concentration rises particle convection becomes most important. Note for the case considered, 100 μm particle diameter with a 30 cm mean residence length, the influence of radiation on particle convection is negligible. One extreme case is shown: the clusters are assumed to be opaque with the surface adjacent to the wall radiating as a black body. This is only valid for very

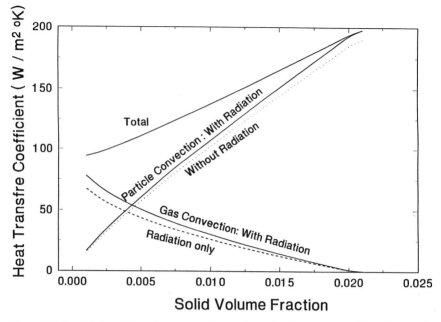

Figure 8.10 Predicted particle and gas convection at elevated bed temperature. The influence of radiation is shown for the portion of the surface covered by the clusters and dilute phase, respectively. T_B, 800°C; T_w, 100°C; d_p, 100 μm; U_g, 7 m/s; L_c, 30 cm. Cluster is assumed opaque, for dilute phase radiation incident on the wall is at core temperature.

high cluster solid fraction. At lower cluster solid fraction, the contribution of radiation is much larger; after 1 s at the wall, radiation is roughly one-half its initial value.

Although most models are based on a renewal mechanism for a continuum at the wall, there are several exceptions. Molerus (1993) contends that particle convection is negligible and gas convection in a rather thick dilute space between the wall and the clusters controls the heat transfer from the clusters to the walls. This is unlikely given the detailed measurements of clusters within one particle diameter of the wall (Lints and Glicksman, 1993) and the independence of heat transfer with superficial gas velocity at a constant cross-sectional particle voidage (e.g. Ebert *et al.* 1993; Tuzla *et al.*, 1991).

Mattmann (1991) has taken heat transfer data at very high pressures, up to 50 bar and high Reynolds numbers, where gas convection enhances the value of h_w, similar to the results of Decker and Glicksman (1983) for bubbling beds.

Martin (1984) proposed a model based on temperature change of only the particles adjacent to the wall. This requires that the cluster time constant is less than one-third the thermal time constant of a particle. For typical contact times of modest sized particles at the wall (Table 8.2) and particle time constants (Table 8.3) it can be seen that this does not hold. The renewal model with wall resistance is preferable since it holds for both short and long contact periods.

8.5 Advanced considerations

8.5.1 Radiation heat transfer

At elevated bed temperature radiation heat transfer becomes an important mechanism that acts in parallel with particle convection and gas convection. Radiation can be dealt with in a straightforward fashion for several limiting conditions. At very low particle concentrations the wall is relatively free of clusters while the bed can still be considered opaque. Radiation can be dealt with as the heat flux between two opaque bodies. At higher solids concentration, a thermal boundary layer has been observed in the downflowing region near the wall of large combustors. This region of intermediate temperature particles is due to radiative heat transfer from the particles to the cold wall. The quantitative treatment of this would involve an energy balance on particles in the annulus. The energy balance requires knowledge of the radial exchange of particles between the annulus and the core, which has only recently been explored (Westphalen and Glicksman, 1995; Zhou et al., 1995).

The downflowing region consists of alternate regions of high-density clusters and dilute solids–gas flow. When a cluster in the downflowing region is in front of the wall, direct radiation between the wall and the core is substantially reduced. When a dilute zone is adjacent to the wall a substantial amount of direct radiation between the core and the wall takes place. At present, a definitive estimate of the net heat transfer from the uncovered wall to core is lacking.

When clusters are at the wall, direct radiation between the wall and the core is prevented. There is radiation between hot particles in the cluster and the wall. If the cluster has a tight solids packing approaching a packed bed or a dense phase of a bubbling bed, only particles in the first one or two rows radiate directly to the wall. The temperature of these particles rapidly drops due to both radiation and conduction to the adjacent wall. In this case radiation rapidly becomes insignificant and for typical cluster transit times at the wall, 1 s or more, radiation has a secondary role compared to particle convection.

Recent experiments (Lints, 1992) have shown that the volumetric solids concentration for clusters at the wall is only 0.1 to 0.2. The particles are more widespread so that particles 5 or 10 rows from the wall can contribute direct radiation to the wall. Particles this far from the wall retain a high temperature for a much longer period than the particles directly adjacent to the surface. In this case the radiant flux to the wall remains significant, even after one or more seconds of contact time.

To estimate the radiant flux to the wall under these conditions, the Mickley type model can be used, assuming that the cluster is a homogenous semi-infinite medium. The cluster is separated from the wall by a gas layer with an equivalent distance of $d_p/3$ through which heat can be transferred by conduction and radiation. Initially the cluster is assumed to be at the bulk mean temperature of the bed. After the cluster contacts the wall, transient heat transfer takes place by conduction and radiation through the cluster. The conduction can be dealt with by use of an effective conductivity. The medium absorbs and re-emits radiation characterized by an absorption coefficient that can be given by the ratio of the surface area of particles per unit volume:

$$K = \frac{1}{4} \frac{A_{particles}}{\text{Volume}} = \frac{2}{3} \frac{\epsilon_c}{d_p} \qquad (8.22)$$

The extinction coefficient is the reciprocal of the mean free path. For an optical thickness Ky of one, y must exceed 6 mean particle spacings at a solids concentration of 0.2.

The energy balance for a layer within the cluster at a distance y from the wall becomes,

$$\rho c \frac{\partial T}{\partial t} = k_e \frac{\partial^2 T}{\partial y^2} + \frac{dq_r}{dy} \qquad (8.23)$$

The first two terms in the energy balance are the standard internal energy change and conduction in the medium. The last term represents the radiation contribution to the layer due to radiant exchange between the layer and all other layers as well as the walls. q_r can be found from a variety of techniques, e.g. Hottel and Sarofim (1967). Equation (8.23) is solved for the temperature distribution as a function of time and location y. The resulting temperature distribution can then be used to calculate the conductive and radiative flux to the wall.

Assuming that the cluster is a homogenous absorbing medium with a unidirectional temperature field, the radiant flux can be expressed as:

$$q_{rad} = \epsilon_{wall} \left[\int_0^\infty \sigma T^4(y) E_2(ky) d(Ky) - \sigma T_{wall}^4 \right] \qquad (8.24)$$

Figure 8.11 shows the solution for typical particle parameters, assuming that the wall and the particles are black bodies. Note that the radiative flux has only fallen by about 50% after one second. At this time radiation and conduction are both important. The time-average total heat transfer coefficient after 1 s is 247 W/m²K; this drops to 164 W/m²K when radiation

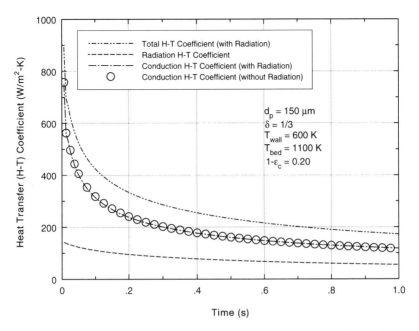

Figure 8.11 Instantaneous radiation and conduction heat transfer coefficients from a cluster to the wall. d_p, 150 μm; ϵ_c, 0.15; δ, $d_p/3$; T_{Bed}, 1100 K; T_{Wall}, 600 K.

is omitted. Further, the conduction contribution is substantially unchanged by the presence of the radiation.

Thus, as a first estimate the average heat transfer from the clusters to the wall can be approximated by the superposition of conduction and one-half to three-quarters of the initial radiant flux between the cluster and the wall. Further work is underway to refine these estimates.

8.5.2 Wall coverage

Measurements at the wall of circulating beds have revealed that the wall does not contain a continuous layer (Lints and Glicksman, 1993; Wu *et al.*, 1991; Rhodes *et al.*, 1992). Rather, the wall is partially covered by a series of clusters that move downward along the surface. The appearance of clusters or large wall density fluctuations correspond to enhanced instantaneous heat transfer coefficients (Wu, 1989; Dou *et al.*, 1992). At higher bed densities the wall becomes fully covered with clusters; however, the densities and the heat transfer coefficients reported for typical commercial circulating bed combustors suggest that the upper portions of the riser will not be fully covered with clusters. For example, Leckner and Andersson (1992) measured temperature fluctuations in the boundary layer of a 1.7 m square combustor, which indicate the periodic appearance of hot particle clusters. Thus, the average heat transfer from the bed to the wall is a function of the wall coverage. It is important to relate the fraction of wall coverage to the hydrodynamic conditions of the bed.

8.5.3 Deposition rate to the wall

The wall coverage is determined from a balance between the deposition of clusters from the core to the wall and the shedding or stripping of clusters from the wall. Rhodes et al. (1992) studied videos of the wall region and concluded that swarms, or clusters, were formed as particles were ejected from the core and thrown against the wall. The wall concentration should be determined by a balance between the rate at which clusters or particles transported to the wall region from the core, less the rate at which clusters shed from the wall, rejoin the core. If the mean time for shedding is t_w and the shedding process is random in time, then the fraction of surface coverage lost due to shedding in time dt is

$$df = -\frac{f \, dt}{t_w} \tag{8.25}$$

For an element of the wall with a radial volumetric flux of particles from the core per unit area J_r, conservation of mass for the wall layer requires

$$\frac{d(\delta(\epsilon_c)f)}{dt} = J_r - \frac{f}{t_w}\delta_w(\epsilon_c) \tag{8.26}$$

where δ_w is the mean layer thickness upon shedding. At steady state with a uniform coverage along the wall, the left-hand side is zero and equation (8.26) can be solved for the wall coverage if δ_w is known.

The wall coverage is also a function of the distribution of clusters at the wall. If the deposition is a random process then some of the clusters will combine and overlap existing clusters rather than filling dilute zones at the wall. Two extremes are random spacial distribution of clusters deposited at the wall and uniform distribution of clusters at the wall. If the coverage is random, i.e. the location of incoming clusters that strike the wall is independent of the locations of clusters already at the wall, then the coverage can be described by a Poisson distribution, i.e. the probability of finding a layer at the wall n particles thick is equivalent to having a two-dimensional surface with n particle centers within a diameter of d_p. The probability of a layer n diameters thick is

$$P_n = e^{-\bar{n}}\left(\frac{\bar{n}^n}{n!}\right) \tag{8.27}$$

where $\bar{n}$ is the mean wall layer thickness. The probability of coverage at least one particle layer thick is

$$P_{1+} = 1 - e^{-\bar{n}}\left(\frac{\bar{n}^0}{0!}\right) = 1 - e^{-\bar{n}} \tag{8.28}$$

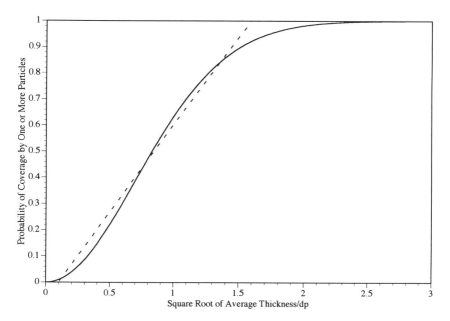

Figure 8.12 Probability of wall coverage as a function of average thickness of particle layers at the wall.

Figure 8.12 shows P_{1+} versus $(\bar{n})^{1/2}$. Note that in the random case, the coverage varies approximately with the square root of the average total thickness of material at the wall.

Rewriting $\bar{n}$ as $f\delta_w$

$$f \approx (f\delta_w)^{1/2} \tag{8.29}$$

when the left hand side of equation (8.26) is zero,

$$J_r \approx \frac{f\delta_w}{t_w} \approx \frac{f^2}{t_w} \tag{8.30}$$

If J_r varies linearly with the cross-sectional concentration, then from equation (8.30), f varies with the square root of the cross-sectional concentration. This may help to explain why the measured heat transfer in most studies varies with the square root of the cross-sectional concentration.

Bolton and Davidson (1988) proposed a model for a circulating bed analogous to that for gas–liquid annular flow where the rate of deposition per unit surface area is proportional to the mean concentration in the core. However, as Bolton and Davidson pointed out, a diffusion model seems inappropriate in light of the positive radial flux from the core to the wall in the presence of an increasing particle concentration gradient in the radial direction. Particles in the dilute phase have a high dispersion rate while clusters move radially at a much slower rate. The dilute phase may have a radial gradient opposite to the cluster radial gradient. It is expected that the deposition rate should be a function of the local solids density in the core. The deposition rate and the corresponding removal by cluster shedding are the key mechanisms linking the wall and core regions. Data and correlations from annular gas–liquid flow are not directly applicable if the radial deposition is due to radial movements of clusters. The gas–liquid results apply to much smaller elements, a maximum particle size of 260 μm, with fully developed flow in pipes of 10 to 20 mm diameter. Large clusters have relaxation times well outside the range of current models and the turbulent momentum to mass transfer analogy breaks down.

8.6 Thermal and dynamic scaling

8.6.1 Hydrodynamic scaling

There is persistent concern about the validity of applying results based on laboratory or pilot plants to larger circulating fluidized beds. As we have seen, heat transfer is controlled to a large extent by the bed hydrodynamic behavior and is dependent on the cross-sectional averaged concentration. Heat transfer parameters, such as wall coverage, cluster solids concentration, and cluster-to-wall spacing are all dependent on the hydrodynamics.

Because of the uncertainties in bed behavior as the riser size increases or the operating conditions shift, the hydrodynamics and bed-to-wall heat transfer of larger beds cannot be confidently predicted from empirical relationships or computational models.

A comparatively new technique has emerged to aid the designer: the use of experimental scale models of commercial beds. By non-dimensionalizing the governing equations of gas and solid momentum transfer along with a representative model of gas–solid force interaction, a set of governing dimensionless parameters has been identified (Scharff et al., 1978; Glicksman, 1984):

$$\frac{U_o^2}{gD}, \frac{\rho_f}{\rho_s}, \frac{\rho_f U_o D}{\mu}, \frac{\rho_f U_o d_p}{\mu}, \frac{L}{D}, \frac{G_s}{\rho_s U_o}, \Phi_s, \text{particle size distribution} \quad (8.31)$$

This set neglects electrostatic forces and assumes that the particle coefficients of restitution and sliding friction has a negligible influence (Litka and Glicksman, 1985).

If a laboratory model is constructed so that the dimensionless parameters are the same as a larger commercial bed, possibly operated at elevated temperature and pressure, the small model and the larger bed will exhibit identical hydrodynamic behavior. A large body of experimental results have shown the validity of these scaling parameters for bubbling beds (Nicastro and Glicksman, 1984; Newby and Keairns, 1986; Roy and

Davidson, 1989; Almstedt and Zakkay, 1990), as well as for circulating beds (Glicksman *et al.*, 1991; Chang and Louge, 1992; Glicksman *et al.*, 1993).

One drawback of this scaling technique is a lack of flexibility. Once the fluidizing gas for the model is chosen, e.g. air at STP, the model dimensions are fixed relative to the commercial bed dimensions. If the model is fluidized with air at standard conditions, the model of a commercial atmospheric bed operating at 800°C has linear dimensions roughly one-quarter those of the commercial riser. The air fluidized model of a commercial bed at 800°C and 12 bar must have linear dimensions roughly equal to those of the commercial bed.

Recently it has been shown (Glicksman *et al.*, 1993) that the number of dimensionless parameters can be reduced in the limit of high particle Reynolds number, where gas inertial forces dominate the gas–solid force, as well as in the limit of low particle Reynolds number, where gas viscous force dominates. Both limits reduce to an identical simplified form of the dimensionless parameters.

$$\frac{U_o^2}{gD}, \frac{\rho_f}{\rho_s}, \frac{U_o}{U_{mf}}, \frac{L}{D}, \frac{G_s}{\rho_s U_o}, \Phi_s, \text{particle size distribution} \qquad (8.32)$$

Furthermore, at intermediate values of the particle Reynolds number, the simplified form, equation (8.32), yields results that are approximately valid. Glicksman *et al.* (1993) experimentally verified the simplified form for a circulating fluidized bed with glass and plastic particles. In addition, with the simplified form, an atmospheric circulating fluidized bed combustor was closely simulated by a cold model with linear dimensions 1/16 of the combustor dimensions.

Recently Glicksman *et al.* (1995a,b) have shown close agreement between a pressurized bubbling bed combustor at 9 bar and a 1/4 scale cold atmospheric pressure model and between a pressurized circulating bed combustor at 14 bar and a 1/2 scale cold atmospheric pressure model, respectively. In both cases the models were designed using the simplified scaling rules (equation (8.32)).

The small-scale experimental models of a larger commercial bed can be used to determine hydrodynamic parameters that influence bed-to-wall heat transfer. For example, current research is underway to determine the influence of column diameter on the fraction of the wall covered by clusters.

8.6.2 Thermal scaling

The scaling concept can also be used to experimentally measure heat transfer components. If the scale model is kept near ambient temperature, radiation cannot be properly scaled relative to convection. Particle convection can be simulated with a small-scale model. An accurate heat transfer simulation requires both hydrodynamic scaling, using equation (8.32), as well as proper thermal scaling. The governing equations for particle convection have been non-dimensionalized using either a continuum model, equation (8.10), or a model of discrete particles, equation (8.4), together with the expressions for effective conductivity (8.8), and wall resistance, equation (8.12). The continuum and discrete particle model give the same result. The governing thermal parameters for particle convection are

$$Nu = \frac{hd_p}{k} = f\left[\frac{U_0 d_p^2 \rho_p c_p}{Lk_f}, \frac{k_s}{k_f}\right] \quad (8.33)$$

In addition, the hydrodynamic parameters, equation (8.32), must be simulated so that parameters such as wall coverage and cluster voidage have the same value as in the commercial bed. The first thermal parameter in equation (8.33) can also be expressed as the ratio of the thermal time constant of the particle at the wall to the transit time of the cluster descending a distance L along the wall. The second factor, the ratio of solid to fluid conductivity, has a comparatively weak influence on the effective cluster to fluid conductivity ratio. Glicksman and Hyre (1994) carried out experiments on two geometrically similar beds with equal values of hydrodynamic

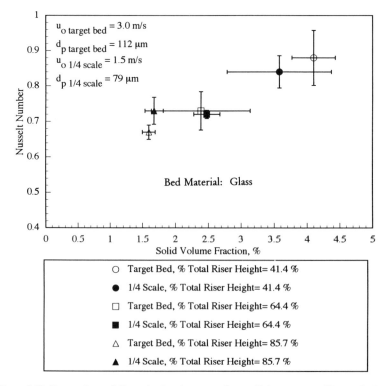

Figure 8.13 Comparison of dimensionless heat transfer coefficient measured in two hydrodynamically and thermally similar CFB units. The linear dimensions of the units differ by a factor of 4 (Glicksman and Hyre, 1994).

parameters, equation (8.32), and the thermal parameters, equation (8.33). The two columns, fluidized with air at ambient conditions, had linear dimensions that differed by a factor of 4. To maintain hydrodynamic similarity the particle sizes were different for the two beds. The results of the wall heat transfer for the two beds show close agreement when compared in dimensionless form (Figure 8.13).

Ackeskog *et al.* (1993) compared heat transfer results from two pressurized beds, the smaller one scaled using the full set of scaling laws, equation (8.31). They found agreement with the large-scale heat transfer within 19%.

The reader is referred to a recent review of scaling in fluidized beds (Glicksman *et al.*, 1994a) for a more in-depth discussion.

8.7 Heat transfer measurement techniques

Before reviewing experimental results it is helpful to review heat transfer measurement techniques. The overall heat transfer coefficient can be determined from measurements of the rate of heat transfer, the wall temperature and the cross-section averaged temperature. In ambient temperature beds a portion of the surface is heated; at low temperature levels the direction of heat flow (to or from the wall) should be immaterial. If only a small area is heated, heat loss from the periphery of the element to the surrounding surface must be limited or accounted for. As pointed out above, results for short active surfaces can differ considerably from long surfaces. A heated dummy section above and below the actively monitored section will give results typical of a long surface if all sections are kept at the same temperature and the junction between the sections is kept perfectly smooth. (A step as small as one particle diameter in height can cause substantial disruption of the clusters along the wall.)

For measurements in hot combustors with cold membrane walls, thermocouple locations in the walls are limited. Fin effects in the membrane wall must also be considered by careful calibration and simulation (Andersson and Leckner, 1992; Jestin *et al.*, 1992). Use of chordal thermocouples to estimate wall temperatures and heat flux may introduce substantial uncertainty in the measurements due to uncertainties in temperature measurement, thermocouple location and heat transfer variation along the tube and fin surface, as well as uncertainties in the heat transfer coefficient for two-phase gas–liquid flow inside membrane wall tubes.

8.8 Heat transfer results – laboratory-scale beds

Almost all heat transfer data currently available have been taken from laboratory beds of modest diameter, 0.2 m or less. Dou *et al.* (1992) and

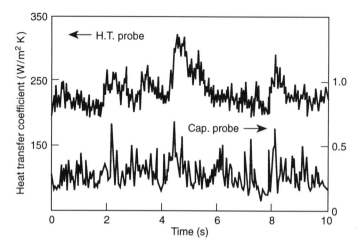

Figure 8.14 Typical trace of instantaneous heat transfer and solids concentration recorded simultaneously. 152 mm diameter circulating bed with 171 μm Ottawa sand, cross-section averaged solid density 47 kg/m³ (Wu et al., 1991).

Wu et al. (1991) have carried out simultaneous measurements of time-resolved heat transfer at a local wall section along with solids concentration close to the same location. Typical results are shown in Figure 8.14. Because of its mass the heat transfer probe will not respond as rapidly as the concentration probe. The close correspondence between heat transfer and concentration peaks is a strong indication of particle convective effects.

The average heat transfer coefficient is found to rise as the mean concentration increases (Dou et al. 1992; Chen and Chen, 1992). There is a strong correlation between these two parameters over two orders of magnitude of the solid fraction, with h varying as the square root of the solids density at the wall as predicted by the renewal model, confirming this modeling approach.

Wu et al. (1989) measured the variation of the local heat transfer coefficient with vertical distance on a 1.59 m long heat transfer surface. The results (Figure 8.15) show a monotonic decrease of heat transfer coefficient with distance down the heater. The results of a renewal model, described in section 8.4, are shown for a bed temperature of 407°C using a cluster residence length of 1.5 m. One-half of the isothermal radiant flux is added to the particle convection. Note that the model predicts local values of h, while the measurements give an average value over a finite distance, explaining the discrepancies of the points near the top, i.e. at small z.

Figure 8.16 shows the results of five different investigations in small low temperature laboratory risers of diameter 100 to 150 mm. Least squares linear fits on logarithmic scales are shown. All of these tests were carried out with a short, 10 to 25 mm long, actively heated section. Most results

HEAT TRANSFER IN CIRCULATING FLUIDIZED BEDS 295

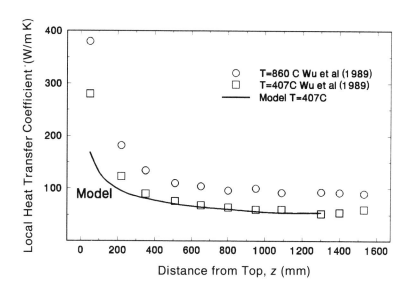

Figure 8.15 Measured local heat transfer along a long membrane wall at two suspension temperatures. 241 μm mean particle size, 152 mm square combustor (Wu *et al.*, 1989). Prediction based on model given in section 8.4.

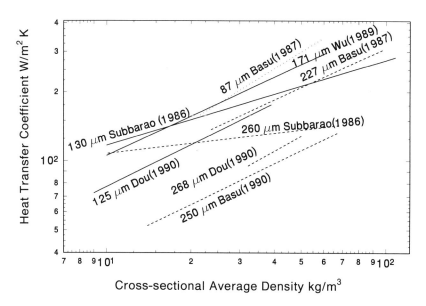

Figure 8.16 Heat transfer coefficient for different particle diameters with active heat transfer surface length of 1–2.5 cm for laboratory units at ambient conditions.

indicate that the heat transfer coefficient varies approximately with the square root of the average cross-sectional suspension density. There is a large variation of heat transfer with particle diameter, tests with small particles yielding higher heat transfer coefficients. Since the clusters are only in contact with the small heat transfer surface for a short time the cluster temperature does not change appreciably. The heat transfer resistance is confined to the cluster-to-wall interface, which is proportional to the particle diameter. Figures 8.17a, b and c show the detailed data for each particle size range. Also shown are the values predicted by the renewal model with surface resistance summarized in sections 8.3 and 8.4.

The increase in overall heat transfer with solids density is caused by the increased wall coverage. At low densities only a small fraction of the wall area is covered by clusters, this area fraction increases with the cross-sectional average solids concentration. The cluster heat transfer rate is much larger than the heat transfer on the surface area covered by the dilute phase. The model captures the correct trend in the data, adding confidence in the basic physical assumptions used. Note that for similar conditions, e.g. Figure 8.17c, there is considerable disagreement between the results of different investigators. However, the overall trend of increasing heat transfer coefficient with small particle sizes is clear when comparing Figures 8.17a, b and c.

When a longer active heat transfer surface is used, of the order of 1 m, as shown in Figure 8.18, the influence of particle size is substantially reduced. In this case, the cluster temperature changes appreciably if the cluster remains at the surface for a good portion of the heater length. The thermal resistance of the cluster is larger than the cluster-to-wall interface resistance. Each line represents the best fit to the data of that particular investigation and particle size.

Cross-plotting the heat transfer coefficient versus particle diameter at a fixed density of $20 \, \text{kg/m}^3$ (Figure 8.19) clearly confirms that small heater lengths have higher heat transfer coefficients than long heaters. The short heaters are also more sensitive to particle size. The model for the longer heater assumes $L_c = 0.5 \, \text{m}$, half of the overall heater length. When the heater is long and less than fully covered by clusters, fresh clusters can be added to the surface and leave the surface at any vertical location. The average cluster contact distance is less than the heater length. Except for one data point of Furchi (1988) at $109 \, \mu\text{m}$, the test results for the long heated surface shows a weak influence of particle diameter. The model captures the average trend of the data from the different studies.

8.8.1 Elevated temperature

Andersson *et al.* (1987a, b) present heat transfer results for a 2.5 MW circulating fluidized bed combustor with a bed cross-section of 70 cm × 70 cm and

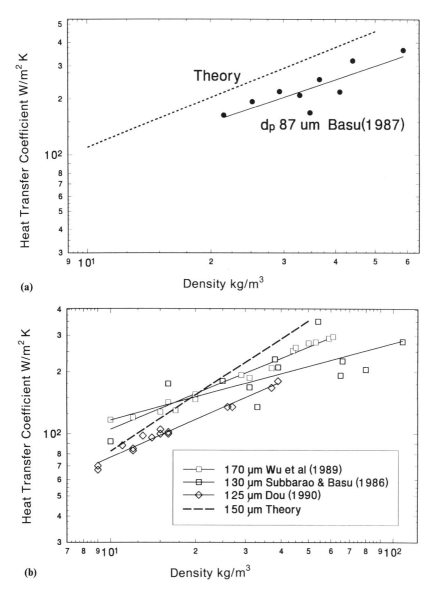

Figure 8.17 Data and prediction at ambient conditions, laboratory units, short active heat transfer surface length; (a) $d_p = 87\,\mu m$, (b) $d_p = 125$–$170\,\mu m$, (c) $d_p = 227$–$260\,\mu m$.

8.5 m high. They used 240 micron Olivine sand. Over the temperature range of the data, the increase of the overall heat transfer coefficient corresponds closely to the increase in the radiation heat transfer from the particles at the bulk bed temperature to the wall. The increase of heat transfer with temperature remains about the same for all densities. These experiments

(c)

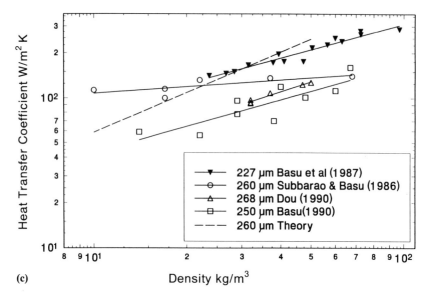

Figure 8.17 continued

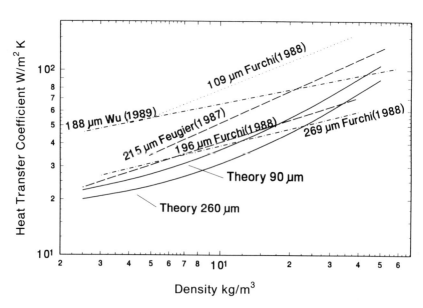

Figure 8.18 Heat transfer coefficient for different particle diameters, long active heat transfer surface length ~1 m, laboratory beds at ambient conditions.

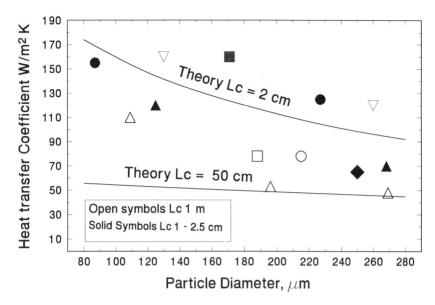

Figure 8.19 Heat transfer coefficient versus particle diameter for different active heat transfer surface lengths, laboratory beds at ambient conditions and density of 20 kg/m^3. Data: Furchi et al. (1988) △; Wu (1987) □; Feugier (1987) ○; Basu and Nag (1987) ●; Wu (1989) ■; Subbarrao and Basu (1986) ▼; Dou (1990) ▲; Basu (1990) ◆.

were conducted with a small active surface so that clusters did not cool substantially while they were in contact with the active surface.

Mattmann (1991) and Xianglin et al. (1990) carried out measurements of bed-to-wall heat transfer in circulating beds at different pressure levels from atmospheric to 0.6 MPa (Xianglin), and 1 to 50 bar (Mattmann, 1991). Xianlgin found a substantial increase at the same cross-sectional density as the pressure was raised for large 0.77 mm particles. A much smaller effect was observed for 0.25 mm particles. With the large particles the wall resistance is more predominant and the pressure increase can substantially reduce the resistance. Mattmann measured a 100% increase of h as the pressure went up by a factor of 4 to 5 for particles of 165 and 194 μm.

Basu (1990) compared heat transfer measurements of various investigators in high temperature beds, all with roughly the same particle diameter and cross-section averaged density. The results of three investigations using a very short active heat transfer surface at the wall showed a marked increase with temperature. In contrast, the results using a 1.5 m long active surface (Wu et al., 1989), had a more modest increase with temperature. For a long surface the cluster temperature has decreased due to its lengthy contact with the wall reducing the influence of radiation.

8.9 Large units

It is doubtful if heat transfer data taken on small diameter circulating bed risers can be directly applied to large beds. Heat transfer coefficients measured in small units have been correlated as a function of cross-section average particle concentration. This correlation collapses experimental data taken from one unit to a single line irrespective of gas velocity or solids recirculation rate. Of course these parameters are imbedded in the value of the solids concentration. The correlation will not be identical for units of different geometry. It has been shown that short active heat transfer surfaces give higher values of h at a given particle concentration than long surfaces. The limited data available for different riser diameters indicates that the heat transfer coefficient also varies with diameter at a fixed solids concentration. Thus, obtaining heat transfer data for large beds is doubly important.

To obtain confident data for a large bed design several courses of action are available. Data from other large units can be used if the surface geometry, particle size and temperature level are similar. Data from a series of different sized risers can be extrapolated to the diameter in question. However, the expected functional relationship is not well known. Alternatively, a room temperature scale model of the large unit can be built using hydrodynamic and thermal scaling. Data on particle convection and wall coverage is available from the model, and radiation augmentation can be calculated as well as the influence of elevated temperature on gas conductivity.

In this section the limited heat transfer data published for large units are reviewed. Werdermann and Werther (1993) measured heat transfer and temperature distribution near the walls of two large combustors, one with a diameter of 8 m and a rated power of 226 MW_{th} while the second has an upper cross-section of 5.13 m × 5.13 m and a rated power of 109 MW_{th}. In the latter plant a thermal boundary layer of approximately 0.2 m thickness was found adjacent to the membrane tube wall (see Figure 8.20). The thickness and relative opacity of the layer suggests that there is substantial radiative heat transfer in this layer between the particles and the wall. The layer also shields the wall from direct radiation from the core. If the lateral exchange between the wall region and the core is modest, the energy loss by radiation will cause a temperature decrease in this region. The authors estimated the radiative contribution based on the measured temperature distribution and an assumed concentration distribution. They found that the radiant flux at the wall was far less than the flux from an isothermal bed to the wall.

Werdermann and Werther compared their results to different heat transfer models. The *closest* model disagreed with the data by almost 50%.

Leckner *et al.* (1991) carried out measurements on a 12 MW atmospheric circulating bed with dimensions 1.7 × 1.7 × 13 m. They found a thermal

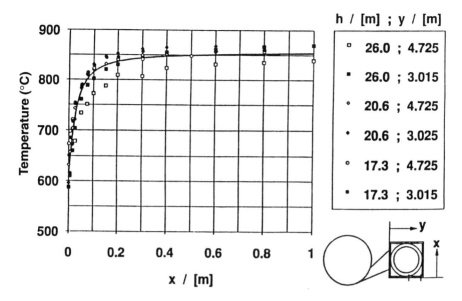

Figure 8.20 Temperature profiles close to the membrane tube wall inside the Flensburg 5.13 × 5.13 m, 109 MW$_{th}$ combustor (x = distance from the fins between the tubes of the membrane wall; h = height above distributor) (Werdermann and Werther, 1993).

boundary layer roughly 0.1 m thick adjacent to a membrane tube wall. They also measured the approximate thickness of the downflowing layer by use of a mechanical 'flag'. The downflowing layer thickness was comparable to the thermal boundary layer. Interestingly, the thermal boundary layer thickness was invariant with distance from the top of the column, in contrast to a single-phase thermal boundary layer, which continuously grows with distance along the wall. The thermal boundary layer in the circulating bed is probably due to the balance between radiation loss from the layer and energy flux by particle transfer between the layer and the core and between the layer and the cold wall. Estimating the radiative mean free path based on the average local solids volume fraction, 0.0025, and mean particle diameter, 260 microns, yields a mean free path of 70 mm comparable to the thickness of the thermal boundary layer. This reinforces the notion that radiative cooling to the wall is responsible, in part, for the measured temperature profile.

The heat transfer coefficients were measured on the membrane walls and neighboring refractory-lined walls by Leckner. The large differences observed are due in part to the thermal boundary layer on the cooled wall and its absence on the refractory-lined surface. Temperature fluctuations were measured near the cooled wall due to hot clusters striking the probe (Figure 8.21). The fluctuations are the result of the non-steady nature of the cluster concentrations within the wall region and the nonuniformity of cluster and

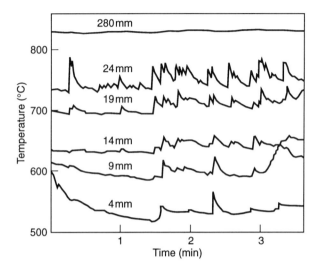

Figure 8.21 Characteristic temperature variations in the boundary layer 11.2 m above the bottom of a square 1.7 m × 1.7 m combustor (Leckner et al., 1991).

dilute phase temperature. The relatively slow response time of the probe masks higher frequency fluctuations.

Heat flux measurements made at the wall of the Chalmers unit approach a constant value corresponding to radiation alone as the bulk density is decreased below 1 kg/m^3. Note that because of the relatively large riser width the calculated bed emissivity due to the heated particles ranges from 0.7 to 0.95 even at these low densities. Gas radiation alone, from the combustion products, should give a bed emissivity in the range of 0.1 to 0.3. The overall heat transfer coefficient for high temperature operation shows the usual increase with suspension density.

Jestin et al. (1992) carried out measurements of heat transfer to the membrane wall, fin and tube, of a 125 MWe circulating bed with upper dimensions 8.6 m by 11 m. They measured the temperature of the tube and fin and determined the value of the heat transfer coefficients on the tube and fin surface that gave the best agreement between the measured temperatures and those predicted numerically. They found that existing correlations did a poor job of predicting the heat transfer rates. Measured heat transfer coefficients were also found to be much higher than those found in a small combustor at the same average concentration. In the course of research on this large boiler, a layer of decreased temperature was also found within 150 mm of the cooled wall. The wall temperature was found to fluctuate, indicative of periodic contact with clusters followed by coverage by a dilute phase.

Divilio and Boyd (1993) have pointed out that heat transfer data from laboratory-scale beds were obtained at suspension densities much higher than those observed in large-scale commercial boilers. Overall heat transfer coefficients for a 110 MWe circulating bed combustor were in the range of 150 to 180 W/m²K (Westsila, 1991). For the upper half of the riser the suspension density was in the range of $5\,\text{kg/m}^3$ or less. Divilio and Boyd contend that at these low particle concentrations, corresponding to a solids volume fraction of about 0.002, the suspension density has a small effect on the overall heat flux since the wall coverage is modest and most of the heat transfer is due to radiation. For radiation between two black bodies at 1550°F and 550°F, the calculated heat transfer coefficient is 148 W/m²K.

Werdermann and Werther present convective heat transfer data of several large beds as a function of the solids volume concentration. The larger beds have significantly higher Nusselt number, based on particle diameter, than small beds with the same solids concentration. Werdermann and Werther explain the increase as due to larger eddies that increase the solids transport to the wall. An alternative explanation is that a larger fraction of the wall is covered by clusters for larger riser diameter, similar to the trends in Figure 8.3. It is interesting to note that the upper portion of the large commercial systems are operated at very low solids concentration, 0.002 to 0.01 in this case. If the concentration was increased, the wall coverage should approach 100% and the influence of unit diameter on heat transfer coefficient would be reduced.

The total heat transfer coefficient versus cross-sectional concentration is shown for five different combustors in Figure 8.22.

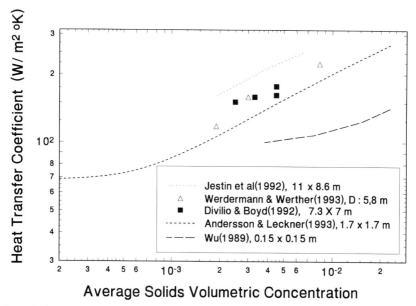

Figure 8.22 Total heat transfer coefficient versus cross-sectional averaged solids concentration for large combustors with long active heat transfer surfaces.

Most of these data were taken on long membrane wall surfaces. All of the results show an increase of h with particle concentration. Note that at a given concentration larger risers give higher heat transfer than small ones. Also, h approximately varies as the square root of the solids concentration. There is a wide difference between commercial units and laboratory size combustors, as exemplified by the data of Wu. At very low concentrations heat transfer should approach a constant of radiation plus gas convection, with the former dominating. The larger units give higher heat transfer because of the larger wall coverage at a given cross-sectional average concentration. At still higher concentrations the walls are completely covered, reducing the influence of riser size. Note that although large combustors run at low average concentrations, the wall coverage is sufficiently large to give a noticeable dependence of h on concentration.

There is a need to develop a working expression of h with riser diameter that can be used with confidence over the range of sizes typical of commercial designs. Additional heat transfer data for large and intermediate size units are needed. A model that explicitly deals with the effects of riser diameter would also be helpful.

8.10 Fins and heat transfer augmentation

As the riser diameter increases the ratio of bed volume to surface area rises. To maintain the bed temperature constant for a combustor the heat transfer coefficient at the wall must be increased or external heat transfer surface must be used. Two possible augmentation techniques for the circulating bed are modification of the surface geometry to increase the heat transfer or the use of fins or wing walls that extend from the wall into the bed interior. In all cases the possibility of enhanced erosion must be carefully evaluated.

Hyre and Glicksman (1995) carried out experiments in which a closely spaced array of very thin horizontal wires were attached to the surface of a plane wall. The wires, with a height of one particle diameter, serve to dislodge the downflowing clusters from the wall. The results, for a small bed, show that up to a 25% increase in heat transfer can be obtained. This technique may be particularly helpful for a membrane wall, a geometry which tends to cause lower surface heat transfer coefficients. The size of the roughness elements may be small enough to minimize erosion effects.

Cao et al. (1994) placed large horizontal rings on a vertical cylindrical surface suspended in a circulating bed. They found the rings increased the heat transfer by 100%. However, the large rings would cause erosion under most circumstances.

In an effort to increase overall heat transfer to walls, fins that extend into the bed interior have been attached to the surface. The gain in heat transfer was generally about 70 to 90% of the increase in surface area due to the fins

(Basu and Ngo, 1993). Nag and Ali Moral (1993) found a modest increase in total heat transfer when pin fins were added to the surface.

With all surface enhancement schemes the incidence of erosion on the extended surface as well as the surrounding wall must be considered. A continuous vertical fin would present the least concern. If substantial internal surface is added to the bed, the overall flow conditions will be affected since the fins will interfere with the downflowing annular region which, in turn, influences conditions in the core.

8.11 Conclusions and recommendations

Although attention to circulating bed heat transfer has increased in the last decade there are still substantial gaps in knowledge. In particular, for large units there is a dearth of experimental data and application of results obtained from small units should be used with caution. Similarly, results for small heat transfer surfaces and smooth planar surfaces differ considerably from those for long surfaces and membrane walls, respectively.

Wall heat transfer is closely tied to the hydrodynamics: the wall coverage by clusters, rate of cluster renewal at the wall and local solids properties near the wall. Confident prediction of heat transfer depends on understanding the hydrodynamics. Factors that influence the hydrodynamics also affect heat transfer. Any heat transfer model or empirical correlation that does not include these hydrodynamic factors will be of limited utility.

There are some limiting cases that should bracket the correct magnitude of the heat transfer. At very low solids concentration the heat transfer from hot beds should be dominated by thermal radiation from bed to wall. At high solids concentration, the wall should be fully covered by clusters, and particle convection and radiation are both important for hot beds. If the solids concentration in the cluster at the wall approaches that of dense bubbling beds, radiation to long heat transfer surfaces will become insignificant.

To determine particle convection, appropriately scaled small experimental models are useful. Such models will also help to explore the importance of proposed design changes on hydrodynamics and, in turn, bed-to-wall heat transfer.

Acknowledgements

Research activities by the author on circulating bed heat transfer have been sponsored by the National Science Foundation, the Electric Power Research Institute and the US Department of Energy. A number of graduate students at Massachusetts Institute of Technology have participated in these studies. Their efforts have been very helpful to the author. Particular thanks to Paul Farrell, Matt Hyre, Michael Lints and Peter Noymer who assisted in recent studies and calculations used in this chapter.

Nomenclature

A	surface area
c	specific heat
C	cross-sectional averaged solids volume fraction
c_p	specific heat of particles
D	bed diameter or width
d_p	particle diameter
E_2	second exponential integral
f	fraction of surface covered by clusters
f_W	mean time cluster remains at wall
G_s	solids recycle rate per unit area
g	acceleration of gravity
h	heat transfer coefficient
h_c	overall cluster-to-wall heat transfer coefficient
h_g	dilute phase heat transfer coefficient
h_H	time averaged heat transfer coefficient of homogenous cluster
h_W	heat transfer coefficient at cluster–wall interface
J_r	solids flux radially to wall
K	extinction coefficient
k	thermal conductivity
L	length
L_m	mean beam length
M	mass
N	number of clusters
Nu	Nusselt number
n	number of particles per unit length
P	probability
q	rate of heat transfer
T	temperature
t	time
t_W	mean time cluster remains at wall
U_o	superficial gas velocity
u	velocity
v	time varying cluster velocity
x	vertical distance
y	coordinate normal to wall
δ	dimensionless gap thickness; cluster layer thickness

ϵ	void fraction
ϵ_c	volume fraction of solids in cluster
ϵ_{eff}	effective emissivity of cluster
δ_W	mean thickness of cluster at wall
μ	viscosity
ρ	density
σ	Stefan–Boltzmann constant
τ	thermal time constant
Φ_s	sphericity

Subscripts

B	bed
c	cluster
e	effective
f	fluid
g	gas
max	maximum
mf	minimum fluidization
$overall$	overall coefficient
p	particle
rad	radiant
s	solid
t	instantaneous
W	wall

References

Ackeskog, H.R.B., Almstedt, A.E. and Zakay, V. (1993) An investigation of fluidized bed scaling: heat transfer measurements in a pressurized fluidized-bed combustor and a cold model bed. *Chem. Eng. Sci.*, **48**, 1459–1473.

Almstedt, A.E. and Zakkay, V. (1990) An investigation of fluidized-bed scaling – capacitance probe measurements in a pressurized fluidized-bed combustor and a cold model bed. *Chem. Eng. Sci.*, **45**, 1071–1078.

Andersson, B.A. and Leckner, B. (1992) Experimental methods of estimating heat transfer in circulating fluidized bed boilers. *Int. J. Heat Mass Transfer*, **35**, 3353–3362.

Andersson, B.A. and Leckner, B. (1993) Local lateral distribution of heat transfer on tube surface of membrane walls in CFB boilers, in *Circulating Fluidized Bed Technology IV* (ed. A. Avidan), AIChE, New York, pp. 311–318.

Andersson, B.A., Johnson, F.J. and Leckner, B. (1987a) A probe for heat flow measurements in fluidized bed boilers. Presented at IEA AFBC Technical Meeting.

Andersson, B.A., Johnsson, F.J. and Leckner, B. (1987b) Heat flow measurement in fluidized bed boilers. Presented at International Conference Fluidized Bed Combustors, Finland.

Bader, R., Findlay, J. and Knowlton, T.M. (1988) Gas/solid flow patterns in a 30.5 cm-diameter circulating fluidized bed, in *Circulating Fluidized Bed Technology II* (eds. P. Basu and J.F. Large), Pergamon Press, Oxford, pp. 123–137.

Basu, P. (1990) Heat transfer in high temperature fast fluidized beds. *Chem. Eng. Sci.*, **45**, 3123–3136.

Basu, P. and Nag, P.K. (1987) An investigation into heat transfer in circulating fluidized beds. *Int. J. Heat Mass Transfer*, **30**, 2399–2409.

Basu, P. and Ngo, T. (1993) Effect of some operating parameters on heat transfer to vertical fins in a circulating fluidized bed furnace. *Powder Technology*, **74**, 249–258.

Bolton, L.W. and Davidson, J.F. (1988) Recirculation of particles in fast fluidized risers, in *Circulating Fluidized Bed Technology II* (eds P. Basu and J.F. Large), Pergamon Press, pp. 139–146.

Brereton, C.M.H. and Grace, J.R. (1993) Microstructural aspects of the behavior of circulating fluidized beds. *Chem. Eng. Sci.*, **48**, 2565–2572.

Burki, V., Hirchberg, B., Tuzla, K. and Chen, J.C. (1993) Thermal development for heat transfer, in *Circulating Fluidized Beds*. Presented at 1993 AIChE Annual Meeting, St. Louis, MO.

Cao, C.C., Bai, D., Jin, Y. and Yu, Z. (1994) Mechanism of heat transfer between an immersed vertical surface and suspension in a circulating fluidized bed, *5th China–Japan Symposium*, Nagoya, Japan, Chemical Industry Press, Beijing, pp. 180–187.

Chang, H. and Louge, M. (1992) Fluid dynamic similarity in circulating fluidized beds. *Powder Technology*, **70**, 259–270.

Chen, C.C. and Chen, C.L. (1992) Experimental study of bed-to-wall heat transfer in a circulating bed. *Chem. Eng. Sci.*, **47**, 1017–1075.

Decker, N.A. and Glicksman, L.R. (1981) Conduction heat transfer at the surface of bodies immersed in gas fluidized beds of spherical particles. *AIChE Symposium Series*, **77** (No. 208).

Decker, N.A. and Glicksman, L.R. (1983) Heat transfer in large particle fluidized beds. *Int. J. Heat Mass Transfer*, **26**, 1307–1320.

Divilio, R.J. and Boyd, T.J. (1993) Practical implications of the effect of solids suspension density on heat transfer in large-scale CFB boilers, in *Circulating Fluidized Bed Technology IV* (ed. A. Avidan), AIChE, New York, pp. 334–339.

Dou, S. (1990) Experimental study of heat transfer in circulating fluidized beds. Doctoral thesis, Lehigh University, Bethlehem, Pennsylvania.

Dou, S., Herb, B., Tuzla, K. and Chen, J.C. (1992) Dynamic variation of solid concentration and heat transfer coefficient at wall of circulating fluidized bed, in *Fluidization VII* (eds O.E. Potter and D.J. Nicklin), Engineering Foundation, New York, pp. 793–801.

Ebert, T., Glicksman, L.R. and Lints, M. (1993) Determination of particle and gas convective heat transfer components in a circulating fluidized bed. *Chem. Eng. Sci.*, **48**(12), 2179–2188.

Fang, Z.H., Grace, J.R. and Lim, C.J. (1995a) Local particle convective heat transfer along surfaces in circulating fluidized beds. *Int. J. Heat Mass Transfer*, **38**, 1217–1224.

Fang, Z.H., Grace, J.R. and Lim, C.J. (1995b) Radiative heat transfer in circulating fluidized beds. *J. Heat Transfer*, **117**, 963–968.

Feugier, A., Gaulier, C. and Martin, G. (1987) Some aspects of hydrodynamics, heat transfer and gas combustion in circulating fluidized beds, in *Proceedings 9th International Conference Fluidized Bed Combustion*, ASME, New York, pp. 613–618.

Furchi, J.C.L., Goldstein Jr, L., Lombardi, G. and Mohseni, M. (1988) Experimental local heat transfer in a circulating fluidized bed, in *Circulating Fluidized Bed Technology II* (eds P. Basu and J.F. Large), Pergamon Press, Oxford, 263–270.

Gelperin, N.I. and Einstein, V.G. (1971) Heat transfer in fluidized beds, Chapter 10 in *Fluidization* (eds J.F. Davidson and D. Harrison), Academic Press, New York, p. 471.

Gidaspow, D., Tsuo, Y.P. and Luo, K.M. (1989) Computed and experimental cluster formation and velocity profiles in circulating fluidized beds, in *Fluidization VI* (eds J.R. Grace, L.W. Shemilt and M.A. Bergougnou), Engineering Foundation, New York, pp. 81–88.

Glicksman, L.R. (1984) Scaling relationships for fluidized beds. *Chem. Eng. Sci.*, **39**, 1373–1379.

Glicksman, L.R. (1988) Circulating fluidized bed heat transfer, in *Circulating Fluidized Bed Technology II* (eds P. Basu and J.F. Large), Pergamon Press, Oxford, pp. 13–29.

Glicksman, L.R. and Decker, N.A. (1982) Heat transfer from an immersed surface to adjacent particles in a fluidized bed; the role of radiation and particles packing. Presented at 7th International Conference on Fluidized Beds Combustion.

Glicksman, L.R. and Farrell, P. (1995) Verification of simplified hydrodynamic scaling laws for pressurized fluidized beds, Part I: Bubbling fluidized beds. *13th International Conference on Fluidized Bed Combustion* (ed. Heinschel, K.J.), ASME, New York, pp. 981–990.

Glicksman, L.R. and Hyre, M. (1994) Scaling of bed to wall heat transfer in a circulating fluidized bed. Presented at AIChE Annual Meeting, San Francisco, California.

Glicksman, L.R. and Westphalen, D. (1994) Lateral solid mixing measurements in circulating fluidized beds. *Powder Technology*, **82**, 153–167.

Glicksman, L.R., Westphalen, D., Woloshun, K., Ebert, T., Roth, K., Lints, M., Brereton, C. and Grace, J.R. (1991) Experimental scale models of circulating fluidized bed combustors, in *11th International Conference on Fluidized Bed Combustion* (ed. Anthony, E.J.), ASME, New York, pp. 1169–1175.

Glicksman, L.R., Hyre, M. and Woloshun, K. (1993) Simplified scaling relationships for fluidized beds. *Powder Technology*, **77**, 177–199.

Glicksman, L.R., Hyre, M. and Farrell, P. (1994a) Dynamic similarity in fluidization. *Int. J. Multiphase Flow*, **20** (supplement), 331–386.

Glicksman, L.R., Hyre, M., Lints, M.C. and Decker, N. (1994b) The role of surface resistance in fluidized bed heat transfer. Presented at 10th International Heat Transfer Conference, Brighton, England.

Glicksman, L.R., Hyre, M., Torpey, M. and Wheeldon, J. (1995) Verification of simplified hydrodynamic scaling laws for pressurized fluidized beds, Part II: Circulating fluidized beds, *13th International Conference on Fluidized Bed Combustion* (ed. Heinschel, K.J.), ASME, New York, pp. 991–1000.

Gloski, D., Glicksman, L.R. and Decker, N. (1984) Thermal resistance at a surface in contact with fluidized bed particles. *Int. J. Heat and Mass Transfer*, **27**, 599–610.

Golriz, M. and Leckner, B. (1992) Experimental studies of heat transfer in a circulating fluidized bed boiler, in *Proceedings International Conference on Engineering Applications of Mechanics*, Volume 3, Sharif University of Technology, Teheran, Iran, pp. 167–174.

Han, G.Y., Tuzla, K. and Chen, J.C. (1992) Radiative heat transfer from high temperature suspension flows. AIChE Annual Meeting, Miami, FL.

Hartge, E.U., Rensner, D. and Werther, J. (1988) Solids concentration and velocity patterns in circulating fluidized beds. *Circulating Fluidized Bed Technology II* (eds P. Basu and J.F. Large), Pergamon Press, Oxford, pp. 165–180.

Hirschberg, B. (1992) Einfluß der Lange der Austauschflache auf die Warmeubertragung in einer zirkulierenden Wirbelschicht. Diploma Thesis, Institut fur Thermische Verfahrenstechnik, Fakultat fur Chemieingenieurwesen, Universitat Karlsruhe.

Horio, M., Morishita, K., Tachibana, O. and Murata, N. (1988) Solid distribution and movement in circulating fluidized beds, in *Circulating Fluidized Bed Technology II* (eds P. Basu and J.F. Large), Pergamon Press, Oxford, pp. 147–154.

Hottel, H.C. and Sarofim, A.F. (1967) *Radiative Transfer*, McGraw-Hill, New York.

Hyre, M. and Glicksman, L.R. (1995) Experimental investigation of heat transfer enhancements in circulating fluidized beds, in *Proceedings 4th ASME/JSME Thermal Engineering Conference*, Maui, Hawaii.

Jestin, L., Meyers, P., Schmitt, G. and Morin, J.X. (1992) Heat transfer in a 125 MWe CFB boiler, in *Fluidization VII* (eds O.E. Potter and D.J. Nicklin), Engineering Foundation, New ork, pp. 849–856.

Katoh, Y., Glicksman, L.R. and Lints, M.C. (1990) Visualization of the particle behavior near the wall of a circulating fluidized bed. Video Presentation, *2nd ASME-JSME Fluid Engineering Joint Conference*.

Kobro, H. and Brereton, C. (1986) Control of fluidity of circulating fluidized beds, in *Circulating Fluidized Bed Technology* (ed. P. Basu), Pergamon Press, Toronto, pp. 263–272.

Kunii, D. and Levenspiel, O. (1991) A general equation for the heat-transfer coefficient at wall surfaces of gas/solid contactors. *Ind. Eng. Chem. Res.*, **30**, 136–141.

Leckner, B. and Andersson, B.A. (1992) Characteristic features of heat transfer in circulating fluidized bed boilers. *Powder Technology*, **70**, 303–314.

Leckner, B., Golriz, M.R., Zhang, W. and Andersson, B.A. (1991) Boundary layers – first measurements in the 12 MW CFB research plant at Chalmers University. *Proceedings 11th International Conference on Fluidized Bed Combustion*, **2**, 771–776.

Li, J., Xia, Y., Tung, Y. and Kwauk, M. (1991) Micro-visualization of two-phase structure in a fast fluidized bed, in *Circulating Fluidized Bed Technology III* (eds P. Basu, M. Horio and M. Hasatani), Pergamon Press, Oxford, pp. 183–188.

Lints, M. (1992) Particle-to-wall heat transfer in circulating fluidized beds. Doctoral Thesis, Massachusetts Institute of Technology, Cambridge MA.

Lints, M. and Glicksman, L.R. (1993) The structure of particle clusters near the wall of a circulating fluidized bed. AIChE Symposium Series, **89** (No. 296), 35–47.
Litka, T. and Glicksman, L.R. (1985) The influence of particle mechanical properties on bubble characteristics and solid mixing in fluidized beds. *Powder Technology*, **42**, 159–167.
Lockhart, C., Brereton, C.M.H., Lim, C.J. and Grace, J.R. (1995) Heat transfer, solid concentration and erosion around membrane tubes in a cold model circulating fluidized bed. *Int. J. Heat Mass Transfer*, **38**, 2403–2410.
Louge, M., Lischer, J. and Chang, H. (1990) Measurement of voidage near the wall of a circulating fluidized bed riser. *Powder Technology*, **62**, 269–276.
Mahalingam, M. and Kolar, A.J. (1991) Emulsion layer model for wall heat transfer in a circulating fluidized bed. *AIChE Journal*, **37**, 1139–1150.
Martin, H. (1984) Heat transfer between gas fluidized beds of solid particles and the surfaces of immersed heat exchanger elements. *Chem. Eng. Proc.*, **18**, 157–223.
Mattmann, W. (1991) Konvektiver Wärmeübergang in vertikalen Gas-Feststoff-Strömungen. Dr.-Ing. Dissertation, Universität Erlangen-Nürnberg.
Maxwell, J.C. (1892) *A Treatise on Electricity and Magnetism*. 3rd edn, vol. I, Clarendon, Oxford.
Mickley, H.S. and Fairbanks, D.F. (1955) Mechanisms of heat transfer to fluidized beds. *AIChE Journal*, **1**, 374–384.
Molerus, O. (1993) Fluid dynamics and its relevance for basic features of heat transfer in circulating fluidized beds, in *Circulating Fluidized Bed Technology IV* (ed. A. Avidan), AIChE, New York, pp. 311–318.
Nag, P.K. and Ali Moral, M.N. (1993) An experimental study of the effect of pin fins on heat transfer in circulating fluidized beds. *Int. J. of Energy Research*, **17**, 863–872.
Newby, R.A. and Keairns, D.L. (1986) Test of the scaling relationships for fluid-bed dynamics, in *Fluidization V* (eds K. Ostergaard and A. Sorensen), Engineering Foundation, New York, pp. 31–38.
Nicastro, M.T. and Glicksman, L.R. (1984) Experimental verification of scaling relationships for fluidized bed. *Chem. Eng. Sci.*, **39**, 1381–1391.
Nowak, W., Mineo, H., Yamazaki, R. and Yoshida, K. (1991) Behavior of particles in a circulating fluidized bed of a mixture of two different sized particles, in *Circulating Fluidized Bed Technology III* (eds P. Basu, M. Horio and M. Hasatani), Pergamon Press, Oxford, 219–224.
Rhodes, M.J., Mineo, H. and Hirama, T. (1991) Particle motion at the wall of the 305mm diameter riser of a cold model circulating fluidized bed, in *Circulating Fluidized Bed Technology III* (eds P. Basu, M. Horio and M. Hasatani), Pergamon Press, Oxford, pp. 171–176.
Rhodes, M., Mineo, H. and Hirama, T. (1992) Particle motion at the wall of a circulating fluidized bed. *Powder Technology*, **70**, 207–214.
Roy, R. and Davidson, J.F. (1989) Similarity between gas-fluidized beds at elevated temperature and pressure, in *Fluidization VI* (eds J.R. Grace, L.W. Shemilt and M.A. Bergougnou), Engineering Foundation, New York, pp. 293–300.
Scharff, M.F., Goldman, S.R., Flanagan, T.M., Gregory, T.K. and Smoot, L.D. (1978) Project to provide an experimental plan for the Merc $6' \times 6'$ fluidized bed test model. Final report J77-2042-FR. US Dept. of Energy, Contract EY-77-C-21-8156.
Sekthira, A., Lee, Y.Y. and Genetti, W.E. (1988) Heat transfer in a circulating fluidized bed. Presented at 25th National Heat Transfer Conference, Houston, Texas.
Soong, C.H., Tuzla, K. and Chen, J.C. (1994) Identification of particle clusters in circulating fluidized beds, in *Circulating Fluidized Bed Technology IV* (ed. A. Avidan), AIChE, New York, pp. 615–620.
Subbarrao, D. and Basu, P. (1986) A model for heat transfer in circulating fluidized beds. *Int. J. Heat Mass Transfer*, **27**, 487–489.
Tuzla, K., Dou, S., Herb, B.E. and Chen, J.G. (1991) Experimental study of heat transfer for circulating fluidized bed combustors. ETCE Energy-Sources Tech. Conf. and Exhibit, 3rd ASME Fossil Fuel Comb. Sam., Houston, Texas.
Werdermann, C.C. and Werther, J. (1993) Heat transfer in large-scale circulating fluidized bed combustors of different sizes, in *Circulating Fluidized Bed Technology IV* (ed. A. Avidan), AIChE, New York, pp. 428–435.
Westphalen, D. and Glicksman, L. (1995) Lateral solid mixing measurements in circulating fluidized beds. *Powder Technol.*, **82**, 153–167.

Westsila, J.W. (1991) Colorado UTE Nucla Station circulating-fluidized-bed demonstration. EPRI GS-7483, **2**.

Wirth, K.E. and Seiter, M. (1991) Solids concentration and solids velocity in the wall region of circulating fluidized beds. *Proceedings 11th International Conference on Fluidized Bed Combustion* (ed. E.J. Anthony), ASME, New York, **1**, 311–316.

Wirth, K.E., Seiter, M. and Molerus, O. (1991) Concentration and velocities of solids in areas close to the walls in circulating fluidized bed systems. *VGB Kraftwerkstechnik*, **10**, 824–828.

Wu, R.L. (1989) Heat transfer in circulating fluidized beds. Doctoral Thesis, University of British Columbia, Vancouver, Canada.

Wu, R.L., Grace, J.R., Lim, C.J. and Brereton, C.M.H. (1989) Suspension-to-surface heat transfer in a circulating fluidized bed combustor. *AIChE Journal*, **35**, 1685–1691.

Wu, R.L., Grace, J.R. and Lim, C.J. (1990) A model for heat transfer in circulating fluidized beds. *Chem. Eng. Sci.*, **45**, 3389–3398.

Wu, R.L., Lim, C.J., Grace, J.R. and Brereton, C.M.H. (1991) Instantaneous local heat transfer and hydrodynamics in a circulating fluidized bed. *Int. J. Heat Mass Transfer*, **34**, 2019–2027.

Xianglin, S., Naijun, Z. and Yigian, X. (1990) Experimental study on heat transfer in a pressurized circulating bed, in *Circulating Fluidized Bed Technology III* (eds P. Basu, M. Horio and M. Hasitani), Pergamon Press, Oxford, pp. 451–456.

Zhang, W., Tung, Y. and Johnsson, F. (1991) Radial voidage profiles in fast fluidized beds. *Chem. Eng. Sci.*, **46**, 3045–3052.

Zhou, J., Grace, J.R., Lin, C.J., Brereton, C.M.H., Qin, S. and Lin, K.S. (1995) Particle crossflow, lateral momentum flux and lateral velocity in a circulating fluidized bed. *Can. J. Chem. Eng.*, **73**, 612–619.

9 Experimental techniques
MICHEL LOUGE

9.1 Introduction

Circulating fluidized beds pose considerable challenges to our understanding. Because it is not yet possible to produce *ab initio* predictions of their complex fluid dynamics, transport and chemical behavior, experimentation remains essential. Because many phenomenological models require empirical input, their quality depends on the accuracy of measurement techniques. Process control and monitoring also require precise experimental data.

Because the relatively high concentrations of circulating fluidized beds generally prohibit diagnostic techniques commonly used with clear fluids, the choice of commercial instruments is limited. The environment of industrial and pilot plants further restricts the range of available instrumentation. Another consequence of high particle concentrations is that, in order to study the interior of typical flows, the experimentalist must often accept some degree of intrusion and attempt to establish how the probe affects the dynamics of the suspension.

The quality of experiments often derives from ingenious new instruments. Grace and Baeyens (1986) reviewed experimental techniques used in fluidization. Cheremisinoff (1986) provided exhaustive descriptions of state-of-the-art instrumentation for gas–solid suspensions and discussed the limitations and advantages of each technique in detail. Saxena *et al.* (1989) gave an extensive review of heat transfer measurement techniques at high temperatures and discussed the design, limitations and advantages of the corresponding probes. Marcus *et al.* (1990) focused on pneumatic transport, in particular surveying commercial and experimental solids flow rate meters. Werther *et al.* (1993) discussed measurements involving pressure, optical fibers and sampling probes. In a review of recent heat transfer research conducted in China, Saxena *et al.* (1992) provided information on relevant instrumentation. Turlier and Bernard (1992) presented a short review of radioisotope techniques and γ-ray tomography, as well as sampling, optical and capacitance probes. In his survey of circulating fluidization, Yang (1993) listed the principal experiments and their instrumentation. Larachi and Chaouki (1994) focused on non-invasive, especially radioactive, techniques for multiphase flows. Soo *et al.* (1994) reviewed instrumentation for dilute flows, while Yates and Simons (1994) provided a comprehensive account of instruments for gas–solid suspensions.

This chapter describes the principal devices and experimental techniques used in circulating fluidization with emphasis on the challenges involved in their implementation and data interpretation.

9.2 Visualization

One way to study circulating fluidization is to observe the flow across a transparent riser wall. Thus many laboratory facilities are built of clear materials like glass or plastic. Because these materials commonly promote considerable electrostatic charges, it is generally essential to add moderate quantities of fine salts to the suspension (Chang and Louge, 1992), coat dielectric walls with a clear conductive wax (Glicksman, 1994), introduce strongly polar vapor molecules such as water or ammonia (Myler *et al.*, 1986), or employ a combination of the above. Moderate use of these methods is required. Excessive introduction of fine salts may produce a film on clear windows; changes in gas density or agglomeration of fine particles may accompany the introduction of a condensable vapor; excessive coating of the wall may alter the descending curtain of solids.

At concentrations typical of riser suspensions, visualization is generally limited to a thin layer near the wall, as a simple calculation illustrates. We assume that monodisperse spheres of diameter d_p are placed along a cubic lattice of uniform solid volume fraction $(1 - \epsilon)$. For moderate $(1 - \epsilon)$, light originally of intensity i_0 is extinguished to a value i after penetrating a distance z into the suspension:

$$i/i_0 = \exp[-(3/2)\kappa_1(1 - \epsilon)z/d_p] \equiv \exp[-z/l], \qquad (9.1)$$

where the extinction coefficient κ_1 is of order 2 when the wavelength is much smaller than the particle diameter (Cutolo *et al.*, 1990). At typical riser concentrations, the extinction distance l is very small.

Because of the considerable light scattering of the suspension, photographic observations through a clear vessel often reveal only the curtain of descending particles near the riser surface. Nevertheless, these observations are useful, as they help understand such features as erosion and heat transfer at the wall. Rhodes *et al.* (1992) studied the formation of swarms and strands with a high-speed video camera. Wirth *et al.* (1991) tracked the velocity of phosphorescent particles after tagging these with a flash introduced by an optical fiber. Arena *et al.* (1989, 1992) employed a thin, 'two-dimensional' riser to visualize flow structures with a video system capable of 2000 frames per second. The corresponding structures were assumed to represent qualitatively the flow in the interior of a more typical riser.

Other studies employed intrusive probes to observe the interior of the suspension. Using the method of Saxena *et al.* (1987) and Saxena and Patel (1988) for bubbling fluidized beds, Li *et al.* (1991) introduced a boroscope composed of a coherent bundle of optical fibers through the riser. With this technique, Hatano et al. (1994) inferred solid volume fraction and velocity from the resulting microscopic images, while Zou et al. (1994) extracted cluster size and probabilities.

For very dilute conditions, the inner core of the flow becomes visible. Tadrist et al. (1994) and Horio and Kuroki (1994) employed a cylindrical lens to introduce a planar sheet of laser light at several orientations through the riser. The light scattered by the particles exposed changing patterns as the coherent radiation sliced through clusters. Although the solid volume fraction was below 0.1%, these observations revealed all characteristic structures attributed to circulating fluidization, including a thin curtain of descending solids at the wall. Larachi and Chaouki (1994) suggested that this visualization technique would be further enhanced by using fluorescent particles.

Others have visualized gas–solid flows via X-ray imaging, thereby extracting quantitative measurements of solid volume fraction. This technique is discussed in section 9.4.3.

9.3 Pressure and stresses

Perhaps the most ubiquitous measurement in circulating fluidization is that of static gas pressure. As long as simple precautions are taken to exclude particles from the lines through filters or purges, this measurement is relatively straightforward. In steady, fully developed flows, ignoring the weight of the gas, the balance of forces on the suspension in a cross-section of the riser is

$$\frac{\partial p}{\partial x} = -\frac{2S_g}{R} - \frac{2S_p}{R} - g\rho_p(1 - \bar{\epsilon}) \tag{9.2}$$

Because they are inversely proportional to R, the contributions of the gas and solid stresses are negligible in circulating fluidized beds unless R is unrealistically small or the velocities approach those of pneumatic transport. Therefore, in regions where vertical gradients of velocities and volume fractions are small, the vertical gas pressure gradient is a direct measure of the cross-sectional average solid volume fraction:

$$\frac{\partial p}{\partial x} \approx -g\rho_p(1 - \bar{\epsilon}) \tag{9.3}$$

Because derivatives are more susceptible to uncertainty than primary variables, the extraction of reliable volume fractions requires uniformly accurate calibrations of successive pressure transducers along the height of the riser. A convenient way to avoid calibration discrepancies is to bring all pressure lines to a single transducer through a multiplexing valve. Because bed operations are generally held steady over long time periods, it is possible to carry out meaningful averages of several successive pressure scans along the column, even though each individual measurement must be time-averaged for several seconds to smooth pressure fluctuations.

The interpretation of time-average pressure measurements is less straightforward in regions where solid velocity or volume fraction vary rapidly. At the base of the riser or near sharp exits, the momentum balance involves convective acceleration terms in both phases that render equation (9.2) incomplete (Weinstein and Li, 1989). Because solids are generally injected in directions not aligned with the main gas flow, the interpretation of gas pressure measurements may require three-dimensional treatment of the flow near the feed points.

Another difficulty arises in the transition between a dense bottom region and a more dilute suspension higher in the riser. If in this transition the cross-sectional average volume fraction varies rapidly, then the vertical pressure drop includes a contribution from the corresponding voidage gradient (Louge and Chang, 1990),

$$\frac{\partial p}{\partial x} = -\rho_p g(1 - \bar{\epsilon}) - \frac{G_s^2}{\rho_p} \frac{\mathrm{d}}{\mathrm{d}x}\left(\frac{1}{1-\bar{\epsilon}}\right) \tag{9.4}$$

Quick-closing valves can reveal discrepancies between the pressure gradient and the true cross-sectional average volume fraction. The idea is to outfit the riser with a number of slide valves simultaneously actuated by fast solenoids. In this way, solids and gases are trapped in sections of the riser delimited by two consecutive valves closing in unison. The weight of the collected solids yields an unequivocal measurement of ensemble-averaged volume fraction in the section. Users of quick-closing valves have included Arena *et al.* (1985, 1988), Mok *et al.* (1989) and Van der Ham *et al.* (1992). The technique is only suitable for small risers and hard to retrofit in an existing unit. It is important to ensure that the assembly does not interfere with the descending curtain of solids, since Lints and Glicksman (1993) reported that the flow at the wall may be disturbed by obstructions as small as one particle diameter.

Fluctuations of static pressure are also exploited to compare the behavior of dimensionally analogous risers (Chang and Louge, 1992; Glicksman *et al.*, 1993), to identify flow regime transitions (see Chapter 1), or to carry out more detailed analyses (Schnitzlein and Weinstein, 1988; Brue *et al.*, 1994).

To record pressure fluctuations without significant attenuation it is essential to consider the dynamic responses of the transducer and the lines connecting it to the riser. Clark and Atkinson (1988) analyzed the frequency response and phase lag of twin static pressure probes in bubbling fluidized beds. For a tube of length l_t and diameter d_t connecting the riser to the transducer, they extracted the dynamic response of the system from the mass and momentum balances of the gas assuming laminar, fully developed flow with negligible exit and entrance effects:

$$\frac{\rho l_t V}{\pi d_t^2 p_0}\frac{d^2 p_t}{dt^2} + \frac{128 \mu l_t V}{\pi d_t^4 p_0}\frac{dp_t}{dt} + p_t = p_b \qquad (9.5)$$

where V is the 'dead volume' of the transducer, and p_0, p_t and p_b are the mean, transducer and bed pressures, respectively. Turton and Clark (1988) generalized the analysis for entrained suspensions, while Stiles and Clark (1992) refined it by considering the onset of turbulence in the line.

Dynamic gas pressure measurements are more difficult to interpret in gas–solid suspensions than in incompressible flows. In this measurement, a tube connected to a gas pressure transducer is aligned with the direction of the main flow. A pseudo-Bernoulli approach considers Eulerian equations along coincident streamlines of the gas and particle phases. Assuming constant voidage along the streamline, eliminating the drag terms by adding the momentum balances of the two phases and neglecting gravity produces a total dynamic pressure p_d that is conserved along the streamline:

$$p_d = p + \tfrac{1}{2}\rho \epsilon u^2 + \tfrac{1}{2}\rho_p(1 - \epsilon)v_p^2 \approx p + \tfrac{1}{2}G_s v_p \qquad (9.6)$$

The idea is that the total gas pressure contains a contribution from the momenta of the two phases that may be obtained by subtracting static pressure. Unless the flow is very dilute, the gas momentum contribution is negligible. In regions where the flow exhibits frequent reversals, pressure tubes must be pointed in both the upward and downward directions. Azzi et al. (1991b) subtracted the corresponding momenta using a differential pressure probe to capture the net value of the product $G_s v_p$. Zhang et al. (1994, 1995) used a similar device in a 12 MW CFB boiler. From a practical standpoint, it is essential to flush particles that accumulate in the probe. One possibility is to use the same tube periodically for solid sampling.

The principal failures of the analysis leading to equation (9.6) are that gravity and collisions can dissipate the momentum of the particles in the tube, that the particle concentration generally increases as the suspension is brought to rest, that gas and particles streamlines do not always coincide, and that the flow is neither irrotational nor inviscid in general. Thus, to exploit the data from a Pitot tube inserted in the riser, Bader et al. (1988) adopted the empirical expression of Van Breugel et al. (1969):

$$(p_d - p) = aG_s v_p \qquad (9.7)$$

with $a \approx 0.75$ instead of equation (9.6), and they independently measured the local particle flux G_s to infer particle velocity. Others have introduced empirical constants in equation (9.6), determining these through calibrations (Azzi et al., 1991b; Zhang et al., 1994, 1995).

Other instruments are designed to record stresses directly. Qian and Li (1994) inserted a strain gauge in the riser to record the combined 'dynamic pressure' of the gas and particle phases. A second gauge provided temperature compensation. The instrument distinguished upward and downward movement. It was used in conjunction with converging optical fibers for a simultaneous record of voidage. Van Swaaij et al. (1970) measured the total shear stresses exerted on a test section by isolating it mechanically from the rest of the riser and measuring the total shear force with a calibrated balance. This ingenious design was sealed using Teflon bellows, while particle contamination was avoided by purging the assembly with air.

9.4 Solids volume fraction

The distribution of solids in the riser has important consequences for mixing, heat transfer and reaction rates. Because riser suspensions are highly heterogeneous and unsteady, measurements of solid volume fraction should be local, fast, and quantitative. Researchers have devised several techniques involving capacitance probes, optical fibers, and transmission of energetic photons through the suspension.

9.4.1 Capacitance instruments

Two closely-spaced conductive surfaces held at different voltages accumulate charges with surface density proportional to the electric field that they shed toward each other. The total charge on each surface is proportional to the voltage difference through a constant 'capacitance'. If the medium between the surfaces is a dielectric, solutions of the field equations show that the capacitance is the product of the dielectric permittivity of the medium γ and a length scale L characteristic of the geometry of the capacitor, i.e.

$$\tilde{C} = \gamma L \qquad (9.8)$$

With two identical parallel plates, L is the ratio of the area of the plates and their separation. For a non-conductive gas, γ nearly equals the dielectric permittivity of free space $\gamma_0 = 8.854 \times 10^{-12}$ F/m. The ratio $K = \gamma/\gamma_0$ is called the dielectric constant of the medium.

Capacitance instruments are used to infer the volume fraction of a suspension by measuring its effective dielectric constant K_e. If the suspension is homogeneous with particles much smaller than L, then the electric field and, consequently, L are independent of γ. For powders with negligible imaginary part of K_e (practically, for non-conductive riser suspensions), the constant is obtained by comparing the capacitance $\tilde{C}$ in the presence of the suspension to the corresponding value $\tilde{C}_0$ in its absence, i.e.

$$K_e = \tilde{C}/\tilde{C}_0 \qquad (9.9)$$

Because K_e is uniquely related to the suspension density, this technique provides quantitative measurements of solid volume fraction.

Traditionally, capacitances are recorded by balancing a bridge circuit. However, this may fail to provide quantitative measurements for several reasons. First, the cables connecting the bridge to the instrument often exhibit a larger capacitance than the probe itself. Second, nearby electrical surfaces may produce 'stray capacitances' that interfere with the measurement. Third, instruments of small dimensions may yield capacitance magnitudes far below the detection threshold of traditional bridges. Finally, the electronics may exhibit drift or instability.

Despite these difficulties, early capacitance probes provided useful qualitative information on the dynamics of fluidized beds in various hydrodynamic regimes (e.g. Lanneau, 1960; Geldart and Kelsey, 1972; Chandran and Chen, 1982; Abed, 1984). Werther and Molerus (1973) devised the needle probe, an instrument consisting of a small active wire protruding from a concentric cylinder held at a different voltage. This device caused remarkably little intrusion in the suspension. Brereton and Stromberg (1986), Li and Kwauk (1980), Hartge et al. (1986) and Brereton and Grace (1993) employed similar designs for circulating fluidized beds. Herb et al. (1989) and Chen et al. (1990) chose instead to traverse two small parallel plates in the riser.

Acree Riley and Louge (1989) virtually eliminated the problems of stray and cable capacitances by incorporating a third conductor in their probes. This 'guard' conductor led to unprecedented accuracy and a more reproducible and quantitative signal. Figure 9.1 shows the amplifier system. A 16 kHz oscillator supplies the 'sensor' with a current of constant amplitude. A separate circuit maintains the guard surface at precisely the same sinusoidal voltage as the sensor. Because it absorbs most distortions of the electric field caused by external interferences, the guard protects the sensor from stray capacitances and confines the measurement volume approximately to the region bounded by the extreme field lines shed from the

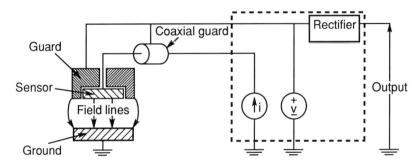

Figure 9.1 Principle of the capacitance amplifier system. A current of constant amplitude $\bar{i}$ supplies the sensor. The voltage $\bar{v}$ is common to the guard surface and the coaxial guard surrounding the sensor cable. It is held equal to the sensor voltage. The dashed lines mark the boundaries of the amplifier. Its output is the rectified guard voltage.

sensor to the ground surfaces. In addition, because the sensor cable is connected to the amplifier circuits through a guarded coaxial cable, the cable capacitance does not participate in the measurement. By eliminating virtually all stray and cable capacitances, this technique can detect capacitances three to six orders of magnitude smaller than conventional bridges.

The output of the amplifier is the rectified guard voltage. It may follow variations of the suspension density as rapid as 3 kHz. Because the guard and sensor voltages are equal, the rectified output is proportional to the amplitude of the sensor voltage. Because the sensor is supplied with a current $\tilde{i}$ of constant amplitude and $\tilde{v} = Z\tilde{i}$, the output is proportional to the magnitude of the impedance $Z = 1/j2\pi f\tilde{C}$ between sensor and ground. Thus, the effective dielectric constant of the medium being measured is

$$K_e = \frac{\tilde{C}}{\tilde{C}_0} = \frac{\tilde{V}_0}{\tilde{V}} \qquad (9.10)$$

where $\tilde{V}_0$ is the rectified output of the probe in air and $\tilde{V}$ is the output in the presence of the dielectric suspension of interest. Commercial amplifiers implement the circuit of Figure 9.1. They are quite stable unless excessive electrostatic charges are present near the probes.

Inaccuracies in earlier interpretation of capacitance probe signals generally resulted from erroneous models relating the effective dielectric constant to solid volume fraction. Often a linear response of the instrument between clean gas and closely packed particles was wrongly assumed. Many experiments and models have successfully captured the effective properties of various suspensions (e.g. Maxwell, 1892; de Loor, 1956; Tinga et al., 1973). For example, dielectric spheres with moderate permittivity (e.g. glass, plastic) satisfy the semi-empirical model of Meredith and Tobias (1960):

$$\frac{K_e}{K_h} = \frac{B_1 - 2(1-\epsilon) + B_2 - 2.133B_3}{B_1 + (1-\epsilon) + B_2 - 0.906B_3} \qquad (9.11)$$

where $B_1 \equiv (2K_h + K_p)/(K_h - K_p)$; $B_2 \equiv 0.409(1-\epsilon)^{7/3}(6K_h + 3K_p)/(4K_h + 3K_p)$; $B_3 \equiv (1-\epsilon)^{10/3}(3K_h - 3K_p)/(4K_h + 3K_p)$, and K_p and K_h are the dielectric constants of the material of the particles and the host fluid, respectively.

On occasion, however, the effective properties of arbitrary powders may be difficult to establish. For example, mildly conductive solids may exhibit values of K_p that depend on electric field frequency. Because of the high dielectric constant of water, humidity also affects the observed dielectric constant of powders. Values of K_p at ambient conditions differ from those at high temperatures. Because the effective dielectric constant of a suspension is sensitive to particle shape, spherical and crushed powders of the same material do not normally exhibit the same effective dielectric properties. In contrast, particle size distribution is relatively unimportant. Thus, before

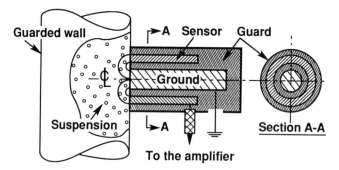

Figure 9.2 'Wall' capacitance probe. Dimensions are not to scale. In this configuration, the wall is held at the guard voltage.

undertaking measurements with a new powder, it is judicious to measure its effective dielectric constant at several volume fractions using a procedure like that of Louge and Opie (1990). However, in most cases a simpler approach may be sufficient. From the model of Böttcher (1952):

$$\frac{K_e - K_h}{3K_e} = (1 - \epsilon)\frac{K_p - K_h}{K_p + 2K_e} \tag{9.12}$$

K_p can be estimated from a single measurement of K_e with the closely packed powder. K_e is then obtained at other volume fractions using equation (9.12). The empirical estimate of K_p inferred in this way may differ substantially from its true value. Nevertheless, our experience is that the corresponding interpolation is sufficiently accurate for most purposes.

As long as they include a guard, a sensor and a ground, quantitative capacitance instruments can be built in many convenient geometries. Louge et al. (1990) used the non-invasive 'wall probe' shown in Figure 9.2 to record the solid volume fraction near the inside surface of a conductive vessel. By connecting the riser section around the instrument to the guard voltage, they confined the measurement volume near the probe face. The central ground can also incorporate another instrument for simultaneous non-invasive measurements, e.g. a small heat transfer sensor (Mohd. Yusof, 1992) or optical fiber (Lischer and Louge, 1992). Louge (1995) described another non-invasive 'wall probe' design that locates ground at the probe periphery. This geometry is particularly attractive for industrial applications.

In another guarded arrangement, Dou et al. (1992) combined small parallel plates protruding into the riser with a measurement of convective wall heat transfer. Other guarded capacitance designs have been used to record radial profiles of volume fraction. For example, Qi and Farag (1993) traversed a 'wall probe' through the riser, while Soong et al. (1994) employed a far less intrusive needle capacitance probe.

Capacitance instruments can also be adapted to hot environments. Almstedt and Olsson (1982, 1985) and Almstedt and Ljungström (1987) reported successive refinements of a water-cooled, needle capacitance design traversed through a pressurized, bubbling, coal-burning fluidized bed. Although this unguarded probe encountered the same difficulties as earlier room temperature instruments, it provided useful results on bubble characteristics. Acree Riley (1989) demonstrated a water and air-cooled version of her guarded 'parallel-plate' configuration (Acree Riley and Louge, 1989) in a coal-burning circulating fluidized bed near 850°C. Although she observed occasional instabilities attributed to electrostatics, the signal was promising.

9.4.2 Optical fibers

Because they are simple, yield high signal-to-noise ratios, and create minimum disturbances in the flow, optical fiber sensors are often used to record profiles of particle volume fraction in dense suspensions. They are either based on forward light scattering between emission and detection fibers separated by a short distance, or on backscattering onto an optical fiber system. Using the former method, Cutolo et al. (1990) correlated measurements of solid volume fractions below 16% using forward scattering theory. However, as equation (9.1) shows, forward scattering is often impractical for denser systems.

Because backscattering sensors are simpler and less intrusive, they have received wider attention, e.g. Horio et al. (1988), Hartge et al. (1986), Zhang et al. (1991). However, because it is difficult to create stable suspensions at an arbitrary voidage, calibration is challenging. To this end, Matsuno et al. (1988) produced streams of particles falling through a vibrating sieve at a known velocity. From a measurement of the flux, they inferred the concentration of particles passing in front of the optical sensor. This calibration, however, was limited to dilute systems. For greater concentrations, Hartge et al. (1986) suspended particles in a water-fluidized bed exhibiting homogeneous expansion over a range of solid volume fractions. Unfortunately, because for transparent particles the signal increases with the ratio of the indices of refraction of the sphere and the suspending medium, suspensions in the gas produce signals of considerably different character than similar suspensions in a liquid. Lischer and Louge (1992) warned that immersing optical fiber sensors in water may provide a misleading calibration for gas suspensions.

Other calibrations techniques may be carried out in the riser itself. Lischer and Louge (1992) inserted an optical fiber through the center of a quantitative capacitance wall probe (Figure 9.3). The calibration was then obtained by comparing the simultaneous time histories of the capacitance and optical signals. Because it is rapid, this method can be executed before every traverse

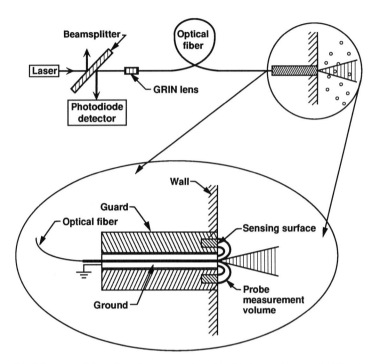

Figure 9.3 Schematic of the calibration setup adopted by Lischer and Louge (1992). Dimensions are not to scale. The semi-toroidal measurement volume shown is that of the capacitance probe of Figure 9.2, penetrating approximately 2 mm into the suspension. © Optical Society of America, reproduced with permission.

of the optical fiber through the suspension. This solves problems of signal drift associated with, for example, degradation of the fiber tip. Tung *et al.* (1988) and Zhang *et al.* (1991) calibrated their optical probes by traversing them through the bed and comparing the resulting measurements with average volume fractions inferred from static pressure gradients. To this end, they assumed that the voidage is a polynomial function of the signal voltage $\tilde{V}$, i.e.

$$\epsilon = \sum_{i=0}^{n} B_i \tilde{V}^i \qquad (9.13a)$$

Thus, on average across the riser,

$$\bar{\epsilon} \equiv \int_0^R 2\frac{r}{R}\epsilon(r)\frac{dr}{R} = \sum_{i=0}^{n} B_i \overline{\tilde{V}^i} \qquad (9.13b)$$

Successive measurements of $\bar{\epsilon}$ and the radial voltage profile produced a redundant set of linear equations (9.13b) to determine the coefficients B_i from multiple regression.

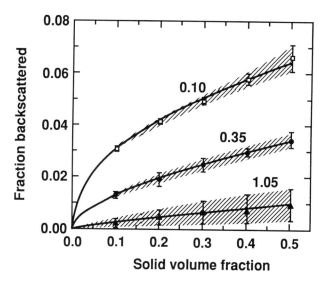

Figure 9.4 Monte-Carlo simulations of the fraction of light backscattered versus solid volume fraction for suspensions of glass spheres in air and a fiber of $0.37NA$. Ratios of particle to fiber diameters are 1.05, 0.35, and 0.10. Error bars represent the sample standard deviation for ten randomly simulated particle placements (Lischer and Louge, 1992). © Optical Society of America, reproduced with permission.

Despite their apparent simplicity, optical fiber signals do not always yield an unequivocal measurement of particle concentration. Using Monte-Carlo simulations, Lischer and Louge (1992) showed that the output and accuracy of single fiber sensors in a suspension of spheres increase with decreasing sphere to fiber diameter ratio (Figure 9.4) and with increasing numerical aperture (NA) of the fiber. Thus it is essential to employ optical fibers with a diameter much larger than that of the particles. In short, there should be enough particles in the measurement volume to average over a greater ensemble, thereby minimizing the effect of particle location on the output. Similarly, the signal should be low-pass filtered to remove spikes associated with individual particles.

Optical fiber systems fall into two main categories. In the first, light is guided through a single fiber (Figure 9.3). The light backscattered from the suspension returns through the same fiber and is directed to the detector via a beam splitter (Werther et al., 1993) or another light segregation technique (Werther and Hage, 1995). In this case, the detector signal always increases with the number of particles causing the backscatter. In this system, proper optical alignment is crucial. In the second design, one or several bundled fibers guide light to the suspension, while other distinct fiber(s) return it directly to the detector. Because the latter method avoids complicated optical paths, it is easier to implement.

However, Louge (1994) warned that systems consisting of separate emission and detection fibers are prone to ambiguous signals, which may rise with solid volume fraction at low concentrations, but then fall in denser suspensions. To illustrate this point, he reformulated in an integral form the analysis of Rensner and Werther (1993), who predicted the behavior of single optical fibers. In this simplified model, the fraction of light returned to the fiber of core diameter d_c is

$$\frac{i}{i_0} = \int_0^\infty \chi\left(\frac{z}{d_c}; NA\right) \exp\left\{-\frac{3}{2}\kappa_1(1-\epsilon)\frac{d_c}{d_p}\frac{2z}{d_c}\right\} \left[\frac{3}{2}\kappa_2(1-\epsilon)\frac{d_c}{d_p}\right]\frac{dz}{d_c} \quad (9.14)$$

The integrand is the contribution to the overall signal of an infinitesimal slice of the suspension at a distance z from the tip. The exponential is the fraction of light transmitted to and from the slice, see equation (9.1). The term in straight brackets represents the fraction of light backscattered by the slice, where κ_2 is a parameter that accounts for the scattering properties of the powder. In an isotropic, homogeneous, absorbing, non-diffusive optical medium, the radiation shape function χ captures the divergence of the beam and the partial acceptance of light returned from the slice. It may be viewed as the fraction of light returned to the fiber from an ideal diffuse screen placed a distance z away from the tip. In a scattering suspension, the shape function is not quite so simple (Amos et al., 1996). However, to achieve an elementary understanding of the optical fiber system, it is convenient to ignore the role played by particle scattering in modifying the shape function.

Note that the transmission and backscattering terms in equation (9.14) have competing dependences on $(1-\epsilon)$. Because transmission decays exponentially with distance from the tip, but backscattering does not, a vanishing shape function near the tip can exacerbate the role of transmission and thus produce an ambiguous signal that first increases, then decreases, with concentration. As equation (9.14) indicates, this effect depends strongly on particle size. Rensner et al. (1991) and Beaud and Louge (1996) reported such behavior for probes with distinct emission and detection fibers and for fiber bundles. This predicament can be avoided by ensuring that the shape function decreases monotonically with distance in the suspension. Practically, the designer should avoid 'blind spots' which either fail to be illuminated by the emission fiber or to be seen by the detection fiber. To this end, a window can be inserted in front of the tip to keep blind spots out of the suspension.

Another difficulty with single optical fibers is that the penetration of light into the suspension and, consequently, the extent of their measurement volume, depends upon the solid volume fraction. Under dilute conditions, the signal may be sensitive to structures deep in the flow, causing the calibration to depend upon riser conditions. To remedy this problem, Reh and Li (1991) proposed a converging arrangement of separate emission and detection fibers, while Tanner *et al.* (1993) used a small lens to achieve convergence with a single fiber. In the two-fiber design, judicious placement of a protective window at the tip avoids the potential for signal ambiguity from 'blind spots' (Cocco *et al.*, 1994).

In general, because of the uncertain extent of their measurement volume and the possible sensitivity of their response to particle placement, instantaneous signals from single optical fibers are not always accurate. However, it is more likely that their time-averaged values, if properly calibrated, give good representations of the local mean volume fraction. Caution should be exercised when higher moments of the volume fraction statistics are sought.

Finally, because optical fibers can potentially withstand high temperatures, these instruments can also be used in fluidized bed combustors (Johnsson and Johnsson, 1992). The temperature of the probe tip must be such that the optical elements exposed to the bed do not soften; for example, the maximum operating temperatures of fused silica and sapphire are 900° and 1800°C, respectively. Sapphire windows or lenses are recommended for their resistance to abrasion. Werther and Hage (1995) exposed water-cooled fibers with a 1000 μm core of raw quartz to a CFB at 850°C. For these, they observed the rapid deterioration of polished tips in the hot CFB, but reported good resilience of cleaved fiber ends.

In hot applications, the designer must consider the different thermal expansion of the optics and their metallic armature, and contend with any fouling. It is also essential to distinguish backscattered light from the radiation produced by the burning solids. To achieve proper discrimination, Johnsson and Johnsson (1992) employed a laser light source and a narrow bandpass optical filter. In addition, by modulating the source at a frequency of 20 kHz, they enhanced the backscattered signal using phase-sensitive detection with a lock-in amplifier.

9.4.3 Transmission densitometry

In many industrial applications it is awkward to introduce a probe in the riser. Fortunately, useful information may be obtained by measuring the transmission of radiation along a line of sight across the entire riser. Because they require optical windows and are quickly extinguished, low energy photons in the visible or infrared are seldom used, unless the path is small or the volume fraction low. Klinzing (1980) recorded extinction from a halogen lamp in a 25 mm pneumatic transport line at loadings (ratio of solids-to-gas mass flow rates) below four. When individual particles are opaque to radiation, Beer's law (equation (9.1)) depends on particle diameter, so smaller solids extinguish light through shorter paths.

In contrast, photons of higher energies such as γ-rays can penetrate solid materials over distances far greater than the diameter of individual particles. In this case, the reciprocal of the fractional absorption per unit length of the solid material replaces particle diameter in equation (9.1), and the measurement becomes essentially independent of particle-size-distribution. In circulating fluidization, the solid volume fraction is not constant across the riser. Neglecting gas absorption, integration of the equation of transfer for a collimated beam yields

$$\frac{i}{i_0} = \exp\left[-\alpha\rho_p \int_0^L (1-\epsilon)\,dz\right] \qquad (9.15)$$

where the path of integration is along a chord of length L from wall to wall in the riser. Note that this integral information differs substantially from the average volume fraction $(1-\bar{\epsilon})$ defined in equation (9.13b).

Although the mass absorption coefficient α is only a function of the photon energy and the composition of the absorbing material, it is wise to establish its practical value through *in situ* calibrations. For example, the source may not be perfectly monoenergetic or highly collimated. A straightforward uncertainty analysis of equation (9.15) indicates that the volume fraction integral is least sensitive to detection errors when the product in brackets is unity. Thus an optimum measurement strategy is to choose a source and a path length that keep this product near one under typical conditions. After reviewing the dependence of α on radiation energy and atomic number of the absorbing material, Bartholomew and Casagrande (1957) recommended sources with a long half-life and energy in the 0.5 to 1.5 mev range like cobalt-60 or cesium-137, which can penetrate the walls of typical risers. Kohl *et al.* (1961) provided exhaustive discussions on radioisotopes and their applications.

Weimer *et al.* (1985) examined several practical issues involving γ-ray measurements. In particular, they noted that conventional γ-ray densitometers are too slow for time-dependent measurements in fluidized suspensions. However, by modifying the electronics of a relatively strong 500 mCi source of ^{137}Cs, they achieved a response as fast as 2 ms. In general, because enough photons must excite the scintillation detector to provide accurate Poisson statistics, the response time decreases with increasing source strength and transmission through the system. Using simultaneous γ- and X-ray densitometry, Seo and Gidaspow (1987) measured the volume fractions of two powders of different absorptivities in a bubbling bed.

In order to obtain radial density profiles, Schuurmans (1980) traversed through the riser a small point source of ^{137}Cs mounted at the extremity of a thin probe, with detection outside the vessel. The ratios of intensities at successive probe steps produced time-averaged profiles of solid volume fraction. This intrusive method is a viable alternative to optical fibers. Wirth *et al.* (1991) traversed a similar γ-ray probe in a CFB riser, while Miller and Gidaspow (1992) did so with an X-ray source.

Because it provides integral information along a straight line-of-sight, the transmission densitometry of energetic photons is well-suited for tomographic reconstruction of the distribution of solid volume fractions in a cross-section of the riser. Much can be learned from medical imaging. Herman (1980) discussed the fundamentals of computerized tomography, including calibration, data acquisition, reconstruction, and difficulties associated with polychromatic beams, photon statistics, etc.

In a computerized tomographic (CT) scanner, an object with a distribution of density is interrogated through several line-of-sight measurements between a source and a detector. In order to cover an entire cross-section of the object, a traditional strategy is to execute a series of N_1 parallel measurements along chords of the riser, and to repeat similar series along N_2 distinct directions around the object. Because this procedure requires multiple translations and rotations of the source/detector assembly, it is slow and thus can only provide time-average information on a system with reasonably steady behavior like a CFB riser. Similarly, its spatial resolution is limited by the time required to record all $N_1 \times N_2$ projections in succession, as well as computing costs.

Azzi *et al.* (1991a) implemented such a system in a catalytic cracker unit (Figure 9.5). Their strategy for numerical reconstruction was to divide the

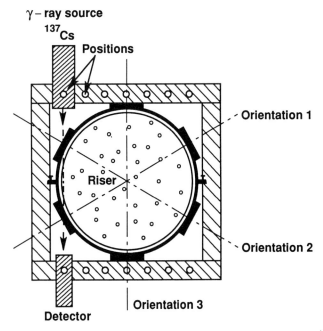

Figure 9.5 Implementation of a CT scanner on an industrial riser (Azzi *et al.*, 1991a). The ^{137}Cs γ-ray source and detector assembly may be fixed at seven transverse positions. The entire support system is rotated to scan three independent directions in the horizontal plane. Yates and Simons (1994) have published a photograph of a similar system. © Éditions Entropie, reproduced with permission.

riser cross-section into a series of adjacent strips covering the paths of the γ-ray beams. For each strip, they defined a basis function that vanishes everywhere but in the strip. Projecting the two-dimensional volume fraction distribution along the basis functions and substituting the result into the transmission equations (9.15) derived from the $N_1 \times N_2$ measurements, they reduced the problem to the inversion of a matrix. However, because the matrix was singular, they employed a regularization technique appropriate for ill-posed problems. To improve the sensitivity of the diagnostic, they also exploited known information about the system through numerical constraints applied to the basis functions. Desbat and Turlier (1993) described efficient algorithms for reconstruction with 'interlaced' sampling schemes that require fewer projections.

Another CT technique interrogates the riser through the 'fan-beam' geometry, where a single source is viewed by detectors placed at several locations across the riser, and the source/detectors assembly is rotated to achieve a sufficient cross-sectional coverage. To reconstruct the density distribution, Bartholomew and Casagrande (1957) employed polynomial basis functions. A promising, albeit more costly, variant of the fan-beam strategy has the potential to yield small enough scanning times to detect real-time features in riser flows. The idea, used in most medical CT scanners, is to rotate continuously a fan-beam source around the riser inside a ring of fixed detectors. Because, unlike the scanning procedure of Azzi et al. (1991a), this technique employs several detectors simultaneously, its time resolution is limited by detection noise. Schlosser et al. (1980) investigated its feasibility for multiphase flows and calculated the source strength that achieves minimum density error for desired time and spatial resolutions in a given geometry.

Other investigators have projected high energy photons across the vessel to record a transverse photographic image of the flow. Weinstein et al. (1984, 1986) traversed a commercial X-ray system along the riser elevation. They produced movies for relatively small solids concentrations and, using longer exposures, obtained images of the mean distribution of solid volume fraction. The images were analyzed with an optical densitometer, with correction for absorption through the plexiglas walls. The mean radial density was assumed to be an axisymmetric polynomial of the radial distance, with coefficients calculated from a set of transmission equations (9.15) obtained at different positions on the photographs. Contractor et al. (1992) observed intensified images from this system through a set of phototransistors. Berker and Tulig (1986) used a γ-ray camera in a similar way. Assuming that the solid distribution is axisymmetric, they employed the Abel transform to extract the radial voidage distribution from the observed transmission:

$$1 - \epsilon(r) = \frac{1}{\alpha \rho_p \pi} \int_r^R \frac{\partial \ln(i/i_0)/\partial y}{\sqrt{y^2 - r^2}} \, dy \tag{9.16}$$

where the integration is along the distance y to the vertical image axis. To minimize noise amplification, they filtered the lateral distribution of $\ln(i/i_0)$ before differentiating it in equation (9.16). Another, more qualitative, use of X-ray imaging was employed by Gajdos and Bierl (1978). More recently, Gamblin *et al.* (1994) observed the interior of a rectangular riser using a rotating anode that emits pulses of X-rays of 1 to 10 ms duration. After image intensification, they captured the flow on a video camera.

9.4.4 Capacitance tomography

This sophisticated non-invasive technique is a promising new development for rapid measurement of solid volume fraction in gas–solid suspensions. It was pioneered in bubbling fluidized beds at the US Department of Energy's Morgantown Energy Technology Center (METC) by Halow *et al.* (1992, 1993), and at the University of Manchester (Huang *et al.*, 1989). The principle is to reconstruct the two-dimensional distribution of effective dielectric constant using capacitance measurements between pairs of N electrodes. In the Manchester design, twelve electrodes of 10 mm height are mounted beneath a 15 mm thick dielectric liner that forms the inner wall of a cylindrical pipe of radius 76 mm. A small guard plate extending 9 mm into the liner from the junction of two adjacent electrodes avoids excessive capacitance between these. In the METC design, four consecutive cross-sectional layers of a 150 mm ID cylinder are probed by rings of 32 electrodes of 25 mm height; guard electrodes extending 150 mm above and below the ring assembly minimize effects of stray capacitances (Figure 9.6). With this layered design, the METC instrument also captures three-dimensional features of the flow.

In the Manchester design, during an individual measurement, fast CMOS switching electronics hold one electrode at a positive voltage for short periods, while the other electrodes are grounded. Then, the rapid discharges of the excited electrode through each grounded electrode are recorded in parallel to provide a measurement of their mutual capacitances (Huang *et al.*, 1992). Because the switching is rapid, the imaging can be repeated at a rate of 100 Hz. The different electronic strategy of the METC instrument permits a similar rate.

Unlike line-of-sight tomography, the forward problem is not confined to a narrow beam. Instead, because Poisson's equation is elliptic, the entire suspension participates in each measurement. Without electrostatics, this equation is homogeneous:

$$\nabla \cdot [K \nabla \tilde{V}] = 0 \qquad (9.17)$$

In the reconstruction strategy employed in Manchester, the pipe cross-section is divided into a number N_3 of finite volume elements called

330　　　　　　　　CIRCULATING FLUIDIZED BEDS

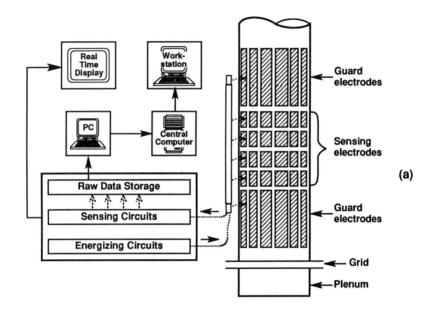

(a)

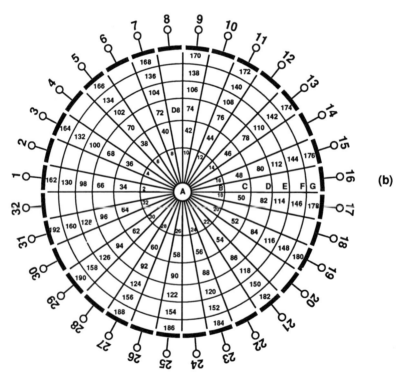

(b)

'voxels'. In the forward problem, Xie et al. (1992) used equation (9.17) to calculate the 'sensitivities' $S_{ij}(k)$ of each electrode pair (i,j) to the presence of a material of dielectric constant K_p in voxel k, while the rest of the pipe is filled with gas at K_h,

$$S_{ij}(k) \sim \left(\frac{\tilde{C}_{ij}(k) - \tilde{C}_{ij}^h}{\tilde{C}_{ij}^p - \tilde{C}_{ij}^h} \right) \qquad (9.18)$$

where $\tilde{C}_{ij}^p$ and $\tilde{C}_{ij}^h$ are the capacitances of the pair (i,j) when the entire pipe is filled with the material at K_p and with the gas at K_h, respectively, and $\tilde{C}_{ij}(k)$ is the capacitance of (i,j) when only voxel k is filled with the material. Although cylindrical symmetry reduces the forward problem to the determination of only $N/2$ cross-sectional distributions of S_{ij}, it is computationally intensive, since each $S_{ij}(k)$ requires a separate solution of equation (9.17).

In the actual suspension, symmetry limits the number of independent measurements to $N(N-1)/2$ capacitance pairs $\tilde{C}_{ij}^m$. The two-dimensional distribution of dielectric constant is reconstructed through the 'gray level' function

$$gr(k) \equiv \frac{\sum_{i=1}^{N-1} \sum_{j=i+1}^{N} \left(\frac{\tilde{C}_{ij}^m - \tilde{C}_{ij}^h}{\tilde{C}_{ij}^p - \tilde{C}_{ij}^h} \right) S_{ij}(k)}{\sum_{i=1}^{N-1} \sum_{j=i+1}^{N} S_{ij}(k)} \qquad (9.19)$$

A homogeneous suspension of effective dielectric K_e yields a constant gray level $gr = (K_e - K_h)/(K_p - K_h)$. From the gray level, the local solid volume fraction may be inferred using a model of K_e like equations (9.11) or (9.12).

In general, K_e is unknown in all N_3 voxels. In order to raise instrument resolution, Xie et al. (1992) chose $N_3 \sim 900$ to be greater than the number of measurements. The resulting indeterminacy manifests itself as a 'leakage' of the gray level into adjacent voxels and as the failure to converge successive iterations between forward and inverse algorithms. Any independent information on the suspension structure (e.g. core/annulus, axisymmetry) helps overcome this indeterminacy. In contrast, because the METC instrument has 2.5 times fewer voxels than the number of independent measurements, it is not indeterminate but instead produces a blurring of images if flow structures are finer than individual voxels.

Figure 9.6 (*previous page*) Capacitance tomography system at METC; (a) electronics and probe assembly; (b) placement of the 32 electrodes and division of the bed cross-section in 193 voxels. Reproduced from Halow et al. © 1993, with kind permission from Elsevier Science Ltd, The Boulevard, Langford Lane, Kidlington OX5 1GB, UK.

Despite these difficulties, the technique is likely to become an important diagnostic for circulating fluidized beds with limited electrostatics. Beck et al. (1993) summarized the early stages of its development in Europe. Brodowicz et al. (1993) used capacitance tomography in horizontal pneumatic conveying. Hayes et al. (1993) envisioned capacitance tomography as a device to record solid mass flow rates from the cross-correlation of two adjacent instruments in a pipe. With its four adjacent ring layers, the METC instrument is a serious contender for meeting that important industrial challenge.

9.5 Local particle flux

The instantaneous particle flux vector G_s is the product of the local bulk density and particle velocity v_p,

$$G_s = \rho_p(1 - \epsilon)v_p \qquad (9.20)$$

Its measurement is therefore a useful complement to the determination of the local solid volume fraction and the particle velocity.

In circulating fluidization, the particle flux is not always directed along the main flow. In the curtain of descending solids near the wall, its time-averaged value is negative. In the core, it may point up or down as entrained packets of solids successively rise and fall. Therefore, its measurement is interpreted differently in the CFB riser than in more steady, homogeneous or unidirectional flows.

The most robust diagnostic technique of particle flux consists of a cylindrical tube connected to a vacuum and separation system (Figure 9.7). In this form, it closely resembles isokinetic probes used in atmospheric sampling. The analysis of these probes has been widely studied (e.g. Lutz and Bajura, 1982). Isokinetic probes draw a specimen that retains the same proportions of gases and solids as the suspension under study. Because their aim is to record accurately the ratio of solid to gas flow rates (loading), the suction rate is tuned in such a way that the velocity in the tube matches that in the suspension. If suction is excessive, then the measured gas and solid flow rates generally yield an apparent loading lower than expected (Belyaev and Levin, 1972, 1974; Fuchs, 1975). The mismatch is exacerbated at higher values of the Stokes number, which is the ratio of the hydrodynamic relaxation time of an individual particle and the time scale characteristic of the flow in the tube. For small particles, this ratio is

$$St = \frac{\rho_p d_p^2}{18\mu} \bigg/ \frac{d_t}{u_\infty} \qquad (9.21)$$

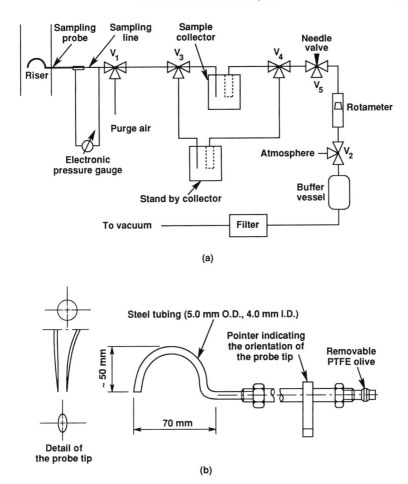

Figure 9.7 Anisokinetic suction probe system of Rhodes and Laussmann (1992b): (a) suction system; (b) details of the probe tip. © Elsevier Science S.A., Lausanne, Switzerland, reproduced with permission.

where u_∞ is the free-stream velocity of the gas. For more massive particles, empirical corrections of the particle relaxation time capture effects of fluid inertia at other than small particle Reynolds numbers (Clift et al., 1978).

At high Stokes numbers, particles have such inertia that the gas suction hardly affects their trajectories. In this case, any mismatch of the suction rate produces an identical relative error in the loading. Such disproportionate sampling is called 'anisokinetic'. In circulating fluidization, the Stokes number is often large (10 to 10^3). Consequently, sampling probes are not generally appropriate for recording solid loading, unless elaborate precautions are taken to ensure isokinetic conditions. To this end, Van Breugel *et al.* (1969) matched the static pressures of the gas inside and outside the tube using miniature purged taps.

For particles of high Stokes number, no such care is strictly necessary if the parameter of interest is solid flux, rather than loading. In this case, the measurement is limited to a record of the mass of solids collected in a given time period, while the gas flow rate is largely ignored. Because the particles are relatively massive, their penetration of the probe is insensitive to suction over a range of gas flow rates. In a CFB riser, Monceaux *et al.* (1986) indeed observed that the mass of solids recovered first increased with suction, then stayed constant over a wide range, and finally climbed again. At low suction, the particle momentum is quickly dissipated in the tube and the collected mass is small. At high suction, the probe disturbs the flow considerably as it draws remote particles into the tube. Using a small riser with a uniform upward flow of 285 μm sand particles, Herb *et al.* (1992) tested a probe with miniature pressure taps similar to that of van Breugel *et al.* (1969). They confirmed the existence of a range of suction velocities where the inferred flux is constant.

This technique has been widely used in the laboratory, e.g. by Bader *et al.* (1988), Werther *et al.* (1990), and Miller and Gidaspow (1992). Rhodes and Laussmann (1992a) and Aguillón *et al.* (1995) evaluated several probe geometries. Leckner *et al.* (1991) and Zhang *et al.* (1994, 1995) employed this technique in a 12 MW CFB boiler, while Werdermann and Werther (1994) did so in CFB powerplants. Samples can also provide information on the solid composition or size distribution in the riser (e.g. Dry, 1987).

Another issue, essential in flows that exhibit significant fluctuations, is the relative orientation of the tube and the instantaneous particle flux G_s. Because the probe can only sample the component of fluxes in the direction of the inward normal n to its opening of area A, the mass M collected over a period $\tilde{T}$ is

$$M = A\tilde{T}\langle(G_s \cdot n)H(G_s \cdot n)\rangle \quad (9.22)$$

where the outer brackets denote time averaging and $H(\zeta)$ is the Heaviside function, unity for $\zeta > 0$ and zero otherwise. Thus, to extract the net flux along the axis of the riser, it is necessary to add the measurements with the probe turned in both the upward and downward directions. Although the downward component is smaller in the core, it does not generally vanish there (Herb *et al.*, 1992). Another consequence of equation (9.22) is that a probe pointed horizontally can sample a significant mass of particles (Qi and Farag, 1993; Zhou *et al.*, 1995), even if the net flux is zero in that direction.

Measurements of net flux from suction probes pointed in the upward and downward directions can be combined with a differential measurement of dynamic pressure to yield an estimate of the local particle velocity and concentration, e.g. Zhang *et al.* (1994, 1995). Temporal correlations of the fluctuations in flux and solid volume fraction are ignored in the corresponding time-averaging of equation (9.7). It is unclear how this simplification affects the accuracy of the estimate.

Because the gas velocity u_∞ vanishes right at the wall, the Stokes number is considerably smaller in the descending curtain of solids than in the core. Consequently, the particles are more readily entrained by suction, and, as observed by Gajdos and Bierl (1978), Bader *et al.* (1988) and Herb *et al.* (1992), collection becomes a stronger, nearly linear, function of the suction rate. Under these conditions, the appropriate value for the suction rate should be zero. However, because the suction is finite in a practical measurement, both upward and downward components of the flux are inferred by carrying out successive measurements at finite rates and extrapolating the results to zero suction.

Bolton and Davidson (1988) introduced a scoop protruding about 12 mm into the riser to collect the descending curtain of solids at the wall by gravity. A small leakage of air through the collection bottle maintained a suction rate sufficient to prevent blockage of the device.

Another, less direct, method can also yield the local particle flux. Plumpe *et al.* (1993) and Gidaspow *et al.* (1989) introduced a small metal sphere in a vertical riser (Figure 9.8). When particles of a different material come into its contact, the surface equilibrium distorts the electronic energy levels creating a contact potential. Upon rapid separation, the two surfaces retain opposite charges (Soo, 1989). The contact is detected as an individual pulse appearing

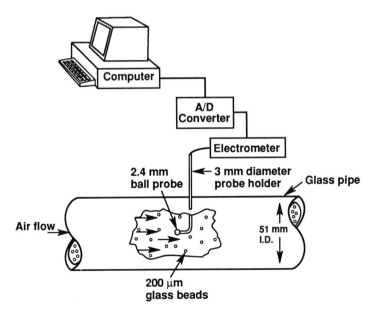

Figure 9.8 'Ball probe' of Plumpe *et al.* (1993). © American Institute of Physics, reproduced with permission.

at the output of a charge-to-voltage converter connected to the sphere. The pulse frequency f produced by a dilute steady flow of particles of diameter d_p and uniform velocity v_p striking a sphere of diameter $d_s \gg d_p$ is

$$f = n\frac{\pi}{4}d_s^2 v_p = \frac{3}{2}\frac{d_s^2}{d_p^3}\frac{G_s}{\rho_p} \tag{9.23}$$

Because the pulse detection is rapid, the main advantage of the 'ball probe' is its ability to record the instantaneous flux, so statistical analyses of its fluctuations may be performed. In particular, it should be noted that, if there is a significant temporal correlation between the velocity and the volume fraction, the time-average flux is in general not equal to the product of the time-average velocity and bulk density:

$$\langle G_s \rangle = \rho_p \langle (1-\epsilon)v_p \rangle \neq \rho_p \langle (1-\epsilon) \rangle \langle v_p \rangle \tag{9.24}$$

For denser suspensions, Zhu and Soo (1992) provide a collision theory that relates the total current detected by the probe to the particle flux.

The ball probe has two main limitations. The first is that it cannot identify the direction of impacting particles. Thus signal interpretation is difficult in flow regions that exhibit frequent reversals. The second has to do with the validation of individual impacts. Because the latter are very rapid, it is difficult to distinguish collisions that produce a small charge transfer from high frequency noise. An impact is accepted if its signal lies above a given threshold. Because the frequency count depends strongly upon the threshold, which cannot be set unequivocally, the probe must be calibrated against a more robust instrument like the sampling probe.

9.6 Solids mass flow rate

Because the overall recirculating solid flux is a basic operating variable of the CFB riser, its determination is essential in experimental facilities. Burkell et al. (1988) reviewed the leading measurement methods. Assuming negligible escape of solids from the cyclones, the flux is generally inferred from the solids flow rate traversing the downcomer.

A simple technique is to observe the flow of identifiable particles through a window located in the dense region of the downcomer. Unless the suspension fails to travel as a homogeneous plug of solids, velocity gradients are small in the downcomer, and the overall flow rate can be estimated from the packing fraction and the observed velocity.

Another method is to close a permeable butterfly valve in the downcomer and to record the accumulation of the fluidized solids above the valve using a differential pressure transducer. The overall riser flux is then

$$\overline{G_s} = \frac{A_d}{A_r}\frac{d(\Delta p)}{g\,dt} \tag{9.25}$$

where A_d and A_r are the cross-sectional areas of the downcomer and riser, respectively, and Δp is the differential pressure recorded. In this method, the accumulation of solids above the butterfly valve may also be determined visually. Another common method is to divert the flow of solids to another vessel for a known period and to record the mass accumulated in that vessel. Burkell et al. (1988) also discussed an impact flowmeter or a calibrated orifice, but fluctuations in the flow rendered these methods largely inoperable.

Marcus et al. (1990) reviewed commercially available solid mass flow meters, which are often limited to relatively small pipe diameters. They mentioned a U-tube that records the Coriolis force imparted by the flowing solids ($D \leq 15$ cm), as well as impact flowmeters and screw conveyors, which are not subject to stringent size limitations, but disrupt the flow considerably. They also cited systems of two cross-correlated capacitance probes ($D \leq 20$ cm).

The cross-correlation of triboelectric signals naturally present in the suspension is also exploited commercially to infer the mean flow velocity in unidirectional flow ($D \leq 8$ cm). However, because electrostatics depends upon the properties of the solids fractal surface, its absolute level is difficult to control (Dybeck et al., 1995) and, consequently, triboelectric instruments alone cannot record concentration with sufficient reliability.

In larger industrial facilities, measurements of solid flow rates remain a considerable challenge. A common method is to carry out an energy balance in a cooled section of the downcomer. The heat extracted is evaluated from measurements of the mass flow rate, specific heat and temperature increase of the cooling fluid. A thermocouple record of the corresponding drop in the solid temperature yields an estimate of the solid mass flow rate. As noted by Burkell et al. (1988), uncertainties of this calorimetric method arise from unknown heat losses and from temperature inhomogeneities as the solids pass through the cooled downcomer section.

Another, equally questionable, method is to infer the flux from the superficial gas velocity and a measurement of the pressure gradient in the upper region of the riser (Patience et al., 1990). In dilute suspensions, a common assumption is that the slip velocity between gas and solids is related to the terminal velocity of individual particles.

Davies et al. (1991, 1992) proposed an ingenious device comprising a chamber where flowing solids are dropped and allowed to escape through lateral vertical slots. The weight of solids accumulating in the chamber is correlated with mass flow rate.

9.7 Particle velocity

For conditions typical of CFB risers, particle velocity measurements are challenging. The most common method is to cross-correlate signals from two optical fiber probes a short distance apart. Laser-Doppler anemometry (LDA) may be used in dilute regions of the riser. Further diagnostics include tracers and the measurement of impact forces.

9.7.1 Cross-correlation

This technique employs two optical fiber probes as described in section 9.4.2. The cross-correlation function $\phi_{12}(\tau)$ of the two voltages $\tilde{v}_1(t)$ and $\tilde{v}_2(t)$ can be estimated as

$$\phi_{12}(\tau) \approx \frac{1}{\tilde{T}} \int_0^{\tilde{T}} \tilde{v}_1(t-\tau)\tilde{v}_2(t)\,dt \tag{9.26}$$

where $\tilde{T}$ is the duration of signal acquisition. A measure of the mean velocity of the particles is $v_m = L/\tau_m$, where L is the distance between the probes and τ_m is the value of τ at which ϕ_{12} reaches a maximum.

If both signals were narrow impulses separated by a fixed time delay, $\phi_{12}(\tau)$ would peak sharply at τ_m. Because this ideal situation seldom occurs in practice, the cross-correlation function broadens considerably, increasing the uncertainty in τ_m. Therefore, the main challenge is to design a system with as narrow a cross-correlation function as possible.

Two principal mechanisms are responsible for the broadening of the cross-correlation. The first is that coherent structures in the riser widen the auto-correlation of each signal. The auto-correlation of $\tilde{v}_1$,

$$\phi_{11}(\tau) \approx \frac{1}{\tilde{T}} \int_0^{\tilde{T}} \tilde{v}_1(t-\tau)\tilde{v}_1(t)\,dt \tag{9.27}$$

peaks at $\tau = 0$. A measure of its temporal coherence is its standard deviation σ_{11}. The second reason for broadening is that the particle concentration evolves as it travels from one fiber to the next. It is convenient to model this evolution as an axial dispersion with diffusion coefficient D_M in a uniformly open system (Kipphan and Mesch, 1978). If it is assumed that voltage is proportional to concentration, the two signals are related by the convolution integral

$$\tilde{v}_2(\tau^\dagger) = \int_{-\infty}^{+\infty} \tilde{v}_1(t^\dagger - \tau^\dagger) g_d(t^\dagger)\,dt^\dagger \tag{9.28}$$

where the dagger indicates time made dimensionless through τ_m. The Fourier transform of equation (9.28) is

$$\tilde{V}_2(f^\dagger) = \tilde{V}_1(f^\dagger) G_d(f^\dagger) \tag{9.29}$$

where the dimensionless response function is

$$G_d(f^\dagger) = \exp\left[\frac{Pe}{2}\left(1 - \sqrt{1 + \frac{8\pi j f^\dagger}{Pe}}\right)\right] \tag{9.30}$$

and $Pe \equiv v_m L/D_M$ (Nauman and Buffham, 1983, p. 107). It follows from equation (9.29) that the Fourier transforms of the correlation functions are related by

$$\Phi_{12}(f^\dagger) = \Phi_{11}(f^\dagger) G_d(f^\dagger) \tag{9.31}$$

Assuming that $\phi_{11}(\tau)$ is an even function, it follows from equations (9.30) and (9.31) that the standard deviation of ϕ_{12} is (Bracewell, 1986, p. 143)

$$\sigma_{12} = \tau_m \sqrt{\frac{2}{Pe} + \left(\frac{\sigma_{11}}{\tau_m}\right)^2} = \sqrt{\frac{2LD_M}{v_m^3} + \sigma_{11}^2} \tag{9.32}$$

Hence the breadth of the cross-correlation and, consequently, the uncertainty in τ_m grow with increasing distance between the two sensors, with diffusion, and with wider auto-correlation, i.e. with greater coherence of the signal, while narrowing with increasing mean velocity. Because the actual dispersion of the particles is three-dimensional, it is likely that the uncertainty grows faster with L than this simple one-dimensional model predicts. Finally, because

$$\int_{-\infty}^{+\infty} \phi_{12}(t)\, dt = \Phi_{11}(0) G_d(0) = \Phi_{11}(0) \tag{9.33}$$

and $\Phi_{11}(0)$ is independent of L, the peak value of $\phi_{12}(t)$ decreases with larger σ_{12} and thus with larger L.

In order to reduce the temporal coherence of the signal, it is useful to design a probe that distinguishes random passage of individual particles. To this end, the measurement volume of each fiber should be small, containing few particles. Note that this contradicts the guidelines for measuring volume fraction, where the measurement volume had to contain many particles to render the output insensitive to their placement, and the signal had to be filtered to avoid detection of their individual passage.

Another difficulty with cross-correlation is that it performs poorly when velocity changes direction. In this case, the averaging may return an ambiguous peak of ϕ_{12} from equation (9.26). For example, this predicament can arise near the boundary between the wall layer and core regions. To alleviate the problem, it is desirable to minimize the acquisition time $\tilde{T}$. Because it is prudent to sample the signal over durations permitting calculation of ϕ_{12} in the range $[\tau_m - \sigma_{12}, \tau_m + \sigma_{12}]$, we recommend sampling for a duration $\tilde{T}$ of order $(\tau_m + \sigma_{12})$. Hartge et al. (1988) provided insight on practical acquisition times in a riser. For a distance $L = 4.4\,\text{mm}$ and velocity v_m of order 5 m/s, they selected $10 \leq \tilde{T} \leq 30$ ms.

The accurate determination of velocity from the transit time τ_m requires knowledge of the effective distance L between the two probes. Because the measurement volume extends beyond the optical fibers and varies with concentration, the effective distance is not necessarily the distance separating the axes of the two fibers. Deviations may occur if the optical axes are not strictly parallel, or the two fibers have slightly asymmetric emission. The relative uncertainty decreases with increasing L. Patrose and Caram (1982) carefully established the effective distance empirically before carrying out measurements.

Transit time correlation anemometers have many variants. Ohki and Shirai (1976) described an optical probe with two detection fibers adjacent to a single emission fiber. Oki et al. (1977) preferred two pairs of fibers. They used a convenient normalization of the cross-correlation integral through the mean $\bar{\tilde{v}}$ and standard deviation σ of each signal,

$$\phi_{12}^*(\tau) \approx \frac{1}{T}\int_0^T \left(\frac{\tilde{v}_1(t-\tau) - \bar{\tilde{v}}_1}{\sigma_1}\right)\left(\frac{\tilde{v}_2(t) - \bar{\tilde{v}}_2}{\sigma_2}\right) dt \qquad (9.34)$$

This normalized integral is more powerful than equation (9.26) for signals that exhibit a step. For example, the cross-correlation of two delayed Heaviside functions does not exhibit any sharp peak, whereas their normalized cross-correlation does.

Horio et al. (1986, 1988) employed the design of Ohki and Shirai (1976) to measure particles velocities in dilute regions and the design of Oki et al. (1977) for the velocity of clusters. Ishii and Horio (1991) proposed an arrangement with vertical axes of each of two sets of optical fibers. Because in this case the long axis of the measurement volume is aligned with the main flow, it is unclear how to define the effective distance L.

Cross-correlation is also used in conjunction with instruments other than optical fibers. Soo et al. (1989) measured the flow velocity in a horizontal transport system by cross-correlating the voltages from two ball probes. In more dilute conditions, non-invasive optical techniques become feasible. Zhu et al. (1991) cross-correlated the transmitted light from two laser beams traversed through a pneumatic transport line.

9.7.2 Laser–Doppler anemometry

Durst et al. (1981) provided a comprehensive review of this mature technique, which several companies offer as a turn-key, albeit expensive, package. Because it is a complex instrument seldom used in relatively dense suspensions, the user should always examine critically the data generated by its elaborate software.

In its basic version, the principle is to create a small measurement volume at the intersection of two laser beams. In the conceptually simplest description, the coherence of the radiation creates a pattern of fringes aligned with the vectorial sum of the directions of the two beams. Particles passing through the crossing region scatter light to one or several detectors. As they traverse alternately light and dark regions of the interference pattern, the particles produce a burst signal with frequency f equal to the ratio of the particle velocity v_p and the known fringe spacing ω,

$$f = v_p/\omega = 2v_p \sin\phi/\lambda_l \tag{9.35}$$

where ϕ and λ_l are, respectively, the half-angle between the two beams and the laser wavelength.

In this configuration, the instrument records the magnitude of the velocity component perpendicular to the fringes, but it fails to distinguish the direction of the velocity. In order to do so, the beams are often passed through a system, often a Bragg cell, that shifts their frequencies by an amount Δf. The result is a moving pattern of fringes that produces bursts of frequency

$$f = (v_p/\omega) + \Delta f \tag{9.36}$$

which permit unambiguous determination of velocities greater than $-\omega\Delta f$. This feature is particularly important for CFB measurements where local velocities may change direction.

In gas flows, seed particles of small Stokes number (e.g. 1 μm alumina particles, oil smoke) are introduced as tracers to travel with the gas. In the presence of a solid phase of larger particles, it is essential to distinguish these particles from the seed. Lee and Durst (1982) discriminated between these by comparing the amplitudes of their respective signals. Signal validation is not always straightforward. For example, large particles can occasionally produce a small burst by intersecting only a limited portion of the crossing region. Modern commercial instruments employ sophisticated software for discrimination.

Because the Doppler bursts from the gas and solid phases are discrete, mutually exclusive events, specific averaging is used to produce unbiased continuum variables and temporal correlations for comparison with theoretical models (Durst et al., 1981). The burst rate can also yield the mean particle flux and an estimate of the local solid volume fraction. Van de Wall and Soo (1994) discussed temporal and spatial conditions necessary to perform this measurement.

To monitor two velocity components simultaneously, some LDA systems intersect two coherent beam pairs of distinct wavelengths and so create two perpendicular fringe patterns. Narrow bandpass filters permit the separate detection of the two signals. Unfortunately, refraction across the vessel wall can prevent the beams from converging at the same point. To avoid this problem, it is best to employ a flat wall perpendicular to the fringe orientation. Thus, LDA measurements are easier to perform in risers of rectangular cross-section. To record velocity profiles in immovable risers, the entire optical train must be traversed. A convenient alternative is to bring laser radiation to a relatively small optical probe through flexible single mode optical fibers.

Because of its requirement for a clear line of sight, LDA is limited to small solid volume fractions in CFB risers. Nevertheless, because the suspension condenses into relatively dense strands of small volume suspended in a larger dilute phase, the laser beam can penetrate over distances longer than expected in a homogeneous suspension of the same overall concentration. However, it is unclear whether this convenient segregation introduces a bias in the resulting velocity statistics.

Arastoopour and Yang (1992) measured one component of particle velocity in a CFB riser of 50 mm diameter. They used a flat optical window to ensure consistent refraction of the laser beams while traversing the riser. Wang et al. (1993) performed a similar measurement in a square riser of 222×222 mm cross-section and 2.5 m height with sands of 530 μm and 718 μm mean sizes. For fluxes and superficial velocities of $G_s = 16$ kg/m^2 s, $U = 5$ m/s and $G_s = 25$ kg/m^2 s, $U = 6$ m/s, they reported profiles of the three components of the mean and fluctuating velocities of the particles. For these respective conditions, the mean solid volume fractions were 1.5 and 1.1%.

Several groups have modified LDA systems or interpreted the burst signals to estimate particle sizes (Cheremisinoff, 1986). For example, in the freeboard of a fluidized bed combustor, Berkelmann and Renz (1989) correlated the particle size with the pedestal amplitude of the Doppler bursts. A recently available commercial instrument permits simultaneous measurement of the diameter and velocity of spheres that scatter light as they traverse the measurement volume. When the Mie parameter, $\alpha_M = \pi d_p / \lambda_l$, is greater than about 10, the principle of this technique is captured by geometrical optics. Because of differences in optical paths, there is a phase shift η_s between a ray directed at the center of the sphere and another ray striking it elsewhere (Bachalo and Houser, 1984). In the far field,

$$\eta_s = 2\alpha_M (\sin \theta - Nm \sin \theta') \qquad (9.37)$$

where θ and θ' are the angles between the plane tangent to the sphere at the point of incidence and the incident and transmitted rays, respectively, and m is the refractive index of the sphere. The integer N is zero if the ray is reflected directly and one if it is transmitted once through the sphere. The phase shift is recorded on detectors viewing the measurement volume from different angles. The resulting determination of α_M yields the sphere diameter. Note that this instrument requires spherical particles. Tadrist et al. (1994) employed it in the dilute upper region of a square glass riser of 200×200 mm cross-section and 2 m height. They fluidized glass spheres and seeded the gas with smoke particles. The mean solid volume fraction in the entire column was less than 2%. The volume fraction at the test location was probably much smaller than 1%. Cattieuw (1992) extracted the velocity statistics of both phases by comparing the distribution of burst velocities in the presence and absence of seeded smoke and by fitting the joint histogram of velocities with Gaussian distributions.

9.7.3 Other velocimetry

Wirth et al. (1991) introduced phosphorescent particles in a riser. After tagging these with a flash via an optical fiber, they inferred the solids velocity at the wall from a video record of the luminous trajectories. In the interior of dense suspensions, Louge et al. (1991) envisioned similar measurements using an integrated fiber optic probe. Fiedler et al. (1994) traversed a probe that illuminates particles and captures images on a charge-coupled device (CCD) array. A particle traveling past successive CCD elements produces a periodic interruption of the signal. Like LDA, the resulting frequency is proportional to the particle velocity. Werther et al. (1993) reviewed similar imaging techniques for dilute pneumatic transport, where the CCD array is replaced by a grid interposed in the optical path. McDonnell and Johnson (1980) recorded particle velocities in a pneumatic transport line by intersecting multiple laser beams to form sharp intense spots separated by dark areas. Their analysis indicates that the method provides a less ambiguous deconvolution of the signal than LDA for particles with a wide distribution of velocities.

Fasching et al. (1992) patented a capacitance probe consisting of four sensors surrounded by a common guard. The cross-correlation of signals from two adjacent sensors provided a measure of the mean solids velocity. Because this design did not control the electric field, it could not yield a quantitative record of solid volume fraction. By joining two grounded capacitance probes side by side and cross-correlating their respective signals, Louge et al. (1995) recorded simultaneously solid velocity and concentration near the wall.

Zheng et al. (1992) employed an ingenious stroboscopic system to monitor the trajectories of white tracers in a suspension of mostly opaque particles. The tracers were illuminated with successive flashes of red, blue and yellow. By recording the three images of different colors on a single photograph, the tracer trajectory became unambiguous.

Heertjes et al. (1970) measured the impulse from groups of small particles impacting a piezoelectric crystal. For relatively large spheres, the peak normal force F_m exerted by the impact of a single, nearly elastic sphere onto a thick, flat plate can be estimated from the Hertzian contact theory,

$$F_m = 2^{-1/5} \left(\frac{5\pi}{24}\right)^{3/5} \frac{\rho_p^{3/5} d_p^2 v_\perp^{6/5}}{\Delta^{2/5}} \qquad (9.38)$$

where the parameter $\Delta \equiv (3/4)[(1-v_P^2)/E_Y + (1-v_P'^2)/E_Y']$ combines Poisson's ratio and Young's modulus for the sphere with the corresponding quantities for the plate. Because the peak signal from individual particles decreases with the square of their diameter, it is generally too small for typical CFB solids. Therefore instead of measuring F_m, Heertjes et al. (1970) recorded the total normal impulse $\int F_\perp dt$ exerted by a series of N

particles of mass M and average velocity $\bar{v}_\perp$ using glass and silica gel particles as small as 60 μm and 105 μm, respectively, with $1 \leq v_\perp \leq 30$ cm/s. Conservation of momentum dictates that, for spheres of restitution coefficient e,

$$\int F_\perp \, dt = (1+e) M N \bar{v}_\perp \tag{9.39}$$

so the probe calibration reveals a signal proportional to velocity. However, because in practical measurements the instantaneous force on the probe of area A is the product of the individual impulse and the collision frequency, it is quadratic with velocity and depends upon the solid volume fraction

$$F_\perp = (1+e)\rho_p(1-\epsilon) A \bar{v}_\perp^2 \tag{9.40}$$

where A is the area of the exposed crystal surface. This dependence complicates the determination of velocity from the probe record. Further, it is unclear how the intrusion of the probe modifies the velocity field. Ettehadieh et al. (1988) employed a force probe to record solid recirculation rates in a jetting fluidized bed.

The direct measurement of gas velocity is challenging for dense flow in a riser. Horio et al. (1992) injected ozone 20 mm upstream of two adjacent optical probes and recorded the resonant absorption of O_3 near 250 nm. Each probe consisted of two optical fibers facing each other across a 5 mm long measurement volume (Figure 9.9). The transmission signals from the two probes, separated by 5 mm, are cross-correlated to extract ozone velocity. In order to minimize light scattering, particles are excluded using wire gauze. Hatano et al. (1993) exploited a similar idea, but with NO_2, a gas tracer that absorbs in the visible wavelength range.

9.8 Heat and mass transfer

Because it dictates the size of a unit and its configuration, the rate of heat transfer is crucial information for the design of CFB boilers. At high temperatures, both forced convection and radiation contribute to the overall rate (see Chapter 8). In cold laboratory facilities, only the former can be evaluated.

9.8.1 Forced convection in cold risers

Because in forced convection the heat flux and energy equations are linear in the temperature of both phases, the convective heat transfer rate can be evaluated for any convenient temperature difference between the wall and the suspension. If the dimensionless numbers characterizing the fluid dynamics and convective heat transfer are appropriately matched (Glicksman et al., 1994), a warm probe immersed in a cold unit can simulate convection in industrial risers. In the absence of radiation, the heat transfer coefficient h is a

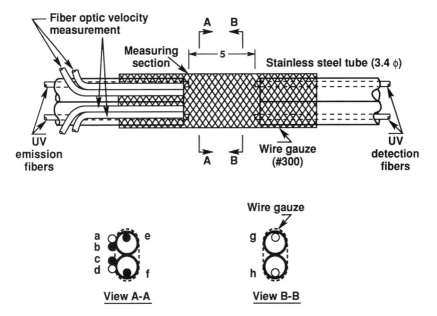

Figure 9.9 Simultaneous measurement of particle and gas velocities (Horio et al., 1992). The particle velocity is inferred from the cross-correlation of the backscattered laser light detected by optical fibers a and d and brought by fibers b and c. Ultraviolet light brought by fibers e and f is detected by fibers g and h, respectively. Ozone absorption in the region between fibers f and h is cross-correlated with a similar signal from the region between e and g. © Engineering Foundation, reproduced with permission.

convenient engineering measure of the convective heat q crossing an elementary wall surface of area A per unit time,

$$q = hA(T_w - T_b) \quad (9.41)$$

Because this definition of h depends upon the choice of the mixed mean temperature T_b, it is essential to define and measure this temperature consistently. This is particularly crucial when much heat is added to a relatively small riser. For example, in the absence of chemical reactions, a constant heat flux q through the wall causes T_b to increase with elevation according to (Louge et al., 1993):

$$\frac{dT_b}{dx} \approx \left(\frac{4}{D_h}\right)\left(\frac{q}{\rho U c_g + G_s c_p}\right) \quad (9.42)$$

The techniques for measuring local heat flux generally fall in three main categories. In the most common, a constant power is supplied to the probe. If conduction losses are negligible between the probe and the surrounding wall, measurement of the probe face temperature yields h. For example, Ebert et al. (1993) employed this principle.

The second method maintains the face temperature constant. Here, h is obtained from the power supplied to the probe. In principle, the heat transfer coefficient at constant flux differs from that evaluated at constant temperature. However, it appears that the difference diminishes with increasing loading (Louge et al., 1993). Because the temperature difference between the active element and the surroundings of the probe is constant, this method permits a more straightforward accounting for conduction losses. The probes often consist of a small platinum element, the resistance of which is a linear function of temperature. By controlling the resistance, temperature is kept constant. Wu et al. (1989) implemented this method with computer feedback control in a cold CFB, achieving an instrument response time as small as 20 ms. Mohd. Yusof (1992) employed a control system similar to that used in hot-wire anemometry. Renganathan and Turton (1989) reviewed the use of thin film heat gauges in fluidization.

One way to maintain the wall temperature constant over larger areas is to condense a fluid behind the wall. Goedicke and Reh (1992) designed a 'heat-pipe calorimeter' around a 700 mm section of a riser of 410 mm diameter (Figure 9.10), with steam raised and circulated in an enclosed space outside the riser. Condensing steam produces a known equilibrium temperature at the wall. The electrical power required to vaporize the condensate is controlled by a feedback loop that maintains the cavity at the desired

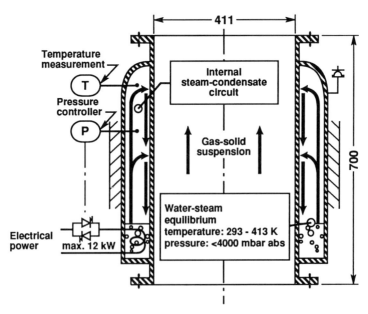

Figure 9.10 Heat pipe calorimeter of Goedicke and Reh (1992). A controller maintains the pressure of the cavity constant by adjusting the electrical power supplied to the boiling coils. Reproduced with permission from the American Institute of Chemical Engineers. © 1992 AIChE. All rights reserved.

vaporization pressure. In the absence of heat losses, the measurements of electrical power and wall temperature yield the average heat transfer coefficient over the section.

The third method infers heat flux from thermocouple measurements located at regular intervals in an insulated cartridge perpendicular to the riser wall. One end of the cartridge is mounted flush with the wall, while the other end is exposed to a hot source or cold sink. If the surrounding insulation of the cartridge is sufficient, the heat flow is one-dimensional, and Fourier's law yields the flux from the known thermal conductivity and the constant temperature gradient. However, thermal inertia prevents instantaneous measurements. Basu and Nag (1987) inserted such an instrument in a CFB riser and circulated water at incipient boiling to maintain a stable temperature at the hot end of the cartridge.

Because CFB risers exhibit rapid fluctuations of voidage and heat flux, a probe should have low thermal inertia and thus rapid response. The temptation is to use as small a probe as possible. However, as Burki *et al.* (1993) demonstrate, thermal entry lengths in a riser are substantial at the wall. It is clear that the wall heat transfer is larger at the leading edge of a vertical heat transfer surface, in a manner reminiscent of the growth of a thermal boundary layer in fluid flow. This increase in h also occurs at the trailing edge of the heated wall surface, probably because of flows periodically upwards and downwards. Therefore, small probes yield abnormally large heat transfer coefficients, unless they are surrounded by guard heaters maintained at the same temperature or average heat flux. Guard heaters are also useful to reduce conduction losses from the probe (Wu *et al.*, 1989).

In order to inform a 'renewal' model of convective heat transfer, Lints and Glicksman (1993) measured the clusters' residence time and coverage of the wall by translating a phonograph needle in the near vicinity of the riser surface. They determined that there is a thin layer with thickness of order 0.2 to 1 particle diameter that is largely devoid of solids, and that obstructions as thin as a particle diameter can alter the curtain of solids at the wall. Therefore, the surfaces of heat flux probes must be made as smooth and non-invasive as possible.

In order to demonstrate the strong correlation between local solid volume fraction and heat transfer rate, Dou *et al.* (1992) combined a constant flux probe with a small, intrusive, parallel plate capacitance instrument driven by the amplifier method of Figure 9.1. The response times of the capacitance probe and the thermopile recording the surface temperature were 2 and 85 ms, respectively. Similarly, Mohd. Yusof (1992) inserted a small platinum coil in the central grounded rod of the non-invasive 'wall' capacitance probe shown in Figure 9.2. The coil was maintained at a constant temperature and, to avoid intrusion, buried beneath a thin, grounded metal layer flush with the wall surface.

Although erosion makes it generally ill-advised to extract heat from within an industrial CFB riser, several studies focus on heat transfer to bed internals. Liu *et al.* (1991) supplied constant electrical power to a horizontal tube spanning the riser cross-section and inferred heat flux from thermocouples on the periphery of the tube. Dou *et al.* (1991) carried out similar measurements with a vertical tube of height 2.6 m heated over a 1.3 m section and immersed along the centerline of an 11 m tall riser. They observed a thermal entry length at the leading edge of the tube. However, unlike their measurement at the wall (Burki *et al.*, 1993), they did not distinguish a thermal development at the trailing edge of the tube. Because solid downflow is very small in the core, this observation further confirms the role played by counter-current solid flow in the development of a thermal boundary layer at the trailing edge of wall heat transfer surfaces.

Wang *et al.* (1989) introduced copper spheres of 11, 15 and 18 mm diameter previously cooled by liquid nitrogen. They calculated the average convective heat transfer coefficient from the sphere temperature using a transient heat balance:

$$\frac{\pi}{6} d_s^3 \rho_s c_s \frac{dT_s}{dt} = \pi d_s^2 h (T_b - T_s) \tag{9.43}$$

The test spheres had a Biot number small enough that their interior temperature relaxed on a time scale $\tau_1 \sim d_s^2 \rho_s c_s / \lambda_s \sim 1$ s, faster than the characteristic time of their cooling in the flow $\tau_2 \sim \rho_s c_s d_s / 6h \sim 30$ s. However, τ_1 was such that the sphere temperature could be sensitive to fluctuations in h associated with the passage of clusters. Because of their peculiar shape, it is not clear that spheres provide useful data for practical CFB applications.

9.8.2 High temperatures

Heat transfer measurements are more ambiguous in CFB boilers. Neither temperature nor flux is usually constant along or through the wall. Further, because of the membrane 'water wall' geometry (see Chapter 8), it is unclear which transfer area is involved in equation (9.41). Finally, the flux is often measured indirectly from a heat balance on the fluid flowing in the tubes. Wu *et al.* (1987) reported such measurements in a pilot CFB riser operating at temperatures from 150 to 400°C. Jestin *et al.* (1991) placed thermocouples at judicious positions in the tube and the fin and used a two-dimensional conduction model of the membrane wall, insulation and welds to infer the heat flux from the thermocouple measurements. Similarly, Andersson and Leckner (1992) observed the transverse distribution of heat flux through steam tubes in a 12 MW CFB boiler.

Radiation is a substantial part of the total wall heat transfer in CFB combustors. At elevated temperatures, the challenge is to distinguish it from the convective component of the flux. For simplicity, we may assume that the suspension and probe are parallel, plane, opaque, gray, diffusely reflecting surfaces following Lambert's cosine law and that the suspension is hotter than the probe. The total heat flux through the probe is

$$\frac{q}{A} = \frac{\sigma_{SB}(T_b^4 - T_w^4)}{\frac{1}{\tilde{\epsilon}_b} + \frac{1}{\tilde{\epsilon}_w} - 1} + h(T_b - T_w) \qquad (9.44)$$

Unlike forced convection, radiation is a non-linear function of temperature. Through its dependence on T_w^4, it is sensitive to the rate of cooling of the wall. In addition, it is affected by temperature gradients in the wall region (Werdermann and Werther, 1994). Similarly, its interpretation requires knowledge of the surface emissivity of the probe face.

Saxena et al. (1989) provided a comprehensive review of fluxmeters at elevated temperatures. They discerned two broad classes of instruments. In the first, the radiative and convective components of the flux are distinguished using two probes of widely different emittances. For example, Botterill et al. (1984) employed an oxidized copper surface with $\tilde{\epsilon}_w \sim 0.8$ and a polished fine gold surface with $\tilde{\epsilon}_w \sim 0.1$. Careful measurements of q/A and T_w for the two probes yield two independent equations (9.44) that may be solved for the unknowns $\tilde{\epsilon}_b$ and h. The principal difficulties with this technique are uncertainties in the emittances of the probes and possible non-uniformities in surface temperatures.

In the second method, the radiative and convective components are separated with optical windows. A fluxmeter located some distance behind the windows measures radiation alone. The total flux is determined by other means. Ozhaynak et al. (1984) emphasized the importance of selecting windows that transmit a wide band of wavelengths and possess a high thermal conductivity. The latter allows vigorous cooling of the window exposed to the hot suspension so that radiation reaching the fluxmeter is not corrupted by additional emission from excessively hot window material. With these criteria in mind, Ozhaynak et al. utilized windows made of zinc selenide rather than quartz. Their probe design is shown in Figure 9.11. They also calculated the total heat flux from a heat balance on the cooling streams of this carefully insulated probe. Han et al. (1992) employed the same probe in a CFB riser operated at 200 to 600°C.

Mathur and Saxena (1987) used two distinct instruments to record the radiative component and the total flux through a horizontal tube in a bubbling fluidized bed. Basu and Konuche (1988) carried out a similar measurement in a CFB. The radiation probe was coated with carbon black to achieve high emittance and a quartz window was used to isolate it from the convective flux. The total heat flux probe was covered with a similar coating, which did not erode substantially over an eight-hour test.

Andersson et al. (1989) measured the total heat transfer flux to the wall of a 16 MW CFB boiler by recording the temperature gradient through a

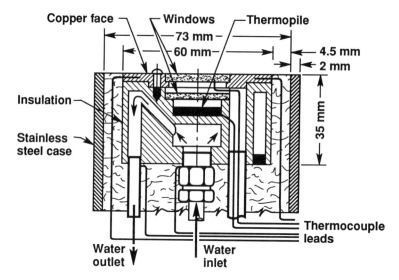

Figure 9.11 Radiometer probe of Ozhaynak *et al.* (1984). The thermopile records the radiation flux through the two ZnS windows. The outer window serves as the external heat transfer surface and is cooled by air forced through the gap. © Engineering Foundation, reproduced with permission.

water-cooled probe. In order to maximize $\tilde{\epsilon}_w$, they covered the probe face exposed to the suspension with temperature-resistant black paint. They calibrated the measurement of heat flux using a black-body furnace. Because paint erosion could compromise accuracy, the layer of paint was renewed periodically. A shortcoming of this measurement was that the radiation component could not be evaluated with certainty.

Temperature measurements are also challenging in hot fluidized beds. Leckner and Andersson (1992) traversed thermocouples through the wall region of their CFB boiler and used a shield to minimize radiation losses from the thermocouples to the cold wall. For the gas temperature, Rink *et al.* (1994) used a suction pyrometer drawing hot gases over a thermocouple protected by a radiation shield. Flamant *et al.* (1992) measured separately the temperatures of the gas and particle phases. The gas temperature was derived from a thermocouple protected against particle contact using a perforated tube, while the particle temperature was obtained by comparing intensities of radiation at 1.3 and 1.5 μm brought to a radiometer via an optical fiber. This radiometric technique has considerable potential in circulating fluidized bed boilers.

9.8.3 Mass transfer

Ebert *et al.* (1993) sublimated naphthalene from a cartridge inserted flush with the wall of a cold riser. They observed a smaller Sherwood number for the sublimation rate of the cartridge than the corresponding Nusselt for wall-to-bed heat transfer. Because chemical adsorption by the bed particles was negligible, the solids failed to enhance the naphthalene concentration gradient at the wall and, consequently, the mass transfer rate there. In contrast, under the same operating conditions, they augmented the wall-to-bed heat transfer through, perhaps, steeper temperature gradients induced by gas–solid heat transfer. To characterize mass transfer rates in suspension, De Kok et al. (1986) introduced naphthalene beads in a bubbling fluidized bed of inert glass beads. A similar technique may be used in circulating fluidized beds.

9.9 Tracers and sampling

Tracers are often used to evaluate the mixing characteristics or residence time distribution (RTD) of a reactor (see Chapters 3 and 4). The idea is to inject a detectable gas or powder that closely matches the hydrodynamic behavior of the suspension. The passage of the tracer is interpreted with a model that incorporates a balance equation for its concentration C:

$$\frac{\partial C}{\partial t} + \boldsymbol{u} \cdot \nabla C = \nabla \cdot (D_M \nabla C) \qquad (9.45)$$

where D_M is a diffusion coefficient or, more generally, a tensor to be determined and $\boldsymbol{u}$ is the velocity vector of the phase carrying the tracer. Boundary and initial conditions are prescribed to represent the form of the vessel and the process of injection.

In this technique, modeling is central to the interpretation of the measurement and knowledge of the underlying flow field is crucial. Because the solution of equation (9.45) is generally three-dimensional, assumptions are invoked to simplify the problem. The leading models have been reviewed by Wen and Fan (1975) and Nauman and Buffham (1983). Further details on solids motion and mixing, including the interpretation of solid measurement techniques, are provided in Chapter 4. In the present review, tracers are classified in four categories.

9.9.1 Particle swarms

In this method, solids are rapidly injected as a group or swarm. Their detection at the exit of the reactor yields the particle RTD, while their passage elsewhere provides a measure of mixing rates. It is tempting to reduce the data by assuming one-dimensional axial dispersion in a uniformly open system. In this case, the constant velocity makes it possible to eliminate the spatial coordinate and to replace equation (9.45) by a single ordinary differential equation. However, this approach is quite approximate in the

CFB riser, where the core-annulus structure and vigorous internal solid recirculation with relatively large length scales dominate the flow.

Tracer solids may be radioactive. Information on radioisotopes is provided by Kohl et al. (1961). Ambler et al. (1990) injected particles labeled with gallium-68 near the base of a CFB riser and recorded their passage near the inlet of the primary cyclone using a scintillation detector. Although the complexity of the underlying flow rendered their results largely inconclusive, they interpreted trends in the bimodal RTD using a core-annulus model. Helmrich et al. (1986) carried out a similar measurement using radioactive $^{24}Na_2CO_3$ tracers. Patience et al. (1991) used sand irradiated in a nuclear reactor and reduced the data with a simpler, less realistic one-dimensional axial dispersion model.

Because the handling of radioactive isotopes is a concern, other solid tracers have also been used. Avidan and Yerushalmi (1985) injected ferromagnetic particles in an expanded bed and monitored their passage through horizontal inductance loops around the bed. Another technique is to introduce a swarm of hot solids into a cold bed. Temperature measurements at other locations then provide data on solid mixing. Westphalen and Glicksman (1995) injected a preheated batch of particles in a riser and measured the spread of the hot tracer using an array of thermistors. Because the hot particles exchanged sensible energy with the surrounding suspension, these authors wrote coupled energy equations for the tracer, the suspension solid and the gas, and they modeled the response of the thermistor. The complexity of the analysis and underlying assumptions was a price paid for the relative simplicity of the measurement.

In another technique, Chesonis et al. (1990) impregnated batches of alumina bed particles with $CaCl_2$. Pulses of this tracer were injected near the base of the riser, and samples were extracted at regular time intervals with an anisokinetic probe and their calcium content analyzed by atomic absorption. The data were analyzed using a model that incorporates a core, a wall region, and the solid return leg. The average residence time in the return leg and a mass exchange coefficient between the core and wall regions were fitted to the time history of the tracer concentrations at different elevations.

Bader et al. (1988) introduced pulses of ordinary table salt in a riser of fluid cracking catalyst. A known mass of sampled solids was then mixed with water and the mass fraction of the dissolved salt was determined from the electrical conductivity in solution. Rhodes et al. (1991) employed the same technique to study longitudinal solid mixing in the riser.

Kojima et al. (1989) impregnated fluid cracking catalyst with a fluorescent dye and injected pulses of the resulting tracer in the bed. An optical fiber guided ultraviolet radiation to the measurement location, where the visible fluorescent emission of the tracer at 520 nm was detected by an adjacent optical fiber. To avoid the complicated interpretation of the absolute signal level from a single probe, Kojima et al. assumed a signal proportional to concentration and one-dimensional axial dispersion in a uniformly open system. With these assumptions, using the transfer function (9.30), they could evaluate the first and second temporal moments of the signal as $\langle t \rangle = z/v_m$ and $\sigma^2 = 2D_M z/v_m^3$, where z is the distance of the probe from the injection point. From changes in these moments between two optical fiber systems a known distance apart, Kojima et al. inferred the mean tracer velocity v_m and its diffusion coefficient D_M. Because observations of a bimodal output signal in the riser contradicted the basic assumptions of the analysis, the latter was convenient, but approximate.

9.9.2 Individual particles

Another tracer technique permits non-invasive tracking of an individual radioactive particle in the flow. Although it has yet to be applied to relatively large experiments, it could be useful for CFB risers. The idea is to triangulate the position of the tracer by recording its irradiance on several uncollimated scintillation detectors. The spreading of the tracer radiation and its attenuation yield a signal that decreases monotonically with distance. The relationship between signal and distance is established through calibration. In principle, the location of the tracer is determined without ambiguity from four detectors. In practice, inhomogeneities of the suspension, secondary low energy gamma emissions, and anisotropy of the detector response contribute to errors in the measured distances. The uncertainty decreases with closer proximity to the detectors and with greater redundancy. Thus, the tracer location is computed from a redundant set of recorded intensities using a least-squares procedure that puts greater weight on nearby detectors. The resulting trajectories are used to construct Lagrangian statistics for the flow. Eulerian information is extracted by subdividing the vessel into a number of bins. In order to build meaningful ensemble averages, the data acquisition must last a considerable time.

Kondulov et al. (1964) first employed this technique in a bubbling fluidized bed. Using advanced computers, Lin et al. (1985) extracted more information from a bed of 140 mm diameter and 160 mm height, while Moslemian et al. (1989) improved spatial resolution with sixteen detectors and a digital photon counting technique. Larachi et al. (1994) detected high energy γ-rays emitted by ^{46}Sc using eight NaI crystals. By simulating the detailed interactions of the radiation with the suspension and the crystals, they achieved a spatial resolution of 5 mm in a cylindrical vessel of 100 mm diameter and 600 mm height for a tracer velocity of 1 m/s and a sampling period of 30 ms. This method is promising in circulating fluidized beds. However, its implementation will likely require many detectors, and experiments may be long, particularly if the tracer spends relatively little time in the riser.

9.9.3 Transient gas injection

Brereton et al. (1988) introduced helium in the windbox and detected its passage at the riser outlet with a gas chromatograph. Because of the short residence time in the column, they found it more practical to interrupt a steady flow of helium with a solenoid valve, rather than to inject a single pulse. They extracted the RTD by differentiating and deconvolving the signal in the frequency domain (Luyben, 1973).

Li and Wu (1991) injected a pulse of hydrogen in a riser of fluid cracking catalyst. Like Kojima et al. (1989), they deconvolved the response of two detectors a known distance apart, assuming one-dimensional axial dispersion to fit a Péclet number to the data. They also characterized the degree of gas back-mixing by monitoring hydrogen upstream of the injection level. Similarly, White and Dry (1989) and Dry and White (1989) injected pulses of argon and detected these with two probes connected to mass spectrometers.

Dry (1986) injected pulses of hot gas near the distributor. An aspiration probe protected by a filter was traversed through the riser at three elevations. A rapid response thermocouple recorded the temperature history of the gas withdrawn by the probe. Using this instrument, Dry et al. (1987) and Dry and White (1992) inferred a qualitative 'contact efficiency' comparing heat transfer with and without solids. The principal shortcoming of this technique is that it employs a tracer dissipating in ways hard to quantify. Because Dry et al. largely ignored heat losses to the aspiration probe and the riser wall, as well as complexities of the underlying gas–solid heat transfer, it is unclear how to interpret their data.

In heterogeneous reactors like the CFB riser, the RTD of an adsorbing gas tracer can yield useful information on contact times. In this context, Krambeck et al. (1987) and Avidan et al. (1985) introduced sulfur hexafluoride in a bed of catalyst particles and varied its adsorptivity by adjusting the moisture content of the catalyst.

9.9.4 Continuous gas injection

Because the diffusion equation (9.45) involves at least two independent axial and radial coordinates in addition to time, a complete transient model of tracer injection is challenging, even in axisymmetric risers. In order to capture both components of gas diffusion, it is convenient to eliminate time from the governing equation. Unlike solid tracers, it is relatively straightforward to do so with gases. A steady flow of tracer is introduced as a point source. The tracer concentration is then monitored at several points of the flow and diffusion coefficients are fitted using a steady version of equation (9.45).

Yang et al. (1984) introduced helium continuously in an air-fluidized riser and monitored the resulting concentration profiles using a thermal conductivity detector. Adams (1988) injected methane at the centerline of a CFB riser and detected it using a total hydrocarbon analyzer. Bader *et al.* (1988) introduced helium at the superficial gas velocity in a riser of fluid cracking catalyst and proposed models for the axial and radial diffusion of the tracer. Werther *et al.* (1991) proposed a core-annulus model to reduce CO_2 tracer data. Martin *et al.* (1992) injected helium isokinetically at the riser centerline and detected it with a gas chromatograph. Their steady, axisymmetric model incorporated a radial profile of mean velocity based on an empirical power law. Amos *et al.* (1993) employed sulfur hexafluoride. Their model assumed uniform gas velocity and radial diffusion in the core, with recirculation in the annulus captured by a hypothetical region surrounding the riser.

This technique can also serve to evaluate gas backmixing. However, because the latter depends on the recirculation of solids, its modeling is less forgiving than that of downstream diffusion. Cankurt and Yerushalmi (1978) injected methane in the riser and sampled its relative concentration profiles upstream. In a similar experiment with catalyst particles, Li and Weinstein (1989) employed helium to avoid adsorption on the surface of the solids.

9.9.5 Gas and solids chemical sampling

The study of reaction rates in circulating fluidized bed reactors requires chemical sampling. In a study of the CFB as a potential waste incinerator, Chang *et al.* (1987) described sampling from the effluents of a high temperature CFB pilot plant. Extraction from the riser is more challenging. Brereton *et al.* (1991) sampled gases and solids at several radial and axial positions in a CFB pilot plant burning coal. Rink *et al.* (1994) described another CFBC pilot facility, together with its gas and particle sampling trains.

For the local measurement of oxygen concentrations, zirconia probes provide a signal proportional to temperature and to the logarithm of the relative partial pressure of oxygen (Stubington and Chan, 1990). Bergqvist *et al.* (1993) introduced air- or water-cooled zirconia cell probes in the 12 MW CFB boiler at Chalmers University.

9.10 Closure

Our understanding of circulating fluidized beds generally advances with the introduction of new instrumentation specifically designed for conditions in dense suspensions, e.g. optical fibers, capacitance sensors, sampling probes, tomography. However, because the general character of riser hydrodynamics is relatively well established, it is no longer satisfactory for instruments to produce qualitative indications of the flow; instead they must provide quantitative data.

Unfortunately, in some cases, inventors emphasize the novelty of an instrument without critically assessing its performance. A strong signal does not necessarily guarantee straightforward calibration or interpretation. Optical fibers, for example, do not always produce an unequivocal measurement of local solid volume fraction. The vast amounts of data generated by modern computer acquisition make it all the more imperative to understand instrument response.

In this context, the present survey has focused equally on the performance and limitations of each instrument. Although experimentalists can now choose from a wide array of techniques, considerable work remains to refine available instruments or develop new ones.

Nomenclature

A	area
a	empirical coefficient in equation (9.7)
a_d	CCD array pixel spacing
B_i	constants
C	concentration
$\tilde{C}, \tilde{C}_0$	capacitances
$\tilde{C}_{ij}^h, \tilde{C}_{ij}^p, \tilde{C}_{ij}(k)$	computed capacitances
$\tilde{C}_{ij}^m$	measured capacitance
c_g	gas mass heat capacity
c_p	solid material heat capacity
c_s	specific heat of test sphere
D	column diameter
d_c	fiber core diameter
D_h	riser hydraulic diameter
D_M	diffusion coefficient
d_p	particle diameter
d_s	ball probe or test sphere diameter
d_t	tube diameter
e	coefficient of normal restitution
E_Y, E_Y'	Young's moduli
$F_\perp$	normal force
f, f_0	frequencies
$f^\dagger$	dimensionless frequency $(= \tau_m f)$
F_m	peak impact force
g	gravitational acceleration
$G_d(f^\dagger)$	dimensionless transfer function (frequency domain)
$g_d(t^\dagger)$	dimensionless transfer function (time domain)
$gr(k)$	gray level of voxel k
G_s	net solids flux

h	heat transfer coefficient
$H(\zeta)$	Heaviside function
$\tilde{i}$	instantaneous current
i, i_0	light intensities
j	$j^2 = -1$
K, K_h, K_p	dielectric constants
K_e	effective dielectric constant
L	length scale
l	extinction distance
l_t	tube length
M	mass
m	refractive index
n	number density
$\boldsymbol{n}$	normal unit vector
N, N_1, N_2, N_3	integers
NA	optical fiber numerical aperture
Nu	Nusselt number
p	gas pressure
Pe	Péclet number
q	heat flux
r	radial coordinate
R	riser radius
S_g	gas wall shear stress
S_p	solid wall shear stress
$S_{ij}(k)$	tomographic sensitivity in (9.18)
St	Stokes number (9.21)
$\tilde{T}$	period
t	time
T_b	mixed mean temperature
T_g	gas phase temperature
T_p	particle phase temperature
T_s	test sphere temperature
T_w	wall temperature
$t^\dagger$	dimensionless time
U	superficial gas velocity
u	gas velocity
V	volume
$\tilde{v}$	instantaneous voltage
$v_\perp$	normal impact velocity component
v_m	cross-correlation peak velocity
$\tilde{V}, \tilde{V}_0$	rectified voltages
$\tilde{V}_1(f), \tilde{V}_2(f)$	voltages in frequency domain
v_p	particle velocity
x, y, z	coordinates, with x in vertical direction

x	vertical
Z	impedance
α	mass absorption coefficient
α_M	Mie parameter
Δ	parameter in equation (9.38)
Δf	Bragg frequency shift
ϵ	voidage
$\tilde{\epsilon}_w$	wall or probe emissivity
$\tilde{\epsilon}_b$	bed emissivity
ϕ	half angle
$\Phi_{11}(f)$	Fourier transform of $\phi_{11}(\tau)$
$\phi_{11}(\tau)$	auto-correlation function
$\Phi_{12}(f)$	Fourier transform of $\phi_{12}(\tau)$
$\phi_{12}(\tau)$	cross-correlation function
$\phi_{12}^*(\tau)$	normalized cross-correlation, equation (9.34)
γ, γ_0	dielectric permittivities
η_s	spatial phase shift
κ_1	extinction coefficient
κ_2	back-scattering coefficient
λ_g	gas conductivity
λ_l	laser wavelength
λ_s	test sphere conductivity
μ	gas viscosity
ν_P, ν'_P	Poisson's ratios
θ, θ'	angles
ρ	gas density
ρ_p	particle material density
ρ_s	density of test sphere
σ_{SB}	Stefan–Boltzmann constant
$\sigma, \sigma_{11}, \sigma_{12}$	standard deviations
τ	cross-correlation time delay
τ_1, τ_2	characteristic times for heat transfer
τ_m	cross-correlation peak time delay
ω	fringe spacing
$\chi(x/d_c; NA)$	radiation shape function

References

Abed, R. (1984) The characterization of turbulent fluid bed hydrodynamics, in *Fluidization* (eds D. Kunii and E. Toei), Engineering Foundation, NY, pp. 137–144.

Acree Riley, C. (1989) Quantitative Capacitive Measurements of Voidage in Dense Gas–Solid Flows, M.S. Thesis, Cornell University, Ithaca, NY.

Acree Riley, C. and Louge, M.Y. (1989) Quantative capacitive measurements of voidage in dense gas–solid flows. *Particulate Science & Tech.*, **7**, 51–59.

Adams, C.K. (1988) Gas mixing in fast fluidized beds, in *Circulating Fluidized Bed Technology II* (eds P. Basu and J.F. Large), Pergamon, Oxford, pp. 299–306.

Aguillón, J., Shakourzadeh, K. and Guigon, P. (1995) Comparative study of non-isokinetic sampling probes for solids flux measurement in circulating fluidized beds. *Powder Technol.*, **83**, 79–84.

Almstedt, A.E. and Ljungström, E.B. (1987) Measurement of the bubble behavior and oxygen distribution in a pilot scale pressurized fluidized bed burning coal, *Proceedings 9th International Fluidized Bed Combustion Conference,* Vol. 1 (ed. J.P. Mustonen), ASME, NY, pp. 575–585.

Almstedt, A.E. and Olsson, E. (1982) Measurements of bubble behavior in a pressurized fluidized bed burning coal, using capacitance probes, *Proceedings 7th International Fluidized Bed Combustion Conference,* Philadelphia, Pennsylvania, pp. 89–98.

Almstedt, A.E. and Olsson, E. (1985) Measurements of bubble behavior in a pressurized fluidized bed burning coal, using capacitance probes – Part II, *Proceedings 8th International Fluidized Bed Combustion Conference,* Houston, Texas, pp. 865–877.

Ambler, P.A., Milne, B.J., Berruti, F. and Scott, D.S. (1990) Residence time distribution of solids in a circulating fluidized bed: experimental and modeling studies. *Chem. Eng. Sci.*, **45**, 2179–2186.

Amos G., Rhodes, M.J. and Mineo, H. (1993) Gas mixing in gas–solid risers. *Chem. Eng. Sci.*, **45**, 943–949.

Amos, G., Rhodes, M.J. and Benkreira, H. (1996) Calculation of optic fibres calibration curves for the measurement of solids volume fractions in multiphase flows. *Powder Technol.*, **88**, 107–121.

Andersson, B.Å. and Leckner, B. (1992) Experimental methods of estimating heat transfer in circulating fluidized bed boilers. *Int. J. Heat and Mass Transfer*, **35**, 3353–3362.

Andersson, B.Å., Johnsson, F. and Leckner, B. (1989) Use of conductivity heat flow meter in fluidized bed boilers. *Trans. Inst. MC.*, **11**, 108–112.

Arastoopour, H. and Yang, Y.F. (1992) Experimental studies on dilute gas and cohesive particles flow behavior using a laser-doppler anemometer, in *Fluidization VII* (eds O.E. Potter and D.J. Nicklin), Engineering Foundation, NY, pp. 723–730.

Arena, U., Cammarota, A. and Pistone, L. (1985) High velocity fluidization behavior of solids in a laboratory scale circulating bed, in *Circulating Fluidized Bed Technology* (ed. P. Basu), Pergamon, Toronto, pp. 119–125.

Arena, U., Cammarota, A., Massimilla, L. and Pirozzi, D. (1988) The hydrodynamic behavior of two circulating fluidized bed units of different sizes, in *Circulating Fluidized Bed Technology II* (eds P. Basu and J.F. Large), Pergamon, Oxford, pp. 223–230.

Arena, U., Cammarota, A., Marzochella, A. and Massimilla, L. (1989) Solids flow structures in a two-dimensional riser of a circulating fluidized bed. *J. Chem. Eng. Japan*, **22**(3), 236–241.

Arena, U., Marzochella, A., Massimilla, L. and Malandrino, A. (1992) Hydrodynamics of circulating fluidized beds with risers of different shape and size. *Powder Technol.*, **70**, 237–247.

Avidan, A.A. and Yerushalmi, J. (1985) Solids mixing in an expanded top fluid bed. *AIChE J.*, **31**, 835–841.

Avidan, A.A., Gould, R.M. and Kam, A.Y. (1985) Operation of a circulating fluid-bed cold flow model of the 100 B/D MTG demonstration plant, in *Circulating Fluidized Bed Technology* (ed. P. Basu), Pergamon, Toronto, pp. 287–296.

Azzi, M., Turlier, P., Bernard, J.R. and Garnero, L. (1991a) Mapping solid concentration in a circulating fluidized bed using gammametry. *Powder Technol.*, **67**, 27–36.

Azzi, M., Turlier, P., Large, J.F. and Bernard, J.R. (1991b) Use of a momentum probe and gammadensitometry to study local properties of fast fluidized beds, in *Circulating Fluidized Bed Technology III* (eds P. Basu, M. Horio and M. Hasatani), Pergamon, Oxford, pp. 189–194.

Bachalo, W.D. and Houser, M.J. (1984) Phase/doppler spray analyzer for simultaneous measurements of drop size and velocity distributions. *Optical Engineering*, **23**, 583–590.

Bader, R., Findlay, J. and Knowlton, T.M. (1988) Gas/solid flow patterns in a 30.5-cm-diameter circulating fluidized bed riser, in *Circulating Fluidized Bed Technology II* (eds P. Basu and J.F. Large), Pergamon, Oxford, pp. 123–137.

Bartholomew, R.N. and Casagrande, R.M. (1957) Measuring solids concentration in fluidized systems by gamma-ray absorption. *Ind. End. Chem.*, **49**, 428–431.

Basu, P. and Konuche, F. (1988) Radiative heat transfer from a fast fluidized bed, in *Circulating Fluidized Bed Technology II* (eds P. Basu and J.F. Large), Pergamon, Oxford, pp. 245–253.
Basu, P. and Nag, P.K. (1987) An investigation into heat transfer in circulating fluidized beds. *Int. J. Heat & Mass Transfer*, **30**, 2399–2409.
Beaud, F. and Louge, M.Y. (1996) Similarity of radial profiles of solid volume fraction in a circulating fluidized bed, in *Fluidization VIII* (eds C. Laguérie and J.-F. Large), Engineering Foundation, NY, in press.
Beck, M.S., Campogrande, E., Morris, M., Williams, R.A. and Waterfall, R.C. (1993) *Tomographic Techniques for Process Design and Operation*, Computational Mechanics Publications, Southampton.
Belyaev, S.P. and Levin, L.M. (1972) Investigation of aerosol aspiration by photographing particle tracks under flash illumination. *J. Aerosol Science*, **3**, 127–140.
Belyaev, S.P. and Levin, L.M. (1974) Techniques for collection of representative aerosol samples. *J. Aerosol Science*, **5**, 325–338.
Bergqvist, K., Andersson, S. and Johnsson, F. (1993) Zirconia cell probe measurements of oxidizing/reducing conditions in a circulating fluidized bed boiler. Presented at the *Swedish–Finnish Flame Days*, Göteborg, Sweden, September 1993.
Berkelmann, K.G. and Renz, U. (1989) The fluid dynamics in the freeboard of an FBC – the use of LDV to determine particle velocity and size, in *Fluidization VI* (eds J.R. Grace, L.W. Shemilt and M.A. Bergougnou), Engineering Foundation, NY, pp. 105–112.
Berker, A. and Tulig, T.J. (1986) Hydrodynamics of gas–solid flow in a catalytic cracker riser: implications for reactor selectivity performance. *Chem. Eng. Sci.*, **41**, 821–827.
Bernard, J.R., Santos-Cottin, H. and Margrita, R. (1989) The use of radioactive tracers for fluidized cracking catalytic plants. *Isotopenpraxis*, **25**, 161–165.
Bolton, L.W. and Davidson, J.F. (1988) Recirculation of particles in fast fluidized beds, in *Circulating Fluidized Bed Technology II* (eds P. Basu and J.F. Large), Pergamon, Oxford, pp. 139–146.
Böttcher, C.J.F. (1952) *Theory of Electric Polarization*, Elsevier, NY.
Botterill, J.S.M., Teoman, Y. and Yüregir, K.R. (1984) Factors affecting heat transfer between gas-fluidized beds and immersed surfaces. *Powder Technol.*, **39**, 177–189.
Bracewell, R.N. (1986) *The Fourier Transform and Its Applications*, McGraw-Hill, NY.
Brereton, C.M.H. and Grace, J.R. (1993) Microstructural aspects of the behavior of circulating fluidized beds. *Chem. Eng. Sci.*, **48**, 2565–2572.
Brereton, C.M.H., Grace, J.R. and Yu, J. (1988) Axial gas mixing in a circulating fluidized bed, in *Circulating Fluidized Bed Technology II* (eds P. Basu and J.F. Large), Pergamon, Oxford, pp. 307–314.
Berereton, C.M.H., Lim, C.J., Legros, R., Zhao, J., Li, H. and Grace, J.R. (1991) Circulating fluidized bed combustion of a high-sulfur eastern Canadian coal. *Can. J. Chem. E.* **69**, 852–859.
Brereton, C. and Stromberg, L. (1986) Some aspects of the fluid dynamic behavior of fast fluidized beds, in *Circulating Fluidized Bed Technology* (ed. P. Basu), Pergamon, Toronto, pp. 133–144.
Brodowicz, K., Maryniak, L. and Dyakowski, T. (1993) Application of capacitance tomography to pneumatic conveying processes, in *Tomographic Techniques for Process Design and Operation* (eds M.S. Beck, E. Campogrande, M. Morris, R.A. Williams and R.C. Waterfall), Computational Mechanics Publications, Southampton, pp. 361–368.
Brue, E., Moore, J. and Brown, R.C. (1994) Process model identification of circulating fluid bed hydrodynamics, in *Circulating Fluidized Bed Technology IV* (ed. A. Avidan), AIChE, NY, pp. 442–449.
Burkell, J.J., Grace, J.R., Zhao, J. and Lim, C.J. (1988) Measurement of solids circulation rates in circulating fluidized beds, in *Circulating Fluidized Bed Technology II* (eds P. Basu and J.F. Large), Pergamon, Oxford, pp. 501–509.
Burki, V., Hirschberg, B., Tuzla, K. and Chen, J.C. (1993) Thermal development for heat transfer in circulating fluidized beds, Annual AIChE Meeting, St Louis, Missouri, November 7–13, paper 66f.
Cankurt, N.T. and Yerushalmi, J. (1978). Gas backmixing in high velocity fluidized beds, in *Fluidization*, Cambridge University Press, Cambridge, pp. 387–393.
Cattieuw, P. (1992) Etude Éxperimentales des Écoulements Gaz-Particules dans un Lit Fluidisé Circulant. Doctoral Thesis, Institut Universitaire des Systèmes Thermiques Industriels, Université de Provence, Marseille, France.

Chandran, R. and Chen, J.C. (1982) Bed-surface contact dynamics for horizontal tubes in fluidized beds. *AIChE J.*, **28**, 907–914.

Chang, H. and Louge M. (1992) Fluid dynamic similarity of circulating fluidized beds. *Powder Technol.*, **70**, 259–270.

Chang, D.P.Y., Sorbo, N.W., Murchison, G.S., Adrian, R.C. and Simeroth, D.C. (1987) Evaluation of a pilot-scale circulating fluidized bed combustor as a potential hazardous waste incinerator. *JAPCA*, **37**, 266–274.

Chen, J.C., Tuzla, K., Dou, S. and Herb, B. (1990) Measurements of fluid mechanic and heat transfer characteristics in a circulating fluidized bed, in *Proceedings of the 2nd Asian Conference on Fluidized Bed and Three-Phase Reactors* (eds W.M. Lu and L.P. Leu), February 18–20, Kenting, Taiwan, pp. 80–87.

Cheremisinoff, N.P. (1986) Review of experimental methods for studying the hydrodynamics of gas–solid fluidized beds. *Ind. Eng. Chem. Process Des. Dev.*, **25**, 329–351.

Chesonis, D.C., Klinzing, G.E., Shah, Y.T. and Dassori, C.G.. (1990) Hydrodynamics and mixing of solids in a recirculating fluidized bed. *Ind. Eng. Chem. Res.*, **29**, 1785–1792.

Clark, N.N. and Atkinson, C.M. (1988) Amplitude reduction and phase lag in fluidized bed pressure measurements. *Chem. Eng. Sci.*, **43**, 1547–1557.

Clift, R., Grace, J.R. and Weber, M.E. (1978) *Bubbles, Drops and Particles*, Academic Press, NY, p. 111.

Cocco, R., Cleveland, J. and Chrisman, R. (1994) Simultaneous *in-situ* determination of particle loadings and velocities in a gaseous medium, in *Fluidization and Fluid-Particle Systems*, Special Suppl. to the 1994 Annual Meeting of the AIChE, AIChE, NY, pp. 251–256.

Contractor, R.M., Pell, M., Weinstein, H. and Feindt, H.J. (1992) The rate of solid loss in a circulating fluid bed following a loss of circulation accident, in *Fluidization VII* (eds O.E. Potter and D.J. Nicklin), Engineering Foundation, NY, pp. 243–248.

Cutolo, A., Rendima, I., Arena, U., Marzocchella, A. and Massimilla, L. (1990) Optoelectronic technique for the characterization of high concentration gas–solid suspension. *Applied Optics*, **29**, 1317–1322.

Davies, C.E. and Foye, J. (1991) Flow of granular material through vertical slots. *Trans. Ind. Chem. Eng.*, **69**(A), 369–373.

Davies, C.E. and Harris, B.J. (1992) A device for measuring solids flowrates: characteristics, and application in a circulating fluidized bed, in *Fluidization VII* (eds O.E. Potter and D.J. Nicklin), Engineering Foundation, NY, pp. 741–748.

Desbat, L. and Turlier, P. (1993) Efficient reconstruction with few data in industrial tomography, in *Tomographic Techniques for Process Design and Operation* (eds M.S. Beck, E. Campograndi, M. Morris, R.A. Williams and R.C. Waterfall), Computational Mechanics Publications, Southampton, pp. 285–294.

De Kok, J.J., Stark, N.L. and van Swaaij, W.P.M. (1986) The influence of solids specific interfacial area on gas–solid mass transfer in gas fluidised beds, in *Fluidization V* (eds K. Østergaard and A. Sørensen), Engineering Foundation, NY, pp. 433–440.

De Loor, G.P. (1956) *Dielectric Properties of Heterogeneous Mixtures*. Doctoral Thesis, University of Leiden, The Netherlands.

Dou, S., Herb, B., Tuzla, K. and Chen, J.C. (1991) Heat transfer coefficients for tubes submerged in circulating fluidized bed. *Exp. Heat Transfer*, **4**, 343–353.

Dou, S., Herb, B., Tuzla, K. and Chen, J.C. (1992) Dynamic variation of solid concentration and heat transfer coefficient at wall of circulating fluidized bed, in *Fluidization VII* (eds O.E. Potter and D.J. Nicklin), Engineering Foundation, NY, pp. 793–801.

Dry, R.J. (1986) Radial concentration profiles in a fast fluidised bed. *Powder Technol.*, **49**, 37–44.

Dry, R.J. (1987) Radial particle size segregation in a fast fluidised bed. *Powder Technol.*, **52**, 7–16.

Dry, R.J. and White, C.C. (1989) Gas residence time characteristics in a high velocity circulating fluidized bed of FCC catalyst. *Powder Technol.*, **58**, 17–23.

Dry, R.J. and White, C.C. (1992) Gas–solid contact in a circulating fluidized bed: the effect of particle size. *Powder Technol.*, **70**, 277–284.

Dry, R.J., Christensen, I.N. and White, C.C. (1987) Gas–solids contact efficiency in a high-velocity fluidised bed. *Powder Technol.*, **52**, 243–250.

Durst, F., Melling, A. and Whitelaw, J.H. (1981) *Principles and Practice of Laser-Doppler Anemometry*, Second Edition, Academic Press, NY.

Dybeck, K., Nagel, R., Schoenfelder, H., Werther, J. and Singer, H. (1995) The D ring sensor – a novel method for continuously measuring the solid circulation rate in CFB reactors, in *Fluidization VIII* (eds C. Laguérie and J.-F. Large), Engineering Foundation, NY, in press.

Ebert, T.A., Glicksman, L.R. and Lints, M. (1993) Determination of particle and gas convective heat transfer components in a circulating fluidized bed. *Chem. Eng. Sci.*, **48**, 2179–2188.

Ettehadieh, B., Yang, W.-C. and Haldipur, G.B. (1988) Motion of solids, jetting and bubbling dynamics in a large jetting fluidized bed. *AIChE J.*, **54**, 243–254.

Fasching, G.E., Smith, N.S. and Utt, C.E. (1992) Three-axis velocity probe system, US Patent 5,170,670.

Fiedler, O., Labahn, N., Müller, I. and Christofori, K. (1994) Halbleiterbildsensor als Ortsfiltersonde – ein Verfahren zur Messung lokaler Partikelgeschwindigkeiten in zirkulierenden Wirbelschichten, *Chem.-Ing.-Tech.*, **66**, 79–82.

Flamant, G., Fatah, N., Olalde, G. and Hernandez, D. (1992) Temperature distribution near a heat exchanger wall immersed in high temperature packed and fluidized beds. *J. Heat Transfer*, **114**, 50–55.

Fuchs, N.A. (1975) Review papers on sampling of aerosols. *Atmospheric Environment*, **9**, 697–707.

Gajdos, L.J. and Bierl, T.W. (1978) *Studies of Recirculating Bed Reactors for the Processing of Coal*. Final Report to the US Department of Energy, microfiche FE-2449–8.

Gamblin, B., Newton, D. and Grant, C. (1994) X-ray characterization of the gas flow patterns from FCC regenerator air ring nozzles, in *Circulating Fluidized Bed Technology IV* (ed. A. Avidan), AIChE, NY, pp. 494–499.

Geldart, D. and Kelsey, J.R. (1972) The use of capacitance probes in gas fluidized beds. *Powder Technol.*, **6**, 45–60.

Gidaspow D., Tsuo, Y.P. and Luo, K.M. (1989) Computer and experimental cluster formation and velocity profiles in circulating fluidized beds, in *Fluidization VI* (eds J.R. Grace, L.W. Shemilt and M.A. Bergougnou), Engineering Foundation, NY, pp. 81–88.

Glicksman, L.R. (1994) Massachusetts Institute of Technology, personal communication.

Glicksman, L.R., Hyre, M.R. and Woloshun, K. (1993) Simplified scaling relationships for fluidized beds, *Powder Technol.*, **77**, 177–199.

Glicksman, L.R., Hyre, M. and Farrell, P. (1994) Dynamic similarity in fluidization. *Int. J. of Multiphase Flow*, **20**, Suppl., 331–386.

Goedicke, F. and Reh, L. (1992) Particle induced heat transfer between walls and gas–solid fluidized beds. *AIChE Symp. Ser.*, **89** (296), 123–136.

Grace, J.R. and Baeyens, J. (1986) Instrumentation and experimental techniques, in *Gas Fluidization Technology* (ed. D. Geldart), Wiley, NY, pp. 415–462.

Halow, J.S. and Nicoletti, P. (1992) Observations of fluidized bed coalescence using capacitance imaging. *Powder Technol.*, **69**, 255–277.

Halow, J.S., Fasching, G.E., Nicoletti, P. and Spenik, J.L. (1993) Observations of a fluidized bed using capacitance imaging. *Chem. Eng. Sci.*, **48**, 643–659.

Han, G.Y., Tuzla, K. and Chen, J.C. (1992) Radiative heat transfer from high temperature suspension flows, Annual AIChE Meeting, Miami Beach, FL., November 1–6, AIChE, NY, paper 117d.

Hartge, E.U., Li, Y. and Werther, J. (1986) Analysis of the local structure of the two-phase flow in a fast fluidized bed, in *Circulating Fluidized Bed Technology* (ed. P. Basu), Pergamon, Toronto, pp. 153–160.

Hartge, E.U., Rensner, D. and Werther, J. (1988) Solids concentration and velocity patterns in circulating fluidized beds, in *Circulating Fluidized Bed Technology II* (eds P. Basu and J.F. Large), Pergamon, Oxford, pp. 165–180.

Hatano, H. Ogasawara, M., Horio, M., Hartge, E.U. and Werther, J. (1993) Local flow of gas and solids in a circulating fluidized bed, *Proceedings 6th SCEJ Symp. on CFB*, Tokyo, Japan, December 13–14, pp. 107–114.

Hatano, H., Kido, N. and Takeuchi, H. (1994) Microscope visualization of solid particles in circulating fluidized beds. *Powder Technol.*, **78**, 115–119.

Hayes, D.G., Gregory, I.A. and Beck, M.S. (1993) Velocity profile measurement in two-phase flows, in *Tomographic Techniques for Process Design and Operation* (eds M.S. Beck, E. Campogrande, M. Morris, R.A. Williams and R.C. Waterfall), Computational Mechanics Publications, Southampton, pp. 369–380.

Heertjes, P.M., Verloop, J. and Willems, R. (1970) The measurement of local mass flow rates and particle velocities in fluid–solids flow. *Powder Technol.*, **4**, 38–40.

Helmrich, H., Schügerl, K. and Janssen, K. (1986) Decomposition of $NaHCO_3$ in laboratory and bench scale circulating fluidized bed reactors, in *Circulating Fluidized Bed Technology* (ed. P. Basu), Pergamon, NY, pp. 161–166.

Herb, B., Tuzla, K. and Chen, J.C. (1989) Distribution of solid concentration in circulating fluidized bed, in *Fluidization VI* (eds J.R. Grace, L.W. Shemilt and M.A. Bergougnou), Engineering Foundation, NY, pp. 65–72.

Herb, B., Dou, S., Tuzla, K. and Chen, J.C. (1992) Solid mass fluxes in circulating fluidized beds. *Powder Technol.*, **70**, 197–205.

Herman, G.T. (1980) *Image Reconstruction from Projection*, Academic Press, NY.

Horio, M. and Kuroki, H. (1994) Three-dimensional flow visualization of dilutely dispersed solids in bubbling and circulating fluidized beds. *Chem. Eng. Sci.* **49**, 2413–2421.

Horio, M., Nonaka, A., Hoshiba, M., Morishita, K., Kobukai, Y., Naito, J., Tachibana, O., Watanabe, K. and Yoshida, N. (1986) Coal combustion in a transparent circulating fluidized bed, in *Circulating Fluid Bed Technology* (ed. P. Basu), Pergamon, Toronto, pp. 255–262.

Horio, M., Morishita, K., Tachibana, O. and Murata, N. (1988) Solid distribution and movement in circulating fluidized beds, in *Circulating Fluidized Bed Technology II* (eds P. Basu and J.F. Large), Pergamon, Oxford, pp. 147–154.

Horio, M., Mori, K., Takei, Y. and Ishii, H. (1992) Simultaneous gas and solid velocity measurements in turbulent and fast fluidized beds, in *Fluidization VII* (eds O.E. Potter and D.J. Nicklin), Engineering Foundation, NY, pp. 757–762.

Huang, S.M., Plaskowski, A.B., Xie, C.G. and Beck, M.S. (1989) Tomographic imaging of two-component flow using capacitance sensors. *J. Phys. E. Sci. Instrum.*, **22**, 173–177.

Huang, S.M., Xie, C.G., Thorn, R., Snowden, D. and Beck, M.S. (1992) Design of sensor electronics for electrical capacitance tomography. *IEEE Proc.-G*, **139**, 83–88.

Ishii, H. and Horio, M. (1991) The flow structures of a circulating fluidized bed. *Advanced Powder Technol.*, **2**(1), 25–36.

Jestin, L., Chabert, C., Flamant, G. and Meyer, P. (1991) *In situ* measurement of particle concentration, temperature distribution and heat flux in the vicinity of a wall in a CFB, in *Circulating Fluidized Bed Technology III* (eds P. Basu, M. Horio and M. Hasatani), Pergamon, Oxford, pp. 247–252.

Johnsson, H. and Johnsson, F. (1992) *Optical Probes for Porosity Measurements in Fluidized Beds*, Report A 92-204, ISSN 0281-0034, Chalmers University of Technology, Sweden.

Kipphan, H. and Mesch, F. (1978) Flow measurement systems using transit time correlation, in *Flow Measurements of Fluids* (eds H.H. Dijstelbergen and E.A. Spencer), North-Holland Publishing, Amsterdam, pp. 409–416.

Klinzing, G.E. (1980) A simple light-sensitive meter for measurement of solid/gas loadings and flow steadiness. *Ind. Eng. Chem. Process Des. Dev.*, **19**, 31–33.

Kohl, J., Zentner, R.D. and Lukens, H.R. (1961) *Radioisotope Applications Engineering*, Van Nostrand, Princeton, New Jersey.

Kojima, T., Ishihara, K.I., Guilin, Y. and Furusawa, T. (1989) Measurement of solids behaviour in a fast fluidized bed. *J. Chem. Eng. Japan*, **22**, 341–346.

Kondulov, N.B., Kornilaev, A.N., Skachko, I.M., Akhromenkov, A.A. and Kruglov, A.S. (1964) An investigation of the parameters of moving particles in a fluidized bed by a radioisotopic method. *Int. Chem. Eng.*, **4**, 43–47.

Krambeck, F.J., Avidan, A.A., Lee, C.K. and Lo, M.N. (1987) Predicting fluid bed reactor efficiency using adsorbing gas tracers. *AIChE J.*, **33**, 1727–1734.

Lanneau, K.P. (1960) Gas–solids contacting in fluidized beds. *Trans. Inst. Chem. Engrs.*, **38**, 125–143.

Larachi, F. and Chaouki, J. (1994) Application des techniques non-intrusives à la caractérisation hydrodynamique des réacteurs polyphasiques, *Actes de la Première Conférence Maghrébine de Génie des Procédés*, Marrakech, Morocco, May 4–6, 1994, pp. 23–46.

Larachi, F., Kennedy, G. and Chaouki, J. (1994) A γ-ray detection system for 3-D particle tracking in multiphase reactors. *Nuclear Instrum. & Methods in Phys. Res. A*, **338**, 568–576.

Leckner, B. and Andersson, B.Å. (1992) Characteristic features of heat transfer in circulating fluidized bed. *Powder Technol.*, **70**, 303–314.

Leckner, B., Golriz, M.R., Zhang, W., Andersson, B.Å. and Johnsson, F. (1991) Boundary layers – first measurements in the 12MW CFB research plant at Chalmers University, in *Proceedings 11th International Conference on Fluidized Bed Combustion* (ed. E.J. Anthony), Vol. 2, ASME, NY, pp. 771–776.

Lee, S.L. and Durst, F. (1982) On the motion of particles in turbulent duct flows. *Int. J. Multiphase Flow*, **8**, 125–146.

Li, H., Xia, Y., Tung, Y. and Kwauk, M. (1991) Micro-visualization of two-phase structure in a fast fluidized bed, in *Circulating Fluidized Bed Technology III* (eds P. Basu, M. Horio and M. Hasatani), Pergamon, Oxford, pp. 183–188.

Li, J. and Weinstein, H. (1989) An experimental comparison of gas backmixing in fluidized beds across the regime spectrum. *Chem. Eng. Sci.*, **44**, 1697–1705.

Li, Y. and Kwauk, M. (1980) The dynamics of fast fluidization, in *Fluidization* (eds J.R. Grace and J.M. Matsen), Plenum Press, NY, pp. 537–544.

Li, Y. and Wu, P. (1991) A study on axial gas mixing in a fast fluidized bed, in *Circulating Fluidized Bed Technology III* (eds P. Basu, M. Horio and M. Hasatani), Pergamon, NY, pp. 581–586.

Lin, J.S., Chen, M.M. and Chao, B.T. (1985) A novel radioactive particle tracking facility for measurement of solids motion in gas fluidized beds. *AIChE J.*, **31**, 465–473.

Lints, M.C. and Glicksman, L.R. (1993) The structure of particle clusters near the wall of a circulating fluidized bed. *AIChE Symp. Ser.*, **89**(296), 35–52.

Lischer, D.J. and Louge, M.Y. (1992) Optical fiber measurements of particle concentration in dense suspensions: calibration and simulation. *Applied Optics*, **31**, 5106–5113.

Liu, D.C., Yang, H.P., Wang, Y.L. and Lin, Z.J. (1991) Experimental studies of heat transfer in circulating fluidized bed, in *Circulating Fluidized Bed Technology III* (eds P. Basu, M. Horio and M. Hasatani), Pergamon, Oxford, pp. 275–281.

Louge, M.Y. (1994) The measurement of particle concentration with optical fiber probes, in *Fluidization and Fluid-Particle Systems*, Special Suppl. to the Annual Meeting of the AIChE, AIChE, NY, pp. 271–280.

Louge, M.Y. (1995) Guarded Capacitance Probes for Measuring Particle Concentration and Flow, US Patent 5,459,406.

Louge, M. and Chang, H. (1990) Pressure and voidage gradients in vertical gas–solid risers. *Powder Technol.*, **60**, 197–201.

Louge, M. and Opie, M. (1990) Measurements of the effective dielectric permittivity of suspensions. *Powder Technol.*, **62**, 85–94.

Louge, M., Lischer, D.J. and Chang, H. (1990) Measurements of voidage near the wall of a circulating fluidized bed riser. *Powder Technol.*, **62**, 267–274.

Louge, M.Y., Iyer, S.A., Giannelis, E.P., Lischer, D.J. and Chang, H. (1991) Optical fiber measurements of particle velocity using laser-induced-phosphorescence. *Applied Optics*, **30**, 1976–1981.

Louge, M., Mohd. Yusof, J. and Jenkins, J.T. (1993) Heat transfer in the pneumatic transport of massive particles. *Int. J. Heat & Mass Trans.*, **36**, 265–275.

Louge, M., Tuccio, M., Lander, E. and Connors, P. (1996) Capacitance measurements of the volume fraction and velocity of dielectric solids near a grounded wall. *Rev. Sci. Instrum.*, **67**, 1869.

Lutz, S.A. and Bajura, R.A. (1982) *Literature Review of Sampling Probes, Topical Report of Phase V: Probe Inlet Design*, NTIS DE 82 01 73 27.

Luyben, W.L. (1973) *Process Modeling, Simulation and Control for Chemical Engineers*, McGraw-Hill, NY.

Marcus, R.D., Leung, L.S. and Klinzing, G.E. (1990) *Pneumatic Conveying of Solids*, Chapman & Hall, London, pp. 471–506.

Martin, M.P., Turlier, P., Bernard, J.R. and Wild, G. (1992) Gas and solid behavior in cracking circulating fluidized beds. *Powder Technol.*, **70**, 249–258.

Massimilla, L. (1973) Behaviour of catalytic beds of fine particles at high gas velocities. *AIChE Symp. Ser.*, **69**(128), 11–13.

Mathur, A. and Saxena, S.C. (1987) Total and radiative heat transfer to an immersed surface in a gas-fluidized bed. *AIChE J.*, **33**, 1124–1135.

Matsuno, Y., Yamaguchi, H., Oka, T., Kage, H. and Higashitani, K. (1988) The use of optical fiber probes for the measurement of dilute particle concentrations: calibration and application to gas fluidized bed carryover. *Powder Technol.*, **36**, 215–221.

Maxwell, J.C. (1892) *Electricity and Magnetism*, Vol. 1, Clarendon, Oxford.
McDonnell, M. and Johnson, E.J. (1980) Multiple beam laser velocimeter: analysis and applications. *Applied Optics*, **19**, 2934–2939.
Meredith, R.E. and Tobias, C.W. (1960) Resistance of potential flow through a cubical array of spheres. *J. Appl. Phys.*, **31**, 1270–1273.
Miller, A. and Gidaspow, D. (1992) Dense, vertical gas–solid flow in a pipe. *AIChE J.*, **38**, 1801–1815.
Mohd. Yusof, J. (1992) Heat Transfer in the Pneumatic Transport of Massive Particles; Modeling and Diagnostics. M.S. Thesis, Cornell University, Ithaca, NY.
Mok, S.L.K., Molodtsof, Y., Large, J.F. and Bergougnou, M.A. (1989) Characterization of dilute and dense phase vertical upflow gas–solid transport based on average concentration and velocity data. *Can. J. Chem. Eng.*, **67**, 10–16.
Monceaux, L., Azzi, M., Molodtsof, Y. and Large, J.F. (1986) Overall and local characterization of flow regimes in a circulating fluidized bed, in *Circulating Fluidized Bed Technology* (ed. P. Basu), Pergamon, Toronto, pp. 185–191.
Moslemian, D., Chen, M.M. and Chao, B.T. (1989) Experimental and numerical investigations of solids mixing in a gas fluidized bed. *Part. Sci. & Technol.*, **7**, 335–355.
Myler, C.A., Zaltash, A. and Klinzing, G.E. (1986) Gas–solid transport in a 0.0508 m pipe at various inclinations with and without electrostatics I: particle velocity and pressure drop. *J. of Powder & Bulk Sol. Technol.*, **10**, 5–12.
Nauman, E.B. and Buffham, B.A. (1983) *Mixing in Continuous Flow Systems*, Wiley, NY.
Ohki, K. and Shirai, T. (1976) Particle velocity in fluidized beds, in *Fluidization Technology*, Vol 1 (ed. D.L. Keairns), Hemisphere, NY, pp. 95–110.
Oki, K., Walawender, W.P. and Fan, L.T. (1977) The measurement of local velocity of solid particles. *Powder Technol.*, **18**, 171–178.
Ozhaynak, T.F., Chen, J.C. and Frankenfield, T.R. (1984) An experimental investigation of radiation heat transfer in a high temperature fluidized bed, in *Fluidization IV* (eds D. Kunii and E. Toei), Engineering Foundation, NY, pp. 371–378.
Patience, G.S., Chaouki, J. and Grandjean, B.P.A. (1990) Solids flow metering from pressure drop measurement in circulating fluidized beds. *Powder Technol.*, **61**, 95–99.
Patience, G.S., Chaouki, J. and Kennedy, G. (1991) Solids residence time distribution in circulating fluidized bed reactors, in *Circulating Fluidized Bed Technology III* (eds P. Basu, M. Horio and M. Hasatani), Pergamon, NY, pp. 599–604.
Patrose, B. and Caram, H.S. (1982) Optical fiber probe transit anemometer for particle velocity measurements in fluidized beds. *AIChE J.*, **28**, 604–609.
Plumpe, J.G., Zhu, C. and Soo, S.L. (1993) Measurement of fluctuations in motion of particles in a dense gas–solid suspension in vertical pipe flow. *Powder Technol.*, **77**, 209–214.
Qi, C. and Farag, I.H. (1993) Lateral particle motion and its effect on particle concentration distribution in the riser of CFB (ed. A.W. Weimer), *AIChE Symp. Ser.*, **89**(296), 73–80.
Qian, G. and Li, J. (1994) Particle velocity measurement in CFB with an integrated probe, in *Circulating Fluidized Bed Technology IV* (ed. A. Avidan), AIChE, NY, pp. 274–278.
Reh, L. and Li, J. (1991) Measurement of voidage in fluidized beds by optical probes, in *Circulating Fluidized Bed Technology III* (eds P. Basu, M. Horio and M. Hasatani), Pergamon, Oxford, pp. 163–170.
Renganathan, K. and Turton, R. (1989) Data reduction from thin-film heat gauges in fluidized beds. *Powder Technol.*, **59**, 249–254.
Rensner, D. and Werther, J. (1993) Estimation of the effective measuring volume of single-fibre reflection probes for solid volume concentration measurements. *Part. Syst. Charact.*, **10**, 48–55.
Rensner, D., Hartge, E.U. and Werther, J. (1991) Different types of optical probes for investigations in highly concentrated gas/solid flows, in *Proceedings of the International Conference on Multiphase Flows* (eds G. Matsui, A. Serizawa and Y. Tsuji), September 24–27, Tsukuba, Japan, pp. 255–258.
Rhodes, M.J. and Laussmann, P. (1992a) A simple non-isokinetic sampling probe for dense suspensions. *Powder Technol.*, **70**, 141–151.
Rhodes, M.J. and Laussmann, P. (1992b) Characterizing non-uniformities in gas-particle flow in the riser of a circulating fluidized bed. *Powder Technol.*, **72**, 277–284.
Rhodes, M.J., Zhou, S., Hirama, T. and Cheng, H. (1991) Effects of operating conditions on longitudinal solids mixing in a circulating fluidized bed riser. *AIChE J.*, **37**, 1450–1458.

Rhodes, M., Mineo, H. and Hirama, T. (1992) Particle motion at the wall of a circulating fluidized bed. *Powder Technol.*, **70**, 207–214.

Rink, K.K., Kozinski, J.A., Lighty, J.S. and Lu, Q. (1994) Design and construction of a circulating fluidized bed combustion facility for use in studying the thermal remediation of wastes. *Rev. Sci. Instrum.*, **65**, 2704–2713.

Saxena, S.C. and Patel, D.C. (1988) Measurements of solids concentration profile around the periphery of an immersed tube in a three-dimensional gas-fluidized bed. *Part. Sci. & Technol.*, **6**, 145–167.

Saxena, S.C., Patel, D.C. and Kathuria, D. (1987) An image-carrying fiber optic probe to investigate solids distribution around an immersed surface in a gas-fluidized bed. *AIChE J.*, **33**, 672–676.

Saxena, S.C., Srivastava, K.K. and Vadivel, R. (1989) Experimental techniques for the measurement of radiative and total heat transfer in gas fluidized beds: a review. *Experimental Thermal & Fluid Science*, **2**, 350–364.

Saxena, S.C., Qian, R.Z. and Liu, D.C. (1992) Recent Chinese heat transfer research on bubbling and circulating fluidized beds. *Energy*, **17**, 1215–1232.

Schlosser, P.A., De Vuono, A.C., Kulacki, F.A. and Munshi, P. (1980) Analysis of high speed CT scanners for non-medical applications. *IEEE Trans. Nucl. Sci.*, **NS27**, 788–794.

Schnitzlein, M.G. and Weinstein, H. (1988) Flow characterization in high velocity fluidized beds using pressure fluctuations. *Chem. Eng. Sci.*, **43**, 2605–2614.

Schuurmans, H.J.A. (1980) Measurements in a commercial catalytic cracker unit. *Ind. Eng. Chem. Process Des. Dev.*, **19**, 267–271.

Seo, Y.C. and Gidaspow, D. (1987) An X-ray–γ-ray method of measurement of binary solids concentrations and voids in fluidized beds. *Ind. Eng. Chem. Res.*, **26**, 1622–1628.

Soo, S.L. (1989) *Particulates and Continuum, Multiphase Fluid Dynamics*, Hemisphere, NY, p. 28.

Soo, S.L., Baker, D.A., Lucht, T.R. and Zhu, C. (1989) A corona discharge probe system for measuring phase velocities in a dense suspension. *Rev. Sci. Instrum.*, **60**, 3475–3477.

Soo, S.L., Slaughter, M.C. and Plumpe, J.G. (1994) Instrumentation for flow properties of gas–solid suspensions and recent advances. *Part. Sci. Technol.*, **12**, 1-12.

Soong, C.H., Tuzla, K. and Chen, J.C. (1994) Identification of particle clusters in circulating fluidized bed, in *Circulating Fluidized Bed Technology IV* (ed. A. Avidan), AIChE, NY, pp. 615–620.

Stiles, R.D. and Clark, N.N. (1992) Accounting for the effect of turbulent flow in large-bore fluidized bed pressure probes. *Flow Meas. Instrum.*, **3**, 9–15.

Stubington, J.F. and Chan, S.W. (1990) The interpretation of oxygen-probe measurements in fluidised bed combustors. *J. Inst. Energy*, **63**(456), 136–142.

Tadrist, L., Azario, E. and Cattieuw, P. (1994) Analysis of two-phase flow in a circulating fluidized bed, in *Circulating Fluidized Bed Technology IV* (ed. A. Avidan), AIChE, NY, pp. 582–587.

Tanner, H., Li, J. and Reh, L. (1993) Radial profiles of slip velocity between gas and solids in circulating fluidized beds, Annual AIChE Meeting, St Louis, MO, November 7–12, AIChE, NY, paper 66e.

Tinga, W.R., Voss, W.A.G. and Blossey, D.F. (1973) Generalized approach to multiphase dielectric mixture theory. *J. Appl. Phys.*, **44**, 3897–3902.

Tung, Y., Li, J. and Kwauk, M. (1988) Radial voidage profiles in a fast fluidized bed, in *Fluidization '88 Science & Technology* (eds M. Kwauk and D. Kunii), Science Press, Beijing, China, pp. 139–145.

Turlier, P. and Bernard, J.R. (1992) Techniques d'étude 'in situ' des lits fluidisés industriels. *Entropie*, **170**, 24–28.

Turton, R. and Clark, N.N. (1988) Predicting the response of pressure probes in pneumatic conveying and fluidized beds. *J. Powder & Bulk Solids Technol.*, **12**, 13–17.

Van Breugel, J.W., Stein, J.J.M. and de Vries, R.J. (1969) Isokinetic sampling in a dense gas–solid stream. *Proc. Inst. Mech. Engrs.*, **184**(3C), 18–23.

Van de Wall, R.E. and Soo, S.L. (1994) Measurement of particle cloud density and velocity using laser devices. *Powder Technol.*, **81**, 269–278.

Van der Ham, A.G.J., Prins, W. and van Swaaij, W.P.M. (1992) Hydrodynamics of a pilot plant scale regularly packed CFB. *AIChE Symp. Ser.*, **89**(296), 53–72.

Van Swaaij, W.P.M., Buurman, C. and Van Breugel, J.W. (1970) Shear stresses on the wall of a dense phase riser. *Chem. Eng. Sci.*, **25**, 1818–1820.

Wang, T., Lin, Z.J., Zhu, C.M., Liu, D.C. and Saxena, S.C. (1993) Particle velocity measurements in a circulating fluidized bed. *AIChE J.*, **39**, 1406–1410.

Wang, X.Y., Zheng, Q.Y. and Feng, J.K. (1989) Heat transfer in central area of circulating fluidized bed, in *Multiphase Flow and Heat Transfer, Second Int. Symp.*, Vol 2 (eds X.J. Chen, T.N. Veziroglu and C.L. Tien), Hemisphere, NY, pp. 1223–1231.

Weimer, A.W., Gyure, D.C. and Clough, D.E. (1985) Application of a gamma-radiation density gauge for determining hydrodynamic properties of fluidized beds. *Powder Technol.*, **44**, 179–194.

Weinstein, H. and Li, J. (1989) An evaluation of the actual density in the acceleration section of vertical risers. *Powder Technol.*, **57**, 77–79.

Weinstein, H., Shao, M. and Wasserzug, L. (1984) Radial solid density variations in a fast fluidized bed. *AIChE Symp. Ser.*, **80**(241), 117–121.

Weinstein, H., Shao, M., Schnitzlein, M. and Graff, R.A. (1986) Radial variation in void fraction in a fast fluidized bed, in *Fluidization V* (eds K. Østergaard and A. Sørensen), Engineering Foundation, NY, pp. 329–336.

Wen, C.Y. and Fan, L.T. (1975) *Models for Flow Systems and Chemical Reactors*, Marcel Dekker, NY.

Werdermann, C.C. and Werther, J. (1994) Heat transfer in large-scale circulating fluidized bed combustors of different sizes in *Circulating Fluidized Bed Technology IV* (ed. A. Avidan), AIChE, NY, pp. 428–435.

Werther, J. and Hage, B. (1995) A fibre-optical sensor for high temperature application in fluidized bed combustion, in *Fluidization VIII* (eds C. Laguérie and J.-F. Large), Engineering Foundation, NY, in press.

Werther, J. and Molerus, O. (1973) The local structure of gas fluidised beds I – a statistically based measuring system. *Int. J. Multiphase Flow.*, **1**, 103–122.

Werther, J. Hartge, E.-U. and Rensner, D. (1990) Meßtechniken für Gas/Feststoff-Wirbelschichtreaktoren. *Chem.-Ing.-Tech.*, **62**, 605–613.

Werther, J. Hartge, E.U., Kruse, M. and Nowak, W. (1991) Radial mixing of gas in the core zone of a pilot scale CFB, in *Circulating Fluidized Bed Technology III* (eds P. Basu, M. Horio and M. Hasatani), Pergamon, Oxford, pp. 593–598.

Werther, J., Hartge, E.U. and Rensner, D. (1993) Measurement techniques for gas–solid fluidized bed reactors. *Int. Chem. Eng.*, **33**, 18–26.

Westphalen, D. and Glicksman, L. (1995) Lateral solid mixing measurements in circulating fluidized beds. *Powder Technol.*, **82**, 153–167.

White, C.C. and Dry, R.J. (1989) Transmission characteristics of gas in a circulating fluidized bed. *Powder Technol.*, **57**, 89–94.

Wirth, K.E., Seiter, M. and Molerus, O. (1991) Concentration and velocities of solids in areas close to the walls in circulating fluidized bed systems. *VGB Kraftwerkstechnik*, **10**, 824–828.

Wu, R.L., Lim, C.J., Chaouki, J. and Grace, J.R. (1987) Heat transfer from a circulating fluidized bed to membrane waterwall surfaces. *AIChE J.*, **33**, 1888–1893.

Wu, R.L., Lim, C.J. and Grace, J.R. (1989) The measurement of instantaneous local heat transfer coefficients in a circulating fluidized bed. *Can. J. Chem. Eng.*, **67**, 301–307.

Xie, C.G., Huang, S.M., Hoyle, B.S., Thorn, R., Lenn, C., Snowden, D. and Beck, M.S. (1992) Electrical capacitance tomography for flow imaging: system model for development of image reconstruction algorithms and design of primary sensors. *IEEE Proc.-G*, **139**, 89–98.

Yang, W.C. (1993) The hydrodynamics of circulating fluidized beds, Chapter 9 in *Encyclopedia of Fluid Mechanics*, Supplement 2: Advances in Multiphase Flows, Gulf Publishing, Houston.

Yang, G., Huang, Z. and Zhao, L. (1984) Radial gas dispersion in a fast fluidized bed, in *Fluidization IV* (eds D. Kunii and E. Toei), Engineering Foundation, NY, pp. 145–152.

Yates, J.G. and Simons, S.J.R. (1994) Experimental methods in fluidization research. *Int J. Multiphase Flow.*, **20**, Suppl., 297–330.

Zhang, W., Tung, Y. and Johnsson, F. (1991) Radial voidage profiles in fast fluidized beds of different diameters. *Chem. Eng. Sci.*, **46**, 3045–3052.

Zhang, W., Johnsson, F. and Leckner, B. (1994) Characteristics of the lateral particle distribution in circulating fluidized bed boilers, in *Circulating Fluidized Bed Technology IV* (ed. A. Avidan), AIChE, NY, pp. 266–273.

Zhang, W., Johnsson, F. and Leckner, B. (1995) Fluid-dynamic boundary layers in CFB boilers, *Chem. Eng. Sci.*, **50**, 201–210.

Zheng, Z., Zhu, J., Grace, J.R., Lim, C.J. and Brereton, C.M.H. (1992) Particle motion in circulating and revolving fluidized beds via computer-controlled colour-stroboscopic photography, in *Fluidization VII* (eds O.E. Potter and D.J. Nicklin), Engineering Foundation, NY, pp. 781–789.

Zhou, J., Grace, J.R., Lim, C.J., Brereton, C.M.H., Qin, S. and Lim, K.S. (1995) Particle cross flow, lateral momentum flux and lateral velocity in a circulating fluidized bed, *Can. J. Chem. Eng.*, **73**, 612–619.

Zhu, C. and Soo, S.L. (1992) A modified theory for electrostatic probe measurements of particle mass flows in dense gas–solid suspensions. *J. Applied Phys.*, **72**(5), 2060–2062.

Zhu, C., Slaughter, M.C. and Soo, S.L. (1991) Measurement of velocity of particles in a dense suspension by cross-correlation of dual laser beams. *Rev. Sci. Instr.*, **62**, 2036–2037.

Zou, B., Li, H., Xia, Y. and Ma, X. (1994) Cluster structure in a circulating fluidized bed. *Powder Technol.*, **78**, 173–178.

10 Combustion performance
CLIVE BRERETON

10.1 Introduction – history and status

Combustion systems, ranging in size from small-scale transportable incinerators (Anderson *et al.*, 1989) to 300 MWe boilers, are one of the main applications of circulating fluidized bed contacting. Circulating fluidized bed combustion (CFBC) systems first gained popularity in the late 1970s, with almost simultaneous development of technologies by Ahlstrom Pyropower in Finland, Lurgi in Germany, and Studsvik Energiteknik in Sweden. Each brought a slightly different design concept to the table, but all were clearly CFB concepts. Since their introduction, CFB combustors have grown in size so that atmospheric circulating fluid bed combustion systems for power generation can be considered in applications up to 400 MWe (see Chapter 11). Atmospheric pressure CFBC systems also find application as part of advanced power generation cycles such as air-blown gasification (Sage *et al.*, 1995), while pressurized circulating fluid beds are being developed both for standalone power plants (Koskinen *et al.*, 1995; Provol and Matousek, 1995), and integrated into cycles such as IGCC (Integrated Gasification Combined Cycle) (Abdulally and Alkan, 1995).

When first introduced into the market, circulating fluid bed combustors were claimed to have all of the benefits of bubbling bed combustors, which until then had been the focus of fluid bed combustion research. They were also claimed to have some inherent advantages. This has largely proved to be true for utility-scale atmospheric pressure units, with circulating bed units dominating this segment of the market. However, bubbling beds remain important in retrofits, in smaller scale applications, where they enjoy cost advantages over circulating beds, and also in certain specialized applications.

Pressurized fluid bed units are subject to entirely different design constraints than atmospheric units. The existing commercial pressurized fluid bed systems are bubbling beds. As noted above, pressurized circulating beds are under development, but do not have such clear advantages as in many low pressure applications.

Fluid bed combustors of all types compete largely with grate units or pulverized units, depending upon the nature of the fuel being fired and the size of unit under consideration. Before any specific comments on circulating fluid bed combustion are made, it is useful to look at some general aspects of combustion in fluidized beds in comparison with these other technologies.

In a fluidized bed combustor the bed is made up primarily of inert material which may be ash, sorbent, or some other inert material such as sand. Solid fuels which, depending upon the reactivity of the fuel, represent between 0.5 and 5% of the total bed material are burnt surrounded by these inerts. High rates of heat and mass transfer to individual fuel particles allow combustion to occur at temperatures which are low compared to the peak combustion temperature of other types of combustion systems, while high rates of particle mixing in the bed minimize temperature gradients. Combustion temperatures vary according to the type of fuel being burned but are typically between 750 and 900°C. At full load the temperature difference between the coldest and hottest points in a CFB furnace may be as little as 30°C. By comparison, a pulverized coal combustor exhibits peak flame temperatures of the order of 1500°C, with the gas temperature dropping to approximately 1150°C by the time the gas exits the main furnace into the superheater section. In a grate system the temperature within the burning fuel bed may be as high as 1200°C; there are also high local temperatures in flames where secondary air mixes with the volatiles leaving the fuel bed, before the temperature falls in the upper furnace. Thus, while grate and pulverized systems are characterized by large temperature gradients, fluid bed systems, and particularly circulating beds, are more nearly isothermal and have lower temperatures in the main combustion section. Low temperatures, together with the high heat capacity and good mixing in the fluid bed, permit combustion of low-grade fuels with low emissions of NO_x and low-cost sulfur capture.

Other features of fluidized bed systems include high bed-to-surface heat transfer coefficients, 100–400 $W/m^2 K$. Circulating beds operate at the lower end of this range, while bubbling beds work in the upper range of coefficients. These high coefficients allow relatively compact designs in spite of the low combustion temperature. Also, operation at temperatures below the ash fusion temperature in all parts of the unit produces a soft ash that has reduced potential for backpass superheater erosion while simultaneously reducing volatilization of corrosive and toxic alkali metals and other inorganic salts (Yates, 1983).

10.2 Clean fossil fuel combustion in fluid bed systems

The principal concerns associated with burning of solid fuel such as coal for power generation are as follows:

- Low efficiencies of the generating cycle and high greenhouse gas emissions per unit of power production.
- Emissions of acid gases, nitrogen and sulfur oxides.
- Emissions of trace elements and products of incomplete combustion (PICs).

Fluid bed combustion technology, in its various forms, addresses each of these concerns:

Cycle efficiency. For conventional atmospheric fluid beds, cycle efficiency is comparable to that for pulverized coal with emissions control at similar steam conditions (typically 35% based on HHV). Current pressurized fluid bed combustion technology offers higher efficiency (up to 45%). In the future, use of arrangements such as the airblown gasification cycle, formerly the British Coal Topping Cycle, or Integrated Gasification Combined Cycle (IGCC) where fluid beds combustors, gasifiers or both, are one option for an important component of the cycle, offer potential for cycle efficiencies approaching 50% (Holley et al., 1995, Gefken and Huber, 1993). Use of higher efficiency cycles offers the potential to significantly reduce greenhouse emissions per unit of electricity production.

NO_x. Nitrogen oxide (NO_x) emissions for fluid bed systems are generally lower than for pulverized or grate systems because of the low uniform combustion temperature. Fluid bed systems are sufficient to meet US NSPS (new source performance standards); however they require additional deNO_x equipment to meet more stringent standards such as those imposed in California. Circulating fluidized beds typically give lower NO_x emissions than bubbling beds burning similar fuels because of inherent differences in the contacting pattern.

SO_x. One of the most attractive features of FBC, bubbling and circulating, is the potential to use a low-cost sorbent to capture sulfur within the fluidized bed. The sorbent is typically limestone or dolomite (a mineral composed of calcium and magnesium carbonates), which is fed to the bed either together with the fuel or as a separate solids stream. At atmospheric pressure, limestone calcines to CaO at the temperature and carbon dioxide partial pressure of the fluid bed. The CaO then reacts with sulfur oxides in the oxidizing portion of the bed to form calcium sulfate. In reducing areas of the fluid bed, or in dedicated fluid bed gasifiers, the calcium oxide may react with hydrogen sulfide to form calcium sulfide. In this way sulfur can be scrubbed from flue gas within the fluid bed itself, with the bed largely formed of sorbent in various states of sulfation. This ability to capture sulfur within the bed precludes the need for add-on scrubbers and may reduce the overall capital and operating costs of a plant compared to a pulverized unit with an add-on sulfur dioxide scrubber (Gefken and Huber, 1993).

In some cases in which fluid bed combustion is applied, it is not emissions issues that are foremost in selection of the technology. Other advantages of fluid bed combustion include high turndown ratios (3:1 and more), and the ability to burn fuels of high moisture content with minimal supplementary fuel. Examples are:

Table 10.1 A comparison of emissions potential and costs for combustion technologies – modified from Gefken and Huber (1993)

Technology	Cost $/kw (1994)	Potential reductions (% change)			Cycle efficiency
		SO_2	NO_x	CO_2	
Pulverized coal with emissions control	1500–1800	90–99	60	NA	33–35
Atmospheric fluidized bed	700–900	90	60	5	34–36
Pressurized fluidized bed	800–1000	90–99	80	30	38–45

(i) Combustion of deinking sludges, where fluid beds compete with grate boilers. In this case a grate boiler might only be able to fire deinking sludge (HHV = 5811 kJ/kg) mixed 25:75 with a higher heating value hog fuel (HHV = 10 500 kJ/kg), while the fluid bed can burn 100% deinking sludge (Kraft and Orender, 1993; Louhimo, 1993).

(ii) Remediation of contaminated soils containing low concentrations of petroleum hydrocarbons or PCBs (Jensen and Young, 1986; Chang et al., 1987). Such soils have virtually no heating value and the ability of the fluid bed to provide good destruction at low temperatures so as to minimize the cost of auxiliary fuel and the furnace size is a major element in marketing the technology.

10.3 Fluid bed versus alternative combustors

The general ability of fluid bed systems to burn low-grade fuels, with low NO_x emissions and *in situ* sulfur capture, applies equally to bubbling and circulating fluidized beds. To discuss the circumstances under which circulating fluid bed systems have special benefits, it is important to understand the approximate operating conditions of each combustor type. These are presented in Table 10.2, in which fluid beds are compared with one another and with combustion systems.

The different operating velocity ranges of bubbling and circulating beds create different footprints. The bubbling bed, Figure 10.1a, has a wide plan area because of its low velocities, with horizontal in-bed tubes for a compact heat transfer surface. The circulating fluid bed by comparison, Figure 10.1b, has a much smaller footprint, but is considerably taller to provide adequate gas residence time and heat transfer area. This in-furnace heat transfer surface is in the form of vertical waterwalls to minimize erosion.

The circulating fluid bed combustor is a tall furnace typically, but not always fabricated from membrane waterwall panels. Membrane wall combustors have a square or rectangular cross-section. Some units, which are

Table 10.2 A comparison of typical key operating parameters for coal combustion systems – modified from Basu and Fraser (1991)

	Bubbling bed	Circulating bed	Pulverized	Stoker fired
Gas velocity (m/s)	1.5–3	4–8.5	4–6	1.2
Bed pressure drop (in H_2O)	80–100	40–70	NA	NA
Coal feed size (mm)	6–0	6–0	<0.01	32–6
Mean bed particle size (micron)	500–1500	150–500	NA	NA
Bed-surface heat transfer coeff. ($W/m^2 K$)	300	120	NA	NA
Entrainment rate ($kg/m^2 s$)	0.1–1	10–40	NA	NA
Excess air (%)	20–25	10–20	15–30	20–30
NO_x emission (ppmv)	100–300	50–200	400–600	400–600
Combustion efficiency	90–96	85–99	99	85–90
Grate heat release rate $MWth/m^2$	0.5–1.5	3–5	4–6	0.5–1.5

NA indicates that a strict comparison is not applicable.

either incinerators or are combustors without heat transfer surface in the main CFB chamber, may have a circular cross-section.

Primary air, between 30 and 100% of the total, enters through a distributor plate at the base of the furnace into the 'primary' zone. Secondary air is injected at some distance above the primary through penetrations in the water wall. In some systems, where NO_x minimization is paramount (Belin *et al.*, 1988) tertiary air may also be introduced still further up the furnace. The combined gas passes up the riser with a full-load superficial gas velocity between 5 and 8.5 m/s and with solids-to-gas mass ratios of between 3 and 30:1 at the furnace exit. It then passes to a primary separator, typically a cyclone, but in some cases some other type of impingement separator such as a U-beam unit (Kavidass *et al.*, 1994), which separates the bulk of the solids. The gases then exit to a conventional boiler backpass. Solids from the primary separator are returned to the combustor either at an uncontrolled rate through a loop seal, or at a controlled rate using an L-valve, depending upon the type of control system applied to the boiler.

10.4 Selection of operating temperature

By looking at the constraints on the temperature range in which most fluid bed systems operate, 750–950°C, it is possible to gain insight into some of the key issues associated with combustion and pollution control.

- Below 750°C, combustion rates are slow, sulfation rates are slow, and CO and hydrocarbons emissions may be of concern
- In fluid bed combustors using a sorbent for *in situ* sulfur capture reactions, an upper temperature limit may be created by sulfur capture efficiency considerations. Ability to capture sulfur effectively decreases in many atmospheric FBC systems beyond 850°C.
- NO_x emissions increase with increasing temperature.

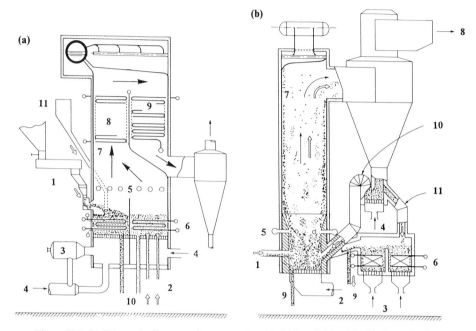

Figure 10.1 (a) Schematic diagram of a conventional bubbling fluidized bed boiler – 1. Coarse fuel feed system (overbed); 2. underbed pneumatic feed system for fine fuel; 3. burner to preheat fluidizing air to raise the bed temperature during start-up; 4. main fluidizing air inlets; 5. overfire (secondary) air inlets; 6. horizontal in-bed tube bundles; 7. membrane (water) wall; 8. superheater bank; 9. convection bank and economizer; 10. bed drain for ash and coarse solids; 11. limestone feed system. (Adapted from Kunii and Levenspiel, 1991.) (b) Schematic diagram of a conventional atmospheric circulating fluidized bed boiler incorporating an external fluid bed heat exchanger – 1. Fuel and limestone feed system; 2. main fluidizing air inlet; 3. fluidizing air to external fluid bed heat exchanger; 4. fluidizing air to non-mechanical seal for cyclone return solids (J-valve); 5. overfire (secondary) air inlets; 6. horizontal in-bed tube bundles in external fluid bed heat exchanger; 7. Membrane (water) wall; 8. cyclone outlet leading to conventional boiler backpass; 9. drains for ash and coarse solids; 10. J-valve overflow for hot solids return to combustor; 11. solids return via external heat exchanger, solids flow rate controlled by a mechanical valve (not shown). (Adapted from Kunii and Levenspiel, 1991.)

- For effective long-term stable operation of fluid bed combustion systems it is essential to either avoid or to deal with particle agglomeration. Particle agglomeration (sintering) begins with small agglomerates, created by softening of ash or other bed materials, with the softened material acting as a bridge to cause other particles to stick together. Agglomerates may grow rapidly, especially in areas which may be either defluidized or operating at low gas velocity due to poor gas distribution. They may trap burning fuel particles and prevent their heat being adequately dissipated. In this way they can be densified and can grow to occupy a significant part of the bed area. They may eventually grow so large as to cause the bed to be shut down because the fluidization quality becomes poor.

Ash softening temperatures vary from as low as 650°C for certain wood ashes to in excess of 1000°C; however, the bulk ash softening temperature may offer only a limited indication of whether sintering will occur. Formation of eutectics at the contact surfaces of particles can reduce the local agglomeration temperature in the presence of salts such as sodium chloride present in hog fuel. Thus, ash characteristics can require reduction of bed temperatures to levels that might be considered lower than desirable for effective combustion. In such cases circulating beds may arguably be slightly preferable to bubbling beds because the more turbulent contacting environment will tend to break up agglomerates as they are formed. Alternatives may be to inject additives such as limestone into the fluid bed so as to influence the bed chemistry and prevent sintering, or to provide a continuous bed classification system that removes bed material, classifies it, and rejects the oversize sinters that act as centres for further agglomeration. Influencing the bed chemistry may be as simple as adding solids, e.g. limestone to a wood-fired unit (Johns and Washcer, 1987), not for sulfur capture but to counter a sodium or potassium–silica eutectic.

Ash softening temperatures create an effective upper operating temperature limit for several fluid bed operations. In the worst cases, the upper limit may be as low as 750°C. In some cases it may be possible to operate up to 1000°C without encountering sintering problems.

Depending upon the specific circumstances, the fuel, and other emissions and economic limitations, any of these constraints may provide an upper or lower limit to operation of the fluid bed combustion system.

In comparisons of bubbling and circulating bed combustors, and of fluid bed combustors with other combustor types, one of the principal distinctions is in the degree of fuel preparation required, and in the cost and complexity of the fuel system. Fluidized bed systems have traditionally competed with stokers or pulverized (PF) units, and bubbling beds have competed favourably with both in this regard. They do not require the expensive pulverizing equipment of PF units, nor are they as sensitive to fuel feed size distribution as stoker units. Bubbling fluidized bed combustors utilize either a simple overbed feeder, or they may utilize complex underbed feed systems – the TVA 160 MWth demonstration plant utilized a total of 120 feed points in underbed mode (Kopetz *et al.*, 1991) together with multiclone recycle of elutriated fines. Overbed feed systems are simple, but result in significant above-bed burning for high-volatiles fuels and lower carbon combustion efficiencies because of rapid elutriation of fine carbon. Once the gases leave the bubbling bed they may be quenched fairly rapidly, and carbon combustion ceases to occur at a rapid rate. The underbed feed system provides for significantly increased in-bed residence time for the fines and improves combustion efficiency, but is much more sensitive to feed size distribution and moisture content and requires increased feed preparation. A discussion of comparative performance of overbed feed and underbed feed systems is given by Castleman (1985) with reference to the TVA 20 MWth demonstration unit. More fundamental information on in-bed volatile combustion and carbon loss by elutriation is given by Park *et al.* (1980, 1981) and Gibbs and Hedley (1978), respectively.

Circulating fluid bed combustors generally provide for considerably simplified feed systems compared to other types of unit. The feed systems are typically simply dump chutes around the furnace periphery often fed by rotary valves. Figure 10.2 indicates spacing values between 9 and 30 m^2 per feeder, corresponding to firing rates between approximately 30 and 130 MWth per feed point (Wang *et al.*, 1993). An alternative feed point location is into the loop seals or fluoseals (see Chapter 7), which may have the advantage of improved mixing of feed with the bed by virtue of the increased solids momentum flux from these ports. In the CFB, maximum feed point spacing arises from the need for high combustion efficiencies and efficient utilization of limestone. Conventional feed point spacings do not provide uniform distribution; however, significant maldistribution in the lower furnace can be rectified by mixing in the large volume of the upper furnace, as practised in most systems. Couturier *et al.* (1989) have given an illustration of gas profiles arising from sorbent and fuel distributions in a commercial furnace and show that they are far from uniform. While the upper furnace rectifies maldistribution, improved distribution can, nevertheless, have a beneficial effect on limestone utilization. This is shown in Figure 10.3 (Wang *et al.*, 1993).

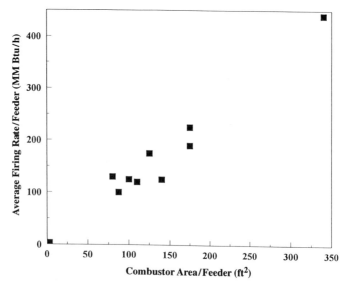

Figure 10.2 Values of combustor cross-sectional area per feeder for commercial ACFBCs (adapted from Wang *et al.*, 1993).

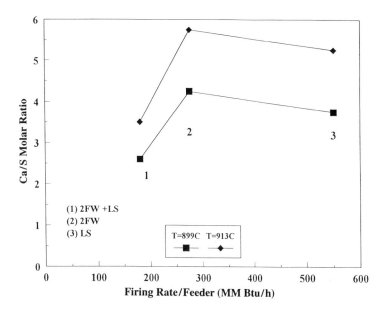

Figure 10.3 Influence of number of feed points in use on sulfur capture for an ACFBC. The unit in question has three feed points, two on the front wall (FW), one in the loop seal (LS) (adapted from Wang et al., 1993).

10.5 Circulating bed combustion fluid mechanics and comparisons with FCC units

In discussion of circulating fluid bed contactors an important distinction must be made between systems using fine Geldart group A particles, and coarser group B particulate systems. Circulating fluid bed combustors utilize relatively course group B solids, and as a result, suspension densities and circulation rates may be an order of magnitude lower than in catalytic contacting operations using group A systems. This is partially due to the importance of minimizing pressure drop (at least in atmospheric systems) so as to decrease the parasitic power for electrical generation, which may be of the order of 10% (Riley et al., 1995). Also, compared to catalytic units, unit sizes for atmospheric pressure boilers may be much larger because of the large production requirements of modern power generation. Furnaces may be 11 m or more in depth and have heights up to 40 m.

Figure 10.4 (Lafanechere and Jestin, 1995) illustrates a typical CFB density profile. The density decays from a turbulent bed in the primary zone (voidage ≈ 0.8, suspension density $\approx 400 \, kg/m^3$) to a value that more closely approximates a dilute transport reactor at the reactor exit (voidage ≈ 0.998, suspension density $\approx 5 \, kg/m^3$). The riser top corresponds to solids-to-gas mass flow ratios as low as 3:1 or a solids circulation flux of $10 \, kg/m^2 s$.

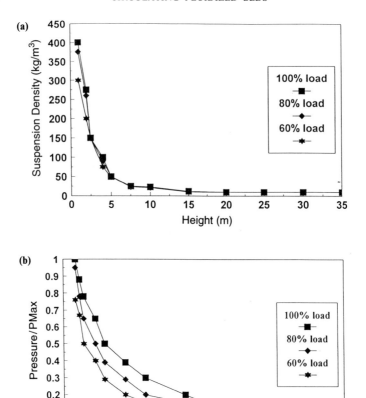

Figure 10.4 Typical commercial CFBC density and pressure profiles from a 125 MWe unit (adapted from Lafanechere and Jestin, 1995).

The vertical density profile is characterized by its sharp decay above the secondary air ports into an 'almost dilute transport' zone. There are enough similarities between the freeboard of a bubbling bed and the secondary zone of a circulating bed that both have been modelled using similar empirical exponential decay approaches (Kunii and Levenspiel, 1991). However, by virtue of the much higher gas velocities, circulation rates in CFBs are orders of magnitude higher than elutriation rates in bubbling beds, and the rate of decay of density at the higher velocity is also lower. Useful expressions for prediction of hydrodynamics in these systems are presented by Werther (1993); their validity to a commercial unit is shown by Lafanechere and Jestin (1995).

The above discussion points to some of the differences between combustors and catalytic systems. These reflect the unique requirements of

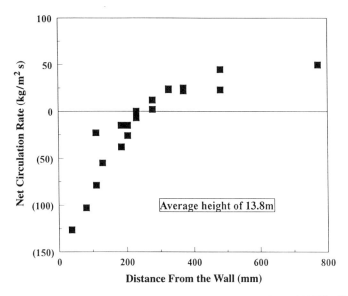

Figure 10.5 Net solids flux as a function of distance from the wall in a 125 MWe CFBC illustrating the downflowing boundary layer and upflowing core. The hydraulic diameter of the unit is 9.65 m (adapted from Lafanechere and Jestin, 1995).

each system. In a catalytic cracking riser rapid circulation is required to maintain catalyst activity; in a combustor the circulation need only maintain temperature uniformity, and in some cases carry heat to an external heat exchanger. The circulation requirement for moving heat is much lower than that for maintaining catalyst activity. The similarity between the units is the presence of the characteristic CFB core-annular structure in both cases. In a CFB combustor the core-annulus structure is pronounced. A dense layer of solids cascades down the furnace walls, and any vertical surfaces, with a thickness of order 0.5 m in large units (Werther, 1993). Inside this downflowing layer is the much more dilute core region in which particle slip velocities may approach their terminal velocity and where only a small fraction of particles travel downwards. These general ideas are illustrated in Figure 10.5, which shows upward and downward fluxes near the wall of a commercial CFBC (Lafanechere and Jestin, 1995).

10.6 Turndown and control strategies

An advantage cited for CFBs versus bubbling bed combustion systems is a potentially simpler turndown strategy. The bulk of solids in a fluid bed are at a relatively uniform temperature because of the good convective solids mixing. As the unit is turned down this temperature must be sustained in order to maintain good combustion and emissions performance; however, in order to reduce the steam generation rate, the rate of heat transfer to the in-furnace heat transfer surfaces must be reduced. This requires either that the heat transfer coefficient be decreased, or that the surface itself be somehow removed from active use. In a bubbling bed unit there are high heat transfer coefficients to horizontal (or vertical) in-bed surfaces which, depending upon the gas velocity, may either increase or decrease as superficial velocity is decreased. There is little or no natural turndown, the process by which the heat transfer coefficient conveniently follows the gas velocity to maintain constant system temperature as load is decreased. Therefore turndown strategies in bubbling beds rely upon:

(i) Mounting tubes in the upper part of the bed, which become uncovered as the gas velocity is decreased, causing reduced heat transfer coefficients.
(ii) Removing part of the bed from service by slumping.
(iii) Recirculating flue gas to cool the bed and to permit the gas velocity to be somewhat independent of load.
(iv) Some combination of the above.

In a circulating fluidized bed, average furnace heat transfer coefficients more closely follow load in a natural fashion as discussed below. The natural turndown is augmented by various methods.

There are effectively two distinct types of CFB systems from the perspective of load following. These can be defined as 'variable' and 'fixed' inventory types, where the inventory refers to the total mass of solids in the circulating bed furnace, but excludes the recycle loop. Figure 10.6 shows a schematic of the first type of unit. In a fixed inventory CFB system, there is no control of the solids recirculation rate, and all of the solids are found in the furnace except for a small and essentially fixed amount in the fluoseals(s) and (if present) external heat exchanger. In Figure 10.6, all of the heat transfer surfaces upstream of the backpass are located in the main furnace; they are located primarily in the upper furnace. As the load is reduced by decreasing primary and secondary air flows simultaneously there is a change in the suspension density profile. Solids move from the upper furnace to the lower furnace as the gas velocity decreases causing the voidage in the lower furnace to decrease and the carryover rate to be reduced. In response, the upper furnace heat transfer coefficients decrease (Tang and Engstrom, 1987), helping to maintain the furnace temperature. The heat transfer from the furnace, and hence its temperature, may be further modified by:

(i) Adjusting the primary-to-secondary air split (i.e. flow ratio). Higher primary-to-secondary air ratios increase the gas velocity in the lower furnace, entraining solids into the upper furnace where they transfer more heat.
(ii) Using flue gas recirculation (FGR) as in a bubbling bed. Higher flue gas recirculation rates again move more solids into the upper furnace

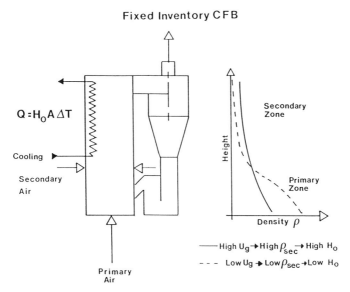

Figure 10.6 Schematic diagram of a fixed inventory CFBC unit and an exaggerated illustration of how changing load in the unit alters the suspension density profile (Brereton, 1987).

thereby increasing heat transfer rates. Flue gas recirculation has additional cooling effects through the sensible heat of the flue gas stream and an additional beneficial impact upon NO_x emissions.

(iii) Increasing the excess air ratio.

An alternative to relying upon variation in the furnace heat transfer coefficient to follow the load is to decouple some or all of the heat transfer from the furnace density profile. This is accomplished by provision of an external heat exchanger (EHE) in the circulating fluid bed return loop (Plass et al., 1986). This is typically a bubbling fluid bed exchanger operating at low superficial gas velocities (0.5–1.0 m/s) (Werdermann and Werther, 1995), where heat transfer coefficients are very high. The exchanger can therefore be relatively compact. Furnace temperature may then be maintained, independent of what happens within the furnace itself, by adjusting the fraction of the return solids passing through this exchanger. Heat transfer coefficients within the exchanger are of order $700 \, W/m^2 K$ (Werdermann and Werther, 1995), and are independent of load. The solids flowing through the exchanger are a slipstream, taken from the main flow circulating through the fluoseal at a rate which is varied by a mechanical valve. The external heat exchanger may contain evaporative, superheat or reheat surfaces, and provides both effective load control and fuel flexibility by extracting up to 65% of the total heat requirement from the primary combustion loop (see Figure 10.7). By decoupling the heat transfer from the combustion loop

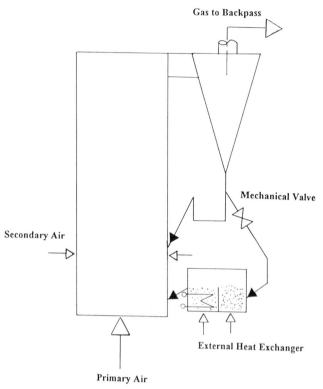

Figure 10.7 Schematic of a fixed inventory CFBC with an external heat exchanger for temperature/load control. Heat removal may be varied independently of other factors by controlling the solids flow rate through the external heat exchanger. This is accomplished by opening and closing the mechanical valve so that the fraction of returning solids passing through the external heat exchanger is changed.

fluid and particle mechanics, the external heat exchanger allows excellent fuel and load flexibility.

An alternative to this 'fixed inventory' type of combustor is a 'variable inventory' unit in which, instead of shifting solids from primary to secondary zone, when increased heat transfer is required, solids are moved from a storage location in the return loop into the main combustion chamber. This configuration requires that the rate of solids return into the combustor be controlled rather than just adjusting itself according to the overall requirements of the system pressure balance. Control cannot be achieved with a fully fluidized fluoseal arrangement, which does not permit control of solids flow but returns everything (see Chapter 7); instead, either a mechanical valve, or a non-mechanical L-valve is needed to control the flow. Figure 10.8 shows how, according to whether the system is at high or low load, the aeration rate to an L-valve may be adjusted to control the solids recycle rate. At

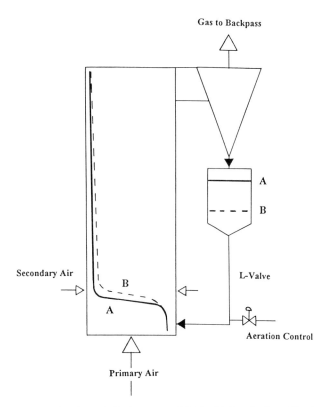

Figure 10.8 Schematic of a variable inventory CFBC with L-valve based load control. As the aeration rate to the L-valve is increased, solids move from the storage hopper into the combustion chamber – the suspension density increases in the combustion chamber and so does the heat transfer rate. The solid line, density profile A and level A in the storage hopper, indicates the situation at low L-valve aeration. The dashed line, density profile B and level B in the storage hopper, indicates the situation at a higher L-valve aeration.

high load, high heat transfer rates, and hence higher solids inventories in the primary combustor chamber are desired. The L-valve aeration flow is therefore increased, and a new solids balance is established with a higher circulation rate and lower inventory on the return side. At low load, the converse is true.

Use of the variable inventory concept allows the density profile and heat transfer rate to be adjusted within limits, independent of either the gas velocity or the primary-to-secondary air ratio. This concept was originally developed by Studsvik Energiteknik (Kobro and Brereton, 1985), and was licensed and modified by Babcock and Wilcox in the United States. It has been used with success on several commercial boilers (Alexander and Eckstein, 1994).

10.7 Temperature profiles and the effect of turndown

Temperature uniformity in a CFB is maintained by virtue of high rates of internal and external solids circulation. These provide a rate of convective heat flow through any zone in the combustor that is an order of magnitude greater than the zonal heat release rate, hence equalizing temperatures. However, there are various circumstances under which temperature may become non-uniform. The first is as gas passes through the cyclone. The suspension density entering the cyclone is typically the lowest in the combustor so that the suspension has a reduced thermal capacity. Added to this, heat release rates per unit volume in the cyclone may be higher than at the top of the furnace, because the cyclone may represent the first opportunity for volatile-rich and oxygen-rich zones of gas to become well mixed in what can otherwise be a relatively poor gas mixing environment. These factors combine to give temperature rises across the cyclone of greater than 50°C for some fuels in some combustors, especially those that do not provide for good mixing of fines or volatiles lower in the furnace. They point to the importance of the cyclone as part of the combustion system. An example of cyclone temperature rise is shown in Table 10.3 (Boemer *et al.*, 1993). While such rises are not uncommon, nor are they generally the case; temperature profiles are strongly dependent on the arrangements of heat transfer surfaces and feed points. For example, a water-cooled cyclone utilized by some manufacturers (Gustavsson and Leckner, 1995; Bernstein *et al.*, 1995) produces a consistent drop in gas temperature.

To further illustrate the importance of the cyclone to the combustion system, Lucat *et al.* (1993) report that the oxygen concentration across the cyclone decreased from 4.8% at the inlet to 3.1% at the outlet. This again shows that the cyclone is needed not just for solids separation, but is also an integral part of the combustor. Development of combustion and pollutant profiles must be considered all the way through the cyclone(s) and even into the convection banks. Both zones represent a significant fraction of the residence time for gas through the system, with the cyclone core offering the first opportunity for gases such as CO and hydrocarbons to burn without being simultaneously generated by combusting solids.

As gas velocity is reduced to follow load in a CFB, circulation rates decrease simultaneously. This leads to increased temperature non-uniformity at reduced load. The extent of the non-uniformity depends strongly upon

Table 10.3 Temperature profile in a CFBC – Data from Boemer *et al.* (1993) from a Lurgi 5.3 m dia × 31 m high CFB with one cyclone and flash recirculation. (Height measurements are height above distributor grid)

Location	In dense bed	10 m	17 m	28 m	After cyclone
Temperature, °C	860	930	880	860	975

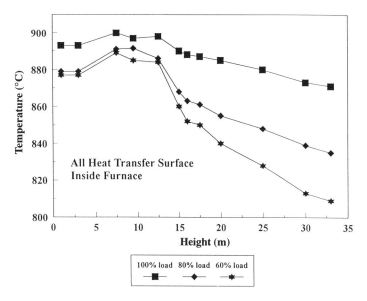

Figure 10.9 Temperature profiles for a 125 MWe CFBC as a function of load (adapted from Lafranchere and Jestin, 1955). Values are from a numerical simulation.

control strategy and heat transfer surface placement, but it would not be uncommon, for example, for the furnace exit temperature to be reduced to 700°C at 50% load (Boemer et al., 1993). Under reduced load conditions, residence times and time–temperature profiles of the gas flowing through the system change, and corresponding changes occur in pollutant formation and combustion efficiency. Figure 10.9 (Lafranchere and Jestin, 1995) shows predicted furnace temperature profiles at full and partial loads for a furnace with all surface within the bed.

In summary, CFB combustion systems are unlike any other type of combustor. At full load they provide for a relatively uniform low temperature over the entire combustor because of high convective particle flows. Also, as shown later, there are combusting fuel particles distributed throughout the CFB. Fine fuel solids may be burnt in a single pass, or may only recirculate once or twice. Coarse solids can have residence times of hours. Coarse combusting solids are present throughout the riser and into the cyclone. The cyclone core represents the first significant opportunity for gases such as CO and hydrocarbons to burn without large-scale simultaneous production of these species. This environment can be contrasted with a bubbling bed, which can be divided into a much more distinct bed region and freeboard. In the latter case volatiles and CO production largely occurs in the bed, with only fine elutriated carbon continuing to produce these species into the freeboard. In this environment it is the freeboard, largely free of combusting solids, that becomes a region where CO burnout and other gas-phase reactions must proceed to completion. In many bubbling bed units the freeboard provides the location for radiant and convective heat transfer surfaces so that temperatures may drop rapidly. However, generalizations are dangerous. In some bubbling bed systems the freeboard may be refractory lined and, with secondary air injection above the bed surface, there can be a large temperature rise due to volatiles combustion.

10.8 Relationship between combustion fundamentals and heat release profiles

A coal particle introduced into a combustion system passes through various stages: drying, devolatilization, ignition and finally combustion of fixed carbon as illustrated in Figure 10.10. An excellent summary of coal combustion in bubbling beds is presented by LaNauze (1985) with basic ideas that are directly applicable to the CFB environment. The combustion processes occur simultaneously with swelling, fragmentation and attrition; these complicate the overall process and make a comprehensive mathematical description extremely difficult. Further, the various stages of the combustion are not distinct, but exhibit some overlap. Despite the problems, each stage of the process has been described by mathematical models, either directly in connection with circulating fluid bed combustion, or related to bubbling beds or pulverized systems. Most of these models contain empirical factors that depend upon the fuel type.

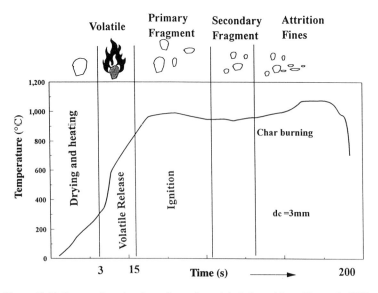

Figure 10.10 Stages of combustion of a coal particle (adapted from Howard, 1989).

To understand CFB combustion it is valuable, at a minimum, to look at the time frames for the various parts of the combustion processes and to compare them with circulation times for solids in the furnace. This gives an indication of where combustion can be expected to occur within the furnace.

Drying and devolatilization have been studied by authors such as LaNauze (1985), Jung and Stanmore (1980), Agarwal (1986), and Anthony et al. (1975). Devolatilization depends upon factors such as heating rate, coal type and particle size. When a coal particle is injected into the bed, volatiles evolution first occurs at a low temperature in the absence of combustion around the particle. As the particle heats up, volatiles continue to be evolved and may burn in a diffusion flame surrounding the particle, with the location of the flame front dependent upon the evolution rate (Agarwal, 1986). Eventually, volatiles are produced at a rate insufficient to sustain the diffusion flame, and though evolution may continue, the volatiles burn directly at the particle surface and the combustion is simultaneous with that of the char. The total amount of volatiles produced will depend upon the heating rate and will not typically be the same as that implied by the normal proximate analysis – for this reason, specific proximate analyses and other mechanical characteristics more suited to predicting behaviour in fluid bed environments have been proposed (e.g. Daw et al., 1991).

The fractional evolution of volatiles versus time in a bubbling fluid bed work has been correlated by La Nauze (1985) using a shrinking core model:

$$\frac{t}{\tau} = 1 - 3(1-x)^{2/3} + 2(1-x) \tag{10.1}$$

where x is the fractional devolatilization at time t, and τ is the time for complete devolatilization.

For coarse coal particles Pillai (1981) observed the time between initiation and extinction of the volatile diffusion flame for 12 different coals. He correlated the results in the empirical form:

$$T_v = a d_0^N \tag{10.2}$$

where T_v is the lifetime of the devolatilization flame in seconds, and is an approximation to τ in LaNauze's expression, d_0 is the initial coal particle diameter in millimetres, N varies from 0.32 to 1.8 and is a function of bed temperature, T_b, and coal type, a, is an empirical constant between 0.22 and 22 which is inversely proportional to $T_b^{3.8}$. Table 10.4 illustrates volatile combustion time as a function of particle diameter for a typical anthracite.

Char combustion in the CFB environment has been studied by Basu and Halder (1989), and by Furusawa and Shimizu (1988). Some representative results are shown in Figure 10.11. The combustion rate is close to the prediction of pure kinetic control, implying that mass transfer is sufficiently rapid to provide little resistance. The results of this figure can be translated

Table 10.4 Predicted devolatilization and char combustion times for an anthracite

Particle diameter (microns)	Estimated devolatilization time (s)^a	Estimated char combustion time (s)^b
10		0.2
100	10	2.5
1000	22	32
10000	50	655

^{a} Devolatilization constants from Pillai (1981)
^{b} Combustion rate constant from Basu and Fraser (1991).

into a char burn-out time which is presented, together with the devolatilization time, in Table 10.4. The char burn-out time depends strongly on the intrinsic reactivity of the char, which may vary by orders of magnitude depending upon the fuel and temperature. Results for a variety of fuels have been compiled by Basu and Fraser (1991).

At full load, circulating fluid bed combustors use superficial velocities of approximately 6 m/s and have heights between 15 and 30 m. In the core, the slip velocity can be approximated by the particle terminal velocity to give a minimum single-pass residence time (for plug flow) varying between 3 and 6 s. Average single-pass residence times are given by the ratio of the total mass of solids in the combustor to the external mass circulation rate and are of the order of 2 minutes. Clearly, for anything other than fine

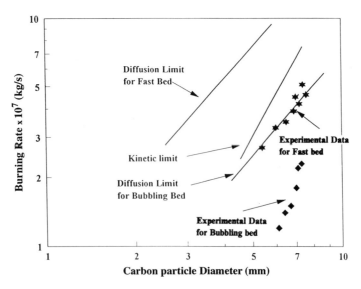

Figure 10.11 Char combustion rates in CFBC. Points indicate experimental measurements for bubbling and circulating beds. CFB points are close to the kinetic limit (adapted from Basu and Halder, 1989).

particles 1 mm or less in size, both devolatilization and char burn-out times are an order of magnitude greater than both the minimum single pass particle residence time, and the mean particle residence time (given by the ratio of the solids mass in the combustor to the external circulation rate). This implies that coarse combusting particles have sufficient residence time to be well mixed with the inerts that form the bulk of the system. Following from this, for a coarse particle feed, release of volatiles, their combustion, and combustion of char, occur throughout the furnace and cyclone. Fine particles may be burnt in a single pass and almost in plug flow. While this simple picture is in practice complicated by factors such as attrition, fragmentation (Walsh and Li, 1993) and very fine or very coarse burning char particles following somewhat different circulation patterns than the bulk of the solids, these simple ideas help explain heat release patterns observed in operating circulating bed systems, and how they change as a function of such factors as fuel size distribution. They also explain why many vendors place restrictions upon the percentage of fine material (e.g. less than 75 micron) in circulating fluid bed feed streams. They are concerned with the potential for reduced combustion efficiency and high rates of heat release in the upper furnace and cyclone. A study by Chelian and Gamble (1995) discusses some of these issues and the concern over temperature increases due to combustion in cyclones. Senior (1992) presented results showing how different sizes of particles segregate in the circulating fluid bed environment. His results imply that segregation is minimal at high load but may be considerable at lower turndown velocities, with preferential circulation of finer particles. Fine particle circulation is significant because it may increase heat transfer in the CFB furnace.

While the circulating fluid bed is a relatively good environment for mixing of solids, so that, for example, coarse char particles are well mixed with inert solids, gas mixing is comparatively poor. This is discussed in Chapter 3 and significantly impacts upon the heat release distribution. In large and even moderately sized pilot units, as in bubbling beds, plumes of volatiles originate from around the feed points; these may persist high into the boiler because of the relatively poor radial gas mixing. The slow mixing of volatiles with combustion air shifts the heat release higher up the furnace, and creates a local reducing atmosphere that influences pollutant formation and reduction processes.

The general concepts of heat release and the development of combustion profiles in CFB systems are illustrated in Figure 10.12. These show the simple core-annulus behaviour in a CFB pilot plant combustor (Zhao, 1992). For fuels such as coals the annulus is a reducing zone with high solids and char loadings. This offers potential for reduction reactions. The core is richer in oxygen with its concentration declining with height, steeply at the base of the unit, and less dramatically in the dilute upper furnace, indicating the progress of the overall combustion. Finally, the cyclone is

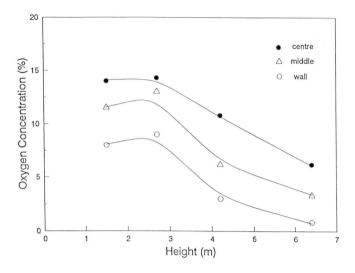

Figure 10.12 Core annulus combustion behaviour in a CFB pilot plant. The core-annulus profile is seen over the entire 7 m combustor height with lower oxygen concentration in the solid-rich wall region (Zhao, 1992).

seen as the region where the bulk of final combustion occurs, although some continued CO generation and destruction continues through the superheater surfaces. The simple core-annulus approach is reasonable for small-diameter pilot plants where volatile plumes mix quickly. In larger units these plumes create dramatic asymmetry, which complicates the profiles. The full-scale results of Couturier et al. (1989) (Figure 10.13) clearly show similar core-annular trends, strong radial profiles, but with dramatic asymmetry due to the plumes and feeding along a single wall.

Differences between pilot plants and commercial units are not limited to plume effects. In the pilot plant the annular zone occupies a much higher fraction of the total unit cross-sectional area, and this reducing zone has a distinct influence upon combustion and pollutant formation behaviour. For example, in a 150 mm ID riser, the downflowing wall layer may be as thick as 25 mm. In commercial units the wall layer may grow to as much as 0.8 m at the base, but the unit may be as wide as 11 m. At the top of the riser the wall layer may be considerably thinner. Proportionally the area occupied by the wall zone is considerably smaller and has less effect on combustion and emissions. In contrast, large units have much smaller height-to-diameter ratios. Thus, the lateral gas mixing, which is easily achieved in pilot units, is much harder to accomplish, and plumes of volatiles and secondary air penetration therefore have a greater effect upon pollutant and combustion results.

Summarizing, the location of heat release depends upon factors such as volatiles content, fines content, fuel size distribution, and fuel reactivity. A

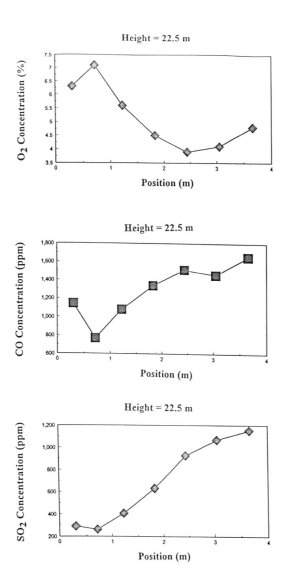

Figure 10.13 Lateral gas profiles in a commercial 22 MWe CFBC. The position in m is the distance from one wall and values are the average of three different lateral positions. Two feed chutes are positioned along the opposite wall (i.e. at 4 m). Profiles show strong radial non-uniformities as well as the importance of the cyclone in finishing combustion – total unit height is 24 m (data from Couturier et al., 1989).

circulating bed design must be sufficiently flexible that large maldistributions in the heat release profile do not cause significant hot spots. Internal and external solids circulation are essential to move heat from the lower furnace, where the bulk of solids lie and the bulk of heat release occurs, to the upper furnace and external circulation loop, where most of the heat transfer surface is located.

10.9 Generation and destruction of pollutants

10.9.1 Hydrocarbons and carbon monoxide

Emissions of hydrocarbons and CO from combustion systems represent losses in the thermal efficiency of the system. At high concentrations they also represent potentially harmful environmental releases.

CO emissions are a complex function of fuel reactivity, fuel distribution, gas mixing and the thermal history of the gas as it passes through the combustion system. Commercial circulating fluid bed designs are such that CO emissions are within the statutory limits for stationary combustion sources. Emissions typically vary from 40 to 300 ppmv and tend to be higher for coals than for more reactive fuels – design is such that the emissions will be within the regulatory standards of the country in question, which are typically in the range from 79–300 ppm at around 6% O_2 (Soud, 1991).

It was noted above that for many fuels, such as coal, CO combustion cannot be completed until the cyclone core is reached. This is because of continued CO generation. It is generally accepted that CO is the main product of char combustion, with subsequent oxidation to CO_2 usually occurring in the zone around the combusting particle. The consequence of CO generation throughout the unit is that the temperature history downstream of the cyclone, as well as the thermal history within the CFB riser and cyclone, determines the final CO levels. Thus, while factors such as increased cyclone temperature tend to reduce CO emissions (Melvin *et al.*, 1993), factors such as soot-blowing frequency in the convection banks also have a strong effect. Table 10.5 (Lucat *et al.*, 1993) shows how emissions of CO and NO_x develop from the cyclone outlet to the baghouse of a

Table 10.5 Development of emissions profiles in and beyond a CFB furnace (Lucat *et al.*, 1993)

NO (ppm)	Mid furnace	Cyclone inlet	Cyclone outlet	Bag filter inlet
O_2 (%)	8.3	14.8	13.1	3.5
CO_2 (%)	11.3	14.2	15.7 (13.1)	(14)
CO (ppm)	360	128	37 (30)	(200)
NO (ppm)	147	129	109	(56)

Numbers in parentheses are corrected to 6% O_2.

250 MWe unit due to char deposition and combustion on the superheater banks.

Throughout the CFB system, prediction of CO oxidation rates are difficult because there is only limited understanding of the oxidation process. There are strong indications that the oxidation occurs dominantly through free radical mechanisms, even at the low temperature of fluid beds, with free radicals present in superequilibrium concentrations (Anthony, 1995). As a result of the various complexities, design to ensure appropriate CO emissions is based largely upon experience.

Emissions from fluid beds, like other systems, are usually based on rolling averages. For homogeneous fuels there is relatively little deviation of the instantaneous CO value from the average. However, with inhomogeneous fuels such as municipal solid waste it may be difficult to ensure feed consistency. In these cases CO spikes and transient phenomena may account for a large portion of CO emissions.

Emissions of other trace organics, measured either individually or as total hydrocarbons, have been less well studied. As with CO, total hydrocarbons are typically within permissible limits and are not problematic except during upsets of the feeding system or when feed inhomogeneity leads to fuel-rich conditions.

Specific measurements of trace organics such as dioxins and furans, and of destruction and removal efficiencies for designated hazardous constituents, have been made in several cases. Ogden Environmental studied a number of wastes including chlorinated benzenes as part of obtaining a permit for a transportable CFB incinerator (Chang *et al.*, 1987). Similar studies have been made by Desai *et al.* (1995) for dichlorobenzene, which was used as a surrogate for PCB. Wong (1994) measured profiles of organic destruction in a pilot CFB system. All these results are for incinerators that are often transportable, and are usually much smaller and provide lower gas residence times than stationary large-scale boilers. The low residence times are compensated, or perhaps more than compensated, by temperatures that tend to be somewhat higher than in full-scale boilers.

With specific reference to dioxins, Desai *et al.* (1995) made measurements of dioxins during combustion of dichlorobenzene. They found flue gas dioxin levels between 3 and 42 ng/m^3, which exceeds the level permitted in many jurisdictions (0.1 to 1 ng/m^3). However, these were measured in the absence of any flue gas scrubbing. Dry scrubbers have been found to remove up to 99.9% of dioxins and furans (Environment Canada, 1986) and similar numbers have been found for wet systems. Hence, the combination of a CFB combustor with an appropriate scrubber is clearly suitable for many incineration applications, with typical destruction and removal efficiencies exceeding the 99.99% requirement of many permitting agencies for destruction of many types of organic wastes (Rickman *et al.*, 1985; Wilbourn *et al.*, 1986).

It may initially seem strange to require a scrubber to follow a fluid bed unit when one of the perceived benefits of the latter is its ability to capture acid gases *in situ* using limestone. However, while limestone is successful in capturing SO_2, thermodynamics and pilot plant operating data (Liang et al., 1991) show that hydrogen chloride cannot be scrubbed *in situ*. At fluid bed temperatures and typical partial pressures of H_2O and HCl associated with waste combustion, the equilibrium of the HCl scrubbing reaction

$$CaO + 2HCl \Leftrightarrow CaCl_2 + H_2O \qquad (10.3)$$

favours the left-hand side. HCl capture can only effectively occur at temperatures below 650°C. Where limestone scrubbing of HCl has been reported (Rickman et al., 1985), it has likely occurred in the cooler baghouse downstream of the fluid bed combustor on elutriated limestone, not in the bed itself.

10.9.2 Sulfur capture

One of the principal incentives for fluid bed combustion is the ability to capture sulfur *in situ* using low-cost sorbents, limestones and dolomites, which are fed as solids to the combustion chamber. At atmospheric pressure, limestone calcines according to

$$CaCO_3 \Rightarrow CaO + CO_2 \qquad \Delta H_{298} = +183\,kJ/g\,mol \qquad (10.4)$$

It can then capture sulfur within the bed according to

$$CaO + SO_2 + \tfrac{1}{2}O_2 \Rightarrow CaSO_4 \qquad \Delta H_{298} = +486\,kJ/g\,mol \qquad (10.5)$$

The overall sulfur capture reaction is mildly exothermic, with the endothermic calcination more than balanced by the exothermic sulfation, but stoichiometric excesses of CaO redress the thermal balance.

Dolomite may calcine to form a particle containing a calcium–magnesium oxide, i.e.

$$CaCO_3 \cdot MgCO_3 \Rightarrow CaO \cdot MgO + 2CO_2 \qquad (10.6)$$

Only the calcium portion subsequently sulfates. However, the calcination of the magnesium portion appears to lead to a more open pore structure, which improves the sulfur capture performance of the calcium component (Senary et al., 1989; Ake et al., 1993).

Under reducing conditions, in gasifiers or in reducing regions of combustion systems, calcium oxide combines with hydrogen sulfide to form calcium sulfide, i.e.

$$CaO + H_2S \Rightarrow CaS + H_2O \qquad (10.7)$$

Sorbent in various states of sulfation and oxidation is one of the main components of the bed material in a CFBC. The fuel particles burn surrounded by this sorbent and other ash components. Gases such as sulfur dioxide released during the combustion follow a tortuous path past these particles on their way through the bed affording a good opportunity for sulfur capture. The spent sorbent particles, together with some unburnt fuel and active sorbent, are ultimately removed as fly ash, entrained from the combustor, and from the main combustion loop. The optimum location of the drain points depends upon the friability of the calcined and sulfated sorbents as well as the amount of coarse inert material in the fuel feed.

The performance of fluid bed sulfur capture is generally measured as a function of the 'calcium-to-sulfur molar ratio Ca:S', the moles of calcium from limestone required to capture one mole of sulfur dioxide released during combustion. Achieving low calcium-to-sulfur ratios is critical to the economics of the fluid bed process, because it affects both limestone costs at the front end, and the amount of ash that must be disposed as residue. As a result of these economics, there has been considerable fundamental and pilot plant work to characterize and understand sulfur capture.

(a) Fundamental aspects of sulfur capture

While the overall reaction for sulfur capture is easily written as equation (10.5) above, this clearly does not describe the mechanism at a molecular scale. Various theories exist, and there is still discussion as to the true mechanism of capture. These have been well summarized by Anthony (1995), with many key ideas discussed by Dennis and Hayhurst (1990). Some competing ideas are:

(i) SO_2 is oxidized to SO_3, which then reacts directly with CaO

$$CaO + SO_3 \Rightarrow CaSO_4 \tag{10.8}$$

(ii) The initial product of reaction is sulfite, which then rapidly disproportionates to sulfide and sulfate (sulfite is unstable above 560°C).

$$4CaO + 4SO_2 \Rightarrow 4CaSO_3 \Rightarrow CaS + 3CaSO_4 \tag{10.9}$$

CaS can subsequently be oxidized to sulfate.

(iii) The sulfation occurs via ionic species that can be represented by the general formula $S_yO_x^{n-}$ (Dennis and Hayhurst, 1990).

Whatever the exact mechanism, it is clear that sulfation is a complex process involving formation of at least two sulfur species, CaS and $CaSO_4$. Under reducing conditions sulfate can be reduced to sulfide, or it can be decomposed directly to the oxide as shown by the phase diagram of Figure 10.14.

$$CaSO_4 \Rightarrow CaS + 2O_2 \tag{10.10}$$

$$CaSO_4 \Rightarrow CaO + SO_2 + \tfrac{1}{2}O_2 \tag{10.11}$$

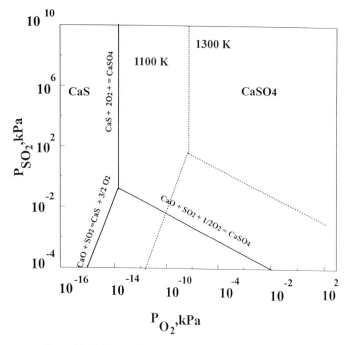

Figure 10.14 Phase diagram for the $CaO-CaSO_4-CaS$ system.

Also, at temperatures above 850°C, solid phase reactions can become important (Davies et al., 1994).

$$CaS + 3CaSO_4 \Rightarrow 4CaO + 4SO_2 \quad (10.12)$$

As a sorbent particle moves around the combustion system, in what may be a residence time of hours, it experiences both oxidizing and reducing conditions, and potentially a series of capture and release reactions with changes in the sulfur species present in the particle. These complex reactions lead to an optimum temperature for sulfur capture in atmospheric circulating fluid bed systems. The exact optimum is a function of sorbent properties and bed operating conditions. There are various theories for the presence of the optimum, which have been summarized by Lin (1994). Perhaps the one that finds most support is that at temperatures beyond the optimum, the reverse sulfation reaction

$$CaSO_4 + CO \Rightarrow CaO + SO_2 + CO_2 \quad (10.13)$$

becomes increasingly important (Lyngfelt and Leckner, 1989a, b; Hansen et al., 1991). Decreased capture at temperatures lower than the optimum is a result of decreased rates of reaction.

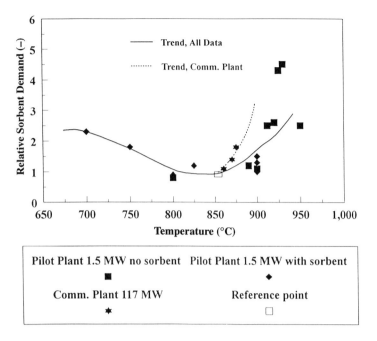

Figure 10.15 Variation of sulfur capture with temperature in atmospheric fluid bed combustion (adapted from Schaub et al., 1989).

The situation is further complicated in pressurized systems where limestone does not readily decompose and where much of the sulfur capture is by direct reaction to SO_2 with carbonate. In such situations no optimum temperature is typically found (Yates, 1983). Figure 10.15 shows the typical variation of sulfur capture with temperature.

The chemical reactions associated with sulfation are only a small part of the total picture of this complex process. Sorbent utilization is also controlled by a series of physical processes both on the level of the individual sorbent particles, and associated with the general unit construction.

A sorbent particle injected into an atmospheric pressure circulating fluidized bed first calcines. Ideally this creates an open porous structure with a good combination of micropores and macropores that allow for effective diffusion of sulfation reactants into the structure, and products out of it. Exposed to sulfur dioxide, hydrogen sulfide, and other gas species, sulfation proceeds from the outside of the particle through to its centre. However, the molar volume of product calcium sulfate (52.2 cc/mol) is greater than that of the parent calcium carbonate (36.9 cc/mol). Therefore the product tends to block the outer pores, preventing further access of SO_2 to the partially unreacted centre (Shearer et al., 1980). Hence, large sorbent particles removed from bubbling and circulating fluidized bed

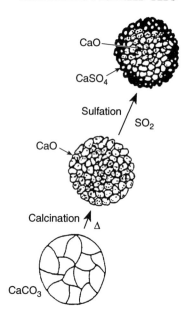

Figure 10.16 Illustration of formation of an impervious sulfate shell in fluid bed sulfur capture (Shearer et al., 1980).

combustors have shells of impermeable sulfate that restrict sorbent utilization (Figure 10.16). Small particles are sulfated more uniformly through the particle, but still suffer from blocking of micropores, which prevents access to unused calcium oxide.

The utilization of sorbent under practical circumstances depends upon many complex and often interacting factors:

- The limestone type, which controls the pore structure developed during calcination and its tendency to decrepitate to form fine reactive particles.
- Attrition of limestone and sulfation products, which determines the extent to which blocked sorbent particles may be broken down by mechanical attrition in the bed to expose fresh calcium oxide surface.
- Hydrodynamics of the bed, and air and fuel distributions, which control the contacting of sorbent with SO_2, and the relative time sorbent spends in oxidizing and reducing zones.
- The mechanical efficiency of the cyclones. Fine partially sulfated sorbent returned to the system enhances sulfur capture. Efficiency of capture of fine particles may be significantly higher than predicted by simple models of cyclones since high particle loadings typical of CFBs enhance fine particle capture – the effect is to increase the residence time of such fines, which may increase both sulfur capture and combustion efficiency (Figure 10.17) (Schaub et al., 1989).

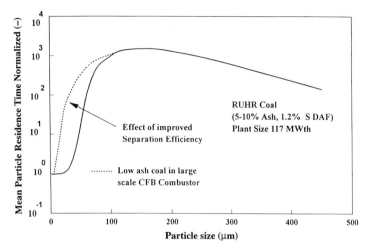

Figure 10.17 The effect of improved separation efficiency in a cyclone on overall solids residence time in CFB systems (adapted from Schaub et al., 1989).

The above factors combine to give sulfur capture performance that is sensitive to limestone type and to system type and operating conditions. Sulfur capture of 90% is usually achieved at molar ratios of calcium to sulfur of between 1.8 and 3:1. Figure 10.18 illustrates how sulfur capture changes with Ca:S for large CFB combustors.

(b) Enhanced sulfur capture
Recently there has been significant work upon improving sulfur capture in CFBs, focused upon removing spent or partially spent sorbent (baghouse or bottom ash) from the system and then reinjecting this material following chemical or physical treatment. The Canadian Electrical Association recently funded a study by Ahlstrom Pyropower (San Diego) to look at the technical and economic feasibility of different 'reactivation' options. Some of the options included (Ahlstrom Pyropower, 1992):

- grinding of spent sorbent to expose fresh unreacted surface;
- slurrying and reinjection of the ash;
- 'dry' rehydration of the ash;
- the ADVACAT process;
- the CERCHAR process.

Each of the final four processes effects reactivation through chemical reaction of the ash with water or steam. In the ADVACAT process the objective is to produce reactive hydrated calcium silicates that are effective absorbers of SO_2 at the cold end of a boiler. In the three other processes the objective is to rehydrate the spent calcium oxide to calcium hydroxide under various water temperatures and stoichiometric ratios. The formation

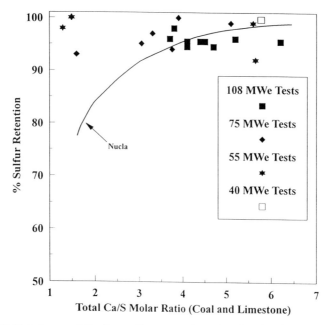

Figure 10.18 Influence of Ca:S on sulfur capture for a CFBC. Points indicate performance of the ACE 108 MWe CFBC at different loads. Solid line indicates the full load (110 MWe) performance of the NUCLA CFBC unit (adapted from Melvin et al., 1993).

of high molar volume hydroxide, or release of steam upon decomposition of this hydroxide upon reinjection into the furnace, apparently breaks open the particle to expose fresh sorbent surface, which can be used when the sorbent is reinjected at the hot end (Couturier et al., 1994).

(c) Uses of fluid bed residues
While in many cases ash from fluid beds is simply landfilled, high disposal costs and shortage of landfill sites provide a strong incentive to find alternative uses for this material. These can at least offset some of the disposal cost. Some examples of uses that have been proposed and found are:

- as a lime supplement for agricultural application making use of the free lime content (Lewnard, 1991);
- as a low grade cement additive (Bland et al., 1995);
- as road base (Bland et al., 1995);
- as a supplement to stabilize sewage sludges and wastes from oil processing, making use of pozzelanic properties;
- as night cover for landfill.

Many of these represent large markets but depend upon case-by-case studies of fly ash, since structural properties such as pozzelanic properties depend upon the specific chemical composition of both coal ash and limestone. Value as an agricultural supplement depends upon the amount of availability of free lime, as well as the impurities present.

(d) Sulfur issues during combustion of gasification residues
The emergence of IGCC and air blown gasification cycles is raising new issues with respect to CFBC sulfur capture. In several of the proposed processes CFBC is used for combustion of a char residue derived from partial gasification in the gasifier side (Sage *et al.*, 1995); this residue contains both unburnt carbon, and calcium sulfide formed by in-bed capture of sulfur using limestone or dolomite. The combustion may occur either at elevated pressure or under atmospheric conditions, depending upon the specific cycle.

There are two issues relating to sulfur during combustion of gasification chars:

(i) Whether sulfur captured as sulfide is released as SO_2 in the combustor.
(ii) Whether sulfide can be effectively oxidized to sulfate so as to minimize residual sulfide in the ash and prevent disposal problems. High sulfide levels prevent landfill of some ashes because of their potential to release H_2S when exposed to water.

Complete conversion is potentially problematic because, as in sulfur capture when firing conventional fuels, an impermeable sulfate layer may prevent reaction with the inner particle core.

Sage *et al.* (1995), Ninomiya *et al.* (1995) and Kudjoi *et al.* (1995), have all studied the oxidation of calcium sulfide in fluid bed environments. It is clear that considerable sulfide residue may remain in the ash (up to 50% of the total sulfur), and that specific measures are needed to control this. It is not clear whether adjusting operating parameters may be sufficient to reduce the sulfide, or whether external treatment followed by reinjection is necessary.

In summary, limestone utilization is a complex subject that has been the subject of considerable research. Predictive tools are now available based on mathematical models and TGA methods (Ulerich *et al.*, 1979; Walsh, 1995). TGA can be used to perform single particle sulfation to determine the limiting sulfation when pore blockage prevents further reaction, providing some basis for performance prediction, although TGA cannot account, for example, for attrition and the complexities of particles moving from oxidizing to reducing zones. Hence, pilot plants and an experience base remain the basis for many commercial guarantees of limestone performance.

In view of the complexities of sulfur capture and dependence upon limestone type, some consideration might be given to optimizing economics by selection of an appropriate stone. However, Ford and Sage (1991) make the important point that due to the significant contribution of transportation costs to the overall costs of the sorbent, particularly coarse grades, it is most likely that the most economic sorbent will be that from the source closest to the plant.

10.9.3 Nitrogen oxide emissions

The ability to produce low nitrogen oxide emissions without add-on pollution control equipment is one of the most attractive features of fluid bed combustion. There are three nitrogen oxide species emitted in significant quantities from fluid bed combustors: NO (nitric oxide), NO_2 (nitrogen dioxide) and N_2O (nitrous oxide). The first two are generally collectively known as NO_x and are regulated by this name in most countries. N_2O emissions have been recognized only recently and are not regulated at the time of writing. However, N_2O emissions may contribute significantly to the greenhouse effect. As a greenhouse gas it is many times more powerful than CO_2 as an absorber of infrared radiation at certain wavelengths so that although three orders of magnitude lower in concentration, its effect is 1/6 that of CO_2 (Elkins, 1989). Hence, there is an increasing interest in the potential regulation of this gas. This is important for fluid bed combustion because N_2O emissions are much higher from low temperature combustion sources, such as fluid beds, than from higher temperature sources such as pulverized coal combustion (50 to 200 ppmv, cf. approximately 10 ppmv or less (Åmand and Andersson, 1989)). As a result there has been considerable recent research upon N_2O formation and potential methods for its destruction in fluid bed environments. In this section, NO_x refers to the combination of NO and NO_2. While coupled to NO_x in terms of emissions, N_2O is considered separately because of its different regulatory status.

NO_x emissions from combustion are generated from two sources, oxidation of nitrogen in the air, 'thermal NO_x', and oxidation of nitrogen from the fuel itself, 'fuel NO_x'. Figure 10.19 illustrates equilibrium concentrations for NO_x as a function of temperature. At fluid bed temperatures thermal NO_x is virtually negligible. This differs from older pulverized coal systems where high local flame temperatures are present, coupled with minimal air staging. Newer staged low NO_x burners give significant NO_x reductions which may approach fluid bed levels. In fluid beds the NO_x is thus all fuel generated from oxidation of nitrogen present in both the volatile matter and char. Oxidation reactions are rapid, forming almost exclusively NO; NO_2 is formed only gradually on cooling the flue gases and represents at most 10% of the total NO_x at the stack, Thus, in fluid bed combustion, the rapid localized combustion of fuel nitrogen generates nitrogen oxides at levels that are beyond the global equilibrium NO_x level for the temperature prevailing in the system. Dissociation of the NO_x that is formed back to equilibrium levels at the prevailing temperature is slow; hence, reduction reactions, catalytic and non-catalytic, are necessary to generate the low

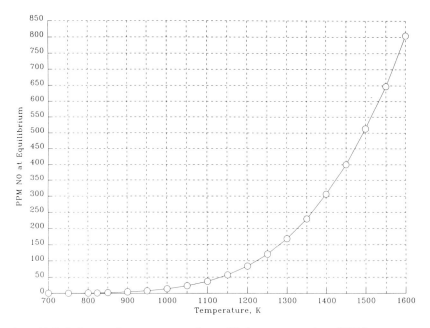

Figure 10.19 Influence of temperature on the equilibrium concentration of NO for an oxygen concentration of 3% by volume and a total pressure of 1 atm.

NO_x emissions that characterize FBC and CFBC systems and may approach the equilibrium values shown in Figure 10.19.

Nitrogen contents in coals are typically 0.5 to 2% by mass. When the coal devolatilizes, this nitrogen is distributed between char and volatiles. Rough proportions for two fuels are shown in Table 10.6, which indicates that some nitrogen may partition to each fraction, but that the amount is fuel dependent. Combustion of char nitrogen proceeds proportionally at approximately the same rate as the carbon (Song et al., 1982), resulting in formation of nitric oxide (NO). Volatile nitrogen is released or decomposes in the gas to form ammonia (NH_3) and hydrogen cyanide (HCN). Ammonia may then be homogeneously oxidized to nitric oxide (NO) or this reaction may be heterogeneously catalysed by either char or calcined limestone. Hydrogen cyanide (HCN) is principally converted homogeneously to nitrous oxide (N_2O).

Table 10.6 Distribution of nitrogen between char and volatile components for two fuels

Fuel type	% Volatile matter (dry basis)	% Fixed carbon (dry basis)	% Nitrogen in original fuel (dry basis)	% Nitrogen in volatile matter	% Nitrogen in devolatilized char
Petroleum coke	10.5	88.5	1.9	2.7	1.8
Sub-bituminous coal	35.9	49.5	0.8	0.3	1.1

The formation mechanisms occur in parallel with complex homogeneous and heterogeneous destruction processes. Among the important reactions are:

- reduction of NO to N_2 by CO catalysed by calcined limestone, i.e.

$$2NO + 2CO \Rightarrow N_2 + 2CO_2 \qquad (10.14)$$

- reduction of NO to N_2 by char
- reduction of NO to N_2O by char nitrogen
- homogeneous reduction of N_2O to N_2 by H and OH radicals
- thermal decomposition of N_2O
- heterogeneous reduction of ammonia to nitrogen by char.

Clearly, formation and destruction of nitrogen oxides are complex processes involving various molecular and radical species. Several authors provide good summaries of these reactions (e.g. Åmand, 1994; Anthony and Preto, 1995), and a large amount of work continues with the aim of providing a predictive capability that can be applied to CFBC. Schematics that show general pathways for formation and destruction of NO and N_2O are provided in Figures 10.20 and 10.21.

While Figures 10.20 and 10.21 provide good indications of the fundamental reactions that underlie nitrogen formation and destruction, they do not provide a tool for estimating NO_x emissions unless they are tied to sophisticated fluid mechanic and kinetic models (e.g. Tsuo et al., 1995). Simple empirical expressions (Zhang and Jones, 1990; Zhao, 1992) have been used based on pilot data but are valid over a limited range. There is also a substantial experience base that allows reasonable prediction of

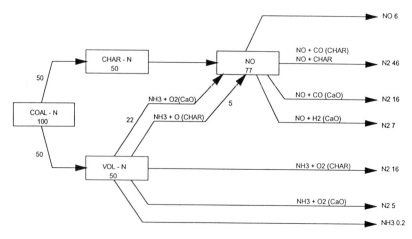

Figure 10.20 Pathways for NO_x formation illustrating typical fractional conversions at each step (adapted from Johnsson, 1989).

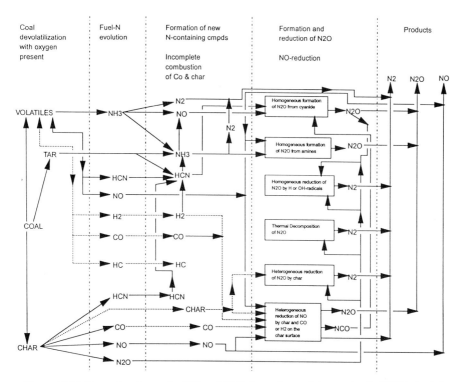

Figure 10.21 General pathways for NO and N_2O formation and destruction in FBC (adapted from Åmand, 1994).

NO_x emissions. Figure 10.22 (Zhao, 1992) shows how, for a pilot plant operating under conditions of reasonably constant temperature and other parameters, NO_x emissions are related to both the fuel nitrogen and to the volatile content of the fuel. Generally, fuels with more volatiles are more reactive, and while they may not form more NO_x immediately on combustion, they lead to significantly lower carbon levels in the furnace so that less NO_x is subsequently reduced.

The following are the generally accepted effects of operating and fuel related variables upon NO_x emission.

Temperature – NO emissions increase with temperature. The rate is strongly dependent upon fuel properties. For example, in pilot plant tests, Brereton et al. (1991) report how for a bituminous coal NO_x emissions increase at a rate of approximately 1 ppm per °C from 60 ppm at 760°C to 200 ppm at 880°C.

Primary-to-secondary air ratio – Decreased primary-to-secondary air ratios decrease NO emissions by increasing NO reduction rates on char in the bottom of the furnace. Also, slow mixing of the secondary air with volatiles plumes from the primary zone provides an almost continuous

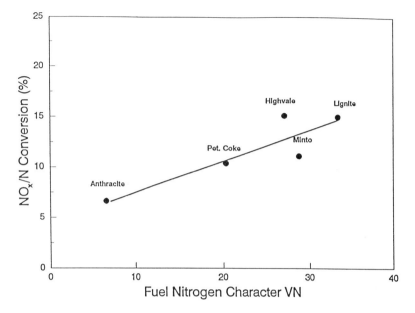

Figure 10.22 Empirical correlation of NO_x as a function of fuel volatile content and nitrogen fraction (Zhao, 1992). The percentage of conversion of fuel nitrogen to NO_x for a variety of fuels of different rank in a pilot CFBC (constant temperature and excess air) is a linear function of VN, the product of (% volatile matter) and (% N in ultimate analysis).

gradual air addition, which reduces volatile nitrogen oxidation rates. This is most pronounced in large furnaces with poor gas mixing; and pilot units may show minimal air staging effects because gas mixing occurs rapidly.

Excess air – Increased excess air increases the oxidation rates of nitrogen species, and limits the extent of zones available for reduction and the amount of carbon available to reduce the NO that has been formed. This has a strong overall effect as shown in Figure 10.23 (Leckner and Åmand, 1987).

Calcium-to-sulfur ratio – Limestone has a complex role in NO_x formation and destruction. While it catalyses the production of NO_x from ammonia, formed from volatile nitrogen, it also catalyses the reduction of NO_x by CO. The relative importance of these effects appears to be a function of fuel reactivity as discussed by Lyngfelt and Leckner (1989a) and Brereton et al. (1991). For reactive fuels such as bituminous or sub-bituminous coals, the oxidation of volatile nitrogen dominates and limestone addition tends to increase NO_x emissions. However, for low volatile fuels such as petroleum cokes, the catalytic reduction can be dominant; in that case limestone addition is beneficial to NO_x emissions.

An overall picture of NO_x formation and reduction reactions in a CFB furnace is shown in Figure 10.24. In this pilot unit there are clearly strong

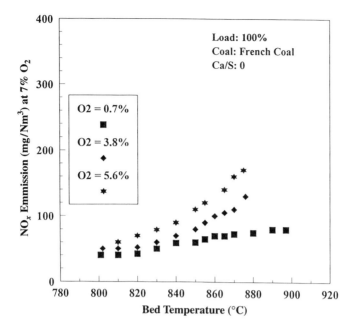

Figure 10.23 Influence of excess air and temperature upon NO_x emissions in a 72 MWth coal fired CFBC (adapted from Boemer et al., 1993).

radial and axial gradients. NO_x is formed primarily in the lower part of the furnace where there is significant char oxidation, due to the high suspension densities, and where the bulk of volatiles are also released and oxidized. This leads to high NO_x levels in the lower furnace. Further up the furnace NO_x production undoubtedly continues because of ongoing char and volatile combustion, albeit at a lower rate because of the reduced suspension density; however, destruction reactions, in particular the reduction of NO on char and limestone, dominate over production reactions creating a net decrease in NO_x along the furnace. In particular, the annulus (wall layer) is a region in which high char and limestone concentrations create reducing or low oxygen zones that favour NO_x reduction reactions.

(a) Factors affecting nitrous oxide emissions
Nitrous oxide emissions are less easily generalized than NO_x emissions. While the same operating variables typically affect N_2O as NO_x, the reaction paths are even more complex. The results from a variety of fluid bed boilers burning a wide variety of fuels have been summarized by Anthony and Preto (1995) from the results of Johnsson (1991), Table 10.7.

The principal findings of the studies are that low rank high volatile coals generate relatively low N_2O emissions, typically around 50 ppmv, while fuels such as petroleum coke may have emissions as high as 150 or

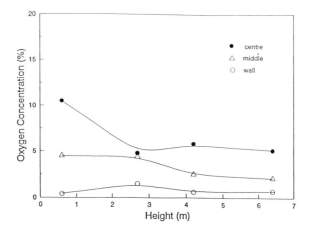

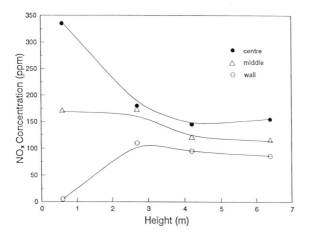

Figure 10.24 NO$_x$ formation and reduction in a 152 mm square × 7.3 m high pilot plant CFB furnace. 'Wall' samples are drawn from the wall layer, 'centre' samples from the centre line, and 'middle' samples from halfway between the two. The oxygen profiles clearly show the core-annulus type structure, with the NO$_x$ profile following (Zhao, 1992).

200 ppmv, a value that exceeds the total NO$_x$ concentration. A further major conclusion is that low N$_2$O emissions are achieved by raising temperatures and decreasing excess air. In particular, high temperatures encourage the thermal decomposition of N$_2$O as shown by Figure 10.25 (Boemer *et al.*, 1993). Thus, the same factors that promote low NO$_x$ promote high N$_2$O, so that, without boiler modifications, there is a trade-off between the two. This is shown clearly by the results of Moritomi (1994), Figure 10.26.

Although as of 1995, N$_2$O emissions are not regulated, various studies have looked at boiler or other system modifications that reduce N$_2$O

COMBUSTION PERFORMANCE 409

Table 10.7 Effect of various factors on N_2O emissions (Anthony and Preto, 1995)

Parameter	Nature of effect	Studies supporting effect
Increasing bed temperature	Decreases N_2O emissions	16 out of 16
NH_3 injection	Increases N_2O emissions	7 out of 7
Increasing fuel volatile content	Decreases fuel-nitrogen conversion to N_2O	10 out of 11
Increasing excess air	Increases N_2O formation	9 out of 11
Air staging	Decreases N_2O formation	3 out of 4
Limestone addition	Decreases N_2O formation	4 out of 7

emissions. Some of these use catalysts such as CaO or olivine sands (Iisa *et al.*, 1991; Miettinen *et al.*, 1991; Johnsson and Dam-Johansen, 1994); however, to date the achievable reductions have been limited. The most promising technique is gas afterburning (Gustavsson and Leckner, 1995), in which gas is burned in the cyclone of a circulating bed boiler to create a high temperature zone promoting thermal decomposition of the N_2O. In the Gustavsson and Leckner (1995) study, a reduction of N_2O concentration from 150 to 30 ppm was attained when the cyclone outlet temperature was raised from 800 to 950°C. Simultaneously the NO emission increased from 70 ppm to 160 ppm.

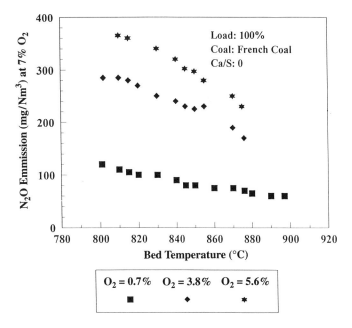

Figure 10.25 N_2O emission as a function of operating temperature and oxygen concentration for a 72 MWth coal-fired CFBC (adapted from Boemer *et al.*, 1993).

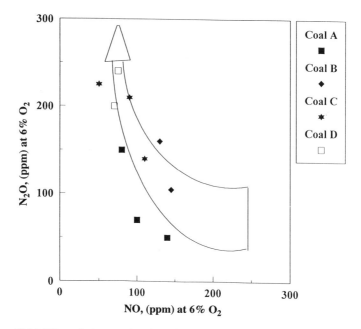

Figure 10.26 NO_x emission as a function of N_2O emission (adapted from Moritomi, 1994).

(b) Ammonia injection for NO_x control

In several CFB applications where extremely low NO emissions are required, the CFB technology has been combined with ammonia injection (SNCR process) to meet prevailing standards (Melvin *et al.*, 1993). This is the same technology as is used in conventional MSW grate boilers and relies on the reduction of NO by NH_3 in a narrow temperature window around 870°C (McInnes and Van Wormer, 1990). Good mixing is required to prevent excess 'slip' of unreacted ammonia through to the stack. A disadvantage of ammonia injection is increased N_2O emissions, as shown, for example, by Hippinen *et al.* (1991), in high pressure systems.

10.10 Summary

There is a wide body of published literature on CFB combustion. This can be combined with a large, but often proprietary, experience base on operating systems, ranging from centimetres to over ten metres in breadth. Within the experience base there is general agreement as to how macroscopic variables qualitatively influence combustion performance, and these trends can largely by explained by a knowledge of the fluid mechanics and reactions that occur within the system. As a result, atmospheric CFB combustion systems can now be quoted and engineered with confidence to large sizes, and they represent a significant portion of newly installed coal-fired utility capacity. Pressurized systems can be expected to develop rapidly.

Atmospheric CFB combustion designs have grown in size, from 150 kW pilot plants to 250 MWe utility units over a period of little more than 10 years. This should be contrasted with bubbling bed designs, which evolved much more slowly and conservatively. The rapid growth owes a great deal to the timeliness of the CFB idea, built upon the broad background knowledge provided by bubbling bed units. It should not be imagined that CFB designs have been developed without some difficulties, but nor have there been any notable utter failures.

Arguably the early success of CFB combustion systems owed little to sound understanding of CFB hydrodynamics, heat transfer or combustion. Rather it says a great deal for the robustness and flexibility of the basic concept that these designs performed acceptably despite a very limited understanding of these fundamental ideas. Only now are fundamentals becoming integrated into designs through modelling, allowing increasingly confident designs of more efficient and cost-effective units. This is especially true for hydrodynamics and heat transfer. With respect to emissions, many CFB units will be designed for combustion of coal, which is not readily chemically characterized. While emissions models are rapidly becoming more developed and are generally qualitatively valid, the complexity of the fuel and its reactions must be added to the complexity of the two-phase fluid and particle mechanics. As a consequence, empirical expressions, representing correlations of experimental results, together with pilot plant studies, will likely remain the basis for many commercial guarantees for some time in the future.

While there has been an attempt to generalize CFBC systems in this chapter, in practice they are far from alike. Philosophical differences make for marked visual and operational differences between different units. These include variations in:

- feed point locations and spacing;
- primary zone geometry;
- usage of external heat exchangers;
- heat transfer surface locations.

Combined, these allow for significant potential differences in performance between different units with respect to combustion and emissions results, turndown and fuel flexibility. These ideas, together with greater detail on design and mechanical features, are discussed further in Chapter 11.

References

Abdulally, I.F. and Alkan, I. (1995) Advanced pressurized circulating fluidized bed system: beyond the demonstration stage, in *Proceedings of the 13th International Conference on Fluidized Bed Combustion* (ed. K.H. Heinschel), Volume 1, ASME, New York, pp. 625–636.

Agarwal, P.K. (1986) A single particle model for the evolution and combustion of coal volatiles. *Fuel*, **65**, 803–810.

Ake, T.R., Dixir, V.B. and Mongeon, R. (1993) High performance test results from Riley Stoker's advanced CFB pilot scale facility, in *Proceedings of the 12th International Conference of Fluidized Bed Combustion* (ed. L. Rubow), Volume 1, ASME, New York, pp. 81–92.

Alexander, K.C. and Eckstein T.G. (1994) Babcok and Wilcox Barberton Ohio, Maintenance experience with circulating fluidized-bed boilers. Presented to Competitive Power Congress '94, Philadelphia, PA, June 1994.

Ahlstrom Pyropower (1992) Activation/reuse of fluidized bed waste; Phase I – Literature review. Report to the Canadian Electrical Association, CEA Project 9131 G 891.

Åmand, L.-E. (1994) Nitrous oxide emission from circulating fluidized bed combustion. PhD Dissertation, Chalmers University, Götheborg, Sweden.

Åmand, L.-E. and Andersson, S. (1989) Emissions of nitrous oxide from fluidized bed boilers, in *Proceedings of the 10th International Conference on Fluidized Bed Combustion* (ed. A. Manaker), ASME, New York, pp. 49–56.

Anderson, B.M. and Wilbourn, R.G. (1989) *Contaminated Soil Remediation by Circulating Bed Combustion, Demonstration Test Results*. Ogden Environmental Services Inc., San Diego, California, November 1989.

Anthony, D.B., Howard, J.B., Hottel, H.C. and Meissner, H.P. (1975) Rapid devolatilization of pulverized coal, in *Proceedings of the 15th International Symposium on Combustion*, Combustion Institute, Pittsburgh, pp. 1303–1316.

Anthony, E.J. and Preto, F. (1995) Pressurized combustion in FBC systems, in *Pressurised Fluid Bed Combustion* (eds M. Alvarez Cuenca and E.J. Anthony), Blackie, Glasgow, pp. 80–120.

Basu, P. and Fraser, S.A. (1991), *Circulating Fluidized Bed Boilers – Design and Operation*, Butterworth-Heinemann, Boston.

Basu, P. and Halder, P.K. (1989) Combustion of single coal particles in a fast fluidized bed of fine solids. *Fuel*, **68**, 1056–1063.

Belin, F., James, D.E., James, D.J., Walker, D.J. and Warrick, R.J. (1988) Waste wood combustion in circulating fluidized bed boilers, in *Circulating Fluidized Bed Technology II* (eds P. Basu and J.F. Large), Pergamon Press, Oxford, pp. 351–368.

Bernstein, N., Goidich, S., Li, S. and Phalen, J. (1995) Influence of fuel type and steam cycle on CFB boiler configuration, in *Proceedings of the 13th International Conference on Fluidized Bed Combustion* (ed. K.J. Heinschel), Volume 1, ASME, New York, pp. 27–37.

Bland, A., Georgiou, D.N., Ashbaugh, M.B., Brown, T.H., Young, L.-J. and Wheeldon, J. (1995) Use potential of ash from circulating pressurized fluidized bed combustors using low-sulphur subbituminous coal, in *Proceedings of the 13th International Conference on Fluidized Bed Combustion* (ed. K.J. Heinschel), Volume 2, ASME, New York, pp. 1229–1242.

Brereton, C. (1987) Fluid mechanics of high velocity fluidized beds, Ph.D Dissertation, University of British Columbia, Vancouver, Canada.

Brereton, C., Grace, J.R., Lim, C.J., Zhu, J., Legros, R., Muir, J.R., Zhao, J., Senior, R.C., Luckos, A., Inumaru, N., Zhang, J. and Hwang, I. (1991) Environmental aspects, control and scale-up of circulating fluidized bed combustion for application in western Canada. Final Report to EMR Canada under Contract 52SS.23440-8-9243, December 1991.

Boemer, A., Braun, A. and Renz, U. (1993) Emission of N_2O from four different large scale circulating fluidized bed combustors, in *Proceedings of the 12th International Conference on Fluidized Bed Combustion* (ed. L. Rubow), Volume 1, ASME, New York, pp. 585–598.

Castleman III, J.M. (1985) Process performance of the TVA 20-MW atmospheric fluidized-bed combustion (AFBC) pilot plant, in *Proceedings of the Eighth International Conference on Fluidized-Bed Combustion*, Volume 1, ASME, New York, pp. 196–207.

Chang, D.P.Y., Sorbo, N.W., Murchison, G.S., Adrian, R.C. and Simeroth, D.C. (1987) Evaluation of a pilot scale circulating bed combustor as a potential hazardous waste incinerator, *JAPCA*, **37**, 266–274.

Chelian, P.K. and Gamble, R. (1995) Combustion of fuel with high fines in Ahlstrom Pyroflow CFB boilers, in *Proceedings of the 13th International Conference on Fluidized Bed Combustion* (ed. K.J. Heinschel), Volume 1, ASME, New York, pp. 535–550.

Couturier, M.F., Doucette, B. and Poolpol, S. (1989) A study of gas concentration, solids loading and temperature profiles within the Chatham CFB combustor. Report to Energy Mines and Resources Canada, November, 1989.

Couturier, M.F., Marquis, D.L., Steward, F.R. and Volmerange, Y. (1994) Reactivation of partially-sulphated limestone particles from a CFB combustor by hydration. *Canadian Journal of Chemical Engineering,* **72**, 91–97.

Davies, N.H., Hayhurst, A.N. and Laughlin, K.M. (1994) The oxidation of calcium sulphide at the temperatures of fluidized bed combustors, in *Twenty Fifth International Symposium on Combustion,* Combustion Institute, Pittsburgh, pp. 211–218.

Daw, C.S., Rowley, D.R., Perna, M.A., Stallings, J.W. and Divilio, R.J. (1991) Characterization of fuels for atmospheric fluidized bed combustion, in *Proceedings of the 11th International Conference on Fluidized Bed Combustion* (ed. E.J. Anthony), Volume 1, ASME, New York, pp. 157–165.

Dennis, J.S. and Hayhurst, A.N. (1990) Mechanism of the sulphation of calcined particles in combustion gases. *Chem. Eng. Sci.,* **45**, 1175–1187.

Desai, D., Anthony, E.J., Mourot, P.L. and Sterling, S.A. (1995) Experimental study of CFBC technology for decontamination of highly chlorinated PCB wastes, in *Proceedings of the 13th International Conference on Fluidized Bed Combustion* (ed. K.J. Heinschel), ASME, New York, pp. 65–74.

Elkins, J.W. (1989) State of the research for atmospheric nitrous oxide (N_2O). Paper contributed to the Intergovernmental Panel on Climate Change, Boulder, Colorado, December 1989.

Environment Canada (1986) The national incinerator testing and evaluation program: Air pollution control technology. Environment Canada, Report EPS 3/UP/2.

Ford, N.W.J. and Sage, P.W. (1991) The characterization of limestone for uses as SO_2 sorbents in coal combustion, in *Proceedings of the Institute of Energy's Fifth International Fluidized Combustion Conference,* pp. 159–168, London, UK.

Furusawa, T. and Shimizu, T. (1988) Analysis of circulating fluidized bed combustion technology and scope for future development, in *Circulating Fluidized Bed Technology II* (eds P. Basu and J.F. Large), pp. 51–62, Pergamon Press, Oxford.

Gefken, J. and Huber, D. (1993) A programmatic look at the role of fluidized-bed technology in the clean coal technology program, in *Proceedings of the 12th International Conference on Fluidized Bed Combustion* (ed. L. Rubow), Volume 1, ASME, New York, pp. 327–333.

Gibbs, B.M., and Hedley A.B. (1978) in *Fluidization* (eds J.F. Davidson and D.L. Keairns), pp. 235–240, Cambridge University Press.

Gustavsson, L. and Leckner, B. (1995) Abatement of N_2O emissions from circulating fluidized bed combustion through afterburning. *Ind. Eng. Chem. Res.,* **34**, 1419–1427.

Hamor, R.J., Smith, I.W. and Tyler, R.J. (1973) Kinetics of combustion of a pulverised brown coal char between 630 and 2200 K. *Combustion and Flame,* **21**, 153–162.

Hansen, P.F.B., Dam-Johansen, K., Bank, L.H. and Ostergard, K. (1991) Sulphur retention on limestone under fluidized bed combustion conditions – an experimental study, in *Proceedings of the 11th International Conference on Fluidized Bed Combustion* (ed. E.J. Anthony), ASME, New York, pp. 73–82.

Hippinen, I., Lu, Y., Jahkola, A.J., Laaitainen, J. and Nieminen, M. (1991) Gas emission from pressurized fluidized bed combustion of solid fuels, in *Proceedings of the 11th International Conference on Fluidized Bed Combustion* (ed. E.J. Anthony), ASME, New York, pp. 281–291.

Holley, E.P., Lewnard, J.J., von Wedel, G., Richardson, K.W. and Morehead, H.T. (1995) Four Rivers second generation pressurized circulating fluidized bed combustion project, in *Proceedings of the 13th International Conference on Fluidized Bed Combustion* (ed. K.J. Heinschel), Volume 2, ASME, New York, pp. 919–924.

Howard, J.R. (1989) *Fluidized Bed Technology,* Adam Hilger, Bristol, UK.

Iisa, K., Tullin, C. and Hupa, M. (1991) Heterogeneous formation and destruction of nitrous oxide under fluidized bed combustion conditions, in *Proceedings of the 12th International Conference on Fluidized Bed Combustion* (ed. E.J. Anthony), ASME, New York, pp. 83–90.

Jensen, D.D. and Young, D.T. (1986) *PCB-Contaminated Soil Treatment in a Transportable Circulating Bed Combustor,* Ogden Environmental Services Inc., San Diego, California.

Johns, R.F. and Washcer, R.E. (1987) Conversion of forest waste into energy – the West Enfield project. Presented to ASME Winter Meeting, Boston, MA, December 1987.

Johnsson, J.E. (1989) A kinetic model for NO_x formation in fluidized bed combustion, in *Proceedings of the 10th International Conference on Fluidized Bed Combustion* (ed. A. Manaker), ASME, New York, pp. 1111–1118.

Johnsson, J.E. (1991) Nitrous oxide formation and destruction in fluidized bed combustion – A literature review of kinetics. Presented at the 23rd IEA-AFBC Meeting in Firenze, November 1991.

Johnsson, J.E. and Dam-Johansen, K. (1994) Reduction of N_2O over char and bed material from CFBC. Abstract from the 6th International Workshop on Nitrous Oxide Emissions, Turku, Finland, June 7th–9th.

Jung, K. and Stanmore, B.R. (1980) Fluidized bed combustion of wet brown coal. *Fuel*, **59**, 74–80.

Kavidass, S., Alexander, K.C., Belin, F. and James, D.E. (1994) Operating Experience with High-Ash Waste Coal in a B&W CFB Boiler. Presented to Power-Gen Asia, Hong Kong, August 1994.

Kobro, H. and Brereton, C. (1985) Control and fuel flexibility of circulating fluidised beds, in *Circulating Fluidized Bed Technology* (ed. P. Basu), Pergamon, Toronto, pp. 263–272.

Kopetz, Jr., E.A., O'Brien, W.B. and Tennant, G.L. (1991) Startup and operating experience of TVA's 160 MW Shawnee atmospheric fluidized bed combustion demonstration plant: a 1991 update, in *Proceedings of the 11th International Conference on Fluidized Bed Combustion* (ed. E.J. Anthony), Volume 1, ASME, New York, pp. 271–278.

Koskinen, J., Lehtonen, P. and Sellakumar, K. (1995) Ultraclean combustion of coal in Ahlstrom Pyroflow PCFB combustors, in *Proceedings of the 13th International Conference on Fluidized Bed Combustion* (ed. K.J. Heinschel), Volume 1, ASME, New York, pp. 369–377.

Kraft, D.L. and Orender, N.C. (1993) Considerations for using sludge as a fuel. *Tappi Journal*, **72**, 139–141.

Kudjoi, A.S., Hippinen, I.T., Lu, Y. and Jahkola, A.K. (1995) Combustion of gasification residues in a pressurised fluidised bed, in *Proceedings of the 13th International Conference on Fluidized Bed Combustion* (ed. K.J. Heinschel), Volume 1, ASME, New York, pp. 117–123.

Kunii, D. and Levenspiel, O. (1991) *Fluidization Engineering*, Second Edition, Butterworth-Heinemann, Stoneham, MA.

Lafanechere, L. and Jestin, L. (1995) Study of circulating fluidized behaviour in order to scale it up to 600 MWe, in *Proceedings of the 13th International Conference on Fluidized Bed Combustion* (ed. K.J. Heinschel), Volume 2, ASME, New York, pp. 971–980.

LaNauze, R.D. (1985) Fundamentals of coal combustion, in *Fluidization* (eds J.F. Davidson, R. Clift and D. Harrison), Academic Press, New York, pp. 631–674.

Leckner, B. and Åmand, L.-E. (1987) Emissions from a circulating and a stationary fluidized bed boiler: A comparison, in *Proceedings of the 9th International Conference on Fluidized Bed Combustion* (ed. J. Mustonen), ASME, New York, pp. 891–897.

Lewnard, J. (1991) Agricultural applications of CFBC ashes from Western coals, in *Proceedings of the 1991 CANMET CFBC Ash Management Seminar*, Halifax, Nova Scotia, pp. 135–152.

Liang, D.T., Anthony, E.J., Loewen, B.K. and Yates, D.J. (1991) Halogen capture by limestone during fluidized bed combustion, in *Proceedings of the 11th International Conference on Fluidized Bed Combustion* (ed. E.J. Anthony), ASME, New York, pp. 917–922.

Lin, W. (1994) Interactions between SO_2 and NO_x emissions in fluidized bed combustion of coal. Ph.D Dissertation, Delft University, The Netherlands.

Louhimo, J.T. (1993) Combustion of pulp and papermill sludges and biomass in BFB, in *Proceedings of the 12th International Conference on Fluidized Bed Combustion* (ed. L. Rubow), Volume 1, ASME, New York, pp. 249–264.

Lucat, P., Morin, J-X., Semedard, J-Cl., Jaud, P., Joos, E. and Masniere, P. (1993) Utility type CFB Boilers: 250 MWe and beyond, in *Proceedings of the 12th International Conference on Fluidized Bed Combustion* (ed. L. Rubow), Volume 1, ASME, New York, pp. 9–15.

Lyngfelt, A. and Leckner, B. (1989a) The effect of reductive decomposition of $CaSO_4$ on sulphur capture in fluidized bed boilers, in *Proceedings of the 10th International Conference on Fluidized Bed Combustion* (ed. A. Manaker), ASME, New York, pp. 675–684.

Lyngfelt, A. and Leckner, B. (1989b) Sulphur capture in fluidized bed combustors: temperature dependence and lime conversion. *J. Inst. Energy*, **62**(450), 62–72.

McInnes, R. and Van Wormer, M.B. (1990) Cleaning up NO_x emissions, *Chemical Engineering Magazine*, September, pp. 131–135.

Melvin, R.H., Friedman, M.A. and Divillo, R.J. (1993) Summary of test program at ACE Cogeneration Company's 108 MWe CFB, in *Proceedings of the 12th International Conference on Fluidized Bed Combustion* (ed. L. Rubow), Volume 1, ASME, New York, pp. 511–520.

Miettinen, H., Stromberg, D. and Linquist, O. (1991) The influence of some oxide and sulphate surfaces on N_2O decomposition, in *Proceedings of the 11th International Conference on Fluidized Bed Combustion* (ed. E.J. Anthony), ASME, New York, pp. 999–1011.

Moritomi, H. (1994) Pressurized fluid bed combustion. Asian Pacific Economic Cooperation (APEC) Clean Coal Technology Training Course, Sydney, Australia, November–December 1994.

Ninomiya, Y., Sato, A. and Watkinson, A.P. (1995) Oxidation of calcium sulfide in fluidized bed combustion/regeneration conditions, in *Proceedings of the 13th International Conference on Fluidized Bed Combustion* (ed. K.J. Heinschel), Volume 2, ASME, New York, pp. 1027–1034.

Park, D., Levenspiel, O. and Fitzgerald, T.H. (1980) A comparison of the plume model with currently used models for atmospheric fluidized bed combustors. *Chem. Eng. Sci.*, **35**, 295–300.

Park, D., Levenspiel, O. and Fitzgerald, T.J. (1981) A model for large scale atmospheric fluidized bed combustors *AIChE Symp. Ser.*, **77**, No. 205, 116–126.

Pillai, K.K. (1981) The influence of coal type on devolatilization and combustion of fluidized beds. *J. Inst. Energy*, **54**, 142–150.

Plass, L., Daradimos, G., Beisswenger, H. and Lienhard, H. (1986) *VGB Kraftwerkstechnic 66*, Vol. 9, pp. 801–807.

Provol, S. and Matousek, W. (1995) A comparison of the Ahlstrom pyroflow pressurized circulating fluidized bed combustor and competing pressurized bubbling bed combustion technologies, in *Proceedings of the 13th International Conference on Fluidized Bed Combustion* (ed. K.J. Heinschel), Volume 1, ASME, New York, pp. 379–390.

Rickman, W.S., Holder, N.D. and Young, D.T. (1985) Circulating bed incineration of hazardous wastes. *Chem. Eng. Progr.*, **81**(3), 34–38.

Riley, K.P., Cleve, K. and Tanca, M. (1995) Large CFB power plant design and operating experience, Texas–New Mexico Power Company 150 MWe (Net) CFB Power Plant, in *Proceedings of the 13th International Conference on Fluidized Bed Combustion* (ed. K.J. Heinschel), Volume 2, ASME, New York, pp. 1501–1504.

Sage, P., Welford, G.B., Brereton, C. and Julien, S. (1995) Development issues for the char combustor component of an integrated partial gasification combined cycle system. *Proceedings of the 13th International Conference on Fluidized Bed Combustion* (ed. K.J. Heinschel), Volume 1, ASME, New York, pp. 49–56.

Schaub, G., Reimert, R. and Albrecht, J. (1989) Investigation of emission rates from large scale CFB combustor plants, in *Proceedings of the 10th International Conference on Fluidized Bed Combustion* (ed. A. Manaker), ASME, New York, pp. 685–691.

Senary, M.K. and Pirkey, J. (1989) Limestone characteristics for AFBC applications, in *Proceedings of the 10th International Conference on Fluidized Bed Combustion* (ed. A. Manaker), ASME, New York, pp. 241–250.

Senior, R.C. (1992) Circulating fluidised bed fluid and particle mechanics; modelling and experimental studies with application to combustion. Ph.D. Dissertation, University of British Columbia, Vancouver, Canada.

Shearer, J.A., Smith, G.W., Moulton, D.S., Smyk, E.B., Myles, K.M., Swift, W.M. and Johnson I. (1980) Hydration process for reactive spent limestone and dolomite sorbents for reuse in fluidized bed coal combustion, in *Proceedings of the Sixth International Conference on Fluidized Bed Combustion*, Atlanta, GA. DOE/METC, Washington DC. pp. 1015–1027,

Song, Y.H., Beer, J.M. and Sarofim, A.F. (1982) Oxidation and devolatilisation of nitrogen in coal char. *Combustion Science and Technology*, **28**, 177–183.

Soud, H. (1991) *Emissions Standards Handbook: Air Pollutant Standards for Coal Fired Plants*, IEA Coal Research publication IEACR/42.

Tang, J.T. and Engstrom, F. (1987) Technical assessment on the Ahlstrom pyroflow circulating and conventional bubbling fluidized bed combustion systems, in *Proceedings of the 9th International Conference on Fluidized Bed Combustion* (ed. J.P. Mustonene), pp. 38–54, ASME, New York.

Tsuo, Y.P., Lee, Y., Rainio, A. and Hyppanen, T. (1995) Three dimensional modelling of N_2O and NO_x emissions from circulating fluidized bed boilers, in *Proceedings of the 13th International Conference on Fluidized Bed Combustion* (ed. K.J. Heinschel), Volume 2, ASME, New York, pp. 1059–1069.

Ulerich, N.H., O'Neill, E.P., Alvin, M.A. and Keairns, D.L. (1979) Criteria for the selection of SO_2 sorbents for atmospheric pressure fluidized-bed combustors. EPRI FP-1307, Electric Power Research Institute, Palo Alto, CA.

Walsh, P.M. (1995) A descriptive model for sulphur capture in atmospheric pressure fluidized bed combustors, in *Proceedings of the 13th International Conference on Fluidized Bed Combustion* (ed. K.J. Heinschel), Volume 1, ASME, New York, pp. 341–350.

Walsh, P.M. and Li, T. (1993) A technique for assessment of char fines formation under fluidized bed combustion conditions, in *Proceedings of the 12th International Conference on Fluidized Bed Combustion* (ed. L. Rubow), Volume 1, ASME, New York, pp. 289–302.

Wang, S.I., Tsao, T.R., Gagliardi, C.R., Herb, B.E., Cox, J. and Parham, D. (1993) Technical challenges in scale up and design of a 250 MW ACFB for the York County Energy Partners Project, in *Proceedings of the 12th International Conference on Fluidized Bed Combustion* (ed. L. Rubow), Volume 1, ASME, New York, pp. 1–8.

Werdermann, C.C. and Werther, J. (1995) Solids flow pattern in an industrial-scale fluidized bed heat exchanger, in *Proceedings of the 13th International Conference on Fluidized Bed Combustion* (ed. K.J. Heinschel), Volume 2, ASME, New York, pp. 985–990.

Werther, J. (1993) Fluid mechanics of large-scale CFB units, in *Circulating Fluidized Bed Technology IV* (ed. A.A. Avidan), AIChE, New York, pp. 1–14.

Wilbourn, R.G., Sterling, S.A. and Vrable, D.L. (1986) Destruction of hazardous refinery wastes by means of circulating bed combustion. Presented at Haztech International Conference, Denver, Colorado.

Wong, P. S-C. (1994) Incineration of industrial organic wastes in a circulating fluidized bed combustor. MASc Thesis, University of British Columbia, Vancouver, Canada.

Yates, J.G. (1983) *Fundamentals of Fluidized-Bed Chemical Processes*, Butterworths, London.

Zhao, J. (1992) Nitrogen oxide emissions from circulating fluidized bed combustion. PhD Dissertation, University of British Columbia, Vancouver, Canada.

Zhang, J.Q. and Jones W.E. (1990) Evaluation of SO_2 and NO_x emission in fluidized bed combustion. Report to CANMET, Energy Mines and Resources Canada, August.

11 Design considerations for CFB boilers
YAM Y. LEE

11.1 Introduction

Since the 1970s, circulating fluidized bed (CFB) technology has been applied to combustion and steam generation. It has become the technology of choice for clean firing of solid fuels. The success of CFB boilers is mainly due to their fuel flexibility and environmental factors. The advantages include high combustion efficiency, low NO_x and SO_2 emissions, and the ability to burn a wide variety of fuels including very low grade fuels. In developed countries, stringent environmental regulations have helped promote the use of CFB technology, while in developing countries, the flexibility of burning a variety of fuels has attracted CFB customers.

Currently, there are about 232 CFB boiler units either under construction or in operation worldwide (excluding small industrial CFB boilers). These circulating fluidized bed combustors (CFBC) units are operating in approximately 20 countries, with additional countries focusing on this technology to solve environmental, waste and fuel problems. Worldwide CFB sales by country are illustrated in Figure 11.1 (Wert, 1993) with the total capacity in excess of 9000 MWe. Simbeck *et al.* (1994) also reported that the overwhelming majority of total steam capacity for fluidized bed combustors (FBC) projects worldwide is in co-generation and utility applications, comprising 55% and 30% of the total capacity, respectively. Small power, district heating, and process steam applications comprise the remaining 15% of the total worldwide capacity equally. The CFB boiler market has been in the 25–250 MW_e range. However, unit capacities into the 250–400 MW_e range, depending on the vendor, are now offered with full commercial guarantees. There are more than ten primary vendors (i.e. developers and licensors) of CFBC technology worldwide. The ranking of CFBC vendors worldwide, indicating type of boiler design sold, is shown in Table 11.1 (Simbeck *et al.*, 1994). In this chapter, the salient features in CFB boiler design and the philosophy of scale-up in CFB boilers are discussed.

The many types of fuels that can be burned effectively and efficiently in a CFB unit are listed in Table 11.2 (Wert, 1994). Almost half of the total worldwide FBC capacity is primarily fueled by bituminous coals. However, other fuels such as coal wastes, petroleum coke, biomass and municipal wastes are

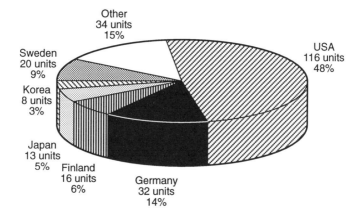

Figure 11.1 Worldwide sales of CFB boilers (Wert, 1993).

gaining in popularity. Table 11.3 describes the properties of the waste fuels including bituminous coal waste, petroleum coke, anthracite coal waste and tire-derived fuel (Makansi, 1994).

In addition to fuel flexibility, CFBC technology has environmental advantages over other boiler designs such as pulverized coal or other combustion systems. CFB boilers inherently produce low NO_x emissions due to lower operating temperatures and staged combustion. For very stringent NO_x emission control, selective non-catalytic reduction (SNCR), such as by injection of ammonia, can be used to reduce NO_x to very low levels as shown in

Table 11.1 Summary of CFBC market by vendor (Simbeck *et al.*, 1994)

Vendor	No. of units sold	
	Atmospheric circulating	Atmospheric hybrid*
Ahlstrom Pyropower	116	0
Tampella Power	21	0
Lurgi	27	0
ABB Combustion Engineering	21	0
Foster Wheeler Energy Corp.	12	0
Deutsche Babcock	0	26
Babcock Enterprise	31	0
Combustion Power Co.	0	12
Babcock & Wilcox	10	0
Kvaemer Generator AB	16	0
Riley Stoker	7	0
Mitsui Engineering & Shipbuilding	4	0
L & C Steinmuller GmbH	3	0
Others	7	2
Total	275	40

* Hybrid design is a combination of bubbling bed and CFB.

Table 11.2 Fuels burnt in CFB combustion systems (Wert, 1993)

Agriculture waste/biomass
Bark
Bitumens and asphaltines
Coals (all grades):
  anthracite
  bituminous
  sub-bituminous
  lignite
  bituminous gob
  anthracite culm
Diatomite
Gases:
  natural
  'off' gases
Gasifier fines
Oil
Oil shale
Peat
Petroleum coke:
  delayed
  fluid
Refuse-derived fuel (RDF)
Residual and waste oils
Sludge
  de-inking
  municipal
  paper mill
Shredded tires
Washery waste and rejects
Wood and woodwaste

Table 11.4, which gives emissions performance data for several CFB units (Makansi, 1994). With staged combustion, the NO_x emission is nominally less than 250 ppm and less than 100 ppm with SNCR. When the NO_x emission requirement is extremely stringent, selective catalytic reduction (SCR) of NO_x may also be used.

Nitrous oxide (N_2O) is another product of the oxidation of fuel-nitrogen during combustion in CFB boilers. The typical values of N_2O emission from

Table 11.3 Comparison of typical waste-fuel analyses (Makansi, 1994)

Component	Bituminous coal waste	Petroleum coke	Anthracite coal waste	Tire-derived fuel (TDF)
Fixed carbon (%)	16.8	83.9	22.0	22.0
Volatile matter (%)	15.2	9.8	6.8	63.1
Ash (%)	61.5	0.7	70.1	12.3
H_2O (%)	6.5	5.6	1.1	2.6
HHV (kJ/kg)	14 700	33 090	7570	32 620
Sulfur (%)	0.8	5.5	0.34	1.3

Table 11.4 Emissions performance for several CFB units (Makansi, 1994)

	NO_x		SO_2		CO		Flyash	
	Permit	Actual	Permit	Actual	Permit	Actual	Permit	Actual
100 MW unit firing low-sulfur coal[1] (kg/MJ)	0.013	0.0043	0.012	0.0021	0.038	0.0074		
Two 160 MW units firing lignite (kg/MJ)		0.021		0.052				5–6% opac.
23 MW unit firing high-sulfur coal (kg/MJ)		≤0.016		≤0.105		≤100 ppm		
48 MW unit firing bituminous coal waste (kg/MJ)		<0.008		<0.08[2]		<0.026		

[1] Includes ammonia injection system for NO_x control.
[2] Calcium-to-sulfur ratio is 1.8 to 2.1.

CFB boilers is about 30–100 ppm (Hiltunen et al., 1991). Lee et al. (1994) summarized the research findings of various researchers on the formation and destruction mechanism of N_2O and reported that the significant factors that affect N_2O emission are fuel type and operating conditions such as the combustion temperature and excess air level. It has been found that N_2O emission increases with excess air level and decreases with an increase in fuel oxygen or volatile content and combustion temperature. For details on N_2O formation and destruction mechanisms and control methods, see Chapter 10.

Sorbents such as limestone are used to capture SO_2 *in situ* in the combustor and eliminate the need for a scrubber downstream of the combustor. For 90% sulfur retention, the required Ca/S molar ratio for feed sorbent is about 0–4, depending on the fuel type and sorbent type. For fuels with a high alkali metal oxide content in the ash, the inherent sulfur capture may be sufficient for the sulfur retention requirement and no addition of limestone is necessary. CO and total gaseous hydrocarbon emission are also low. Typically, in operating commercial boilers, the CO emission has been 20–100 ppm at 850–900°C and the total gaseous hydrocarbon emission is below a few ppm methane equivalent. In addition, due to the relatively low operating temperature and the addition of limestone to the combustor, heavy metals emissions from CFB units are less than from pulverized coal boilers. For some low-grade fuels that contain high amounts of chlorine and fluorine, excess lime from the combustor has also been found to be effective in capturing the HCl and HF emitted during combustion, in the cooled section of the boiler and in the baghouse. With the use of baghouse or electrostatic precipitator at the outlet of the boiler, particulate emissions usually can be controlled to be less than 6.4 mg/MJ.

11.2 Boiler configuration

The main components of a typical CFB boiler are shown in Figure 11.2. These consist of a combustion chamber, a cyclone separator, a return leg and loop seal or L-valve for recirculation of the bed particles (Tang and Engstrom, 1987). The combustion chamber is enclosed with water-cooled tubes and a gas-tight membrane. The lower section of the combustor is covered with refractory, with openings for introducing fuel, limestone, secondary air, recycled ash, one or more gas or oil burners for start-up and bottom ash drains. Most of the combustion occurs in the lower section while heat transfer to the furnace wall is achieved mainly by particle convection and radiation in the upper section of the combustor. The cyclone is refractory-lined and is designed to separate the entrained solids from the hot flue gas and return them through the return leg and loop seal. The loop seal prevents backflow of gas from the riser up the standpipe and has no movable mechanical parts. The gas velocity employed in CFB is usually in the range 4.5 to 6 m/s. Air is fed to the unit as primary air, secondary air, transport air for fuel and limestone feed, air to the loop seal and fluidizing air to the ash classifier. The bottom ash classifier is designed to remove larger bed particles and recycle small particles back to the combustor for improved heat transfer. The operating bed temperature is usually in the range of 850–900°C. This temperature range is chosen to optimize the sulfur capture efficiency of limestone. The coal and limestone feed sizes are

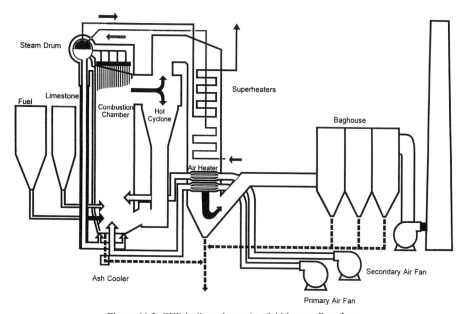

Figure 11.2 CFB boiler schematic of Ahlstrom Pyroflow.

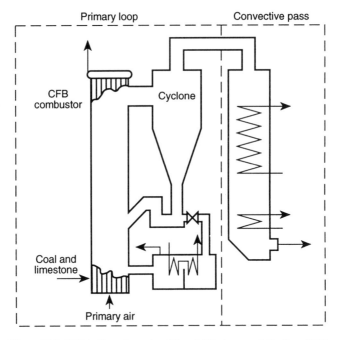

Figure 11.3 CFB boiler schematic of Lurgi (Gottung and Darling, 1989).

typically 6 mm × 0 with mean size of 1–3 mm and 1 mm × 0 with mean size of 0.1–0.3 mm, respectively.

The basic boiler configuration of major CFBC boiler manufacturers such as Ahlstrom, Lurgi and Foster Wheeler are similar, with the main components being the furnace with water-wall, cyclone and return leg, and back pass. The main design differences are in the external heat exchanger, grid design and ash handling systems. The Lurgi design usually features an external heat exchanger, whereas the Foster Wheeler design has an INTREX internal heat exchanger. Figure 11.3 (Gottung and Darling, 1989) and 11.4 (Goidich *et al.*, 1991) illustrate the designs of Lurgi and Foster Wheeler, respectively.

11.3 Combustor design

The combustor design is based on optimizing the combustion efficiency of the fuel and the emission characteristics. The gas velocity and excess air level are normally predetermined, while the combustor height is defined by the operating temperature and the necessary heat transfer surfaces in the combustor. It must also provide enough residence time for combustion of the fuel. CFB boilers normally operate with 20–25% excess air and a superficial velocity of

DESIGN CONSIDERATIONS FOR CFB BOILERS 423

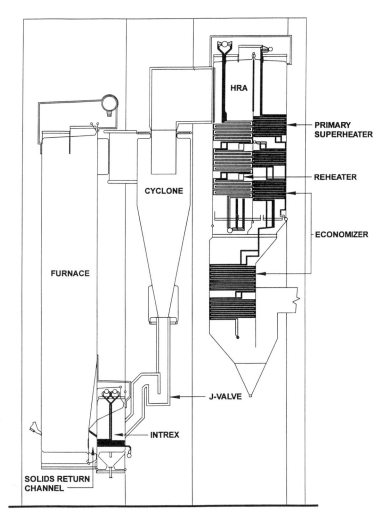

Figure 11.4 CFB boiler schematic of Foster Wheeler (Goidich *et al.*, 1991).

4.5–6 m/s. Steam and feedwater conditions define the split of sensible heat, evaporation and superheat duty. For example, for steam conditions of 103 bars and 513°C, the heat duty for sensible heat, evaporation and superheat is distributed into 30%, 50% and 20% respectively. With consideration of the fuel type, the allocation of heat transfer surfaces in the furnace and in the backpass (or convection pass) are determined. The backpass layout of in-line or over-the-top design is the designer's choice. The in-line design for the backpass is one with the backpass on the same side of the cyclone relative to the combustor as shown in Figure 11.2. On the other hand, the over-the-top design for the backpass is one with the backpass on the opposite side of the cyclone relative to the combustor. Typically, an over-the-top design saves some space but increases the height of the combustor. The boiler furnace is made of insulated, gas-tight, vertical membrane walls with either a top- or bottom-supported structure. When the boiler size becomes sufficiently large, the top-supported structure is normally used. The cyclone and return leg are constructed of steel plate shell with refractory-lining. In order to minimize the height of the combustor for cost control, evaporative surfaces such as wingwalls or omega panels may be used.

The furnace cross-sectional area is determined from the air flow rate, the gas velocity and the fuel type. Typically, the area should be rectangular with the width less than 9 m due to limited secondary air penetration. For example, a 95 MWe CFB burning sub-bituminous coal has an approximate plan dimension of 7 m × 14 m and a height of 33 m. In order to maintain good fluidization in the lower furnace, some designers use a tapered lower section. The taper helps to increase the superficial velocity near the grid, thereby minimizing the chance of segregation and clinker formation (Basu and Fraser, 1991). Some designers use the same cross-section in the lower furnace to maintain a denser bed at the bottom. The lower furnace is refractory-lined to protect it from erosion, with introduction of secondary air and return of solids via a loop seal. The proper sizes and number of openings and their location in this section are basic design parameters.

11.4 Fuel characteristics

One of the key advantages of CFB boilers is fuel flexibility. In order to design properly the boiler and auxiliary equipment for burning various types of fuels, fuel characteristics must first be determined. Proximate, ultimate and heating value analyses are all important. Since the ash and sorbent forms the bed material, the ash composition of the fuel has an impact on maintaining the proper bed inventory for optimum operation. The friability of fuel ash and sorbent affect the split of fly ash and bottom ash, thereby influencing the design of the ash handling equipment. In some cases, where the fuel has a very low ash content or very friable ash, make-up bed material, such as sand, may be needed. Fuel moisture content affects the flue gas volume and leads to an increase in the volume of the boiler for a given operating velocity. Fuel moisture content also affects the handling of the fuels. The fuel nitrogen and sulfur content together with the local emission regulations dictate the emission control strategy. The alkali content (sodium, potassium) and chlorine content of the fuel also need to be determined since they profoundly influence the potential for agglomeration or fouling. In addition, the form of the alkali (organically or inorganically bound) is important. The fuel particle size varies with fuel reactivity and for coal, the top size varies from 6 to 25 mm. Typically, the fuel particle size for more reactive fuels such as sub-bituminous coals and lignite can be bigger than for less reactive fuels such as bituminous coal and anthracite.

11.5 Refractory

Refractory is used in CFB boilers to protect areas susceptible to erosion, to provide insulation and to maintain bed temperature for varying fuel heating values. Due to the high turbulence in the combustor, significant metal wastage tends to occur if the solid particles are allowed to impact on the tubes. Abrasion-resistant refractory with low insulating characteristics is generally used for erosion protection. Refractories can be divided into a number of groups such as preformed (e.g. brick, tile, special shapes), unformed (castables, plastics, gun mixes, etc.) and special materials (e.g. ceramic fiber) (Crowley, 1991). Current designs rely on the combined usage of firebrick and castable refractory for hot face working linings, in conjunction with insulating products for backup linings (Linck and Tietze, 1993). The material options are listed in Table 11.5. The typical hot face lining is 114 to 127 mm thick, whereas the backup lining is about 178 to 190 mm thick. There are two common types of anchoring materials, alloy and ceramic. Alloys commonly used are the 300-series stainless steels and Incoloy 800. They are normally formed in the shape of a 'V' or 'Y' and are connected to a threaded stud welded to the steel shell. For most refractory linings, a minimum stud diameter of 95 mm is required.

The main areas of refractory lining are in the lower combustion chamber, the area where in-furnace heat transfer surfaces penetrate the combustion chamber, flue gas exit to cyclone, cyclone, return leg and loop seal. Proper

Table 11.5 Refractory material options (Linck and Tietze, 1993)

Material	Density (kg/m^3)
Dense	
Firebrick	2200–3000
Conventional castables and gun mixes	2000–3000
Low-cement castables and gun mixes	2300–3100
Ultra-low-cement castables	2400–3000
Phos-bonded plastics	2400–3000
Special medium density	
Conventional castables and gun mixes	1700–2000
Low-cement castables	1860–2300
Insulating – Backup	
Castables and gun mixes	20–40

selection and installation of refractories can increase the service life of linings for CFB boilers. The loss of refractory material can be due to erosion, thermal stresses caused by thermal cycling or thermal shocks and chemical attacks by alkalis. Alkali attack is caused by the reaction of alkalis with alumina silicates in the refractory forming new compounds of higher thermal expansion coefficient. Surface spalling can then occur. Regular inspection with preventive maintenance of the refractories would ensure safe and reliable operation of the boiler.

11.6 Air and solids feed system

Figure 11.5 shows a typical CFB boiler air and flue gas schematic. Air supply to the CFBC boiler is divided into primary and secondary air. Primary air is supplied by a fan through the grid to fluidize the bed material and for

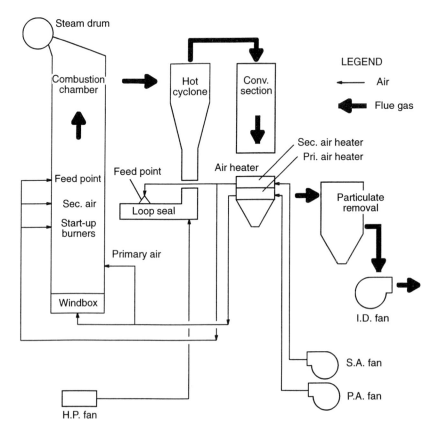

Figure 11.5 Air and flue gas flow schematic of Ahlstrom Pyroflow.

combustion in the lower part of the combustor. The grid is composed of nozzles that have to be properly designed to avoid problems of bed material leakage into the windbox, pluggage or erosion of nozzles. Common nozzle types include cap nozzles, 'pig-tail nozzles' and directional nozzles. For some designers, in order to reduce the impact of change in boiler load on the pressure drop across the grid, a portion of the primary air can be injected above the grid from the combustor walls. Secondary air is supplied by another fan and is injected into the combustor through the side walls in order to create staged combustion leading to lower NO_x emission. The number of secondary air injection nozzles and the injection levels vary among boiler designers. Nonetheless, the primary air to total air split varies from 50%–75% depending on fuel type and boiler design. The removal of flue gas from the combustor is assisted by an induction draft fan system. In addition, high pressure air is used to fluidize the loop seal area and for classifying of particles in the bottom ash drain classifier or ash cooler system.

For start-up of CFB boilers, two types of burners burning oil or gas can be used. A duct burner located in the primary air duct at the entrance to the combustion chamber is used to preheat the incoming air to the windbox. In addition, one or more start-up burners can be installed in the front and the side walls of the combustion chamber to raise the furnace temperature to the ignition temperature of the fuel. The capacity of the start-up burner is normally in the range of 30% to 50% of fuel heat input. When the bed temperature is above the fuel ignition temperature (for coal, approximately 550°C), fuel is fed into the combustor. The fuel feed system typically includes gravimetric feeders, chain conveyors and screw feeders. The types of equipment used for conveying various types of fuels must be chosen carefully to ensure proper operation. Drag chain, belt and screw conveyors present different problems for different fuels (Raskin *et al.*, 1990). In general, belt conveyors are not as maintenance-intensive as drag chain conveyors and are especially suitable for conveying fuels that are particularly abrasive, sticky or potentially corrosive. Fuel is usually fed into the front wall and the loop seal. Feeding into the loop seal has an advantage of covering more area than feeding into the front wall due to better distribution. The number of fuel feed points per unit area is determined from the fuel characteristics and the degree of lateral mixing in the specific design of combustor. Typically, one feed point per 7 to 38 square meters can be used. The rate of fuel feed is automatically controlled in response to the main steam header pressure.

Limestone is usually employed as sorbent for SO_2 removal. The limestone is crushed to a fine powder and is fed pneumatically into the combustor through the front wall, side wall and/or loop seal. The rate of limestone feed is automatically controlled in response to the SO_2 emissions and the required level.

11.7 Separator and return system features

Cyclones are typically used to retain particles in circulation in the CFB combustor. These can be water-cooled, steam-cooled or without cooling. In other CFBC designs, various kinds of impact separators are used such as U-beam (Alexander and Eckstein, 1994) and louver-type separators (Wang et al., 1991). However, the efficiency of these separators varies. The efficiency of the particle separator affects the combustion efficiency and limestone utilization. The particle separator separates entrained particles from the flue gas stream and returns the particles to the combustor. Hence, the solids mean residence time in the combustor is increased by improving the cyclone efficiency. The recycle system (L-valve or J-valve) works on the principle of pressure balance between the solid return leg and the furnace pressure above the solid recycle point (Basu and Fraser, 1991).

Studies have shown that the exit configuration of the combustor has an effect on the internal solids circulation in the combustor (Brereton and Grace, 1993; Zheng and Zhang, 1993; Allison, 1993). For an abrupt exit, a higher percentage of solids reaching the cyclone fall back to the combustor than for a smooth exit. Typically, commercial CFB units have abrupt exits to the cyclone, which facilitates internal solids recirculation leading to longer solids residence times. An extended top section may increase the internal solids recirculation.

Although cyclones have performed well in CFB units, they have been high cost items. In the development of second generation CFB boilers, designers have tried to replace the cyclone. Substitute candidates for high temperature operation can be divided into three main categories, which include separators built inside the furnace, traditional high temperature cyclones made of water/steam-cooled membrane walls and various kinds of particle settling chambers and impact labyrinths designed to collect solids without the use of centrifugal force. For one designer (Gamble et al., 1993), a new kind of centrifugal separator consisting of flat walls has been developed to replace the traditional cyclone. This square separation chamber is equipped with a vortex finder and has a collection hopper underneath, which provides for solids return to the lower combustor. Figure 11.6 shows the new Pyroflow Compact design (Gamble et al., 1993).

11.8 Ash handling system

Final particulate clean-up is achieved by using a baghouse or electrostatic precipitator. Fly ash particles leaving a cyclone are typically less than $100\,\mu$m in diameter and have a mean size of approximately $30\,\mu$m. It has

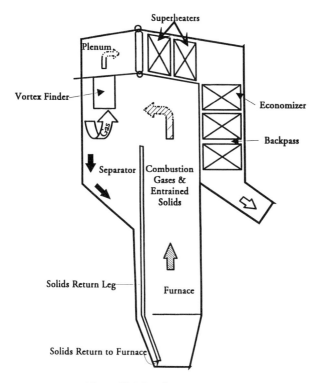

Figure 11.6 Pyroflow compact concept.

been found that the majority of carbon loss is in fly ash particles smaller than 15 μm (Engstrom and Lee, 1990). Captured fly ash particles are pneumatically carried to the ash silos for storage. In some cases, the fly ash stream may be diverted to an ash reinjection system. Ash to be reinjected passes through air locks where it is conveyed to the combustor through a pneumatic transport line. Fly ash reinjection is used in some cases for improved combustion efficiency and limestone utilization.

Bed ash is removed from the bottom ash drain system. In order to maintain a constant bed pressure and good bed quality, large ash particles are removed from the bottom ash drain system. The bottom ash drain system can be in the form of a fluidized bed cooler or ash drain pipe with water-cooled screw conveyor. For fluidized bed coolers, the superficial velocity is used to control the classification of particles. High pressure air is used to classify the material flowing out of the combustor ash drain pipe systems. The hot bed material is fed into a variable-speed water-cooled screw conveyor before discharging to the ash handling system.

For disposal or utilization, fly ash and bottom ash can be removed from the storage silo by either a dry unloading spout or through a twin-paddle mixer ash conditioner. In general, fly ash has a higher unburned carbon and free lime content than bottom ash (Young and Cotton, 1994). For example, the unburned carbon content for coal-fired CFB fly ash is approximately 4–9% whereas the bottom ash is about 0.2–4%. The free lime content as $Ca(OH)_2$ is about 20–40% for fly ash and 10–30% for bottom ash. Landfill is the main ash disposal method. In order to reduce the volume of waste, ash utilization methods for agriculture, sludge stabilization, and cement and concrete production have been developed.

11.9 Control system

Plant control system design for CFB boiler applications is dependent on the system architecture. There are a number of available options (Swartz and Utt, 1990). Distributed Control Systems (DCS) can perform both digital and analog functions, while Programmable Logic Control (PLC) can be used to perform the digital functions with the DCS performing the analog control, display, alarm, and operator interface functions.

Typically, CFB boilers have the ability to change load at approximately 5% maximum capacity rating (MCR)/min or faster. In most applications, the boiler is not the limiting factor in the rate of load change, but the turbine metal temperature is (Lyons, 1993). The boiler can turn down to approximately 30% load without auxiliary fuel. The minimum load without auxiliary fuel is dependent upon maintaining the required adiabatic combustion temperature in the combustor.

11.10 Scale-up

The largest CFB boiler currently in operation is the 165 MWe boiler supplied to Nova Scotia Power Corp. at Pt. Aconi, Nova Scotia. A 250 MW_e CFB boiler is being installed in France, while a 225 MW_e unit is being designed for installation in the United States and two 230 MW units are being designed for installation in Poland. Several CFB boiler suppliers can offer 400 MW_e CFB boilers with full guarantees. Several designers suggest that the 400 MW_e design can be scaled easily to 600 MW_e and beyond. Thus, CFB technology is competing with pulverized coal technology for utility-scale applications. In the scale-up of CFB boilers, there are a number of factors to be considered (Darling and Andrews, 1994) including:

- Heat transfer surfaces
- Air distribution
- Fuel distribution
- Cyclone design

11.10.1 Heat transfer surfaces

In CFB boilers, heat transfer to the furnace wall is mainly accomplished by particle convection and radiation. As the height increases, particle suspension density decreases. Therefore, in scale-up of furnaces, the benefits from increasing the overall height may not compensate for the increase in cost. The combustor height is usually less than 43 m. Designers have selected different approaches to cope with this issue: multiple furnaces with a common heat recovery pass; wingwalls, platens, or partial division walls in the furnace; waterwall surface in the external heat exchanger (EHE); 'pant-leg furnace' design; a full division wall in the furnace; or a combination of these techniques (Makansi, 1993). The wingwalls and platens, also called Omega-surfaces, are placed in the low solid density region of the combustor where erosion is reportedly not a problem. Currently, new designs of heat exchangers utilizing internal or external circulating solids are also being developed by different designers. Utility plants are designed with reheat steam to improve plant efficiency. The reheat steam design operates with approximately 10% better thermal efficiency than a non-reheat design (Wert, 1993). Reheat options for CFB boilers include steam/water-cooled cyclones, external heat exchangers (EHE), tube bundles in the convective backpass section, and/or Omega panels and wingwalls in the upper section of the combustor (Makansi, 1994). For additional material on heat transfer, see Chapter 8.

11.10.2 Air distribution

Furnace depth affects air distribution over the furnace cross-section. Secondary air is distributed from the walls of the lower furnace. The penetration of secondary air is important to prevent a locally fuel-rich region at the center of the furnace leading to poor combustion efficiency and higher emissions. At least one designer (Makansi, 1993) emphasizes that beyond a furnace depth of 9 to 11 m, adequate air penetration cannot be ensured. Therefore, in general, the furnace depth is fixed while the width is increased as boiler size increases. However, above 300 MW_e, a limit on furnace width is reached where plant layout and buckstay design become impractical. Dual grid furnace design, which allows air and fuel to be fed into the furnace center, is used to increase the depth of furnace. Figures 11.7 and 11.8 illustrate the dual grid design and the philosophy of scale-up of one designer (Lyons, 1994). For large boilers, a segmented windbox also helps to provide control of air flow to different parts of the combustor.

11.10.3 Fuel distribution

The number of fuel feedpoints affects the distribution of fuel over the furnace cross-section. Without reasonable fuel feed distribution, both combustion

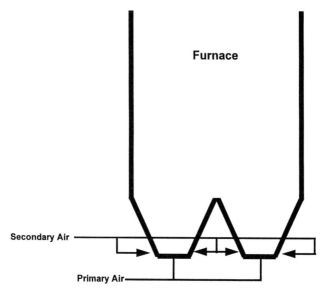

Figure 11.7 Dual grid design (Lyons, 1994).

efficiency and sulfur capture will suffer. The required number of feedpoints depends partly on the characteristics of the fuel. Fuels with high volatile content, such as sub-bituminous coal or lignite, burn quickly on introduction into the combustor. To avoid a fuel-rich region and the resulting reduced performance, such fuels require more feedpoints than low-volatile fuels such as bituminous coal or petroleum coke.

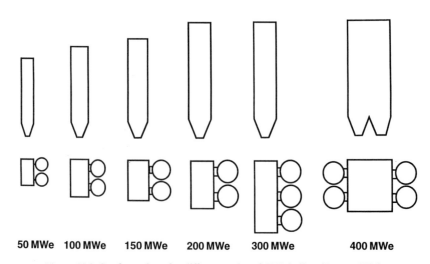

Figure 11.8 Configurations for different scales of CFB boilers (Lyons, 1994).

11.10.4 Cyclone design

The cyclone diameter can affect cyclone efficiency, which in turn influences solids recirculation rates, furnace solids density and furnace temperature. Classical cyclone design theory suggests that as size increases, the efficiency of solids separation decreases. However, the inlet solids loading in CFB cyclones is so high that significant particle interaction occurs, with larger particles causing smaller particles to be carried to the cyclone wall (Darling and Andrews, 1994). Cyclone efficiency appears not to deteriorate for diameters up to about 8 m and the cyclone is limited more by concern for mechanical design than collection efficiency (Makansi, 1993). Multiple cyclones are still used in all designs. Other considerations include the solids return from the cyclone. Different approaches include a split loop seal design providing two solids return ports for each cyclone for designs without EHEs. When cyclones discharge into EHEs, in one design, the solids leaving the EHE pass into a solids-return channel. The solids then re-enter the combustion chamber through multiple openings in the furnace rear waterwalls. Cyclone design is considered in detail in Chapter 6.

11.10.5 Research needs for scale-up

Pilot plant test burns have been used to predict performance of commercial units reasonably well. Nonetheless, due to the small cross-section of pilot plants, solids and gas mixing are more efficient than in a commercial unit. It is important to determine the impact of gas and solid mixing on boiler performance as the boiler size increases. In order to improve the quantitative prediction of boiler performance for scale-up, mathematical modeling work, together with gathering of bench scale, pilot plant and commercial scale test data, need to be developed. Mathematical modeling of CFB units of various areas have been developed. In order to investigate solids and gas mixing phenomena in the CFB, three-dimensional modeling is needed (Hyppanen et al., 1991; Tsuo et al., 1995). For details of fluid mechanics modeling and reactor modeling development, see Chapters 5 and 15. The use of scale modeling for predicting the behavior of commercial units has also been developed (Glicksman, 1988; Horio et al., 1989). The governing parameters of the scaling laws are determined from the non-dimensionalization of the two-component flow equation. Experimental verification of the scaling laws is in progress (Westphalen and Glicksman, 1993). Chaos theory has also been applied to study the fluidization characteristics of fluidized beds (Daw and Halow, 1991; Van Den Bleek and Schouten, 1993). On the other hand, dynamic simulation of a CFB boiler is essential for optimum control of boiler operation during start-up and load change (Hyppanen et al., 1993). Another means of obtaining quantitative predictions of important scale-up parameters is by developing empirical correlations based on data collected from bench scale/pilot scale cold or hot model testing. Neural networks can also be utilized to analyze the data and build non-linear, data-based models. However, the usefulness of these correlations or models need to be verified with commercial unit data. Parameters that are of great interest for scale-up include:

- vertical and horizontal solids concentration profiles
- local solids mass flows
- solids circulation rates
- thickness of the wall downflow zone
- gas mixing coefficient
- solids mixing coefficient
- penetration depth of secondary air

Diagnostic tools have been developed to measure the above parameters in cold model and bench scale or pilot plant CFB units. However, measurement of these parameters in commercial units is difficult so that further development is needed. For details on instrumentation, see Chapter 9.

11.11 Pressurized circulating fluidized bed (PCFB) technology

The success of atmospheric CFBC technology has led to the development of pressurized CFBC technology for higher plant efficiency and lower pollutant emissions. PCFB technology is an attractive option for meeting the needs of both new plant installations and for repowering of existing stations during the next decade. Economic analysis has shown that the cost to repower with the PCFB technology is expected to be much lower than Integrated Coal Gasification Combined Cycle (IGCC) and Pressurized Bubbling Fluidized Bed (PBFB) technologies (Green et al., 1994; Provol and Matousek, 1995).

The PCFB process utilizes a combined cycle that employs a combination of a gas turbine and steam turbine to generate power. For one designer, the gas turbine provides 20–25% of the net power station output and the steam turbine provides the remaining 75–80% of the plant power output. A simplified process flow diagram of the Ahlstrom Pyroflow PCFB process and its subsystems is shown in Figure 11.9. A paste mixture of coal, limestone and water is fed to the PCFB combustor. Air from the gas turbine's compressor enters the boiler through the grid and through secondary air injection points. The gas velocity in the combustor is about 4–5 m/s at a pressure of 10–17 bar (Sellakumar and Lamar, 1994). The combustion temperature is 870–900°C. Heat generated during the combustion of coal is transferred to the combustor waterwalls, and to superheater and reheater heat transfer surfaces located in the middle and upper sections of the combustor. The PCFB boiler operates on the same principles that have been proven with atmospheric CFB boilers. The dust-laden flue gas leaving the PCFB hot

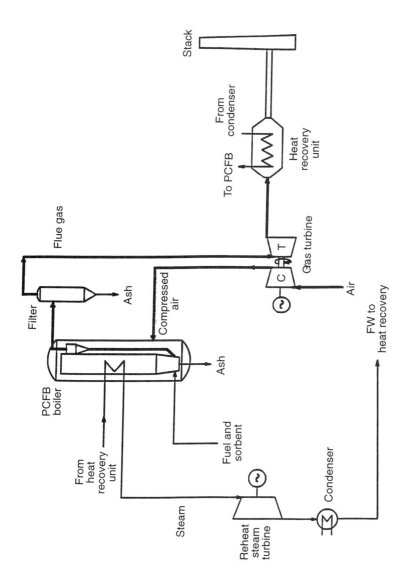

Figure 11.9 Simplified process flow diagram of the Ahlstrom Pyroflow PCFB.

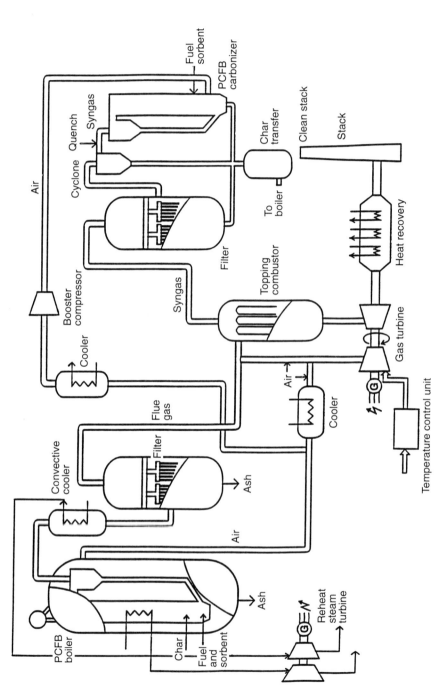

Figure 11.10 Simplified process flow diagram of the Ahlstrom Pyroflow topping-cycle PCFB.

cyclone is cleaned by hot ceramic filter to meet the requirements of the gas turbine as well as the environmental requirements. The gas from the filter enters the expander section of the gas turbine and the energy generated by the expansion of the gas is used to drive the compressor and to generate electric power. For one designer, a 10 MWth PCFB pilot plant was built in Finland and commissioned in mid-1989 (Isaksson *et al.*, 1990). The test results obtained from the pilot plant testing of different coals and limestones have successfully demonstrated that the PCFB technology can provide extremely low emissions meeting current and future needs (Engstrom *et al.*, 1993; Koskinen *et al.*, 1995). One of the key components of the PCFB technology is the high temperature, high pressure ceramic filter system for removal of particulates in the flue gas before entering the gas turbine. Various ceramic candle and cross-flow filters have been tested. Major progress in filter development has been achieved in recent years (Lippert *et al.*, 1995).

A second generation or topping cycle PCFB technology is also under development by Ahlstrom Pyropower and Foster Wheeler (Provol and Ambrose, 1993; Robertson and Van Hook, 1993; Conn *et al.*, 1995). Figure 11.10 shows a simplified process flow diagram of the Ahlstrom Pyroflow topping cycle PCFB technology. This advanced PCFB technology includes all the components of the first generation PCFB technology but also includes a carbonizer or pyrolyzer to partially devolatilize the coal thereby producing a synthetic gas. The synthetic gas is used to fire a topping combustor and raise the gas turbine inlet temperature to between 1100 and 1300°C. In this advanced PCFB technology, the air from the compressor is split by control valves and fed to the combustor, carbonizer and the topping combustor. The flue gas from the combustor after passing through the ceramic filter system is mixed with the air from the compressor and is used as the source of hot combustion air for the topping combustor. The hot synthetic gas from the carbonizer is cooled to about 600–650°C and then cleaned by passing through the ceramic filter before entering the topping combustor. Char collected from the cyclone of the carbonizer and the ceramic filter for cleaning the synthetic gas are transferred to the PCFB combustor for complete combustion. The net efficiencies (on lower heating value (LHV) basis) for the first and second generation PCFB technologies are in the range of 40–43% and 46–48%, respectively (Sellakumar and Lamar, 1995).

11.12 Summary

Circulating fluidized bed boilers have become proven, reliable and cost effective commercial technology for both industrial steam use and industrial/utility power generation. Despite the good performance of the technology, designers have continued to optimize and improve the design of CFBC units. A new generation of CFB boilers is gradually entering the market. At the same time, the CFB boiler market is growing worldwide. The expanding market for both new fuel applications and scale-up to utility scale boilers promises a good future for the application of CFB technology in combustion. Current, CFB boilers of 400 MW_e capacity are being offered by several designers. It is anticipated that the technology can be easily scaled to 600 MW_e. Concurrently, pressurized CFB technology is being developed for improved cycle efficiency and even lower gaseous emissions.

References

Alexander, K.C. and Eckstein, T.G. (1994) Maintenance experience with circulating fluidized bed boilers. Presented at Competitive Power Congress '94.

Basu, P. and Fraser, S.A. (1991) *Circulating Fluidized Bed Boilers: Design and Operations*. Butterworth-Heinemann, Stoneham, MA.

Brereton, C.M.H. and Grace, J.R. (1993) End effects in circulating fluidized bed hydrodynamics, in *Proceedings of the 4th International Conference on Circulating Fluidized Beds* (ed. A.A. Avidan), AIChE, New York, pp. 137–144.

Conn, R., Van Hook, J., Robertson, A. and Bonk, D. (1995) Performance of a second-generation PFB pilot plant combustor, in *Proceedings of the 13th International Conference on Fluidized Bed Combustion* (ed. K.J. Heinschel), AIChE, New York, pp. 75–89.

Crowley, M.S. (1991) *Guidelines for Using Refractories in Circulating Fluidized-Bed Combustors*. Electric Power Research Institute, Palo Alto, CA.

Darling, S.D. and Andrews, N.W. (1994) Design and experience with utility-scale CFB boilers. Presented at Power-Gen Asia '94 Conference.

Daw, C.S. and Halow, J.S. (1991) Characterization of voidage and pressure signals from fluidized beds using deterministic chaos theory, in *Proceedings of the 11th International Conference on Fluidized Bed Combustion* (ed. E.J. Anthony), ASME, New York., pp. 777–786.

Engstrom, F. and Lee, Y.Y. (1990) Future challenges of circulating fluidized bed combustion technology, in *Circulating Fluidized Bed Technology III* (eds P. Basu, M. Horio and M. Hasatani), Pergamon, Oxford, pp. 15–26.

Engstrom, F., Isaksson, J. and Sellakumar, K.M. (1993) Process performance and demonstration of Ahlstrom Pyroflow® PCFB technology, in *Proceedings of the Power-Gen Euro-93*, Paris, France.

Gamble, R., Hyppänen, T. and Kauranen, T. (1993) Pyroflow Compact – A second generation CFB boiler by Ahlstrom Pyropower, in *Proceedings of the 12th International Conference on Fluidized Bed Combustion* (ed. L.N. Rubow), ASME, New York, pp. 751–760.

Glicksman, L.R. (1988) Scaling relationships for fluidized beds. *Chemical Engineering Science*, **43**, 1419.

Goidich, S.J., McGee, A.R. and Richardson, K. (1991) The Nisco Cogeneration Project 100-MWe circulating fluidized bed reheat steam generator, in *Proceedings of the 11th International Conference on Fluidized Bed Combustion* (ed. E.J. Anthony), ASME, New York, pp. 57–64.

Gottung, E.J. and Darling, S.L. (1989) Design considerations for circulating fluidized bed steam generators, in *Proceedings of the 10th International Conference on Fluidized Bed Combustion* (ed. A.N. Manaker), ASME, New York, pp. 617–623.

Green, G., Lamar, T. and Provol, S. (1994) Performance and cost of Ahlstrom Pyroflow® pressurized CFB technology. Presented at POWER-GEN Americas '94, Orlando, FL.

Hansen, P. and Dam-Johnson, K. (1993) Limestone catalyzed reduction of NO and N_2O under fluidized bed combustion conditions, in *Proceedings of the 12th International Conference on Fluidized Bed Combustion* (ed. L.N. Rubow), ASME, New York, pp. 779–787.

Hiltunen, M., Kilpinen, P., Hupa, M. and Lee, Y.Y. (1991) N_2O emissions from CFB boilers: experimental results and chemical interpretation, in *Proceedings of the 11th International Conference on Fluidized Bed Combustion* (ed. E.J. Anthony), ASME, New York, pp. 687–694.

Horio, M., Ishii, H., Kobukai, Y. and Yamanishi, N. (1989) A scaling law for circulating fluidized beds. *Journal of Chemical Engineering of Japan*, **22**(6).

Hyppänen, T., Lee, Y.Y. and Rainio, A. (1991) A three dimensional model for circulating fluidized bed boilers, in *Proceedings of the 11th International Conference on Fluidized Bed Combustion* (ed. E.J. Anthony), Montreal, pp. 439–448.

Hyppänen, T., Lee, Y.Y., Kettunen, A. and Riiali, J. (1993) Dynamic simulation of a CFB based utility power plant, in *Proceedings of the 12th International Conference on Fluidized Bed Combustion* (ed. L.N. Rubow), ASME, New York, pp. 1121–1127.

Isaksson, J., Sellakumar, K.M. and Provol, S.J. (1990) Development of Pyroflow pressurized circulating fluidized bed boilers for utility applications, in *Circulating Fluidized Bed Technology III* (eds P. Basu, M. Horio and M. Hasatani), Pergamon, Oxford, pp. 439–444.

Koskinen, J., Lehtonen, P. and Sellakumar, K.M. (1995) Ultraclean combustion of coal in Ahlstrom Pyroflow® PCFB combustors, in *Proceedings of the 13th International Conference on Fluidized Bed Combustion* (ed. K.J. Heinschel), ASME, New York, pp. 369–377.

Lee, Y.Y., Tsuo, Y. and Khan, T. (1994) Significant factors influencing N_2O emissions from circulating fluidized bed boilers, in *Proceedings of the 6th International Workshop of Nitrous Oxide Emissions* (eds M. Hupa and J. Matinlinna), Abo Akademi, pp. 43–57.

Linck, F.E. and Tietze, W.M. (1993) A review of designs that minimize refractory concerns in circulating fluidized bed boilers. *Proceedings of the 12th International Conference on Fluidized Bed Combustion* (ed. L. Rubow), ASME, New York.

Lippert, T., Alvin, M.A., Bruck, G., Isaksson, J., Dennis, R. and Brown R. (1995) Testing of the Westinghouse hot gas filter at Ahlstrom Pyropower Corporation, *Proceedings of the 13th International Conference on Fluidized Bed Combustion* (ed. K.J. Heinschel), ASME, New York, pp. 251–261.

Lyons, C. (1993) Current comparison of Coal Fired technologies to meet today's environmental requirements. Presented at Power Gen Americas '93.

Lyons, C. (1994) New developments in fluidized bed boiler technology. Presented at Competitive Power Congress '94.

Makansi, J. (1993) Can fluid-bed take on P-C units in the 250- to 400-MW range? *Power*, September, 45–50.

Makansi, J. (1994) Circulating fluidized-bed boiler technology for Asia. *Electric Power International*, McGraw-Hill, New York.

Provol, S.J. and Ambrose, S. (1993) The Midwest Power DMEC-2 advanced PCFB demonstration project Ahlstrom pyroflow advanced pressurized CFB technology, in *Proceedings of the 12th International Conference on Fluidized Bed Combustion* (ed. L.N. Rubow), San Diego, pp. 937–948.

Provol, S. and Matousek, W. (1995) A comparison of the Ahlstrom Pyroflow® pressurized circulating fluidized bed and competing pressurized bubbling fluidized bed combustion technologies, in *Proceedings of the 13th International Conference on Fluidized Bed Combustion* (ed. K.J. Heinschel), ASME, New York, pp. 379–390.

Raskin, N. and Van Winkle, K. (1990) Fuel feed systems for fluidized bed boilers. Presented at CIBO Fluid Bed IV.

Robertson, A. and Van Hook, J. (1993) Circulating pressurized fluidized bed pilot plant, in *Circulating Fluidized Bed Technology IV* (ed. A. Avidan), AIChE, New York, pp. 235–239.

Sellakumar, K.M. and Lamar, T. (1995) Application of pressurized circulating fluidized bed technology for combined cycle power generation. *Heat Recovery Systems and CHP*, **15**, 163–170.

Simbeck, D.R., Johnson, H.E. and Wilhelm, D.J. (1994) The fluid bed market: status, trends and outlook. Presented at CIBO Fluid Bed X.

Swartz, M. and Utt, J. (1990) Optimization of circulating fluidized bed boiler operation through distributed control system design management. Presented at 1990 Joint Power Generation Conference.

Tang, J. and Engstrom, F. (1987) Technical Assessment on the Ahlstrom Pyroflow® CFB and Conventional Bubbling Fluidized Bed Combustion Systems, in *Proceedings of the 9th International Conference on Fluidized Bed Combustion* (ed. J.P. Mustonen), ASME, New York, pp. 38–54.

Tsuo, Y.P., Lee, Y.Y., Rainio, A. and Hyppänen, T. (1995) Three dimensional modeling of N_2O and NO_x emissions from circulating fluidized bed boilers, in *Proceedings of the 13th International Conference on Fluidized Bed Combustion* (ed. E.J. Anthony), ASME, New York, pp. 1059–1069.

Van Den Bleek, C.M. and Schouten, J.C. (1993) Can deterministic chaos create order in fluidized-bed scale-up? *Chem. Eng. Sci.,* **48**, 2367–2373.

Wang, T., Yan, G., Zhang, Z., Pan, Z., Jiang, Z., Jiang, H. and Tang, W. (1991) The circulating fluidized bed combustion with staged solid circulation, in *Proceedings of 3rd International Conference on Circulating Fluidized Bed Technology* Volume 3 (eds P. Basu, M. Horio and M. Hasatani), Pergamon, Oxford, pp. 479–484.

Wert, D.A. (1993) Application of fluidized bed combustion for use of low grade waste fuels in power plants. Presented at Power-Gen Europe '93 Conference.

Westphalen, D. and Glicksman, L. (1993) Experimental verification of scaling for a commercial-size CFB combustor, in *Circulating Fluidized Bed Technology IV* (ed. A.A. Avidan), AIChE, New York, pp. 436–441.

Wu, S. and Allison, M. (1993) Cold model testing of the effects of air proportions and reactor outlet geometry on solids behavior in a CFB, in *Proceedings of the 12th International Conference on Fluidized Bed Combustion* (ed. L.N. Rubow), ASME, New York, pp. 1003–1009.

Young, L. and Cotton, Jr., J.L. (1994) Beneficial uses of CFB ash, in *Proceedings of 11th International Pittsburgh Coal Conference* (ed. S.H. Chiang), vol. 1, University of Pittsburgh, pp. 33–38.

Zheng, Q. and Zhang, H. (1993) Experimental study of the effect of exit (and effect) geometric configuration on internal recycling of bed material in CFB combustor, in *Circulating Fluidized Bed Technology IV* (ed. A.A. Avidan), AIChE, New York, pp. 145–151.

12 Applications of CFB technology to gas–solid reactions
RODNEY J. DRY AND COLIN J. BEEBY

12.1 Introduction

The aim of this chapter is to provide the reader with some insight into the procedure for determining whether a circulating fluidized bed (CFB) system is appropriate for a given process. The approach is to examine a number of commercially significant examples (other than combustion, which is dealt with elsewhere in this book) and to explore the features that make the CFB option more attractive than other forms of reactor for gas–solid reactions. The following applications are examined:

CFB gasification of coal
CFB calcination of alumina
CFB roasting of sulfide ores
CFB treatment of hot smelter offgas
CFB pre-reduction of iron ore for direct smelting
Rotary kiln metallization of ilmenite

This list is not intended to be exhaustive – its purpose is merely to illustrate a number of key aspects of the CFB and how these can be put to good use. The inclusion of a non-CFB process at the end of this list is deliberate in that it illustrates the inappropriateness of CFB technology in some circumstances. Other forms of reactor have features that can lead to competitive advantage when used correctly. It is up to the process engineer to keep an open mind.

There is no absolute definition for a CFB. Bubbling fluidized beds with direct dust recycle have some features similar to those of a CFB, and there is a significant grey area between the two. In this chapter a CFB is considered to be the following:

(i) For a single-stage process such as a circulating fluidized bed combustor, a CFB is regarded as a fluidized bed system in which solids circulation is significant from a process point of view. Solids separated in the primary cyclone (or equivalent hot gas–solid separator) are returned to the main bed without substantial heating or cooling.
(ii) For a two-stage process such as a fluidized catalytic cracking (FCC) unit, a CFB is defined as a system in which fluidized solids are transferred between two different chemical environments to meet a primary process

chemistry objective. This usually occurs in conjunction with an energy balance objective.

12.2 Process considerations

Whether a circulating fluidized bed should be used in an arbitrary application involving a gas–solid reaction is unclear. For some processes a CFB will offer significant competitive advantage while for others it may be a retrograde step. In many cases the choice has been driven by non-technical considerations and incomplete understanding. This has led to a number of inappropriate industrial applications of CFB technology and also to a number of systems which (in hindsight) would have been in a stronger position had they utilized a CFB.

The challenge is to understand when a CFB is appropriate. There is no general answer, since each process application is unique and is driven to various degrees by process chemistry, energy management and economic issues. The CFB represents one type of gas–solid contacting device in which a physical or chemical transformation is conducted. As such it is only one component of an overall process. In economic terms it may not be important whether a particular process is carried out in a CFB, a bubbling fluidized bed, a multiple hearth system, a shaft kiln, a rotary kiln or some other form of gas–solid contacting device. In situations such as this the selection process is usually based on perceived simplicity and/or prior experience with one form of process unit. This is often the case for high-value materials such as refined foods and pharmaceuticals where process intensity and energy efficiency have little impact on profitability.

For treatment of bulk commodity materials such as minerals this is not so. For processes in this category it is common for energy efficiency and process intensity to play major roles in establishing economic viability. In this case, selection of the most appropriate form of gas–solid contacting device can make the difference between success and failure. There are very few rules available to guide this decision. The marriage between process chemistry and a gas–solid reactor system, CFB or otherwise, must be assessed on a case-by-case basis taking into account all relevant factors, both technical and non-technical.

The learning process in this context is slow and expensive. The proverbial wheel has been re-invented many times and finding the correct starting point for development of a new process represents a significant challenge. An examination of historically successful examples is helpful in this regard. If a broadly similar problem has been successfully (or unsuccessfully) dealt with in the context of a different process, then there is significant merit in examining this and understanding how it was done. However, it is dangerous to assume that a solution developed in one process environment will work in another. Even if energy management constraints are similar, moving from one CFB process to another could involve significant changes in thermodynamics, particle reactivity, accretion-forming tendencies, etc. When these changes are fully accounted for and a process environment has been moulded to meet these needs, it is likely that the system will differ substantially from its original form. In other words, each process is unique and use of any 'standard' packaged system for different process objectives is likely to lead to sub-optimal solutions.

12.3 CFB gasification of coal

Gasification involves conversion of coal and air (or oxygen) into a gaseous mixture of mainly CO, CO_2, H_2, H_2O, CH_4 and N_2. It is an old technology that has been used for a variety of industrial and commercial purposes. These include fuel gas generation and production of synthesis gas. Gasification at high pressure is becoming increasingly important in the context of integrated gasification combined cycle (IGCC) power generation, due to the promise of high thermal efficiency and low environmental impact. The emergence of fuel-cell technologies for future 'super-efficient' power generation provides a further long-term incentive for gasifier development.

Since much of the cost and many of the problems associated with coal-based advanced power cycles revolve around hot gas cleaning, it is desirable to minimize the volume of dusty gas that such processes generate. Gasification represents a good way of achieving this, since only part of the air (or oxygen) required for ultimate combustion is added at the gasification stage. The rest is added later, after the fuel has been cleaned. By comparison, all the air needed for combustion is added to the bed in a pressurized fluidized bed combustor and the volume of dusty gas produced is significantly greater. Gasification, therefore, represents a good way of minimizing the hot gas cleaning burden.

Coal gasification may be conducted in a variety of ways. At present, CFB or fluidized bed gasification in general is not the main thrust of development. One of the strongest economic leverage points in the IGCC system is carbon conversion. For advanced IGCC power cycles, oxygen-blown lean-phase or entrained-bed gasification at high temperature (1300–1700°C) and pressure (15–25 bar) is favoured. At these temperatures reaction rates are high and good carbon efficiency can be achieved. Ash in the coal melts and is withdrawn in the form of a molten slag. Examples include the Shell, Dow and Texaco gasification processes (Patterson, 1990). A number of large-scale demonstration plants are currently being built or commissioned to prove this technology.

Use of a bubbling or circulating fluidized bed for coal gasification is constrained by the need to operate at temperatures significantly below that at which the ash melts. Gasifiers of this type are usually operated at 800–1000°C. Molten ash, if present, is likely to lead to accretions and ultimately to loss of fluidization. At these temperatures good carbon conversion is difficult to achieve and losses of micron-sized char fines can be excessive. There is some evidence (Rayner, 1952) that the production of char fines arises from the gasification process rather than mechanical attrition. It is likely that higher-reactivity 'binding phases' in the char are consumed first, leaving low-reactivity sub-particles behind. For this reason it is often necessary to provide a secondary high-temperature (slagging) combustion system. With this complication much of the initial incentive to use a fluidized bed for gasification is lost.

A number of bubbling bed systems such as the IGT/Tampella and KRW gasifiers use controlled slagging/agglomeration with selective withdrawal of coarse solids from the bed as a means of achieving high carbon efficiency (Patterson, 1990). Whilst this approach is effective, it is also sensitive to coal ash chemistry and this can lead to plant control difficulties.

A number of technologies for gasifying coal in a fluidized bed have been developed over the years. Some of these, like the Cogas system (Yerushalmi, 1982) developed in the UK in the 1970s, used a circulating burden of char to provide heat for gasification reactions in much the same way that circulating catalyst provides reaction heat in a catalytic cracker. This system, shown in Figure 12.1, utilizes a slagging combustor to achieve high carbon utilization. A 2 T/h pilot unit was operated successfully at Leatherhead, UK for a number of years. This technology has not subsequently achieved widespread commercial application due to the complexity of the system, coupled with poor economics.

Fluidized bed gasification of low-rank (brown) coals differs significantly from fluidized bed gasification of higher-rank (black) coals. High-reactivity char is obtained from most low-rank coals and high carbon utilization is possible at typical fluidized bed gasification temperatures. This may be due to the catalytic effect on gasification of alkali metal species present in many low-rank coals. In IGCC this is significant since volatilization of sodium, which impacts on hot gas cleaning and gas turbine systems, can be minimized at the relatively low temperature.

The Winkler gasifier (Yerushalmi, 1982) is arguably the most successful fluidized bed gasification system. The original 1920s configuration comprised an atmospheric bubbling bed operating at a gas velocity of around 3 m/s. Under these conditions micron-sized char dust is generated and losses from the bed are a major limiting factor. The system was later developed by Rheinbraun (Bocker and Englehard, 1994) into a high-pressure (up to 30 bar) circulating fluidized bed gasifier through addition of a cyclone and a direct fines return system to the bed as illustrated in Figure 12.2. An industrial-scale plant capable of producing 100 000 T/a of synthesis gas was commissioned in 1986 at Berrenrath near Cologne. An advanced pilot

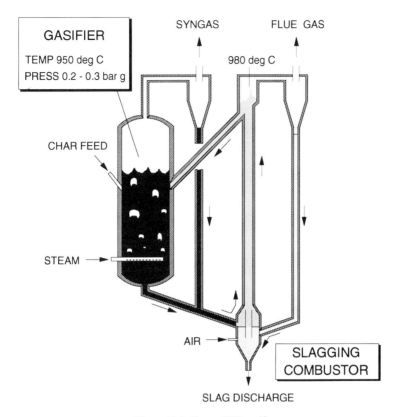

Figure 12.1 Cogas CFB gasifier.

plant for IGCC testing, rated at 6.5 T/h coal feed, was also operated between 1989 and 1992. It is envisaged that this form of technology will be used for commercial lignite-fired IGCC power generation in the near to medium term.

To what extent can the Winkler system be classified as a CFB? What is the feature which makes the CFB a good choice for this application?

The key to a successful fluidized bed gasification system is high carbon efficiency. The upper temperature limit is set by ash stickiness (typically 950–1000°C) and high-reactivity char is therefore desirable. Though the main bed within the Winkler CFB gasifier is not far beyond bubbling (gas velocity around 3 m/s), there is a fines component that passes repeatedly from the bed to the cyclone and back again. The hot circulating fines loop effectively controls the residence-time of char fines in the reaction zone and hence carbon efficiency. In other words, it is the ability of the CFB component of the system to capture and recirculate fines that dominates overall carbon utilization.

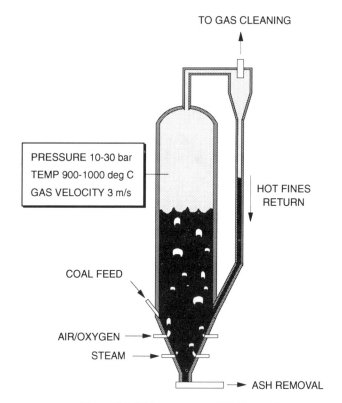

Figure 12.2 High-temperature Winkler gasifier.

Finding methods which enhance the performance of the cyclone and fines return system represents a major technological challenge. If fines retention in this system could be dramatically improved, then higher-rank coals with lower char reactivity could be used with good carbon efficiency. CFB gasification could then become a significant challenger in the race for an efficient IGCC process, rather than a niche technology for low-rank coal.

Cyclone collection efficiencies for fine particles can significantly exceed theoretical limits when some form of in-flight agglomeration takes place. This is known to occur in catalytic cracker regenerator cyclones (King, 1992), and it may be possible to enhance this effect in CFB gasifiers. Improvements of this type can reasonably be expected, and this is likely to shape the future for CFB gasification.

12.4 CFB calcination of alumina

Aluminum is produced by electrolysing alumina, Al_2O_3, in a molten salt bath of cryolite (AlF_3 and NaF) at temperatures around 1000°C. Cryolite is not consumed in the process – its function is to dissolve alumina to allow effective electrolysis. Since the molten salt bath contains fluorine, any atomic hydrogen that finds its way into the smelter leads to production of gaseous HF. This depletes fluorine from the cryolite, and this can upset the chemistry of the bath to the point where alumina dissolution and electrolysis performance are compromised. Emission of HF is also a significant environmental hazard. To manage these effects, there is a need to minimize hydrogen (in all forms) from the alumina feed to the smelter.

Alumina is produced by dissolving bauxite in caustic solution and subsequently recrystallizing gibbsite, $Al(OH)_3$, from solution. It is then calcined at high temperature to yield Al_2O_3. The water removal sequence occurs in stages: the first two water molecules are removed in a short time at around 300°C, while removal of the third molecule requires significantly higher temperatures and residence-times. Commercial calciners operate at temperatures of 900–1150°C to achieve removal of the final water molecule and to convert an appropriate proportion of the alumina to the stable alpha crystalline form. Conversion of at least a proportion to alpha alumina is necessary to minimize rehydration (i.e. to re-form $Al(OH)_3$) when the product is shipped to smelters. Complete conversion is undesirable since this decreases pore volume through recrystallization and renders the alumina a poor adsorber of HF gas. In most modern smelters the incoming alumina is contacted with HF-contaminated air from the pot-room in order to adsorb HF and return it to the electrolysis cell.

Energy efficiency is a key factor in designing a modern alumina calciner. Most current designs incorporate staged countercurrent contacting between gas and solid as illustrated in Figure 12.3. Only 'clean' fuel types such as natural gas can be utilized since, if coal were used, the ash would contaminate the product and compromise the performance of the smelter.

There are currently three main suppliers of fluidized bed alumina calcination technology: Lurgi, Alcoa and FL Smidth. The Lurgi system is the only one that uses a CFB. It was developed by Reh and others in the 1960s (Reh, 1971) and formed the basis for Lurgi's considerable contribution to development of the CFB as a general reactor system for the mineral processing industry.

The Lurgi CFB system, Figure 12.4, has as its centre a CFB in which natural gas is burned at around 1000°C. Apart from the electrostatic precipitator and blowers, the balance of the plant is dedicated to recovery of heat from the two CFB effluent streams: (i) hot flue gas and (ii) hot solids. This system is highly efficient and achieves energy consumption of around 3.0–3.2 GJ/tonne of product.

> To what extent can the Lurgi calciner be classified as a CFB? What is the feature which makes the CFB a good choice for this application?

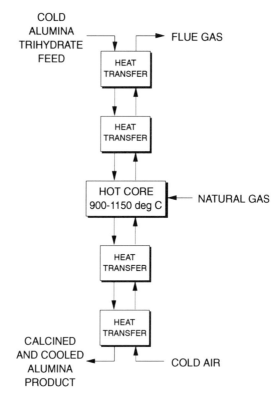

Figure 12.3 Generalized alumina calcination strategy.

The 'hot core' of the calciner circuit is a high-circulation CFB with a loop seal solids return system. The process objective is to combust natural gas in direct contact with hot alumina, then to hold the alumina at temperature for an appropriate period. If the CFB did not recirculate material, solids residence-time at temperature would be too brief to achieve the required level of water removal, causing product quality to deteriorate. Circulation of solids via the CFB cyclone achieves the desired increase in residence-time.

However, this does not explain why a CFB is needed for alumina calcination. Once a particle has been heated to its target temperature of about 1000°C, it merely requires a given residence-time at this temperature for the desired transformations to occur. In a single pass through the CFB riser it is possible to achieve particle heating. It is also feasible to collect hot solids and hold them in a low-velocity bubbling bed for a suitable time. As illustrated in Figure 12.5, this is how the Alcoa fluid flash calciner operates (Fish, 1974). The FL Smidth calciner uses a single-pass heating system with a degree of internal recirculation, Figure 12.5 (Raahauge,

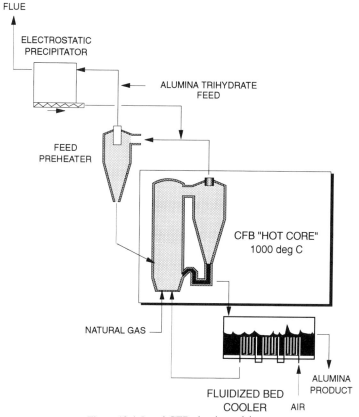

Figure 12.4 Lurgi CFB alumina calciner.

1988). In this case a higher temperature is used to compensate for the lower particle residence-time in the hot zone.

The commercial success of these alternative systems demonstrates that a CFB is not essential for good alumina calcination performance. Alumina calcination therefore does not represent a good example of how a circulating fluidized bed can provide strong technical advantage.

12.5 CFB roasting of sulfide ores

Fluidized bed roasting of metal sulfide ores has been practised for over 50 years. This technology flowed from early developments in fluidized catalytic cracking (late 1930s and early 1940s), when Dorr-Oliver obtained the rights to apply these fluidization methods to non-petrochemical systems (Squires, 1982). Bubbling fluidized bed roasting of ores subsequently became

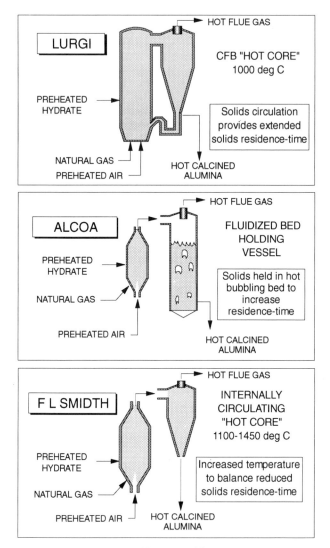

Figure 12.5 Alternate calciner systems.

a standard processing method, and both Dorr-Oliver and Lurgi built numerous plants of this type. The world's largest bubbling fluidized bed roaster is located at Pasminco EZ in Hobart, Tasmania, Australia (Morgan, 1985). This unit roasts 900 T/day of zinc sulfide ore to produce zinc oxide for subsequent leaching. It measures around 16 m in diameter in the freeboard and is considered the limit of how far this technology can reasonably be developed. This size limit is related to the construction of the roof of the vessel. It comprises a large refractory arch built from individual bricks. Any further increase in size would require a complete change in construction method and would involve a significant increase in cost.

The likely first development of CFB sulfide ore roasting occurred in the late 1980s and was led by Lurgi (Pienemann et al., 1992). The application involved roasting a gold-bearing pyrite ore in order to liberate the gold and render it susceptible to cyanide leaching. The main reaction can be summarized as follows:

$$FeS_2 + O_2 \rightarrow Fe_2O_3/Fe_3O_4 + SO_2$$

This reaction is highly exothermic and there is a need to control the ratio of magnetite (Fe_3O_4) to hematite (Fe_2O_3) in the product for optimum gold recovery. It is therefore important that the air-to-sulfur feed ratio be controlled. Process temperature is also critical, with as small a spread as possible required to maximize gold recovery. Target temperature is around 640°C. Feed is introduced as a slurry and the water content is adjusted such that no indirect heat transfer is required. A simple refractory-lined vessel is used.

Bubbling and circulating fluidized bed options were investigated. Despite the fact that commercial CFB sulfide roasting was unproven at the time, the CFB option was selected. This decision was based on a projected advantage in terms of calcine quality, and on additional complexity in the alternative which required multiple bubbling beds due to scale-up limitations. A single CFB was considered more cost-effective, despite the risk. The resulting unit, located at Gidgie in Western Australia, was designed to treat 575 T/day of concentrate (dry basis). This plant was commissioned in 1989 and has performed exceptionally well. A second unit, of similar capacity, was subsequently built at the same location and commissioned in 1990.

> To what extent is this system a true CFB? What is the feature which makes the CFB a good choice for this application?

The Gidgie gold roaster is similar to a low-temperature version of an atmospheric CFB combustion unit (Engstrom and Lee, 1990) in terms of its fluid mechanics. High gas velocities lead to rapid solids mixing, good oxygen distribution and excellent thermal uniformity. Recirculation of solids around the CFB loop further reduces the temperature spread. An ore particle in the CFB system is exposed to a uniform environment in terms of temperature and oxygen potential. This leads to a consistent product with improved gold recovery. In an equivalent bubbling bed a thermal gradient could be established between the bed and the freeboard due to freeboard afterburning. Within the bed a strong oxygen-potential gradient could also exist between gas bubbles and the dense phase. Use of a CFB in preference to a bubbling bed achieves an important process objective in this case. Gold ore roasting is an excellent example of appropriate CFB application.

Similar arguments are likely to apply to other forms of sulfide ore roasting. For zinc sulfide a CFB is likely to provide better tolerance to stickiness through increased turbulence and mixing. Since lead levels in the ore are generally responsible for the formation of sticky phases, a CFB system may be able to accept higher lead contents in the feed. This has significant implications in terms of metal recovery in ore beneficiation circuits and may well provide the impetus for further development.

12.6 CFB treatment of hot smelter offgas

Metallurgical smelting operations often emit hot offgases which contain molten, sticky materials. Examples include smelters for nickel, copper and lead such as Outokumpu (Walker, 1989), Kivcet (Perillo *et al.*, 1989), Isasmelt (Jahanshahi and Player, 1989) and Vanuykov (Bystrov *et al.*, 1992). In some cases the waste heat from these processes has significant value and effective recovery is needed.

The conventional method of dealing with hot, sticky offgases of this type is to pass them into a radiation-cooling chamber which has few internals and water or steam-cooled walls. Molten material can either freeze in-flight, or impact on the cooled walls and freeze in layers. Conventional systems are usually designed to ensure that all molten phases have frozen by the time the gas emerges from the radiant cooling zone. Subsequent offgas cooling is based on conventional waste heat recovery methods for particulate-laden gases, viz. tubular heat exchangers with soot-blowers, etc.

The build-up of frozen material on the walls of the radiant cooler must be dealt with while the plant is in operation. This is usually achieved by some form of rapping (by a hammer system on the outside of the cooled wall), which causes the frozen material on the inside to break loose and fall away. A new layer of frozen material builds up and the cycle is repeated. This type of mechanical rapping occurs more or less continually and is part of normal operation. However, it can stress the wall of the cooler leading to premature failure if the system is not correctly set up and carefully monitored.

The waste heat boiler is often a dominant part of a smelting system in terms of capital cost. Its maintenance requirements can rival those of the smelter itself and this can have a strong negative effect on plant availability and hence economic performance. Advanced smelting technologies are heavily dependent on offgas treatment systems. Hence there is a strong incentive to develop improved offgas handling technologies.

Ahlstrom has pioneered the application of CFB technology to this situation under the name 'Fluxflow' (Hiltunen and Myohanen, 1992). The overall concept, illustrated in Figure 12.6, is to feed the hot offgas containing sticky materials directly into the conical base of a distributor-free CFB. The hot gas meets an array of cool particles in the bed and a rapid thermal quench

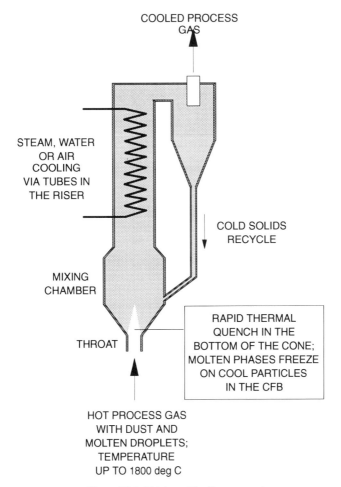

Figure 12.6 Ahlstrom Fluxflow concept.

is achieved. Molten material in the incoming hot gas tends to freeze in-flight or on cold particles in the CFB, rather than on the walls of the system. In this way the need for rapping is minimized (or eliminated) and the 'dirty' wall accretion zone is dramatically reduced in size.

The rapid quench action can also be used to avoid unwanted chemical reactions. For example, in sulfide ore smelting it is desirable to minimize oxidation of SO_2 to SO_3 in the waste heat system. Excessive SO_3 presence can lead to dew-point and corrosion problems. SO_3 formation is solid-catalysed, with the reaction rate typically passing through a peak at around 500–600°C. It is less favoured thermodynamically at typical smelter temperatures (1200–1300°C) and formation is kinetically limited at low temperatures. If the gas is allowed to cool slowly, as in conventional systems,

454   CIRCULATING FLUIDIZED BEDS

then significant SO_3 can form. With the CFB system the time available for SO_3 formation is reduced to the order of milliseconds so the reaction is effectively arrested.

The first commercial application of Fluxflow technology was as a retrofit on a cement kiln at Partek, Finland (Hiltunen and Myohanen, 1992). This kiln had previously been fitted with a cyclone train for energy recovery as shown in Figure 12.7. The problem that arose was build-up of alkalis

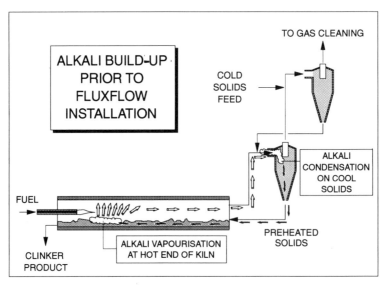

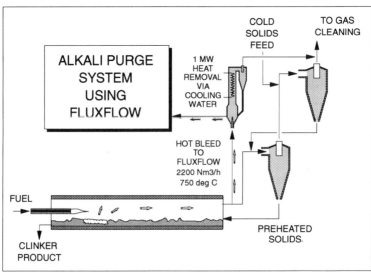

Figure 12.7 Alkali purge using Fluxflow.

(Na_2O and K_2O). These species, even though present at very low levels in the kiln feed, had no escape route from the system and therefore accumulated. They would vaporize at the hot end of the kiln, then migrate to the cold end in vapour form and condense on the incoming solids in the heat recovery cyclones. From here they would return with these solids to the hot end of the kiln to be re-vaporized. The result was a 'heat-pipe' effect, with a flattening of the thermal gradient along the kiln and a loss of overall performance.

The solution was to install a Fluxflow unit as shown in Figure 12.7. A relatively small slip-stream of hot gas (2200 Nm^3/h) was passed from the kiln to the CFB where the quench-cooling effect condensed vapour-phase alkalis. Periodic removal of a portion of the CFB solids inventory is all that is needed for the system to operate continuously. High-efficiency removal of alkalis has been demonstrated and the unit has been operated successfully since 1986.

Currently the world's largest Fluxflow unit is located at Mt Isa, Australia. This unit, illustrated in Figure 12.8, treats around 94 000 Nm^3/h of hot offgas from a 180 000 T/a Isasmelt copper smelter. Minimization of SO_3 formation is a significant advantage in this case. This unit commenced operation in 1992.

> To what extent is it necessary to use a CFB for hot smelter offgas treatment? What is the feature which makes the CFB a good choice for this application?

Fluxflow was originally conceived as a replacement for a conventional waste heat boiler. Low pressure drop is essential, since no change in blower type was envisaged in moving from a conventional system to a Fluxflow unit. To perform the same quench and heat exchange duty at low pressure drop in a bubbling fluidized bed is not possible without a proper gas distributor, which would foul rapidly. In addition, the plan area for such a system would be significantly larger and placement of heat transfer surfaces would pose significant problems.

CFB offgas treatment relies on an ability to feed hot, dirty gas directly into a fluidized bed. There is no gas distributor since fouling would be too rapid. Consequently, when the gas flow stops, any material present in the riser or 'mixing chamber' falls downward through the throat. It is therefore necessary to empty solids from the bed prior to turning the gas off. The easiest way to do this is to stop solids recirculation by collecting solids in the cyclone base and/or return leg system as shown in Figure 12.9. It is the circulating nature of the CFB which makes it possible to start and stop the unit in a practical manner.

The CFB arrangement for treatment of hot smelter offgas containing sticky materials is both natural and critical to success. This is an excellent example of a niche for CFB technology, which provides competitive advantage based on circulating fluidized bed fluid mechanics.

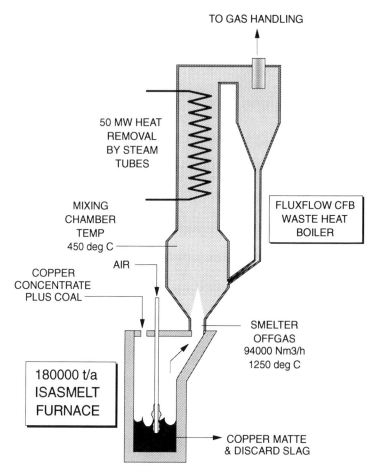

Figure 12.8 Mt Isa copper Fluxflow unit.

12.7 CFB pre-reduction of iron ore for direct smelting

The traditional route for ironmaking is the blast furnace. A gas-permeable bed is required in the blast furnace to allow gas upflow and moving-bed solids downflow. Any fine particles fed to the blast furnace are likely to be blown out with the departing offgas. Material smaller than 4–6 mm is usually classed as 'fines' and excluded from blast furnace feed for this reason.

Iron ore fines are converted into sinter or pellets as a pre-treatment step as shown in Figure 12.10. This involves fusing the material at around 1100–1300°C in a separate unit. Lump ore (larger than 6 mm) can be fed directly into the blast furnace if it has suitable strength characteristics under reduction conditions. However, on average there is significantly

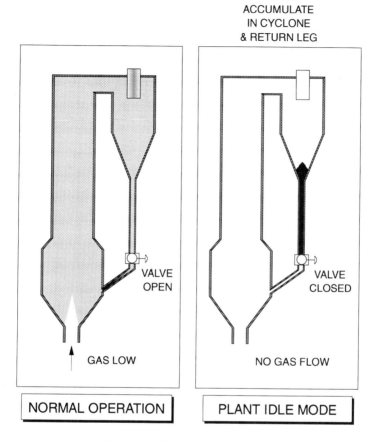

Figure 12.9 Fluxflow start/stop mode.

more fine ore available and a high proportion of blast furnace feed is in the form of sinter or pellet.

The other major feed material is metallurgical coke. Coal cannot be fed directly into the blast furnace since devolatilization in the upper regions would give rise to tar and a range of other undesirable organic compounds. It is also necessary for the reductant to be in lump form and to have enough strength when hot to maintain the gas permeability of the system. Coal is pre-treated in coke ovens for devolatilization and rearrangement of the carbon lattice. It is then fed as lump coke into the blast furnace. Coking coal is carefully selected on the basis of its performance in the blast furnace and is priced at a premium relative to thermal (steaming) coal.

At present there is little incentive for new investment in blast furnace ironmaking. In order to achieve economies of scale it is necessary to install large

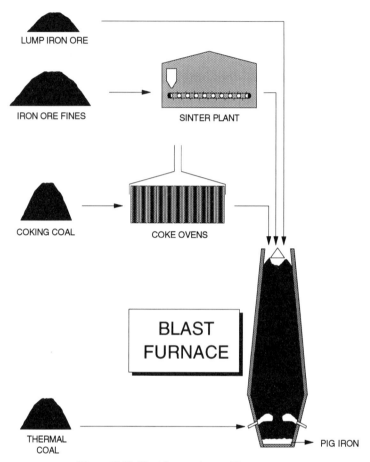

Figure 12.10 Blast furnace ironmaking.

units (typically 2–3 MT/a) and the capital cost per unit of installed capacity is high. Feed materials (coking coal and lump ore) are expensive and the additional cost of sinter or pellet production is significant. Coke ovens are regarded as environmentally undesirable and periodically require costly rebuilding.

The future of blast furnace ironmaking is not in any short-term danger, since installed units will continue to operate as long as they are economically viable. Energy efficiency has been significantly improved by using pulverized thermal coal injection into the bottom of the blast furnace, though the need for coke to keep the burden gas-permeable remains. Oxygen enrichment and pressurized operation have also yielded efficiency benefits and there is every indication that this type of development will continue.

The shortcomings of blast furnace technology (large unit size, expensive feed materials and poor environmental performance) have created an opportunity for a direct bath-smelting alternative. The aim of this alternative is to use iron ore fines and steaming coal directly, with good energy efficiency and environmental performance. In addition, a technology which is viable at capacities of 0.5–1.0 MT/a would be attractive since this offers good market agility and is more in tune with local steel production and consumption patterns in many parts of the world.

The concept of a direct bath-smelting system is shown in Figure 12.11. Thermal coal is fed directly into a molten iron/slag bath at temperatures of 1400 to 1700°C, where it devolatilizes rapidly without formation of tars or other organic compounds. Char dissolves in the melt to act as reductant. Gas leaving the top of the melt is partly combusted to provide process heat and subsequently leaves the vessel. This gas, which contains approximately equal parts of CO and CO_2, is brought into contact with incoming iron ore. Partial reduction takes place according to the reaction:

$$Fe_2O_3 + CO \rightarrow Fe_3O_4/FeO + CO_2$$

at around 800–900°C. Approximately 20–30% of the oxygen in the ore is removed in this way, leading to a coal saving of similar magnitude. With this approach it is possible to compete with the blast furnace in terms of overall coal consumption. This ability to match (or exceed) the energy efficiency of the blast furnace without export gas credits is what distinguishes direct smelting technologies from other less efficient competitors such as Corex (Cusack *et al.*, 1994a).

Though direct smelting technology is not yet commercial, it is considered to have the potential to revolutionize ironmaking in the next century. Two versions, HIsmelt (Cusack *et al.*, 1994b) and DIOS (Inatani, 1991), have reached large pilot-plant status at greater than 50 000 T/a of hot metal production. A third, being developed by the American Iron and Steel Institute (Aukrust and Downing, 1991), is testing the core smelting step (without pre-reduction) at a similar scale. Though the future for this technology is still unclear, it is likely that it will be used commercially early in the twenty-first century. The three current variants are different from one another in process terms and this is likely to translate into different preferred feed materials. The nature of locally available iron ore and coal may have a significant influence on selection of a direct smelting process. All three versions have the potential to establish a significant market share on this basis.

The pre-reduction step is different in each of these technologies:

(i) The American Iron and Steel Institute (AISI) process is based on pelletized feed and aims to use a shaft furnace for pre-reduction.
(ii) The DIOS process is pressurized and oxygen-enriched. Coal is added to the hot offgas to reform it as it leaves the smelting vessel. The aim is to cool it from smelter temperature (1400–1700°C) to pre-reduction temperature (800–900°C) through devolatilization and cracking. This chemically enriches the gas by using a portion of its sensible heat. Gas

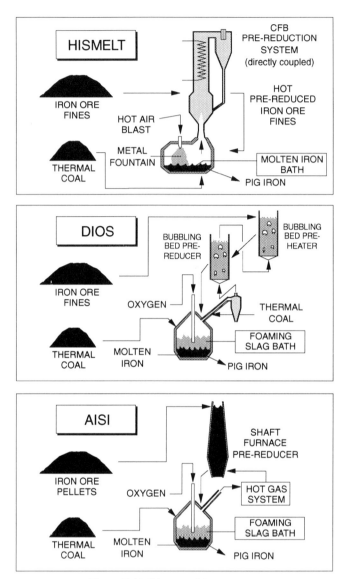

Figure 12.11 Direct smelting processes.

fed to the pre-reduction step is cool enough (800–900°C) to be passed through a cyclone and a gas distributor. A conventional bubbling fluidized bed is used for pre-reduction, and a second bubbling bed is needed for ore pre-heating as depicted in Figure 12.11.

(iii) HIsmelt is an air-based process. Offgas volume is large in comparison with other direct smelting technologies. Though there is scarcely sufficient reduction potential in this gas for pre-reduction, direct coal addition for cooling is not appropriate. For the required amount of cooling (e.g. 1600°C to 850°C) it would be necessary to add a large amount of coal. The offgas would become over-enriched and the residual fuel value in the offgas from the pre-reduction step would be high. This would lead to excessive coal consumption in the overall process. It is therefore necessary to feed hot offgas at 1400–1700°C directly into the pre-reduction unit, with molten/sticky carryover from the smelter present. A CFB directly coupled to the smelting vessel is an appropriate solution for HIsmelt.

The HIsmelt strategy for handling molten, sticky carryover using a CFB is similar to that described in the previous section. HIsmelt has access to Fluxflow technology for this purpose. The HIsmelt system represents a step beyond Fluxflow, with hot quench for accretion control functioning in conjunction with chemical pre-reduction in the CFB. The smelter and the CFB pre-reduction system are closely coupled as illustrated in Figure 12.11.

> To what extent is it necessary to use a CFB for HIsmelt iron ore pre-reduction? What is the feature which makes the CFB a good choice for this application?

There are currently three versions of direct ironmaking technology and, of these, only one uses a CFB for pre-reduction. This suggests that the CFB may be incidental as with CFB alumina calcination. However, there are a number of significant differences in the smelting stage between the HIsmelt, DIOS and AISI processes that influence the choice of a pre-reduction system.

Both DIOS and the AISI process use a foaming slag as the heat transfer medium in the bath. Pre-reduced iron ore and coal are dropped onto a layer of slag and oxygen-enriched air is injected into this layer. Combustion takes place within the slag, as does coal dissolution and reduction. The slag becomes very hot (e.g. 1800–2000°C) in the combustion zone and circulation of hot slag is the means by which heat is returned to the bath. A pool of molten metal is maintained below the slag layer. Due to oxygen enrichment the corresponding offgas volume is small.

HIsmelt is air-blown and uses a molten metal fountain instead of foaming slag as the heat transfer medium. Coal is bottom-injected into an iron bath to create a metal fountain. Hot air (1200°C) is top-injected and partial combustion takes place in the top zone. Metal droplets passing through this hot top zone are heated and transfer energy back to the bath. Since molten metal is easier to contain than molten slag, this approach offers the advantage of reduced refractory wear. It is also less sensitive to coal ash chemistry. However, the offgas volume is high and the hot gas handling problem demanding. Direct coupling into a CFB is seen as an elegant solution for the HIsmelt configuration. The HIsmelt pre-reduction system is an excellent example of problem-solving via CFB technology.

12.8 Rotary kiln metallization of ilmenite

The currently favoured route for production of titanium-based pigment involves hot chlorination in a bubbling fluidized bed. The feed material is rutile, consisting of TiO_2 containing low levels of iron and other impurities. Performance of the final pigment is dramatically downgraded by the presence of impurities. Distillation of the $TiCl_4$ resulting from chlorination is an effective means of achieving the required impurity removal. After purification the $TiCl_4$ is re-oxidized in a specially-designed burner system to produce TiO_2 pigment.

The desired feedstock for chlorination is natural rutile derived from beach sands. However, supplies of this material are limited, and this is reflected in its price. Ilmenite ($FeO.TiO_2$) from beach sands is far more abundant, but is unsuitable as a feedstock without pre-treatment. Each mole of iron in the chlorinator feed leads to production of a mole of iron chloride ($FeCl_3$). This is an undesirable waste material and leads to excessive chlorine consumption. In general, it is desirable to have as little iron as possible present in the feed to a chlorinator.

The price differential between ilmenite and rutile is substantial and this has led to the development of a number of processes for converting ilmenite into synthetic rutile. The Becher process (Palmer, 1992) is the most commonly used these days and this process provides the basis of a world-scale synthetic rutile production industry in Western Australia.

The Becher process involves selective coal-based reduction of the iron component in the ilmenite to metal. This metallization step proceeds according to the reaction:

$$FeO.TiO_2 + C \rightarrow Fe.TiO_2 + CO/CO_2$$

This is followed by a catalysed, aerated leaching step in which metallized iron is effectively 'rusted out' of the matrix. The iron-lean particles can be separated from the precipitated 'rust' by filtration and the result, after some further acid leaching to remove residual iron and other impurities, is synthetic rutile suitable for fluidized bed chlorination.

The ilmenite metallization step is conducted at temperatures of 1100–1200°C in a large rotary kiln. At these temperatures metallized iron would normally be sticky, but in the ilmenite grains the refractory nature of titanium oxides provides a suitable inhibition mechanism as far as sticking is concerned. Free-flowing behaviour is maintained in the kiln.

> This metallization step would seem to be an excellent opportunity for CFB technology. Why is a CFB (or some other form of fluidized bed) not used, and what is the feature of the rotary kiln that makes it more attractive?

In a process such as this, energy efficiency is critical. Strongly reducing conditions are needed to achieve the desired level of metallization. At the same time, a large amount of heat must be supplied to establish the correct operating temperature. This translates into a pair of competing constraints:

(i) Oxygen potential must be as low as possible to achieve high degrees of metallization, and
(ii) The offgas from this process should have as much CO_2 as possible and as little CO as possible to achieve good heat release and hence good energy efficiency.

One of the most attractive features of the CFB is good mixing and near-uniform particle exposure to the gas. In this particular system the process requirement is for something quite different: a segregated arrangement where (a) strongly reducing conditions can be maintained in a solids metallization region, and (b) a highly oxidizing region in which gas-phase combustion of CO can occur for heat release. The heat released by combustion must then be transferred back to the metallization zone without oxidizing the solids.

The solids are contained in a rolling bed in the bottom region of the kiln. Char gasification within this bed provides a 'blanket' of strongly reducing gas which percolates upward and keeps oxidizing gases away from the solids. In the region above the bed air is added and near-complete combustion takes place. Heat transfer to the reducing zone is achieved by a combustion of radiation and heating of the refractory walls of the kiln. The offgas from this system contains a small percentage of carbon monoxide (typically 1%). Most of the offgas fuel value is captured within the kiln and this gives the process good energy efficiency.

This example is one where the features of the rotary kiln are exploited to achieve an efficient process. A single-stage CFB would not be able to meet these process constraints since particle–gas mixing is too great. A segregated dual-stage system (i.e. as per FCC units) would require circulation of some suitable heat-carrying medium at temperatures in excess of 1200°C. Such a system is unlikely to lead to a more efficient process.

The conclusion is that the rotary kiln is preferred for ilmenite metallization. Reduction chemistry and energy management demands are such that a CFB is not easily able to provide a good environment for this process.

12.9 The future

Development of a new process represents an intellectual challenge of the highest order. CFB technology is but one of a number of tools available to the process engineer. 'Getting it right' involves careful screening of alternatives. Decisions of this nature are often subject to pre-conceived notions and political factors within an organization; senior management rarely has the ability or the time to become involved at this level. It is therefore up to the process engineer to 'sell' an assessment of which system is most appropriate. He or she must also be sufficiently knowledgeable in the field to defend this recommendation. This aspect of CFB design is unlikely to change in the short or medium term.

At present there is heavy reliance on simple engineering models, plant observation and 'gut feel'. Though computational fluid dynamic (CFD) models are not yet capable of addressing process design issues in CFB systems at a practical level, this may change over the next 3–5 years. Successful CFD modelling will lead to a major change in design procedures, with improved predictive capability being used to screen options earlier (and more rigorously) in the design cycle. This is likely to improve significantly the quality of decision-making at a technical level.

References

Aukrust, E. and Downing, K. (1991) AISI direct steelmaking program, in *50th Ironmaking Conference*, Washington, AIME, pp. 659–663.

Bocker, D. and Engelhard, J. (1994) R&D efforts for a continuous development of the lignite industry in Germany, in *10th International Conference on Coal Research*, Brisbane, Australia, 9–12 Oct 1994, pp. 121–142.

Bystrov, V.P., Fyodorov, A.N., Komkov, A.A. and Sorokin, M.L. (1992) The use of the Vanuykov process for the smelting of various charges, in *Proceedings International Conference on Extractive Metallurgy of Gold and Base Metals*, Kalgoorlie, Australia, AusIMM, pp. 421–425.

Cusack, B.L., Taylor, I.F. and Hardie, G.J. (1994a) The HIsmelt project – Australia. A coal-based direct smelting process, in *Proceedings 10th International Conference on Coal Research*, Brisbane, Australia, 9–12 October 1994, pp. 71–91.

Cusack, B.L., Wingrove, G.S. and Hardie, G.J. (1994b) Initial operation of the HIsmelt research and development facility, in *Ironmaking 2000*, Mertyle Beach, South Carolina, 2–4 October 1994.

Engstrom, F. and Lee, Y.Y. (1990) Future challenges of circulating fluidized bed combustion technology, in *Circulating Fluidized Bed Technology III* (eds P. Basu, M. Horio and M. Hasatani), Pergamon, Oxford, pp. 15–26.

Fish, W.M. (1974) Alumina calcination in the fluid-flash calciner, in *Light Metals Vol. 3: Alumina, Bauxite and Carbon* (ed. H. Forberg), Metallurgical Society of the American Inst. of Mining, Metallurgical & Petroleum Engineers, pp. 673–682.

Hiltunen, M. and Myohanen, K. (1992) Extremely rapid cooling of hot gas containing sticky components by means of Fluxflow technology, in *Fluidization VII* (eds O.E. Potter and D.J. Nicklin), Engineering Foundation, New York, pp. 841–848.

Inatani, T. (1991) The current status of JSF research on the direct iron ore smelting reduction project, in *50th Ironmaking Conference*, Washington, AIME, pp. 651–658.

Jahanshahi, S. and Player, R. (1989) Process chemistry studies of the Isasmelt lead reduction process, in *Proceedings Non-Ferrous Smelting Symposium*, AusIMM/IEAust, Sept 17–21, Port Pierie, Australia, pp. 33–40.

King, D. (1992) Fluidized catalytic crackers: an engineering review, in *Fluidization VII* (eds O.E. Potter and D.J. Nicklin), Engineering Foundation, New York, pp. 15–26.

Morgan, S.W.K. (1985) *Zinc and its Alloys and Compounds*, John Wiley, New York.

Palmer, R. (1992) Synthetic rutile kiln practice, in *Pyrosem WA* (eds E.J. Grimsey and N.D. Stockton), Murdoch University, 9 November 1990, pp. 68–106.

Patterson, W.C. (1990) Coal use technology in a changing environment. *Financial Times Business Information Report*, FTBE Ltd., London.

Perillo, A., Carlini, G. and Ibba, R. (1989) Portovesme KSS lead smelter sets new technology, in *Proceedings Non-Ferrous Smelting Symposium*, AusIMM/IEAust, Sept. 17–21, Port Pierie, Australia, pp. 21–26.

Pienemann, B., Stockhausen, W. and McKensie, L. (1992) Experience with the circulating fluid bed for gold roasting and alumina calcination, in *Fluidization VII* (eds O.E. Potter and D.J. Nicklin), Engineering Foundation, New York, pp. 921–928.

Raahauge, B.E. (1988) Theory and application of gas suspension calciner. *117th AIME Annual Meeting*, Phoenix, USA.

Rayner, J.W.R. (1952) Gasification by the moving burden technique. *J. Inst. Fuel*, **25**, 50–59.

Reh, L. (1971) Fluidized bed processing. *Chem. Eng. Progress* **67**(2), 58–63.

Squires, A.M. (1982) Contributions toward a history of fluidization, in *Proceedings Joint Meeting of Chemical Industry & Engineering Society of China and AIChE*, Beijing, Sept. 19–22, 1982, pp. 322–353.

Walker, S. (1989) Flash smelting – 40 years on with success. *Int. Mining*, May 1989, pp. 41–46.

Yerushalmi, J. (1982) Applications of fluidized beds, in *Handbook of Multiphase Systems* (ed. G. Hetsroni), Hemisphere, Washington, pp. 8:152–8:240.

13 Fluid catalytic cracking
AMOS A. AVIDAN

13.1 Introduction

Fluid catalytic cracking, or FCC, was the first application of fine-powder fluidization. Interestingly, the first commercial FCC unit (Esso's Model I in 1942) had both the reactor and regenerator in the form of entrained circulating fluid beds. Over 50 years later, FCC is still the major application of fluidization with over 350 FFC units operating worldwide, and with new ones coming on stream every year. FCC units convert heavy fuel oil and petroleum residue to lighter products. Major FCC products are gasoline, diesel fuel, heating oil, heavy cycle oils, which are used as bunker fuels or converted to carbon black, and light gases such as propene and butenes, which are converted to alkylate gasoline, methyl-tertiary-butyl ether, and are used for the production of petrochemicals.

FCC has an enormous economic impact on petroleum refining and hence on worldwide economy. It is an impressive operation in terms of sheer numbers. For example:

- Total worldwide capacity of FCC units is over 2.4 million tonnes per day (16 million barrels per day). The uplift across the FCC unit (difference between the value of products and feed) can be as high as $US10 per barrel.
- The total inventory of FCC catalyst in the average FCC unit is 300 tonnes. Over 450 000 tonnes of FCC catalyst are manufactured every year, representing a $US600 million business worldwide.
- The daily catalyst makeup rate to the typical vacuum gas oil feed FCC unit is 4 tonnes, whereas the daily makeup rate to a resid processing FCC unit can be as high as 30 tonnes.
- The total catalyst manufactured worldwide is approximately 1200 tonnes/day corresponding to a surface area of 380 000 km^2, the size of Egypt, each year.

Fluidization has become established as a well-studied, and economically important unit operation used in tens of different applications because of the advent of FCC, and it is still very important to FCC design, operation, and optimization. This chapter introduces the basics of FCC, with special emphasis on the fluidization aspects.

13.2 Brief history

The first large-scale fluid bed was the Winkler gasifier, commercialized in 1926. However, it only achieved limited use and did not immediately spawn new fluid-bed applications. FCC technology was developed in a brief period in the early 1940s following the commercialization of catalytic cracking in fixed-bed reactors by E. Houdry and the Socony-Vacuum Company (Avidan et al., 1990). The first commercial FCC unit, the ESSO Model I, which was started up in 1942, was a circulating fluid bed (CFB) system, comprised of an upflow reactor and an upflow generator. The empirical realization that Stokes' Law underpredicts the 'blow-out' velocity of fine powders, and that a stationary fluidized bed can be maintained at gas velocities many times the single particle terminal velocity, led to the invention of the low-velocity-fluidized bed, LFB, used in the 'downflow' FCC Model II commercialized in 1943. FCC technology has enjoyed continuous growth and evolution, and has never achieved the status of a mature technology.

Other fluid-bed applications have also used CFBs in preference to LFBs, but we show in Chapter 17 that the use of CFBs is limited to situations where the higher capital and operating costs of higher gas velocity can be justified by significant process advantages. In many applications, a well-designed LFB will suffice and be less costly to construct and operate than a CFB.

The first commercial FCC unit, the upflow Model I (Figure 13.1) was designed and commercialized rapidly. Oil fed to the Model I was vaporized, and heat was recovered from the regenerator via an external catalyst cooler. (This 'non-heat balanced' concept has recently regained popularity with the trend of processing heavier feeds such as residual oil.) The feed met regenerated catalyst and proceeded up a riser, which had an expanded diameter reactor section. (A long residence time was needed because of the low activity of the acid-treated natural clay catalyst used in those days.) While subsequent designs used the downflow LFB (Figure 13.2), the upflow riser design came back into prominence following Mobil's invention of zeolite cracking catalysts in the early 1960s. The cracked products in the Model I were separated from spent catalyst in a cyclone. (Quick separation is still a key FCC design element.) The spent catalyst was lifted into an upflow regenerator. Regenerated catalyst was split into two streams; one diverted to the external cooler and the other contacted with feed.

The capacities of early FCC units were gradually raised to three times their original design values through catalyst and hardware improvements. Hardware improvements included better understanding of standpipe flow and pressure balances, which allowed a significant reduction in unit height (Figure 13.2), and hence in construction and operating costs. Additional information on the evolution and scale-up of the FCC process is available in Chapter 14.

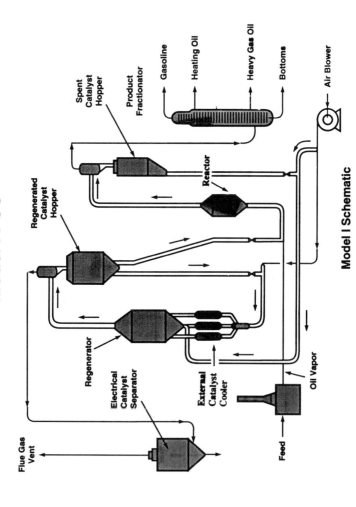

Figure 13.1 The Model I FCC commercialized in 1942. Both reactor and regenerator were upflow circulating fluid beds.

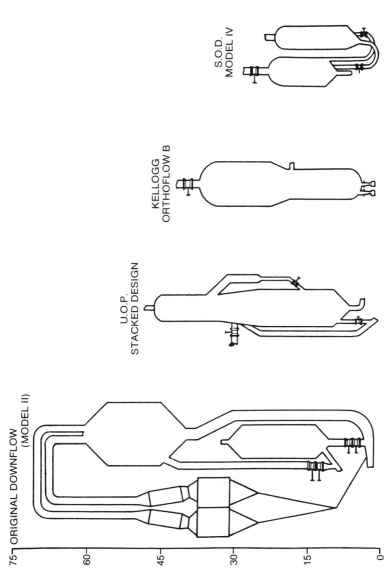

Figure 13.2 FCC evolution from the downflow FFB Model II (1943) to the Esso pressure-balanced FFB Model IV (1952). Note the reduction in size and height for a constant throughput.

13.3 Catalyst flow in modern FCC units

One example of a modern FCC unit is the M.W. Kellogg Heavy Oil Cracker shown in Figure 13.3a. It has evolved from previous designs and has many advanced features. The riser reactor is completely vertical, rather than having curved sections, promoting more uniform radial profiles throughout. The riser diameter is initially the same as the regenerated catalyst standpipe diameter, but is increased after feed introduction. (The number of moles increases during cracking, and the flaring design keeps the gas velocity from increasing too much.) The feed is injected through several radial feed nozzles, mixed with dispersion steam. These nozzles are typically designed to atomize the feed into small droplets, and to distribute the feed over the entire riser cross-section, thereby promoting intimate contact between the oil and catalyst (see Figure 13.4, Miller *et al.*, 1994).

The top of the riser has a 90 degree capped turn to minimize erosion. This turn affects the axial density profile (see Chapters 16 and 17). Solids are quickly separated from gaseous products in a 'closed cyclone' system and are returned via a dipleg to the top of a stripping vessel. (The 'closed cyclone' system provides a continuous path for the gas to leave the reactor without mixing within the reactor vessel freeboard.) The stripper is a standpipe, which typically has horizontal baffles. Steam is injected in one or more locations in the lower section of the stripper and rises countercurrently to the downflowing solids. The steam displaces hydrocarbon vapor and strips hydrocarbons adsorbed on the catalyst, and these leave with reactor product. More importantly, these hydrocarbons are not combusted in the regenerator, where they would deactivate the catalyst and decrease allowed cracking severity, or limit throughput of the FCC unit.

Typical stripper solids residence times are 0.5 to 5 minutes, and typical relative velocities between upflowing gas and downflowing solids are 0.1 to 0.5 m/s. A well-designed stripper has good solids and gas distribution, and good contact between them. One indicator of stripper efficiency is the amount of hydrogen left in the coked, spent catalyst. Strippable hydrocarbons have a hydrogen-to-carbon ratio of up to 2, or a hydrogen concentration of 14 wt%. Well-stripped coke has a hydrogen content of approximately 4 wt%; typical strippers have values of 5 to 8 wt%. Considerable effort is expended in designing reliable and efficient strippers and in their maintenance and troubleshooting.

Stripped solids descend through a vertical standpipe into the regenerator. Many FCC units have angled standpipes to allow lateral movement of solids between side-by-side units. Special care has to be taken in the design and proper aeration of standpipes, especially sloped ones (see Chapter 7). Properly designed FCC standpipes circulate up to 1800 kg/m^{2}s of solids, whereas many improperly designed (sloped, or improperly aerated) standpipes only achieve about 600 kg/m^{2}s. A common way of overcoming

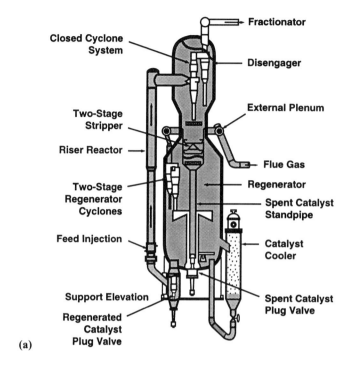

Figure 13.3 (a) An example of a modern FCC unit – the M.W. Kellogg heavy oil converter stacked unit. Note the reactor closed cyclone system, counter-current flow regenerator and dense FFB catalyst cooler. (b) Counter-current regeneration principles and mechanical details of the gas and solids distributors.

poor understanding of standpipe flow has been to design standpipes larger than needed.

The solids flow rate between the stripper and regenerator can be controlled by pressure balance as in the popular Esso Model IV, first commercialized in 1952. This model does not use valves to control the flow, and hence, has the lowest overall height as shown in Figure 13.2. In most designs, solids circulation is controlled by at least one slide valve. The stripper vessel is elevated to provide the static head required to contain a pressure drop across the spent catalyst slide valve. Solids flow out of the regenerator can be controlled by an overflow well in the regenerator or by another slide valve (as in the Exxon 'Flexicracker', for example).

Most other designs provide additional flexibility by using two valves, a spent catalyst valve and a regenerated catalyst valve. In the M.W. Kellogg design (Figure 13.3a), plug valves are used, while slide valves are used in other designs. Non-mechanical valves have only limited applications for well-flowing fine powders, such as FCC catalyst. The Model IV can provide

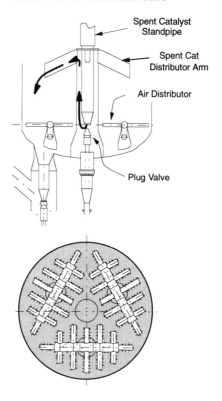

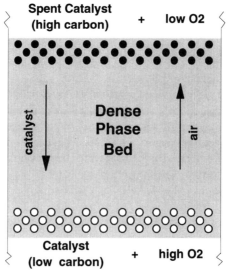

(b)

Figure 13.3 (continued).

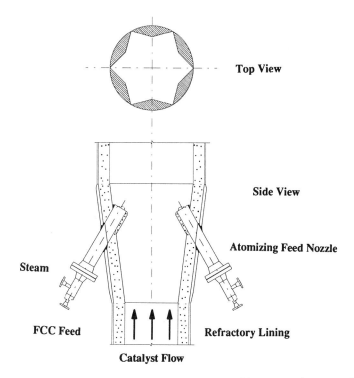

Figure 13.4 FCC feed-catalyst contacting. The feed is atomized with steam and contacts the upflowing regenerated catalyst. Nozzle placement ensures good radial coverage (top view).

some limited control of solids circulation rates by varying aeration to the 'U-based' standpipes.

13.4 Catalyst regeneration

The regenerator in FCC units is a combustor where coke deposited on the catalyst and unstripped hydrocarbons are oxidized by air to carbon monoxide, carbon dioxide, water, sulfur and nitrogen oxides, and other trace compounds. The FCC regenerator can operate in 'partial CO combustion' mode, where the flue gas contains substantial quantities of CO, and is then further oxidized in an external CO boiler, or in 'full CO combustion' mode.

Proper distribution of air and solids is very important to efficient regenerator operation. Solids can enter the regenerator tangentially ('swirl-flow'), laterally ('cross-flow'), with the air through a riser, or from the top (through several distribution arms as in the Kellogg HOC design). Catalyst bed height in the regenerator can vary from a very shallow 2.5 m in some units (to save air blower cost) to 20 m or more in CFB regenerators, as in the UOP 'high-efficiency' regenerator.

There are many regenerator designs of different vintages and this has resulted in a large variance in regenerator efficiency. Efficiencies as low as 20% and as high as 70% have been reported (van Deemter, 1980). Regenerator efficiency is manifested in several important areas. An efficient regenerator can provide values *lower* than:

- Carbon level on regenerated catalyst (higher catalyst activity)   0.05 wt%
- Afterburn in the cyclones (requires less Pt combustion promoter)   10°C
- Required excess O_2 (major operating cost)   1 vol%
- NO_x emissions for typical FCC feed   75 ppm

Some generic regenerator design types are compared in Table 13.1.

One particularly efficient CFB regenerator design developed by Mobil and UOP in the 1970s is the 'high-efficiency combustor' (Figure 13.5). The fluidization regimes in the various zones of this combustor are illustrated on a 'phase diagram' (Squires *et al.*, 1985). The bottom part of the combustor operates in the turbulent fluidization regime, as do most commercial LFB FCC regenerators. A necking down of combustor diameter causes the gas velocity to exceed the 'blow-out' (or transport) velocity in a fashion equivalent to the introduction of secondary air in CFB boilers. The fast fluid-bed denoted by 'C' entrains the inventory at a velocity of approximately 2 m/s. The entrained solids are separated in a simple inertial separator, which can have an efficiency as high as 85%, and then by internal cyclones. The solids are recirculated to the bottom of the combustor, and into the bottom of the riser reactor.

This efficient CFB combustor was developed originally to operate in full CO combustion. More than 30 of these 'riser regenerators' have been built, with an additional 20 CFB regenerators built in China (Chen *et al.*, 1994). However, the commercialization of noble metal CO combustion promoter by Mobil at about the same time provided an alternative to full CO combustion in LFB regenerators. In addition, in the past decade there has

Table 13.1 Relative comparison of generic regenerator types – typical commercial performance relative to LFB with central catalyst distributor base

	Dense bed, LFB 'cross-flow'	Shallow LFB 'swirl'	CFB type	LFB with central catalyst distribution
Inventory	1.5	2.5	2*	1 (base)*
Catalyst makeup	1.5	2	1	1 (base)
NO_x emissions	6	6	1.5	1 (base)
Relative cost	1	1.7	2	1 (base)

*Two times LFB base; A significant portion of the inventory is in the inactive 'upper combustor'.

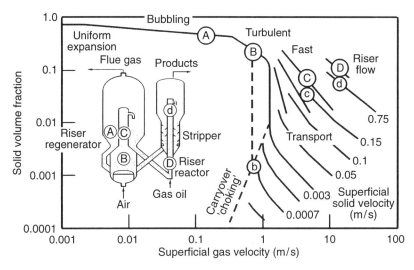

Figure 13.5 'Phase diagram' for a Group A powder and a schematic FCC 'high efficiency' unit showing where various fluidization regimes are found.

been an accelerating trend to process increasing amounts of petroleum residua in FCCs. Most new FCC units are now constructed as resid FCC units, RFCCs, and many units are now designed to operate in partial CO combustion mode where CFB regenerators provide little benefit.

While CFB regenerators have clear advantages over most prior LFB regenerators, it is clear from Table 13.1 that similar, or even better performance can be obtained from a well-designed LFB regenerator. The main inefficiency of the early LFB regenerator designs seems to have resulted from poor distribution of air and carbon laden catalyst. When these problems are solved via:

- efficient and reliable multi-zone air grid,
- central spent catalyst distribution,
- countercurrent flow of coked catalyst and air,

the LFB regenerator can have the same, or better, performance than the CFB regenerator at a lower capital and operating cost. The same conclusion has been reached in other CFB–LFB comparisons; for example, the new Sasol LFB reactor is better and less costly than the Synthol CFB reactor (see Chapters 14 and 17).

The M.W. Kellogg regenerator shown in Figure 13.3 includes many elements of a well-designed LFB regenerator. The air is distributed through a multi-zone pipe grid distributor. The catalyst is lifted in a low velocity riser and distributed radially at the top of the dense bed through several lateral arms. This arrangement encourages good distribution of air and catalyst, which are then contacted in counter-current flow. Some regenerated catalyst is withdrawn to a dense-phase catalyst cooler, where heat is removed by generating medium-pressure steam.

LFB dense-phase catalyst coolers are superior to the CFB dilute-phase catalyst coolers of the 1940s since they afford a much higher heat transfer coefficient and are less prone to erosion. Various designs of dense-phase catalyst coolers are available for license by M.W. Kellogg, UOP and Stone & Webster.

SINOPEC developed several commercial designs of one- and two-stage regenerators, including designs employing a CFB, occasionally termed the fast fluidized bed, with superficial gas velocities higher than LFB regenerators (1.0 to 1.8 m/s). It combines elements of the successful stacked M.W. Kellogg design with a larger CFB regenerator. However, as shown in Table 13.1, it is not clear whether a CFB single-stage or two-stage regenerator has sufficient benefits over a well-designed single-stage LFB regenerator to justify the higher capital and operating costs.

13.5 FCC process basics

FCC is such a widespread, useful and profitable processing unit in a petroleum refinery because it cracks heavy petroleum fractions such as vacuum gas oils, atmospheric gas oils, coker gas oils, lube extracts, slop streams, and residuum, to lighter products. On average, FCC capacity is 38% of the crude run in the refinery.

FCC feed is typically in the 320°C to 600°C boiling range, but residuum processing units can have feeds with higher end point. Some other typical key property ranges of FCC and RFCC feeds are:

- API gravity   10–30
- Sulfur, wt%   0.1–3
- Nitrogen, wt%   0.01–0.5
- Carbon residue, wt%   0.1–7
- Nickel and vanadium, ppm 0.1–50

The most useful FCC products are gasoline, light cycle oil and light olefins. Typical FCC yields are shown in Figure 13.6. Yields can be improved by the use of feed hydrotreating, but they are still far lower than the theoretical yields (taking into account thermodynamic constraints and the amount of available hydrogen in the feed) of approximately 90 wt% (Avidan *et al.*, 1990; Wilson-Thomas, 1994). There are two main reasons for current FCC yields being lower than the theoretical maximum:

- Heavy aromatics in the feed have a low cracking tendency with today's catalysts.
- The useful products (gasoline, light cycle oil, light olefins) can react further.

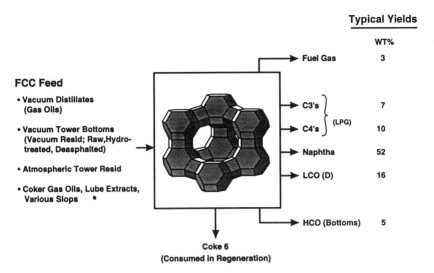

Figure 13.6 Typical FCC feeds, and typical yields in 'gasoline mode' operation. The structure in the box is a schematic representation of Y-zeolite – the key active ingredient in FCC catalyst.

Improvements in catalysts and FCC hardware have increased useful FCC yields by over 50% over the past fifty years. Improvements in FCC are evolutionary, but occasionally there are true revolutions. Chief among them was the invention of zeolite cracking catalysts by Mobil scientists in the early 1960s. The Y-zeolite, which is still the main cracking component of today's catalysts, increased coke selectivity (the ratio of cracking to coking, which is the dominant factor in FCC design and operation), and caused an increase of gasoline yield of up to 50% (at the same coke yield). Zeolite catalysts had an important effect on design. Due to their inherently higher activity (by a factor of about a thousand), a much shorter time, of only a few seconds, is needed to achieve desired yields. This eliminated the need for dense-bed crackers, and resulted in pure riser cracking. Typical modern FCC operating ranges in a riser reactor are:

Feed temperature, °C	150–300
Regenerated catalyst temperature, °C	675–750
Riser top temperature, °C	500–550
WHSV, h^{-1}	10–100
Catalyst/oil ratio, wt	4–10
Dispersion steam, wt%	0–5
Pressure, kPA	150–300
Oil residue time, s	1–5
Solids residence time, s	3–15

The term conversion is used differently in FCC than in other catalytic processes. It is usually defined as one hundred minus the volumetric yield of all products boiling above gasoline end point (typically 220°C), i.e.

$$X = 100 - (\text{vol}\% \text{ LCO} + \text{HCO}) \tag{13.1}$$

Since cracking can be described as a second-order reaction, a term called crackability is commonly used in FCC correlations:

$$\bar{X} = \frac{X}{100 - X} \tag{13.2}$$

These and other chemical reaction engineering aspects of FCC were recently summarized by Avidan and Shinnar (1990).

Today's FCC catalysts are comprised of several components: Y-zeolite crystals, silica alumina, or clay matrix, weighing agents, Pt promoter, bottoms cracking agents, and vanadium traps. Some of these components can be used as separate additives. Other separate additives are: ZSM-5 for octane enhancement, SO_x transfer additive (transfers SO_x from regenerator to reactor where it is converted to H_2S), fluidization aides, nickel passivation agents, and other proprietary additives.

Obviously, the FCC catalyst system is complex and considerable expertise goes into catalyst selection, scale-up and optimization. Major catalyst vendors are Grace-Davison, Akzo-Nobel, Engelhard, and CCIC. FCC catalyst particles are microspheres, typically produced by spray drying. They produce a quintessential Geldart Group A powder, with the following typical fresh properties:

Average particle size, μm	70
Size range, μm	20–150
Sphericity	nearly 1.0
Angle of internal friction	79°
Angle of repose	32°
True density, kg/m^3	2500
Particle density, kg/m^3	1200–1700
Bulk density, kg/m^3	750–1000
Typical U_{mf}, m/s	0.001
Typical V_T, m/s	0.1

Freshly produced FCC catalyst has a log-normal particle size distribution. The natural proportion of fines, taken as particles smaller than 40 μm, for such a powder is approximately 25 wt%. However, catalyst vendors typically reduce this fraction by more than half. This is unfortunate from the point of view of flow properties in standpipes and bubble size in regenerators (average bubble size for an FCC powder with the fines fraction undiminished is smaller than 0.02 m, while the fines-poor powder can have voids of up to

0.15 m in diameter, even in the turbulent fluidization regime). The removal of fines from fresh FCC powder is intended to lower particulate emissions, especially in units with poor solids recovery systems. Some operators have found that recycling third stage separator fines is beneficial.

Another important aspect in FCC catalyst evaluation, as well as any other process variation, is scale-up. It is difficult to simulate contact in FCC risers in a small bench-scale reactor. Various small-scale pilot plants are used, including a fixed-bed ASTM Test Method D3907, an LFB test, small CFBs, etc. The scaling relationships between those units, and between them and commercial operation, are not completely understood, though some progress has been made.

13.6 FCC feed atomization and mixing

The feed injection system of an FCC unit plays a crucial role in yield performance (Miller *et al.*, 1994). Ideally, cracking reactions should take place in the vapor phase on the solid catalyst surface, and rapid and uniform mixing is essential to ensure quick vaporization and intimate contact between the oil and catalyst during their brief residence time in the riser. Poor feed injection creates localized regions of high and low catalyst-to-oil ratios and induces catalyst backmixing. These phenomena generally reduce conversion and lead to increased coking. The feed injection system and the design of the riser should produce the most uniform cracking environment possible. Oil needs to be atomized into small droplets with a narrow size distribution since small droplets vaporize more quickly than larger ones. FCC feed needs to be distributed evenly over the riser cross-section; empirically, it has been found that a uniform catalyst density profile is advantageous. The feed also needs sufficient momentum to effectively penetrate the flowing catalyst without causing erosion of the riser wall or excessive catalyst attrition.

Many commercial FCC feed nozzles have been developed and used in practice. While it has been generally accepted that the best nozzles produce a relatively flat spray of uniform small droplets, preferably with a narrow droplet size distribution, there have been many different ways of approaching this objective. Some feed nozzles rely on high liquid pressure drop, while others rely on atomization via high pressure dispersion steam. Some examples of commercial nozzles and the results of cold flow model testing of them are shown in Figures 13.7 and 13.8 (Miller *et al.*, 1994).

Equally important for the design of the FCC feed mixing section is riser design, and hydrodynamics just below and just above the feed injection point. Tests on several riser configurations conducted in one of the largest CFB cold flow models (riser diameter of 0.3 m and height of 12 m) are described by Miller *et al.* (1994). They used several commercial riser bottom configurations (Figure 13.9), including a 'Y' configuration with a

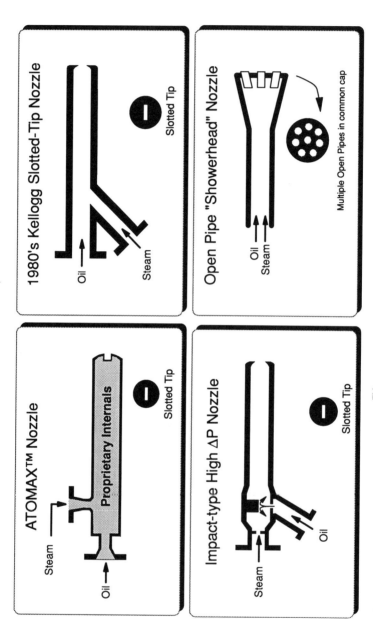

Figure 13.7 Schematics of several nozzle types: (a) ATOMAX™ steam is used to break up the oil into small droplets; (b) older slotted tip nozzle; (c) impact-type high ΔP nozzle, and (d) open pipe 'showerhead' nozzle.

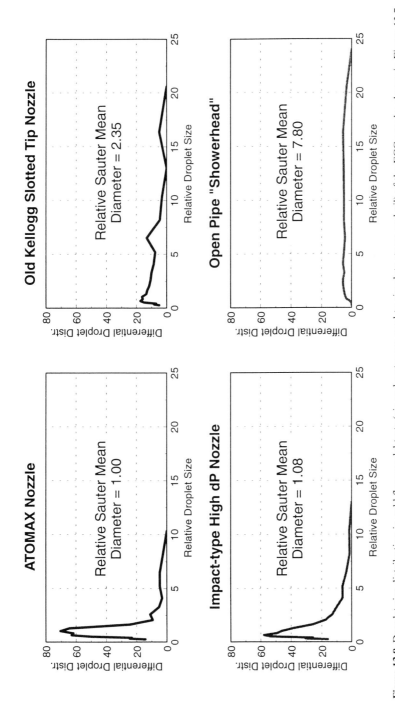

Figure 13.8 Droplet size distribution in cold flow model tests (air and water are used to simulate steam and oil) of the FCC nozzles shown in Figure 13.7. A steep narrow distribution is believed to be advantageous in FCC.

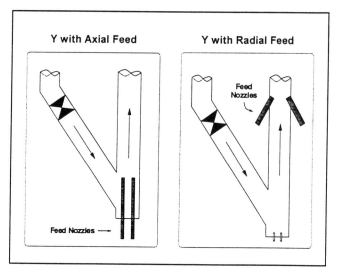
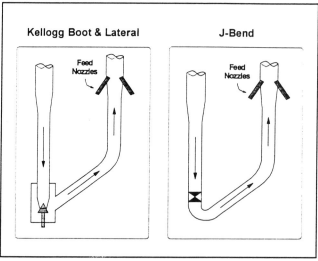

Figure 13.9 Different FCC riser bottom configurations used to test feed-catalyst mixing effectiveness in a large (0.3 m ID, 12 m tall) cold flow model. Air simulated vaporized FCC feed (Miller *et al.*, 1994).

sloping standpipe and axial bottom feed nozzles, a 'Y' configuration with radial nozzles, a 'boot and lateral', and a 'J bend' configuration. Density profiles for those configurations, measured by a gamma-ray densitometer, are shown in Figure 13.10. The 'Y axial' design exhibited the most non-uniform radial density profile, which is generally believed to be disadvantageous in FCC. The other three designs exhibited much flatter radial density

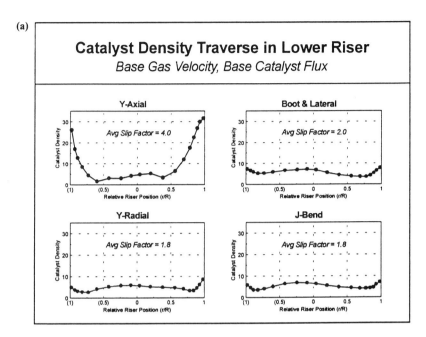

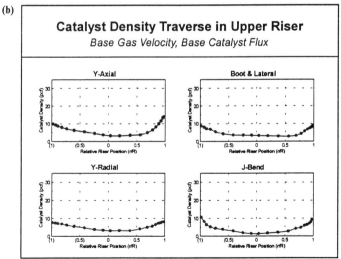

Figure 13.10 Catalyst density profiles for the four FCC riser bottom arrangements of Figure 13.9 measured by a gamma-ray densitometer. (a) Density profiles above the feed injection point in the lower part of the riser, and (b) profiles in the upper riser section. A flat density profile is thought to be advantageous in FCC.

profiles. The 'Y axial' configuration also produced the highest average slip factor, defined as:

$$\text{Average slip factor} = \frac{U_g}{G_s/\rho_s} \cdot \frac{(1-\epsilon)}{\epsilon} \qquad (13.3)$$

Catalyst density profiles in the upper section of the riser are also shown in Figure 13.10. All riser configurations become more similar as the flow approaches fully-developed conditions, although the 'Y-axial' configuration continues to exhibit an asymmetric distribution.

13.7 FCC catalyst–product separation

Separation of FCC products and catalyst has become an imperative for the high cracking temperatures now in use. This rapid separation is needed to eliminate long residence time dilute phase catalytic cracking downstream of the riser (Avidan *et al.*, 1990b). FCC reactor cyclones were traditionally 'negative pressure' cyclones 25 years ago, before the advent of all riser cracking (i.e. the pressure in the primary cyclone was lower than the reactor pressure). The reactor products flowed from the higher pressure reactor freeboard into the reactor cyclones (hence the term 'negative pressure' as applied to the pressure inside the first cyclone separators). With the conversion to all riser cracking, inertial separators have been developed. Many different types of inertial separator devices have been patented, and several have been commercialized. Some examples appear in Figure 13.11. The rough-cut separator is basically a low-efficiency cyclone that relies on momentum

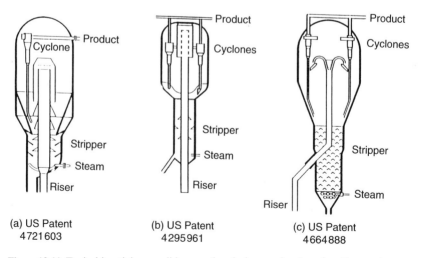

Figure 13.11 Typical inertial gas–solid separation devices used as 'rough-cut' separators.

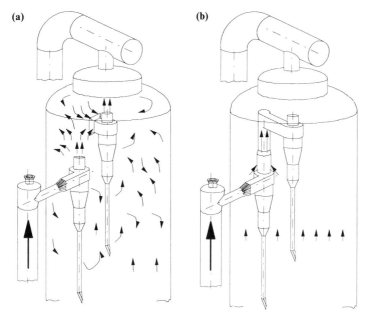

Figure 13.12 Evolution of FCC product–catalyst separation – improvements over 'rough-cut' separations. (a) Riser cyclones, introduced in the 1970s, provided a higher separation efficiency than 'rough cuts', and (b) positive-pressure closed cyclones, introduced in the 1980s to eliminate product residence time in the reactor vessel. A vent in the transfer line provides an outlet for stripper gas.

direction changes rather than centrifugal force as in a cyclone. The inertial force acting on solids particles is typically 10 to 30 times lower than in the cyclone, and separation efficiency is typically about 80% compared to greater than 99.9% achieved in a cyclone.

The introduction of riser cyclones (see Figure 13.12a, from Johnson *et al.*, 1994) has improved FCC reactor performance and changed the primary cyclone pressure balance. It now becomes a 'positive pressure' cyclone – with the pressure inside the primary cyclone higher than in the reactor freeboard. The improvement in reactor performance due to the introduction of riser cyclones has been attributed to a reduction of catalyst holdup in the reactor freeboard. This reduction in holdup reduces the amount of coke deposited on the catalyst, thereby leading to significant reduction in regenerator temperatures, and allowing higher conversions or higher throughputs.

Looking for further improvements in reactor performance, several FCC developers have sought better riser terminations or improvements in stripper efficiency. In the 1970s Mobil developed and commercialized the concept of stripper cyclones. Steam was introduced into the riser cyclone bodies, and internal spirals mimicked the internal structure of a stripper. While no operational problems were encountered, no performance improvement was noted. In trying to understand the reasons for achieving no measurable performance improvement with stripper cyclones, Mobil engineers concluded that cyclones have an inherently low contact efficiency between solids and vapor, and, coupled with short residence times, low stripping efficiency is achieved. It was clear that the main reason for performance degradation was thermal cracking, which continued in the reactor freeboard despite the reduction in solids holdup, since the gaseous products residence times were still high, exceeding 30 s on average.

While many riser exit configurations have been proposed to reduce thermal cracking, only a truly closed, coupled cyclone system can reduce the average product residence time in the reactor from over 30 s to as low as 2 s. There are several methods of achieving a closed, coupled cyclone

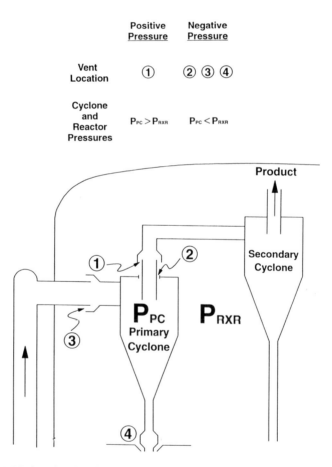

Figure 13.13 FCC riser closed cyclone pressure balance. The vent location determines whether the primary cyclone is 'positive pressure' ($P_{PC} > P_{RXR}$), or 'negative pressure' ($P_{PC} < P_{RXR}$). Vent locations at points ②, ③ and ④ are 'negative', while vent location ① is positive.

system. One main distinguishing design feature is the means provided for removing stripper gas from the system:

- Negative pressure systems, which provide for removal of stripper gas through a vent in front of the primary or riser cyclone (hence changing the pressure balance from positive to negative).
- Positive pressure systems, which preserve the positive pressure feature of the primary riser cyclone by providing a stripper gas vent between the primary and second cyclones.

Cold Flow Model tests by Mobil (Avidan *et al.*, 1990) have shown that positive pressure systems are less susceptible to upsets.

The pressure balance of a positive (Mobil–M.W. Kellogg) and several patented negative pressure systems (Mobil and others) are shown in Figure 13.13.

Nomenclature

FCC	fluid catalytic cracking
LCO	light cycle oil (an FCC product typically boiling between 220°C and 350°C)
LFB	low velocity (bubbling or turbulent) fluidized bed. A bed where a bed level can be discerned
CFB	circulating fluidized bed
G_s	solid flux, kg/m^{2}s
HOC	heavy oil cracker (M.W. Kellogg's RFCC Unit)
HCO	heavy cycle oil (an FCC product typically boiling above 350°C)
P_{PC}	pressure in the FCC primary riser reactor cyclones
P_{RXR}	pressure in the FCC reactor vessel
RFCC	resid (processing) FCC
U_g	superficial gas velocity, m/s
WHSV	weight hourly space velocity (mass of reactant per hour divided by mass of catalyst)
X	FCC conversion (defined in equation (13.1), vol%)
$\bar{X}$	crackability (or second order conversion, defined in equation (13.2))
ϵ	cross-section average voidage
ρ_s	solids particle density, kg/m^3

References

Avidan, A.A. and Shinnar, R. (1990) Development of catalytic cracking technology. A lesson in chemical reactor design. *Ind. Eng. Chem. Res.* **29**, 931–942.

Avidan, A.A., Krambeck, F.J., Owen, H. and Schipper, P.H. (1990) FCC Closed-Cyclone system eliminates post-riser cracking. *Oil and Gas Journal,* March 26.

Johnson, T.E., Miller, R.B., Santner, C. and Avidan, A.A. (1994) FCC reactor product–Catalyst separation – Ten years of commercial experience with closed cyclones. Grace European FCC Technology Conference, Athens, September 27–30.

Chen, J., Cao, H. and Lin, T. (1994) Catalyst regeneration in fluid catalytic cracking, in *Fast Fluidization* (ed. M. Kwauk), Academic Press, San Diego, pp. 389–419.

Miller, R.B., Niccum, P.K., Sestili, P.L., Johnson, D.L., Dou, S. and Hansen, A.R. (1994) New developments in FCC feed injection and riser hydrodynamics. AIChE 1994 Spring National Meeting, April 12–21, Atlanta.

Squires, A.M., Kwauk, M. and Avidan, A.A. (1985) *Science*, **230**(4732), 1329–1337.

van Deemter, J.J. (1980) Mixing patterns in large-scale fluidized beds, in *Fluidization* (eds J.R. Grace and J.M. Matsen), Plenum Press, New York, pp. 69–89.

Wilson-Thomas, M. (1994) Detailed FCC feed characterisation. Grace FCC Technology Conference, Athens, September 27–30.

14 Design and scale-up of CFB catalytic reactors
JOHN M. MATSEN

14.1 Introduction

The scale-up of circulating fluid bed reactors has been a continuing activity in the chemical process industries for over half a century. Despite that record, such scale-up is still not an exact science, but is rather that mix of physics, mathematics, witchcraft, history and common sense that we call engineering. The purpose of scale-up efforts is not to achieve fundamental and total knowledge of the process, but rather to minimize the possibility of expensive errors in reaching commercial operation. Better physical data, more realistic models and more exact equations are always desirable. Nevertheless, the basic goal is not ever-increasing accuracy of calculation but rather the recognition and management of uncertainties and risk.

The term 'circulating fluid bed' (CFB) has been loosely applied to a variety of phenomena. This chapter limits the term to units in which particles are circulated to a riser at a rate controlled for process reasons and carried out of the top of the riser by the process gas stream. Only three catalytic CFB processes have undergone scale-up to commercialization: fluid catalytic cracking (see Chapter 13), the 'Synthol' Fischer Tropsch synthesis practised by Sasol, and duPont's maleic anhydride process, which will reach commercial operation in late 1996. The much broader experience in scale-up of non-circulating fluid beds is instructive and generally applicable to CFBs, however, and many useful examples exist in that literature.

14.2 Scale-up issues

14.2.1 Study design

The term 'scale-up' may be somewhat misleading when applied to the typical progression from test tube, through bench scale and pilot plant stages. 'Scale-down' is often equally appropriate. A detailed study design of the commercial plant is invariably executed very early in development, and design and operation of intermediate-scale equipment is guided to a very great degree by the commercial study design. This design is in sufficient depth to establish dimensions for major process vessels and specifications for significant mechanical components. It becomes the basis for preliminary process economics and also serves to identify which aspects of the process have the greatest economic impact. The study design should identify basic data needs, materials and machinery developments, and any design correlations to be undertaken.

14.2.2 Reactor engineering

For both circulating and non-circulating fluid bed processes, reactor design information is usually an important goal of scale-up studies. The important issues are gas–solids contacting and residence time distributions. As shown in Chapters 3 and 15, some gas is not in close contact with catalyst particles, and there are significant deviations from plug flow. The debits of limited gas–solids contacting and of axial dispersion invariably increase with reactor size, although in some circumstances those debits are of little practical consequence. It is often assumed that gas–solids contacting is very good and that gas flow approaches plug flow in a CFB reactor. This may be a useful approximation in comparison with non-circulating bubbling fluid bed reactors. It is not, however, a fundamental CFB characteristic, as demonstrated by Ouyang et al. (1995) who measured decomposition of ozone in a 250 mm diameter riser at gas velocities up to 8 m/s and solids circulation rates to 240 kg/m^{2}s. They found large ozone concentration differences between the axis and wall at all heights. Overall ozone conversion was less than for a continuous stirred tank reactor (CSTR) with no contacting limitations, and the performance could be expressed as that of a CSTR with a 60% catalyst effectiveness factor. This specific experimental behavior probably does not encompass a quantitative design principle, but it suggests the nature of reactor performance debits that might be encountered in practice.

In the past 30 years a great deal of fundamental understanding of the physics of gas–solids contacting in fluid beds has emerged. Highly detailed reactor models based on fundamental first principles ('learning models') have been widely proposed. In commercial scale-up, however, simpler 'design' models are often used, with critical parameters being determined by tracer experiments on fairly large-scale equipment. Two examples of the latter approach prove instructive. May (1959) performed large-scale cold reactor simulations for the development of the Esso (Exxon) fluid hydroforming process. He used helium decay tests to determine gas residence times and radioactive solids tracer tests to measure solid phase dispersion coefficients. Model parameters determined in the cold unit, i.e. a gas crossflow ratio from bubble to emulsion phase and an emulsion phase gas dispersion coefficient, were sufficient to predict reaction performance. The beneficial effects of high fines content and horizontal baffling or staging were established in the cold unit and proved critical in the satisfactory

DESIGN AND SCALE-UP OF CFB CATALYTIC REACTORS   491

operation of commercial units. A slightly different approach was used by Krambeck *et al.* (1987) in developing Mobil's methanol-to-gasoline (MTG) process. They measured residence times of SF_6 tracer in their cold units. This tracer is adsorbed on the catalyst, and the adsorption coefficient can be changed by changing the moisture content of the gas stream. By determining the tracer decay function for two different adsorptivities, the reactant 'contact time' and hence the extent of reaction can be estimated. The theory and application of this technique have been discussed by Pustelnik and Nauman (1991). The Mobil work found that a turbulent fluidized bed has negligible contacting limitations and again documented the beneficial effect of high fines concentration and horizontal baffling.

14.2.3 Experimental work

On the way from initial concept to commercial realization, several experimental units are needed to address the many questions or concerns which arise. Typically there will be one or two bench-scale units to demonstrate the basic reaction concept. First rate data are often generated in these units since many reactions for which a CFB might be considered cannot be studied by pure kinetics studies in thermogravimetric analysers or differential packed beds. There will likely be vendor tests of machinery and mechanical components. Specialized tests may be warranted for materials of construction, attrition and agglomeration, and solids handling. There will inevitably also be a large pilot plant to provide an integrated demonstration of the critical process components.

It is also common for model tests to be done on critical hydrodynamics aspects of the process. Such tests are usually performed on a reduced scale, using expected commercial particles and air at ambient conditions. The models often have plastic sections for visual observation and may be provided with probes, tracer detection, and other instrumentation that could not be easily incorporated in a commercial unit. Glicksman and his students have published a number of studies demonstrating the correct dimensionless groups to be used in scaling of gas–particle hydrodynamics. The paper by Westphalen and Glicksman (1994) deals with CFB scaling (see Chapters 2 and 8). Two units of differing size can exhibit hydrodynamic similitude if operated so that the following dimensionless groups are the same in both units: Reynolds number, Froude number, ratio of gas-to-solids density, gas-to-solids flux ratio, and ratios of all particle and apparatus dimensions. This assumes the absence of significant interparticle surface forces and ignores significant reaction or heat transfer effects. In practice it is very difficult to perform such an experiment for a large scale-up factor, and such an approach does not seem to have been used in an actual process scale-up. Cold flow modellers should be aware of this rigorous approach, however, when translating model data to commercial design.

14.3 Commercial components in CFB systems

14.3.1 Gas and particle introduction

A great many configurations have been used to introduce gas and particles to the bottom of a riser reactor. For catalytic reactors, two systems are in common use. In the Synthol Fischer–Tropsch reactors, catalyst drops vertically from the slide valve into a horizontally flowing gas stream. The particles are carried in dilute phase transport through a long sweeping bend terminating in a vertical section where the main reaction occurs. The horizontal dilute phase flow requires a relatively high gas velocity and low solids loading. The horizontal component of solids movement may be an important consideration in the arrangement of vessels and piping. Some older FCC units use a similar arrangement on the oil feed riser, but it is believed that the inevitable solids stratification is undesirable for modern short contact time reactors.

In most FCC reactors and in duPont's maleic anhydride process, catalyst is already moving vertically upwards in dense phase riser flow before the hydrocarbon feedstock is introduced. This seems to be the preferred configuration for high particle mass flux rates (500 to 1100 kg/m^{2}s).

Injection of hydrocarbon feed is a critical aspect of the FCC process. The feedstock is a high boiling point oil which becomes vaporized only after injection into the hot catalyst stream. With the very rapid reactions now practiced, it is very important that the oil be atomized into tiny droplets and be dispersed rapidly across the entire riser cross-section. Feed nozzle design is an area of intense proprietary development. For example, Kellogg and Mobil conducted air–water tests at full scale for several nozzle designs, using measurements of drop size and spray patterns as an index of nozzle performance (Miller *et al.*, 1994). These tests are discussed further in Chapter 13, with drop size distributions shown in Figure 13.8. Cold test results were confirmed in operating FCC units. It should be emphasized that this sort of optimization provides incremental but lucrative improvement in a mature process. That degree of refinement should not normally be necessary nor expected in the design of new processes.

Weinstein *et al.* (1995) documented feed injection studies for the maleic anhydride process. They concluded that the best gas distribution and the most rapid approach to fully developed riser flow occurred if gas was injected in a ring midway between the riser wall and centerline.

14.3.2 The riser

The riser is the defining element in a CFB catalytic reactor. Catalyst particles introduced at the bottom of the riser are entirely carried out at the top in co-current flow with reaction vapors. It is necessary to have independent control of particle and vapor rates. Beyond that, risers may have a wide range of configurations and design characteristics, dependent on process requirements. The Sasol Synthol reactor has a section of expanded diameter, where lower gas velocity permits a higher particle holdup. The expanded section also accommodates heat transfer surfaces to remove the heat of reaction. Most CFB reactors, however, have cross-sections which do not vary with height.

Solids mass flow rates in risers can range up to 1700 kg/m²s, although the highest rates in reactor risers seem to be about 1000 kg/m²s. Diameters of at least 3.6 m and heights up to 50 m have been used, and gas velocities up to 25 m/s are encountered. Slip factors of approximately 2 or 3 are often seen in the upper portions of long risers. Patience et al. (1992) proposed an equation for slip factor:

$$\phi = \rho_p U(1-\epsilon)/G_s\epsilon = 1 + \frac{5.6}{Fr} + 0.47 Fr_T^{0.41} \tag{14.1}$$

This shows a clear scale-up effect with the slip factor increasing with increasing riser diameter. Riser hydrodynamics are discussed more fully in Chapter 2.

14.3.3 Riser termination

It has recently been recognized that the configuration of the riser exit piping can have a significant effect on riser hydrodynamics. A sharp bend, a sudden decrease in diameter or an increase in diameter causes increased refluxing of solids downwards along the riser walls. This refluxing causes an increase in solids holdup that can extend upstream (i.e. down the riser) for a number of riser diameters. Thus the termination or exit should be selected to meet catalyst holdup goals. A blind tee has often been used to minimize erosion, and this design causes substantial refluxing.

14.3.4 Solids separation

Riser exit gas with entrained solids may be fed directly to a cyclone or may first pass through a knock-out drum and then to a cyclone. For most catalytic processes, catalyst recovery requirements dictate that two stages of cyclone be used, with collected solids returned to the process.

Cyclone design is usually carried out by equipment vendors. The effects of size and operating conditions are generally well understood and are usually incorporated into a design procedure, such as that presented in Chapter 6.

In FCC practice, cyclone inlet velocities are typically limited to 20 or 25 m/s for erosion considerations. Barrel diameters less than 1.2 m are usually avoided because of difficulty of access and maintenance. Cyclone height is usually 4 or 5 times barrel diameter, and most FCC units specify diameters less than 1.5 m to avoid excessive height. There is, however, no fundamental problem with larger cyclones. Multiple cyclones in parallel may be used in lieu of a single larger cyclone.

14.3.5 Standpipe

Particles circulating through the flow control valve, riser, riser termination, and cyclone travel successively from higher to lower pressure regions. The standpipe is the engine that drives the circulation by generating the necessary increase in pressure to bring the particles back to conditions at the control valve inlet. Standpipes can generate hydrostatic pressure gradients close to those corresponding to the minimum fluidization density of the particles of either type A or type B powders. A more reliable value for design purposes would be 75% to 80% of the minimum fluidization pressure gradient. Typical mass flow rates range up to 1700 kg/m^{2}s, and high rates seem to be somewhat less troublesome than low rates.

In nearly all catalytic reactor standpipes, the net flow of gas is downwards towards the higher pressure. Aeration is therefore provided to counteract gas compression and to maintain good particle fluidization. Aeration taps are often distributed at 2 or 3 m intervals along the length of the standpipe, and such taps must be provided at the bottom of the standpipe. Design aeration rates are calculated from the gas downflow and the pressure gradient. An example is given by Matsen (1973). Provision is made to optimize aeration rates for actual conditions once the unit is operating, and quality of circulation is often very sensitive to small changes in aeration pattern.

Standpipes operate best if straight and vertical. Bends and constrictions can sometimes be the source of problems. Standpipes are discussed further in Chapter 7.

14.3.6 Circulation control

Particle circulation in a CFB unit can be controlled by slide valves, by non-mechanical valves, by differential pressure, or by riser gas flow rate.

Slide valves are used in many FCC units and in the Sasol Synthol reactors. They are a type of gate valve in which the gate or slide can open and shut without interference from stagnant solids. They have good reliability and are used at temperatures up to 750°C. Slide valves offer considerable process flexibility and accommodate differential pressures of at least 5 to 100 kPa. Shingles and Silverman (1986) report that the relationship between the solids flow rate and pressure drop may be expressed by a standard orifice pressure drop equation, with a discharge coefficient of 0.72 based on particles flowing at minimum fluidization density. In evaluating plant flow data, one should note that the reported valve opening is usually a linear valve stem position between minimum and maximum slide travel, and open area is not directly proportional to this measure.

DESIGN AND SCALE-UP OF CFB CATALYTIC REACTORS   495

Slide valve FCC units have two controllers for the circulation system. The slide valve for hot regenerated catalyst is controlled by the temperature of the cracked oil vapor–catalyst mix leaving the top of the reactor riser. The valve for cold spent catalyst is on level control to maintain a desired pressure drop across the fluid bed in the reactor stripper. Slide valves do not have a natural loop seal to prevent reverse flow of riser gas up the standpipe. The controller is therefore set to close the valve if the valve pressure drop becomes too low, signaling imminent flow reversal.

Circulation control by adjustment of differential pressure is practiced in those FCC units that have a solids overflow well for take-off from the regenerator vessel. The overflow well automatically ensures that (at steady state) solids leave the regenerator as quickly as they enter. An increase in reactor pressure causes an increase in circulation from the reactor to the regenerator with a slight shift in inventory to the regenerator. The higher bed level in the regenerator results in a greater flow of catalyst into the overflow well. Catalyst level in the well rises slightly in response to the higher circulation rate and higher reactor pressure. Such a control scheme is used, for instance, in the Exxon Model 4 FCC design. The reactor-to-regenerator differential pressure is controlled based on reactor riser outlet temperature.

In the Model 4 FCC units, a portion of the regeneration air is used to transport spent catalyst in the riser leading to the regenerator. In the original embodiment, this transport air (called control air) was adjusted to change circulation rate, but it was later found preferable to use differential pressure control.

Non-mechanical valves have not yet been used in CFB catalytic processes, but are in common practice in CFB combustion units (see Chapter 7). Non-mechanical valves are especially attractive at temperatures above 750°C, but they are difficult to use with type A powders, which often retain enough aeration to flow rapidly even without the addition of valve control gas.

The circulation rate of catalyst is often calculated from heat balances. It may also be estimated from slide valve pressure drop and open area.

14.4 Development of specific processes

14.4.1 Fluid catalytic cracking

By far the greatest use of CFB reactors is in fluid catalytic cracking in petroleum refining. The process is discussed in greater detail in Chapter 13. The first pilot plant configuration for FCC consisted of a great length of serpentine piping through which a powdered catalyst and reactant vapors were passed at high velocity. This initial concept was a scale-up nightmare and had serious operating problems. W.K. Lewis had the inspiration of operating entirely with gas upflow and reducing velocity so as to augment increased catalyst holdup. This became a cornerstone of the first commercial unit (Model I FCC), which came on stream in May, 1942. In the Model I configuration (see Figure 13.1), both the reactor and the regenerator operated with catalyst fed at a high rate to the bottom of the vessel and entirely carried out at the top, entrained by the process gas streams. We now term this a circulating fluid bed, but it was then called an upflow unit. Three Model I units were built, with one remaining in service until 1959. An enlightening history of early cracking is given by Squires (1986).

Even before start-up of the Model I units, design and construction were underway on five units of the Model II design. These operated at lower superficial velocities with well-defined bubbling fluid beds, and catalyst was withdrawn from the bottom of the beds. Much better control over circulation and catalyst inventory was possible. Cyclones were housed within the fluid bed vessels and returned particles to the bed from which they were entrained. For the low activity catalysts then available, the Model II bubbling bed design was a decided improvement over the Model I CFB. Model II units are still in operation, some with feed rates six times greater than the original design values. Jahnig et al. (1980) has given an excellent account of the scale-up development of these early FCC units.

Reaction kinetics and reaction engineering were in their infancy when the Model I and II FCC units were developed. Scale-up of the reactor consisted simply of maintaining constant space velocity, superficial gas velocity and ratio of catalyst circulation to feed. There was no thought of residence time distributions or gas–solid contacting debits, although it was understood that backmixing would hurt performance. A small scale-up debit appeared in going from the 100 barrel (15 tonnes) per day pilot plant to the 13 000 (1950 tonnes) barrel per day first commercial unit, but that caused no significant problems. Gas–solids contacting is relatively favorable with cracking catalyst, and in any case contacting and backmixing are not critical parameters in low conversion, low selectivity reactions such as FCC. A more critical reactor scale-up problem surfaced in the regenerator, that of afterburning. Both CO and CO_2 are evolved from the catalyst surface during regeneration, and the subsequent gas phase combustion of CO by any remaining oxygen is low at the bed temperatures of the early regenerators, typically 600°C. In the dilute phase, thermal damping due to catalyst is greatly reduced, and the $CO-O_2$ reaction can become autothermal given the right conditions of temperature, residence time, and catalyst concentration. This problem was never encountered or envisioned in the pilot plant, but the critical conditions often existed commercially. Water sprays were therefore installed in the dilute phase and cyclones to prevent temperature runaway.

Until the mid-1970s, virtually all FCC units maintained a dense phase bubbling or turbulent bed in the reactor vessel. A vestigial CFB reactor remained in the feed riser, in which liquid oil feed was sprayed into a transfer line carrying hot regenerated catalyst. This configuration was used mainly because of its effectiveness in vaporizing the heavy oil feedstock, and any reaction in the riser was incidental. With the introduction of zeolites, however, catalyst activity began to increase steadily. FCC operations began reducing catalyst inventory in the dense phase reactor bed (a simple operating change). By 1980 many units were operating without any dense bed at all, and FCC units had reverted back to true CFB operation on the reactor side. While there may have been examples of CFB reactor modeling during this period, the change was primarily an evolutionary development in commercial units. There was little scale-up in this phase of development.

Once reactor dense beds were eliminated, it became widely appreciated in the industry that the low velocity dilute phase region should also be reduced or replaced. The long vapor residence time and backmixing in the dilute phase produced undesirable after-cracking, and that debit became more serious as catalysts became more active and reactor temperatures increased. To counter this problem, riser termination devices were developed to decrease entrainment of catalyst into the dilute phase, and risers were extended upwards to reduce the residence time of gas in the dilute phase. The culmination of this trend saw risers discharging directly into the cyclones, effectively eliminating the low velocity dilute phase. Chapter 13 further documents this development. This trend was again incremental and mainly in existing units. Because mechanical changes were required, it became common practice to build plastic flow visualization models to assist in the development of a suitable geometry. Typical model measurements might include pressure drop, catalyst separation efficiency, and tracer gas residence time distribution. Avidan *et al.* (1990) documented one such development. A great profusion of hardware configurations now exists for FCC riser reactors, partly because development took place simultaneously at many different locations, and partly because many improved riser configurations were retrofitted to existing units and accommodated site-specific design constraints.

14.4.2 Maleic anhydride

The oxidation of butene to maleic anhydride in a CFB catalytic reactor has been developed recently by duPont. This first commercial unit is now under construction in Asturias, Spain and is scheduled to begin operation in late 1996. Development of the process and many of its scale-up aspects has been extensively documented in a remarkable series of papers by Contractor and numerous co-workers. A general description of the process was given by Contractor (1988).

Butene oxidation is catalysed by various vanadium phosphates. As previously practiced, selectivity to maleic anhydride was adversely affected by the presence of gaseous oxygen. The duPont process uses the vanadium phosphate catalyst itself as the source of reactant oxygen, with improved selectivity. The catalyst is reduced in a CFB reactor and is then circulated to a regenerator where it is re-oxidized. The catalyst thus becomes a stoichiometric reactant for butane oxidation. Preparation of the catalyst was discussed by Contractor *et al.* (1987). Attrition resistance is an especially important property for fluid bed catalysts, and that aspect of catalyst development is covered in the above paper and also by Contractor *et al.* (1989).

As is customary for fluid bed process scale-up, a large demonstration plant was built and operated for a year (Contractor *et al.*, 1994). This featured a CFB riser reactor 0.15 m in diameter and 30 m high. The foremost objective was to demonstrate satisfactory reactor yields and production rates. Another part of the operation included the manufacture of tonnage quantities of catalyst, to ensure that the necessary activity, selectivity, and attrition resistance could be achieved in large-scale production. It was found that this semi-commercial catalyst had better attrition resistance than the laboratory catalysts.

The demonstration plant riser operated at solids mass fluxes up to $1100 \, kg/m^2 s$ and gas velocities up to 10 m/s. Engineering data on density profiles were reported by Contractor and Patience (1992) and were well correlated by the model of Patience *et al.* (1992). Slip factors approaching 3 were observed at 4 m/s and decreased to 2 at high gas velocities. Gas tracer tests showed little dispersion or bypassing at high solids mass flux ($590 \, kg/m^2 s$), but some peak broadening at $270 \, kg/m^2 s$ (Contractor *et al.*, 1994). Because of the nature of the chemistry of the reaction as well as the low degree of gas bypassing and dispersion, uncertainty in scale-up of reactor performance was not a critical issue.

A possible dangerous temperature runaway was anticipated in the event of sudden loss of catalyst circulation. In such a case it was necessary to know how rapidly catalyst already in the reactor would be removed by entrainment. This was studied in the laboratory-scale CFB unit at the City College of New York by Contractor *et al.* (1992). Weinstein *et al.* (1995) also used that unit to study the effects of gas feed nozzle configuration. A combination of pressure and X-ray measurements indicated that the best combination of solids holdup and gas–solids contacting was achieved with gas injection between the axis and the wall, rather than at the axis or at the wall.

14.4.3 The Synthol process

In 1955, Sasol in South Africa began commercial synthesis of hydrocarbons from carbon monoxide and hydrogen via the Fischer–Tropsch process. Chemistry and catalysis aspects of the process have been extensively documented by Dry (1981, 1983). An excellent overview of engineering and scale-up issues has been given by Shingles and McDonald (1988). The plant used synthesis reactors of both fixed bed and CFB designs, the latter being termed 'Synthol' reactors.

M.W. Kellogg had done the early development of the CFB process and operated a pilot plant with a 109 mm diameter riser beginning in 1947. From this, the first commercial Synthol represented a 23-fold scale-up in diameter and a 500-fold scale-up in throughput. Two of the Synthol units were built initially. Start-up was difficult, and the process could not be operated on a commercial basis until many mechanical and operating modifications had been made. Major problems included:

- Fluidization of the dense iron catalyst caused severe unit vibration, not encountered in the pilot plant. The shock dampers were totally inadequate and a new damper system had to be designed and installed.
- The Synthol reactors were closely integrated with other units in the process train. They did not have the flexibility to ride out a temporary failure in the other units; even a slight mishap often required a complete re-start of the Synthol units.
- Compressor capacity was underdesigned, and any increase in reactor pressure drop above the design value made operation of downstream equipment nearly impossible.
- Near isothermal operation was required but not initially achieved. The low temperature at the reactor inlet caused wax formation, resulting in sticky catalyst. Changes in feed preheat and reactor internals eventually solved this problem.
- Catalyst bridging often occurred in the standpipe. Backflow of synthesis gas up the standpipe was also a problem. Maintenance of a high percentage of fine catalyst was found to be essential.
- The units were designed for continuous addition of fresh catalyst and withdrawal of equilibrium catalyst. However, valve technology at that time was inadequate for the severe conditions of pressure, temperature and erosion that were encountered. The reactors were therefore operated on a cyclic basis, with shutdown and catalyst replacement every 40 days.
- Fouling and erosion of heat exchangers were serious, necessitating a modification of internals.
- Catalyst life and selectivity were low initially, but were later greatly improved.

The above problems were steadily reduced by a concerted engineering effort, and by 1960 Synthol had become a routine and profitable commercial process. A third reactor of the same basic design was started up uneventfully that year. Further gradual optimization and improvement of these reactors continued for another 14 years.

When OPEC dramatically raised oil prices in 1974, Sasol decided to build the Synthol 2 plant at Secunda. With the assistance of Badger Engineering, concerted development work was undertaken to correct some remaining design shortcomings and scale-up the reactor size. Perhaps the greatest effort was devoted to changing the reactor cooler from a fixed tube sheet design to a cooling coil. This was demonstrated on one of the operating Sasol 1 units, first with a single cooling coil between the shell and tube banks and then with coils replacing the shell and tube exchangers. Erosion and heat transfer coefficients were thoroughly studied (Shingles and Jones, 1986). Shingles and Silverman (1986) studied standpipe and slide valve operation in commercial and cold units. This led to a decrease in the relative size of standpipes in the Sasol 2 design. The vertex angle of the catalyst hopper cones was decreased in Sasol 2 based on studies showing funnel flow in the hoppers at Sasol 1.

Scale-up to the Sasol 2 design was by a factor of 3.5 on feed rate and 2.4 on reactor cross-section. Gas superficial velocity was kept constant and pressure was increased by 40%. Sasol 2 started up in 1980 and Sasol 3, built to the same design, began operation in 1982. These start-ups were smooth, with the reactors quickly achieving design conditions. There are now 19 Synthol reactors in operation, so that this CFB process may be regarded as having achieved a mature state of development.

Although the CFB process has proven successful by any measure, the continuing quest for process economies led Sasol to re-examine low-velocity fluidized bed technology (LFB). CFB and LFB reactors were compared by Silverman *et al.* (1986). A large demonstration plant for FFB synthesis had been operated by HRI in Brownsville, Texas during the early 1950s but was shut down in 1957 for economic reasons. Apparently the reactor suffered from unacceptable gas bypassing. Sasol subsequently conducted extensive studies on the effects of catalyst particle size distribution. They found that with fine type AC powders, fluidization and contacting could be much improved. Pilot plant operation demonstrated FFB reactor performance equivalent to that of the Synthol reactors. Because the FFB reactor has substantially lower capital and operating costs this is now the preferred configuration at Sasol. Shingles (1993) discussed these issues and projected commercial operation in 1995.

14.5 Closure

Beginning in the mid-1970s there was a great increase in the number of research papers published on circulating fluid beds, fast fluidized beds, high velocity fluidization, and turbulent fluidization. It remains apparent that many aspects of the basic physics of these operations are poorly understood. While there has been nothing comparable to the elegance and fundamental simplicity of the bubbling bed models developed for low-velocity non-circulating systems, this state of affairs has had little effect on the path and scale-up of CFB processes. A number of factors contribute to this:

- A large pilot plant or demonstration unit is usually necessary for many reasons besides the establishment of hydrodynamic and reaction engineering data. Such reasons may include demonstration of solids handling techniques; establishment of safe start-up, shutdown, and operating procedures; manufacture of product for testing; operator training, development of mechanical components; and exploration of operability issues such as attrition, erosion, and fouling. The availability of data from large pilot plants ensures that simple design models (e.g. for reaction, hydrodynamics, and heat transfer) can be calibrated at a reasonable scale at realistic operating conditions.
- By far the most important reaction parameters in CFB catalytic reactors are the controlled variables of gas and catalyst feed rate, temperature, and pressure. Dependent variables such as riser density and gas–solids contacting are secondary and often somewhat compensating. For example, riser density generally increases with scale because of greater segregation of gas and particle flows, and poorer gas–solids contacting therefore also results. The greater catalyst inventory somewhat counteracts the effect of poorer contacting on the reaction. Also, while scale-up debits are an unpleasant surprise, in many cases they may be unavoidable within feasible design limits.
- The debits for a process that cannot be started up as scheduled or which has a low service factor far outweigh the debits usually associated with poor reactor performance. Even a routine anticipated maintenance turnaround on an FCC unit may take 5 or 6 weeks. Having a large unit offstream is expensive. The five-year period to achieve routine commercial operation of Synthol must have been a very stressful time for the Sasol engineers.

Better design tools and greater understanding of the effects of scale will be helpful in the optimization of existing processes and in the successful scale-up of new ones. Engineers should not be deterred from CFB scale-up, however, by the current primitive state of knowledge. Enough examples of CFB and low-velocity fluid bed scale-up have been documented for one to say that most problems have surfaced and been conquered. That experience is surely the best teacher.

Nomenclature

Fr	Froude number, $U/(gD)^{0.5}$	
Fr_T	$v_T/(gD)^{0.5}$	
G_s	solids mass flux	kg/m²s
g	gravitational acceleration	9.8 m/s²

U superficial gas velocity   m/s
v_T particle terminal velocity   m/s
ϵ cross-section average voidage
ρ_p particle density   kg/m^3

References

Avidan, A., Krambeck, F.J., Owen, H. and Schipper, P.H. (1990) FCC closed cyclone system eliminates post-riser cracking. *Oil and Gas Journal*, March 26.

Contractor, R.M. (1988) Butane oxidation to maleic anhydride in a recirculating solids riser reactor, in *Circulating Fluidized Bed Technology II* (eds P. Basu and J.F. Large), Pergamon Press, Oxford, pp. 467–474.

Contractor, R.M. and Patience, G.S. (1992) Density profiles in a tall experimental circulating fluidized bed. A.I.Ch.E. Annual Meeting, Miami.

Contractor, R.M., Bergna, H.E., Horowitz, H.S., Blackstone, C.M., Malone, B., Torardi, J.C.C., Griffith, B., Chowdhury, D. and Sleight, D.W. (1987) Butane oxidation to maleic anhydride over vanadium phosphate catalysts. *Catalysis Today* **1**, 49–58.

Contractor, R.M., Bergna, H.E., Chowdhury, U. and Sleight, A.W. (1989) Attrition resistant catalysts for fluidized bed systems, in *Fluidization VI* (eds J.R. Grace, L.W. Shemilt and M.A. Bergougnou), Engineering Foundation, New York, pp. 589–596.

Contractor, R.M., Pell, M., Weinstein, H. and Feindt, H.J. (1992) The rate of solids loss in a circulating fluid bed following a loss of circulation accident, in *Fluidization VII* (eds O.E. Potter and D.J. Nicklin), Engineering Foundation, New York, pp. 243–248.

Contractor, R.M., Patience, G.S., Garnett, D.I., Horowitz, H.S., Siler, G.M. and Bergna, H.E. (1994) A new process for n-butane oxidation in maleic anhydride using a circulating fluidized bed reactor, in *Circulating Fluidized Bed Technology IV* (ed. A. Avidan), A.I.Ch.E., New York, pp. 387–391.

Dry, M.E. (1981) The Fischer Tropsch synthesis, in *Catalysis* (eds J.R. Anderson and M. Boudart), Springer Verlag, Berlin, pp. 159–255.

Dry, M.E. (1983) The Sasol Fischer-Tropsch process, in *Applied Industrial Catalysis*, Vol. 2 (ed. B.G. Leach), Academic Press, New York, pp. 167–213.

Jahnig, C.E., Campbell, D.L. and Martin, H.Z. (1980) History of fluidized solids development at Exxon, in *Fluidization* (eds J.R. Grace and J.M. Matsen), Plenum Press, New York, pp. 3–24.

Krambeck, F.J., Avidan, A.A., Lee, C.K. and Lo, M.N. (1987) Predicting fluid bed reactor efficiency using adsorbing gas tracers. *A.I.Ch.E. Journal*, **33**, 1727–1734.

Matsen, J.M. (1973) Flow of fluidized solids and bubbles in standpipes and risers. *Powder Technology* **7**, 93–96.

May, W.G. (1959) Fluidized bed reactor studies. *Chem. Eng. Prog.* **55**(12), 49–56.

Miller, R.B., Johnson, T.E., Johnson, D.L., Avidan, A.A. and Schipper, P.H. (1994) Third generation FCC feed injection technology: introduction of the ATOMAX feed nozzle. Presentation AM-94-55, Annual NPRA meeting, San Antonio, Texas.

Ouyang, S., Li, X.-G. and Potter, O.E. (1995) Circulating fluidized bed as a catalytic reactor: Experimental study. *A.I.Ch.E. Journal*, **41**, 1534–1542.

Patience, G.S., Chaouki, J., Berruti, F. and Wong, R. (1992) Scaling considerations for circulating fluidized bed risers. *Powder Technol.* **72**, 31–37.

Pustelnek, P. and Nauman, E.B. (1991) Contact time distributions in a large fluidized bed. *A.I.Ch.E. Journal*, **37**, 1589–1592.

Shingles, T. (1993) Why CFBs for exothermic catalytic reactions? Workshop presentation at 4th International Conference on Circulating Fluidized Beds, Somerset PA.

Shingles, T. and Jones, D.H. (1986) The development of Synthol circulating fluidized bed reactors. *Chem SA*, **12**(8), pp. 179–182.

Shingles, T. and McDonald, A.F. (1988) Commercial experience with Synthol CFB reactors, in *Circulating Fluidized Bed Technology II* (eds P. Basu and J.F. Large), Pergamon Press, Oxford, pp. 43–50.

Shingles, T. and Silverman, R.W. (1986) Determination of standpipe pressure profiles and slide valve orifice discharge coefficients on Synthol circulating fluidized bed reactors. *Powder Technology*, 129–136.

Silverman, R.W., Thompson, A.H., Steynberg, A., Yukawa, Y. and Shingles, T. (1986) Development of a dense phase fluidized bed Fischer–Tropsch reactor, in *Fluidization V* (eds K. Ostergaard and A. Sorensen), Engineering Foundation, New York, pp. 441–448.

Squires, A. (1986) The story of fluid catalytic cracking, the first circulating fluidized bed, in *Circulating Fluidized Bed Technology* (ed. P. Basu), Pergamon Press, Toronto, pp. 1–19.

Weinstein, H., Feindt, H.J., Graff, R.A., Pell, M., Contractor, R.M. and Jordan, S.P. (1995) Riser gas feed nozzle configuration effects on the acceleration and distribution of solids, Proceedings 8th International Fluidization Conference, Tours, France, May, in press.

Westphalen, D. and Glicksman, L.R. (1994) Experimental verification of scaling for a commercial size CFB combustor, in *Circulating Fluidized Bed Technology IV* (ed. A.A. Avidan), A.I.Ch.E., New York, pp. 436–441.

15 Reactor modeling for high-velocity fluidized beds
JOHN R. GRACE AND K. SENG LIM

15.1 Introduction

Reactor models are idealized human constructs that attempt to capture the essence of the behaviour of the flow, mixing and contacting of reacting species and phases in order to be able to predict reactor conversions, yields and dynamic responses. Accurate modeling also requires an adequate representation of the chemical kinetics, while other factors such as heat transfer and thermodynamics may also play significant roles. Models are more complex for multi-phase reactors than for single-phase reactors as they must take into account the mixing behaviour within each individual phase, as well as the exchange of reactants and products between the phases. Each of these aspects is directly influenced by the flow patterns throughout the entire reactor.

There has been considerable work on the modeling of low-velocity fluidized bed (LFB) reactors operated in the bubbling and slugging hydrodynamic regimes (see Grace, 1986 for a review). In those regimes it is essential to consider the bed to be comprised of two distinct phases – a dilute (also called dispersed or bubble) phase and a dense (also called continuous or emulsion) phase. It is most common to consider one-dimensional flow in each of these phases. Competing models then make a series of assumptions regarding the axial mixing in each of the phases and the interphase transfer between the phases, usually described in terms of a single mass transfer coefficient. Competing models also rely on different descriptions of hydrodynamic features such as bubbles. Heat transfer is usually considered to be rapid enough for these low-velocity regimes that temperature gradients can be ignored. In some cases, separate allowance is made for the grid region upstream of the bubbling zone and/or the freeboard region downstream.

As interest in high-velocity fluidized beds operating in the turbulent and fast fluidization regimes has grown, attempts have been made to provide appropriate reactor models, often by extending models initially devised for the bubbling regime or derived for simpler reactors. There have now been sufficient experimental work and modeling efforts that some guidance can be provided to those wishing to use models for design, simulation or process control. This chapter gives a brief overview of this body of work.

15.2 Turbulent regime

As noted in Chapter 1, transition to the turbulent regime results in a decrease in the prevalence of large voids and provides a transition to the fast fluidization regime. The turbulent regime is of interest in its own right for a number of fluidized bed processes, but also because the bottom regions of many CFB risers operate in the turbulent fluidization regime.

Models of turbulent fluidized bed reactors have all assumed one-dimensional gas flow (i.e. have ignored lateral gradients), despite the significant radial density gradients reported by Abed (1984). The earliest workers (Van Swaaij, 1978; Wen, 1979; Fane and Wen, 1982) proposed single-phase plug flow of gas. On the other hand, several of the models discussed below for circulating fluidized beds assume that the turbulent lower region of the riser can be treated as a CSTR (i.e. in terms of perfect mixing). Neither of these approaches, however, seems to have been based on direct experimental evidence.

A better approach (Avidan, 1982; Edwards and Avidan, 1986; Baerns et al., 1994; Foka et al., 1994) is to assume axially dispersed plug flow of gas. Some experimentally derived axial Peclet numbers are given in Table 15.1 based on tracer studies. Note that the results of Edwards and Avidan (1986) suggest that axial dispersion is reduced by the addition of ring-type horizontal baffles and by reducing the bed diameter. Li and Wu (1991) correlated their data for the turbulent regime by

$$\mathcal{D}_{g,ax} = 0.84\epsilon^{-4.445} \quad \text{(SI units)} \tag{15.1}$$

where ϵ is the overall voidage in the turbulent regime. Until there are more data available, this equation can be used to estimate the gas axial dispersion coefficient for turbulent beds or for turbulent regions at the base of CFB risers.

As far as the solids are concerned, the particles almost always stay in the bed long enough that perfect mixing of the solids is a good assumption for modeling purposes.

Table 15.1 Values of axial Peclet number $Pe = UL/\mathcal{D}_{g,ax}$ for turbulent fluidized beds

Authors	Particles	D, m	U, m/s	Baffles	Pe
Edwards and Avidan (1986)	catalyst, $\bar{d}_p$ and ρ_p not specified	0.1 0.6 0.6	0.6 0.3 to 1.0 0.3 to 1.0	none none horizontal	10 7 12
Foka et al. (1994)	catalyst, $\bar{d}_p = 196\,\mu\text{m}$ $\rho_p = 1100\,\text{kg/m}^3$	0.1	1.07 to 1.24	none	1.6 to 3.6
Li and Wu (1991)	FCC, $\bar{d}_p = 58\,\mu\text{m}$, $\rho_p = 1575\,\text{kg/m}^3$	0.09	<1.3	none	<8

15.3 Fast fluidization: single-region one-dimensional models

Gas mixing in circulating fluidized beds is considered in detail in Chapter 3. Early reviews of the characteristics of CFB reactors have been given by Grace (1990) and Contractor and Chaouki (1991). A summary of models for CFB combustors was prepared by Sanderson (1993).

The simplest approach for predicting gas mixing in risers operating in the fast fluidization regime is to ignore lateral or radial gradients, thereby treating the entire cross-section of the riser as if it were uniform. Models that make this assumption may then be further subdivided into those that also assume hydrodynamic uniformity in the axial direction (represented schematically by Figure 15.1a) and those which allow for axial gradients in hydrodynamic properties. These two cases are considered below.

15.3.1 Without allowance for hydrodynamic axial gradients

The simplest possible approach, adopted or tested in several reactor models (e.g. Hastaoglu *et al.*, 1988; Gianetto *et al.*, 1990; Ouyang *et al.*, 1993) is to assume single-region plug flow of gas through the entire riser while ignoring axial variations in voidage and other hydrodynamic properties along the riser. While this may be a reasonable approximation for some conditions (Martin *et al.*, 1992), it leads to significant over-prediction of conversions

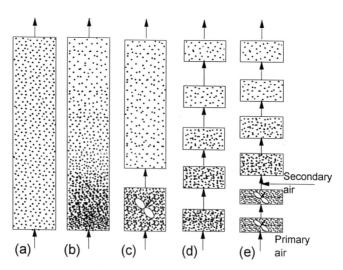

Figure 15.1 Schematic diagram showing various single-region one-dimensional reactor models for CFB risers: (a) without hydrodynamic axial gradients; (b) with axial variation in voidage; (c) with separate turbulent bed region at the bottom; (d) cells in series with gradient in voidage from cell to cell; (e) Arena *et al.* (1995) model with perfect mixing of gas in n_p cells below secondary air injection level and plug flow in n_s cells above this level.

in other cases (e.g. Ouyang *et al.*, 1995b). This approach ignores axial dispersion caused by both non-uniform radial profiles of gas velocity and down-flowing solids at the wall (see Chapters 2 and 3).

Several authors (e.g. Gianetto *et al.*, 1990; Marmo *et al.*, 1995) have introduced axial dispersion of gas in their models to account for deviations from plug flow. Ouyang *et al.* (1993, 1995a) also tested a perfect mixing (i.e. CSTR) model. However, reactor models that assume both one-dimensional flow and axial uniformity of hydrodynamic properties do not give an appropriate description of the actual behaviour in CFB risers, with the result that these approaches have not been found to be very successful. If they are employed, they then require introduction of correction factors, such as an effectiveness factor (Ouyang *et al.*, 1993), to fit reactor data to the models.

15.3.2 With allowance for hydrodynamic axial gradients

As shown in Chapter 2, there tend to be marked axial variations in solids concentration in CFB risers. In addition, axial gradients may be caused (as in most CFB combustion systems, see Chapters 10 and 11) by introducing some of the gas at the bottom and additional (secondary) gas at a higher level, and this can lead to a change in hydrodynamic regime. A number of reactor models have sought to allow for these variations in solids concentrations and/or hydrodynamic regime, while maintaining the assumption of radial uniformity.

The simplest addition to the models of the previous sub-section, indicated schematically in Figure 15.1b, is to retain plug flow while allowing for gradients of solids concentration with height as in the models of Lee and Hyppanen (1989), Pagliolico *et al.* (1992) and Marmo *et al.* (1995). The axial variation in concentration can be derived from experimental pressure profile measurements along the riser, based on the voidage profile expression introduced by Li and Kwauk (1980), or obtained from other empirical expressions as in the model of Pagliolico *et al.* (1992) where an exponential profile was fitted to experimental pressure gradient data.

To allow for the relatively dense bottom region, especially when secondary gas is introduced part way up the riser, a number of models have assumed that the lower part of the riser operates as a turbulent bed (Figure 15.1c). Interestingly, gas flow in this region has then commonly been taken (e.g. Sotudeh Gharebaagh *et al.*, 1994; Marmo *et al.*, 1995) to be perfectly mixed, rather than in plug flow or axially dispersed plug flow as in the models discussed in section 15.2 above. Alternatively (Figure 15.1d), a number of models (e.g. Weiss and Fett, 1986; Muir, 1995, Dersch *et al.*, 1995) have divided the riser into a number of compartments or cells in series, with allowance for different particle concentrations in each cell.

Li and Zhang (1994) outline a model for predicting various features of solid fuel combustion in circulating fluidized beds. The axial voidage profile of Li and Kwauk (1980) is assumed, with lateral gradients ignored. Simplified chemical kinetic expressions are assumed. The effect of temperature, particle size, excess air, primary/secondary air ratio and calcium/sulphur molar ratio on combustion efficiency, SO_2 capture and NO_x emissions are predicted by the model, showing trends that are consistent with operating experience.

In the CFBC model of Arena *et al.* (1995), the cell heights and mixing characteristics are assumed to change once the secondary air injection level is passed on proceeding up the riser. The particles are taken as fully mixed in all cells (i.e. above and below the secondary air nozzles); however, the gas is assumed to be in plug flow for all cells above the secondary air injection level while being perfectly mixed in each cell below the secondary injection level. These mixing patterns are shown schematically in Figure 15.1e. In the model of Zhang *et al.* (1991), the region below the secondary nozzles is treated as a bubbling bed, followed by several well-mixed compartments in the upper part of the riser.

In the case of Jiang *et al.* (1991), the riser is divided into five compartments in series, with the boundaries between compartments corresponding to the heights of four annular-ring horizontal baffles mounted around the periphery of the column to interrupt the downward flow of solids along the outer wall of the column. The solid particles are assumed to be perfectly mixed within each compartment, while two cases were considered for the gas, one with perfect mixing in each compartment and the other with plug flow.

While these models can be used to provide a first approximation of reactor characteristics, they ignore lateral gradients that are major factors in circulating fluidized beds. We turn next to models that make allowance for these gradients.

15.4 Fast fluidization: core/annulus models

Recognizing the finding (e.g. see Chapter 2) that most circulating fluidized beds operating in the fast fluidization regime are subject to predominantly downflow of relatively dense streamers along the outer wall while there is net dilute upflow in the core, a number of modelers have adopted core/annulus two-region models. In these models the gas and solids flows are assumed to be quite different in the outer annular region than in the inner core, with mass transfer back and forth across the interface between the two regions. The flow in each of these regions is commonly (but not always) assumed to be one-dimensional. As with the models treated in section 15.3, some models (covered in the next sub-section) assume longitudinal hydrodynamic uniformity, while others (section 15.4.2) also allow for axial hydrodynamic variations.

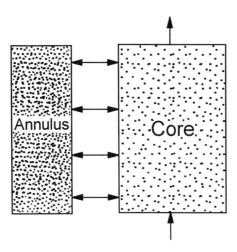

Figure 15.2 Schematic diagram showing basic core/annulus model used to describe gas mixing and gas–solid contacting in CFB risers; axial variations in hydrodynamic properties are ignored.

15.4.1 Without allowance for hydrodynamic axial gradients

The earliest core-annulus model was that of Brereton *et al.* (1988), who assumed that all of the gas passes upwards in plug flow through a central core, this gas being exchanged with an outer annular region where the gas is stationary. A schematic showing this simple two-region model appears in Figure 15.2. Each of the two-regions was assumed to be radially uniform. An inter-region mass transfer coefficient and the radius of the core, both assumed to be independent of height, were obtained by fitting the models to experimental axial gas mixing data. This model was applied to a first-order reaction with steady state operation by Kagawa *et al.* (1991) and Marmo *et al.* (1995) leading to mass balance equations:

core region: $$U\left(\frac{R^2}{r_c^2}\right)\frac{dC_c}{dz} + \frac{2K}{r_c}(C_c - C_a) + k_1(1 - \epsilon_c)C_c = 0 \quad (15.2)$$

annulus region: $$\left(\frac{r_c^2}{R^2 - r_c^2}\right)\frac{2K}{r_c}(C_a - C_c) + k_1(1 - \epsilon_a)C_a = 0 \quad (15.3)$$

where K and k_1 are the cross-flow mass transfer coefficient and first-order kinetic rate constant, respectively. Here the particles are assumed to be small enough that both intra-particle diffusion resistances and external particle mass transfer do not constitute factors to be considered. The boundary condition at $z = 0$ is $C_c = C_0$. Note that this model is formally analogous to the two-phase bubbling bed model (Grace, 1986), with the annulus replacing the (stagnant) dense phase and the dilute core replacing the bubble phase. Solution of these equations is straightforward and leads to a concentration at the exit of the riser of

$$C_H = C_0 \exp\left\{-k_1^*\left[\frac{(1-\epsilon_a)(1-\phi^2)X}{k_1^*(1-\epsilon_a)(1-\phi^2)+X} + (1-\epsilon_c)\phi^2\right]\right\} \quad (15.4)$$

where $\phi = r_c/R$ is the ratio of core radius to riser radius,

$$k_1^* = \frac{k_1 H}{U} \quad (15.5)$$

is a Damkohler number or dimensionless reaction rate constant, and

$$X = \frac{2KHr_c}{UR^2} \quad (15.6)$$

is a dimensionless inter-region mass transfer coefficient.

Some predictions from this model are shown in Figure 15.3 for overall voidages, $\bar{\epsilon}$, of 0.88 and 0.98 with values of ϕ, ϵ_a and ϵ_c as obtained from Kagawa et al. (1991) giving

$$\phi = 0.85; \quad \epsilon_c = 0.4 + 0.6\bar{\epsilon}; \quad \epsilon_a = 2\bar{\epsilon} - 1 \quad (15.7)$$

As expected, the conversion (given by one minus the ordinate in Figure 15.3) increases with increasing kinetic rate constant, increasing inter-region mass transfer coefficient, increasing riser height, decreasing superficial gas velocity and decreasing overall voidage. In the work of Kagawa et al. (1991), the inter-region mass transfer coefficient was varied and found to give a good fit to experimentally measured axial gas concentration profiles for $K = 0.001$ m/s.

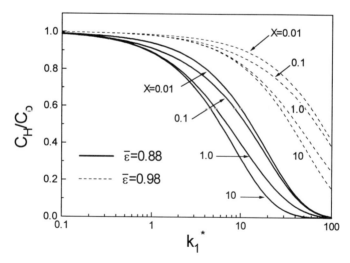

Figure 15.3 Outlet concentration for a first-order reaction in a CFB riser for a simple core/annulus model with $\phi = r_c/R = 0.85$ and core and annulus voidage given by equation (15.7).

White et al. (1992) performed gas mixing studies and fitted their data to a core/annulus model. The assumption of stagnant gas flow in the annulus zone was found to be appropriate. Fitted cross-flow coefficients then ranged from about 0.015 to 0.06 m/s, being higher for 71 μm FCC particles than for 140 μm sand and increasing with increasing solids circulation rate and decreasing gas velocities. The fitted r_c/R values ranged from about 0.53 to 0.95, decreasing with increasing G_s and increasing with increasing U.

Patience and Chaouki (1993) carried out residence time distribution tests using a radioactive tracer gas in a 83 mm diameter riser using a similar model. Both r_c/R and the cross-flow mass transfer coefficient, K, were obtained by fitting, leading to values of 0.72 to 0.99 and 0.031 to 0.11 m/s, respectively. Comparison of their data with other data in the literature led to the conclusion that r_c/R increases with gas velocity and particle density, while decreasing with riser diameter and solids circulation rate.

A similar model has been applied by Zhao (1992), but with more complex chemical kinetics, to predict NO_x variations for fluidized bed combustors. Oxygen concentration profiles were uncoupled from the NO_x reactions and assumed to be determined only by the char combustion, with lower oxygen concentrations in the wall layer due to the higher concentration of char there. The core/annulus mass transfer coefficient was estimated to be 0.10 m/s in order to give 3% oxygen at the outlet, and then this value of K was assumed to apply to the inter-region mass transfer of other gaseous species. All volatile nitrogen was assumed to be released as ammonia. The voidages in the core and annulus regions were taken as 0.98 and 0.60, respectively, while r_c/R was taken as 0.87. Predicted NO profiles showed trends consistent with experimentally measured profiles.

Variants on the above models have been proposed by several others. Werther et al. (1992) allowed for radial dispersion in the central core, whereas radial gradients were ignored in the wall region. In the wall zone, there were assumed to be alternate periods of upflow of a dilute phase and downflow of a dense phase. Based on hydrodynamic studies, the authors assumed $r_c/R = 0.85$, a downflow velocity in the annulus region of 1 m/s and 80% volume fraction of dilute phase in the outer wall region. With these values, tracer gas experiments in a 0.4 m diameter riser of height 8.5 m with 130 μm sand and different values of the dilute phase gas velocity in the annulus led to a volumetric interphase mass transfer coefficient of $0.23 \, \text{s}^{-1}$ between the dense and lean phases. This value was insensitive to the solids circulation rate for the range of conditions studied.

Amos et al. (1993) also allowed for radial dispersion in the core zone. In their case, the core occupied the entire riser cross-sectional area, with the annulus considered as a separate recycling region without radial gradients beyond the containing wall of the vessel. The model was fitted using gas mixing data where tracer was injected in the axis of a 152 mm diameter riser.

Ouyang and Potter (1994) and Ouyang *et al.* (1995b) considered a model with downward flow of both solids and gas in the annulus region, leading to a split boundary value problem whose solution is significantly more difficult to obtain than for the stagnant annulus case. Experimental ozone concentration profiles using FCC particles, impregnated with ferric oxide to catalyse ozone decomposition, were relatively flat in the core, but fell off in the wall region (Ouyang *et al.*, 1995a, b). The authors also found some circumferential variations in gas concentration, especially in the bottom section of the reactor near where solids re-enter the riser from the recirculation loop (Ouyang *et al.*, 1995a). One way of dealing with deviations from idealized models was to introduce effectiveness factor corrections, supposedly to account for mass transfer resistances imposed by cluster-like structures.

15.4.2 With allowance for hydrodynamic axial gradients

The next level of sophistication consists of models that recognize both the core/annulus structure of gas–solids flow in CFB risers and the change in solids density (and possibly flow regime) with height. A number of such models have been proposed in recent times.

The simplest examples of such models place one or more zones upstream of (i.e. below) the core/annulus structure, which is assumed to prevail throughout the upper part of the riser. For example, Marmo *et al.* (1995) explored a model where the lower zone of the riser acted as a CSTR, with a core/annulus division above this. The height of the CSTR was kept as a fitting parameter, while the core/annulus mass transfer coefficient was obtained from the results of Patience and Chaouki (1993) and voidages in the core and annulus were estimated from a correlation of Zhang *et al.* (1991). Saraiva *et al.* (1993) considered a bubbling bed zone upstream of a region represented by core/annulus flow in order to model reactions in a CFB combustion unit. This concept was extended a step further by Neidel *et al.* (1995) who inserted an acceleration zone between the bubbling bed region and the upper part of the riser, the latter being subdivided into a wall zone and core region (see Figure 15.4). On the other hand, Pugsley *et al.* (1994) represented both the bottom and top zones of a riser in terms of core/annulus flow structures, with acceleration of solids occurring in the lower zone and all the gas travelling up the core.

A somewhat different concept was utilized by Kunii and Levenspiel (1990). A lower dense zone of overall voidage 0.75 to 0.85 was treated as a combination of a lean quickly rising core region and slow-moving 'clumps' of emulsion and wall region. Their bubbling bed model was extended to cover this combination. The upper zone was treated like the freeboard of a bubbling bed with exponentially decaying solids concentration.

Talukdar *et al.* (1994) considered a turbulent bed zone below the level where secondary air is injected into a CFB combustor, and a fast fluidized

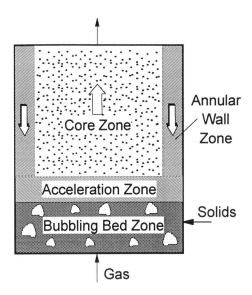

Figure 15.4 Schematic showing zones considered in the model of Neidel *et al.* (1995), an example of a core/annulus model with allowance for hydrodynamic axial gradients.

bed above this level. The lower turbulent zone was divided into three 'cells' in series, while the upper fast bed was divided into ten separate cells, each including a core and an annulus region. With unequal rates of solids transfer inwards and outwards across the core/annulus interface, the size of the wall region changes with height in this model. Further results of this model were given by Talukdar and Basu (1995).

In the CFB reactor model of Pugsley *et al.* (1992), applied to the case of catalytic oxidation of n-butane to maleic anhydride, the authors allow for up to three zones in series – an acceleration zone corresponding to the dense region at the bottom of the riser, a fully developed zone occupying most of the riser, and a deceleration zone at the top. The third of these is omitted for risers with smooth exits. The voidage in the acceleration zone was obtained from an expression of Wong *et al.* (1992), while the voidage in the fully developed zone was estimated assuming

$$\text{Slip factor} = \rho_p U(1 - \bar{\epsilon})/(\bar{\epsilon} G_s) = 2 \qquad (15.8)$$

Exchange of solids from the core to the annulus was based on an equation due to Bolton and Davidson (1988) for turbulent diffusion, whereas the reverse transfer was obtained from an analogy with gas–liquid annular flow proposed by Senior and Brereton (1992). The core/annulus gas exchange coefficient was estimated from the Higbie penetration theory, giving values of the order of 10^{-2} m/s. Predictions of the model suggest that conversion will increase with increasing G_s and with decreasing U. A preliminary catalyst

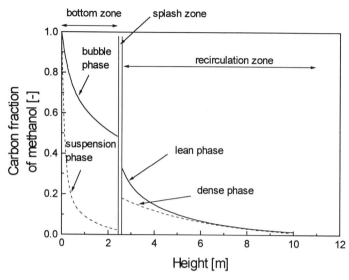

Figure 15.5 Predicted axial concentration profiles of reactant (methanol) predicted by the model of Schoenfelder *et al.* (1994) for the MTO process.

deactivation model suggests that deactivation is not a major factor for the conditions simulated.

Four zones in series are considered in the model of Schoenfelder *et al.* (1994), a bubbling zone at the bottom (treated with a two-phase bubble model), a splash zone (treated as a CSTR), a recirculation zone, and an exit region. The latter merely serves to mix the contents of the riser at the top without further reaction. The recirculation zone, occupying most of the riser length, is considered to be occupied by two regions of one-dimensional flow which, though not explicitly labeled as core and annulus, can be thought of in this light. Hence the model can be represented schematically as in Figure 15.4, only with the splash zone CSTR replacing the acceleration zone. Figure 15.5 shows axial profiles of reactant concentration based on the model for the methanol-to-olefins (MTO) process. Note the abrupt changes at the boundaries between the zones. The model suggests that inter-region mass transfer in the recirculation zone is significantly more rapid than the interphase (bubble-to-dense phase) transfer in the lowest zone.

This model was extended by Kruse *et al.* (1995) to consider two-dimensional flow in both the dilute core and outer dense annular region of the recirculation zone, relabeled as the 'upper dilute zone'. After fitting a radial dispersion coefficient, assumed to be identical in the core and outer annular zones, with the aid of tracer axial injection tests, a crossflow coefficient was fitted to measured concentrations of a tracer introduced through a ring distributor. The volume fraction of solids in the outer dense region of the upper dilute zone was also obtained by fitting tracer concentration profiles. The radial Peclet number, $UD/D_{g,r}$ averaged 387 for test results in a column of diameter 0.4 m with 163 μm sand particles.

Sung (1995) modeled NO_x and N_2O reactions in a circulating fluidized bed combustor unit based on the hydrodynamic model of Senior and Brereton (1992). Above the secondary air level, the model allows for an annulus phase to grow thicker or thinner with height, depending on the relative calculated magnitude of inward and outward solids flows. Sung simulated the solids distribution below the secondary air injection level by means of a core/annulus model. The model also allows the voidage in the core and annulus and the core/annulus interfacial perimeter to be estimated as a function of height. Stagnant gas flow was assumed in the annulus region, with all gas passing through the core. Fifteen reactions were included in the chemical kinetics scheme, and a devolatilization model was also provided. A core/annulus mass transfer coefficient of order 0.1 m/s gave predictions for oxygen close to experimental values, while NO and N_2O profiles were better predicted if their inter-region mass transfer coefficients were an order of magnitude higher.

15.5 Some experimental findings

A number of studies have been performed in recent years showing concentration profiles in CFB reactors. Grace *et al.* (1990) demonstrated that there are substantial lateral concentration gradients in a pilot-scale CFB combustion unit. Because of the substantially higher concentration of char near the walls of the riser, oxygen concentrations were substantially lower there than in the core of the vessel. Lateral concentration gradients of other reacting species like unburnt hydrocarbons, SO_2 and NO_x, could be explained also in terms of the predominance of particles in the wall region. Further detailed lateral profiles of concentration in the same CFBC unit have been reported by Zhao (1992) and Sung (1995).

Similar findings have been reported (Jiang *et al.*, 1991; Ouyang *et al.*, 1995a, b; Schoenfelder *et al.*, 1996) for the catalytic ozone decomposition reaction. One such set of profiles is plotted in Figure 15.6. While the experimental concentration profiles confirm that the reactant concentrations are significantly lower at the wall than in the interior, they suggest that the concept of a sharply defined core/annulus boundary with one-dimensional flow on either side of this boundary is oversimplified. In reality, the concentration profiles are smooth rather than having a sharp transition at a core/annulus boundary, as one would expect from the models described in section 15.4. This is also consistent with radial profiles of solids concentration (e.g. Berker and Tulig, 1986), which show a continuous change with radial position.

Given the pronounced lateral gradients and also the possibility of tangential gradients (Ouyang *et al.*, 1995a) and volatile-rich zones originating from

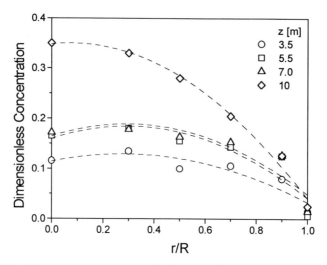

Figure 15.6 Radial ozone concentration profiles determined by Ouyang *et al.* (1955a) in a 0.25 m diameter CFB reactor with $U = 3.8$ m/s, $G_s = 106$ kg/m²s and $k_1 = 57.2$ s^{-1}.

solids feed points in CFBC units (Chapter 10), it is important not to assume that gaseous concentrations measured at one position are representative of concentrations at other locations for the same level, or even that there is symmetry around the axis of the column. Experimental axial concentration profiles that indicate increases in ozone concentration with height near the bottom (Jiang *et al.*, 1991; Ouyang *et al.*, 1995a) presumably reflect such local variations. Lateral gradients and lack of symmetry are likely to be especially prominent in square (or rectangular) columns, where solids travel downwards primarily in the corners (Zhou *et al.*, 1995), or in the vicinity of solids feed-points or return-positions.

15.6 Other models

While most CFB reactor models are covered in the sections above, there are some models which differ fundamentally. We provide a brief introduction to these in this section.

15.6.1 Cluster/gas two-phase model

Fligner *et al.* (1994) presented a model in which there are two phases, a dispersed cluster phase containing all the particles and a surrounding continuous phase consisting only of gas, for FCC risers operated at sufficiently high gas velocities that there is no downflow of solids at the wall. The clusters are assumed to have voidage equal to ϵ_{mf}, to be spherical in shape and to have sizes calculated from their slip velocities and drag as proposed by

Yerushalmi et al. (1978). Radial density distributions were measured by gamma-ray densitometry in a 0.3 m diameter cold flow riser of height 12 m at two levels using typical commercial gas velocities and catalyst fluxes. Solids fluxes were measured at the same positions by sampling. Mass transfer is assumed to take place at the outer surface of the clusters as they travel up the risers. Local and average exit conversion data from a commercial FCC reactor were used for parameter fitting. The model was then applied to predict the effect of changes in solids feed configuration.

15.6.2 Co-existing upflow/downflow model

Schoenfelder et al. (1996a, b) recently proposed a promising model which uses experimentally measured suspension density profiles and lateral profiles of solids flux (both upward and downward) to account for intermittency in the upper part of the riser. A simple CSTR model is applied at the bottom. The model allows upflow of lean suspension to co-exist with downflow of clusters at every point in the upper zone. This avoids the step changes of core-annulus models (see above) and gives realistic lateral concentration profiles.

15.6.3 Monte Carlo model

In the model developed by Ju (1995), the riser is divided into 40 cells, 20 for the core and 20 for the annulus. A series of fuel particles is then considered, one by one, to move through the system subject to the laws of chance, with the probability of being transferred to or from the adjacent wall or core cell, or upwards or downwards to the next cell, simply given by the ratio of mass flow out of the cell in a given direction to the mass of the particles in that cell. The time steps were chosen so that the probability of the fuel particle leaving that cell in each time step is 10%. It is assumed that the particles in the core move only upwards or outwards, whereas particles in the annulus cells can move only downwards or inwards. A fraction of the particles reaching the top of the riser is assumed to be reflected back downwards, the rest entering the cyclone, which is treated as a CSTR. Each fuel particle is assumed to first undergo devolatilization, then char combustion. Keeping track of the cell and time at which heat is released allows the heat release distribution to be compiled. The temperature and oxygen partial pressure distributions must initially be guessed, then updated once the heat release distribution is calculated. The model algorithm is shown in Figure 15.7.

The Monte Carlo method offers advantages in programming relative to models where a series of differential equations must be solved iteratively. For the conditions studied, tracking as few as 100 particles gave reproducible results. A modular structure was adopted so that separate subroutines for devolatilization, plume evolution, char combustion, oxygen profiles and temperature prediction can be adopted. The model was shown to give correct

518   CIRCULATING FLUIDIZED BEDS

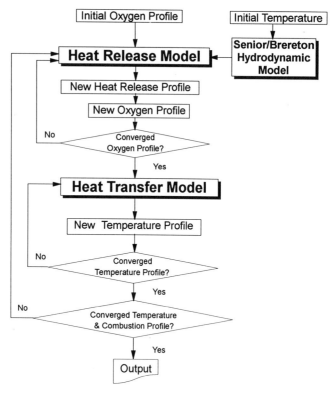

Figure 15.7 Algorithm used to obtain converged profiles of oxygen concentration, temperature and heat release in the Monte Carlo CFB combustion model of Ju (1995).

trends in comparison with a pilot scale CFBC unit and a 2.5 MW$_{th}$ prototype CFBC unit.

15.6.4 Models based on solving fundamental equations

A number of models have been established and applied to CFB risers based on solving fundamental continuity, momentum and energy equations using finite difference or finite element techniques. The approach has been reviewed in Chapter 5 and by Gidaspow (1994). In principle, it is not a major step to add chemical reaction terms to the species mass conservation equations to allow predictions to be made for reactors.

One application of such modeling was reported by Hyppanen et al. (1991), who established a fully three-dimensional model to describe commercial circulating fluidized bed combustors. Results are presented for a 85 MW$_{th}$ boiler obtained using a $8 \times 12 \times 20$ grid with the solid particles divided into five separate size fractions. The calculations show high volatile concentrations and low oxygen concentrations near the outer walls, similar to measured profiles discussed above. Insufficient experimental data were available to allow detailed comparison.

15.6.5 Transient models

All the models described so far have been steady state models, written for reactors operating under steady state conditions. However, it is important to establish unsteady or transient models capable of being used for reactor control, predicting start-up and shut-down, and calculating the response to upset conditions.

Several attempts have been made to provide transient models for CFB combustors. Whereas almost all of the models considered so far in this chapter have considered only what happens within the riser, transient combustion models must also take into account:

- time derivative terms in the various equations;
- the recirculation of solids through the external recirculation loop and the inventory of solids there, which affect the recycle period;
- the capture efficiency and return characteristics of both primary and secondary separators;
- the thermal capacity of refractory materials and the heat transfer to these surfaces, especially for small units which have high surface-to-volume ratios;
- the steam-side characteristics of the boilers;
- the temperature dependence of the kinetic rate constants and possibly of other properties;
- reactions in the cyclones and possibly also in the downcomer(s).

These factors make transient models considerably more cumbersome than steady state models. In addition, few experimental studies have been carried out on the dynamic behaviour of CFB reactors.

Control models can be either mechanistically or empirically based. Mechanistic models incorporate mass and energy balances and generally allow simulation and design as well as control, whereas empirical models may be more helpful when the process is poorly understood and operate within a relatively narrow range of conditions.

The earliest dynamic model for a CFB reactor system was that of Weiss and Fett (1986) who simulated the start-up conditions in a reactor where there is thermal decomposition of sodium bicarbonate. The riser and return loop were both divided into cells and allowance was made for backmixing. The predictions show cycling of the CO_2 concentration after introduction of the feed. The model was later extended (Weiss et al., 1988) to apply to CFB combustion.

Zhang et al. (1991) developed a dynamic model for CFB combustion in which the riser is divided into a dense bubbling bed and a dilute zone, with the latter subdivided into a number of cells in series. The hydrodynamics

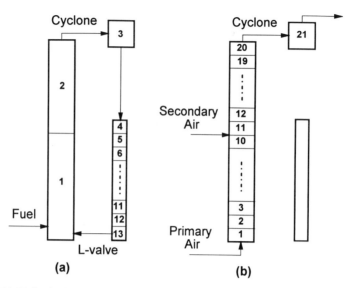

Figure 15.8 Well-mixed cells adopted in dynamic CFBC model of Muir (1994): (a) for solids; (b) for gas.

were considered to reach steady state quickly. Reactions involving NO and SO_2 were included in the reaction scheme. On the other hand, Mori *et al.* (1991) ignored these reactions and divided the riser into three zones (a dense zone, reducing zone and oxidizing zone) in their CFBC model. Both solids and gas were assumed to be perfectly mixed in each of these zones, with solids downflow from zone to zone being an important model parameter. Hyppanen *et al.* (1993) incorporated not only the boiler but also the steam process in their dynamic model. The riser was treated as a number of perfectly mixed cells in series. The suspension-to-wall heat transfer coefficient was correlated as a function of suspension density, and the latter was fitted empirically. Other details were also treated empirically or not described in detail. The model was verified using data from a 125 MWe power plant.

The zones employed in the dynamic model of Muir (1994) are shown schematically in Figure 15.8. Note the allowance for the cyclone and for solids passage through the recirculation loop. In view of the complexity of other aspects of the model, no allowance is made for the core/annulus structure. The solids density profile was fitted using an exponential decay function. Gas flow in the standpipe is ignored. Both mass and heat balances were written, with heat transfer coefficients correlated empirically, allowance being made for heat losses in the standpipe as well as to the reactor refractory and heat transfer tubes. The model was shown to give good predictions of the responses of combustion temperature, flue gas oxygen concentration and heat removal rate resulting from changes in fuel feed rate, primary and secondary air flow and solids circulation rate in a pilot-scale CFBC unit.

15.7 Concluding remarks

It is clear that circulating fluidized bed reactors lend themselves to a variety of models of differing degrees of sophistication. There is no single best approach to modeling. Simple one-dimensional models (section 15.3) may be adequate for some purposes, especially when dealing with overall estimations or process control. Core/annulus models (section 15.4) give a better overall simulation of flow in the riser and are excellent from an educational point of view, but they are also unable to capture all aspects of the complex flow patterns. More complex models (section 15.6) may ultimately be more successful, but generally require much more effort to write and more computer time to solve.

Continuing efforts are needed to improve the models and to provide verification using experimental data. Data giving gas concentration profiles and profiles of other variables like temperature and voidage for large-scale CFB reactors are especially critical for testing CFB reactor models in the future.

Nomenclature

C_a, C_c	gas concentration in annulus, core zone, mol m^{-3}
C_H	gas concentration at $z = H$, i.e. at top of riser, mol m^{-3}
C_0	inlet gas concentration, mol m^{-3}
D	riser column diameter, m
$\mathscr{D}_{g,ax}$	gas axial dispersion coefficient, m^2 s^{-1}
$\mathscr{D}_{g,r}$	gas radial dispersion coefficient, m^2 s^{-1}
d_p	mean particle diameter, m
G_s	solids circulation flux, kg m^{-2} s^{-1}
H	riser total height, m
K	inter-region crossflow mass transfer coefficient, m s^{-1}
k_1	first order reaction rate coefficient, s^{-1}
k_1^*	dimensionless first order reaction rate coefficient, $k_1 H/U$
L	depth of turbulent bed, m
Pe	axial Peclet number, $UL/\mathscr{D}_{g,ax}$
R	column radius, m
r_c	radius of core region, m
U	superficial gas velocity, m s^{-1}
X	dimensionless inter-region mass transfer coefficient defined by equation (15.6)
z	Height coordinate, m
$\bar{\epsilon}$	cross-sectional average voidage
ϵ_a, ϵ_c	voidage of annulus, core region
ρ_p	particle density, kg m^{-3}
ϕ	dimensionless core radius, r_c/R

References

Abed, R. (1984) The characterization of turbulent fluid bed hydrodynamics, in *Fluidization* (ed. D. Kunii and R. Toei), Engineering Foundation, New York, pp. 137–144.

Amos, G., Rhodes, M.J. and Mineo, H. (1993) Gas mixing in gas–solids risers. *Chem. Eng. Sci.*, **48**, 943–949.

Arena, U., Chirone, R., D'Amore, M., Miccio, M. and Salatino, P. (1995) Some issues in modelling bubbling and circulating fluidized bed coal combustors. *Powder Technol.*, **82**, 301–316.

Avidan, A.A. (1982) Turbulent fluid bed reactors using fine powder catalysts, *A.I.Ch.E.-C.I.E.S.C. Meeting*, Chemical Industry Press, Beijing, pp. 411–423.

Baerns, M., Mleczko, L., Tjatjopoulos, G.J. and Vasalos, I.A. (1994) Comparative simulation studies on the performance of bubbling and turbulent fluidized bed reactors for the oxidative coupling of methane, in *Circulating Fluidized Bed Technology IV* (ed. A.A. Avidan), AIChE, New York, pp. 414–421.

Berker, A. and Tulig, T.T. (1986) Hydrodynamics of gas–solid flow in a catalytic cracker riser: implications for reactor selectivity performance. *Chem. Eng. Sci.*, **41**, 821–827.

Bolton, L.W. and Davidson, J.F. (1988) Recirculation of particles in fast fluidization risers, in *Circulating Fluidized Bed Technology II* (eds P. Basu and J.F. Large), Pergamon, Oxford, pp. 139–146.

Brereton, C.H.M., Grace, J.R. and Yu, J. (1988) Axial gas mixing in a circulating fluidized bed, in *Circulating Fluidized Bed Technology II* (ed. P. Basu and J.F. Large), Pergamon, Oxford, pp. 307–314.

Contractor, R.M. and Chaouki, J. (1991) Circulating fluidized bed as a catalytic reactor, in *Circulating Fluidized Bed Technology III* (ed. P. Basu, M. Horio and M. Hasatani), Pergamon, Oxford, pp. 39–48.

Dersch, J., Fett, F.N., Brunne, T. and Bernstein, W. (1995) Modelling of circulating fluidized bed combustors firing wet brown coal, in *Preprints for Fluidization VIII* (eds C. Laguerie and J.F. Large), vol. 1, pp. 565–572.

Edwards, M. and Avidan, A. (1986) Conversion model aids scale-up of Mobil's fluid-bed MTG process. *Chem. Eng. Sci.*, **41**, 829–835.

Fane, A.G. and Wen, C.Y. (1982) Fluidized bed reactors, Chap 8.4 of *Handbook of Multiphase Systems* (ed. G. Hetsroni), Hemisphere, Washington.

Fligner, M., Schipper, P.H., Sapre, A.V. and Krambeck, F.J. (1994) Two phase cluster model in riser reactors: impact of radial density distribution on yields. *Chem. Eng. Sci.*, **49**, 5813–5818.

Foka, M., Chaouki, J., Guy, C. and Klvana, D. (1994) Natural gas combustion in a catalytic turbulent fluidized bed. *Chem. Eng. Sci.*, **49**, 4269–4276.

Gianetto, A., Pagliolico, S., Rovero, G. and Ruggeri, B. (1990) Theoretical and practical aspects of circulating fluidized bed reactors for complex chemical systems. *Chem. Eng. Sci.*, **45**, 2219–2225.

Gidaspow, D. (1994) *Multiphase Flow and Fluidization: Continuum and Kinetic Theory Descriptions*. Academic Press, San Diego.

Grace, J.R. (1986) Fluid beds as chemical reactors, in *Gas Fluidization Technology* (ed. D. Geldart), Wiley, Chichester, U.K., pp. 285–339.

Grace, J.R. (1990) High velocity fluidized bed reactors. *Chem. Eng. Sci.*, **45**, 1953–1966.

Grace, J.R., Zhao, J., Wu, R.L., Senior, R.C., Legros, R., Brereton, C.M.H. and Lim, C.J. (1990) Proc. Workshop on Materials Issues in Circulating Fluidized Bed Combustion. EPRI-GS-6747, Palo Alto, **10**, 1–21.

Hastaoglu, M.A., Berruti, F. and Hassam, M.S. (1988) A generalized gas–solid reaction model for circulating fluidized beds – an application to wood pyrolysis, in *Circulating Fluidized Bed Technology II* (ed. P. Basu and J.F. Large), Pergamon, Oxford, pp. 281–288.

Hypannen, T., Lee, Y.Y. and Raimo, A. (1991) A three-dimensional model for circulating fluidized bed combustion, in *Circulating Fluidized Bed Technology III* (eds P. Basu, M. Horio and M. Hasatani), Pergamon, Oxford, pp. 563–568.

Hyppanen, T., Lee, Y.Y., Kettunen, A. and Riiali, J. (1993) Dynamic simulation of a CFB based utility power plant, in *Proceedings 12th International Fluidized Bed Combustion Conference*, ASME, New York, pp. 1121–1127.

Jiang, P., Bi, H.T., Jean, R.H. and Fan, L.S. (1991) Baffle effects on performance of catalytic circulating fluidized bed reactors. *AIChE J.*, **37**, 1392–1400.

Jiang, P., Inokuchi, K., Jean, R.-H., Bi, H.T. and Fan, L.S. (1991) Ozone decomposition in a catalytic circulating fluidized bed reactor, in *Circulating Fluidized Bed Technology III* (eds P. Basu, M. Horio and M. Hasatani), Pergamon, Oxford, pp. 557–562.

Ju, D.W.C. (1995) Modelling of steady state heat release, oxygen profile and temperature profile in circulating fluidized bed combustors. MASc. thesis, Univ. of British Columbia, Vancouver.

Kagawa, H., Mineo, H., Yamazaki, R. and Yoshida, K. (1991) A gas–solid contacting model for fast-fluidized bed, in *Circulating Fluidized Bed Technology III* (eds P. Basu, M. Horio and M. Hasatani), Pergamon, Oxford, pp. 551–556.

Kruse, M., Schoenfelder, H. and Werther, J. (1995) A two-dimensional model for gas mixing in the upper dilute zone of a circulating fluidized bed, *Can. J. Chem. Eng.*, **73**, 620–634.

Kunii, D. and Levenspiel, O. (1990) Fluidized reactor models. *Ind. Eng. Chem. Res.* **29**, 1226–1234.

Lee, Y.Y. and Hyppanen, T. (1989) Coal combustion model for circulating fluidized bed boilers, in *Proceedings 10th International Conference on Fluidized Bed Combustion*, A.S.M.E., New York, pp. 753–764.

Li, Y. and Kwauk, M. (1980) The dynamics of fast fluidization, in *Fluidization* (eds J.R. Grace and J.M. Matsen), Plenum, New York, pp. 537–544.

Li, Y. and Wu, P. (1991) A study on axial gas mixing in a fast fluidized bed, in *Circulating Fluidized Bed Technology III* (eds P. Basu, M. Horio and M. Hasatani), Pergamon, Oxford, pp. 581–586.

Li, Y. and Zhang, X. (1994) Circulating fluidized bed combustion. *Adv. Chem. Eng.*, **20**, 331–338 (ed. M. Kwauk), Academic Press, San Diego.

Marmo, L., Manna, L. and Rovero, G. (1995) Comparison among several predictive models for circulating fluidized bed reactors, in *Preprints for Fluidization VIII* (eds C. Laguérie and J.F. Large), vol. 1, pp. 475–482.

Martin, M.P., Turlier, P., Bernard, J.R. and Wild, G. (1992) Gas and solid behaviour in cracking circulating fluidized beds. *Powder Technol.*, **70**, 249–258.

Mori, S., Narukawa, K., Yamada, I., Takebayashi, T., Tanii, H., Tomoyasu, Y. and Mii, T. (1991) Dynamic model of a circulating fluidized bed coal fired boiler, in *Proceedings 11th International Fluidized Bed Combustion Conference*, ASME, New York, pp. 1261–1266.

Muir, J.R. (1995) Dynamic modelling for simulation and control of a circulating fluidized bed combustor. Ph.D. thesis, Univ. of British Columbia, Vancouver.

Neidel, W., Gohla, M., Borghardt, R. and Reimer, H. (1995) Theoretical and experimental investigation of mix-combustion coal/biofuel in circulating fluidized beds, in *Fluidization VIII* (ed. C. Laguérie and J.F. Large), Volume I, pp. 573–583.

Ouyang, S. and Potter, O.E. (1994) Modelling chemical reaction in a 0.254 m i.d. circulating fluidized bed, in *Circulating Fluidized Bed Technology IV* (ed. A.A. Avidan), AIChE, New York, pp. 422–427.

Ouyang, S., Lin, J. and Potter, O.E. (1993) Ozone decomposition in a 0.254 m diameter circulating fluidized bed reactor. *Powder Technol.*, **74**, 73–78.

Ouyang, S., Li, X.G. and Potter, O.E. (1995a) Circulating fluidized bed as a catalytic reactor: experimental study. *AIChE J.*, **41**, 1534–1542.

Ouyang, S., Li, X.G. and Potter, O.E. (1995b) Investigation of ozone decomposition in a circulating fluidized bed on the basis of core-annulus model, in *Preprint for Fluidization VIII* (eds C. Laguérie and J.F. Large), vol. 1, pp. 457–465.

Pagliolico, S., Tiprigan, M., Rovero, G. and Gianetto, A. (1992) Pseudo-homogeneous approach to CFB reactor design. *Chem. Eng. Sci.*, **47**, 2269–2274.

Patience, G.S. and Chaouki, J. (1993) Gas phase hydrodynamics in the riser of a circulating fluidized bed. *Chem. Eng. Sci.*, **48**, 3195–3205.

Pugsley, T.S., Patience, G.S., Berruti, F. and Chaouki, J. (1992) Modeling the catalytic oxidation of n-butane to maleic anhydride in a circulating fluidized bed reactor. *Ind. Eng. Chem. Res.*, **31**, 2652–2660.

Pugsley, T.S., Berruti, F. and Chakma, A. (1994) Computer simulation of a novel circulating fluidized bed pressure–temperature swing absorber for recovering carbon dioxide from flue gases. *Chem. Eng. Sci*, **49**, 4465–4481.

Sanderson, W.E. (1993) A review of overall models of circulating fluidized bed combustors. Report of Dept. of Chemical Process Technology, Delft University, Netherlands.

Saraiva, P.C., Azevedo, J.L.T. and Carvalho, M.G. (1993) Modelling combustion, NO_x emissions and SO_2 retention in a circulating fluidized bed, in *Proceedings 12th International Conference on Fluidized Bed Combustion*, A.S.M.E., New York, pp. 375–380.

Schoenfelder, H., Kruse, M. and Werther, J. (1996a) Spatial reactant distribution in a pilot-scale CFB riser: model predictions and measurements, 5th International Circulating Fluidized Bed Conference, Beijing, Preprint MSS4.

Schoenfelder, H., Kruse, M. and Werther, J. (1996b) Two-dimensional model for circulating fluidized bed reactors. *AIChE J.*, **42**, 1875–1888.

Schoenfelder, H., Werther, J., Hinderer, J. and Keil, F. (1994) A multi-stage model for the circulating fluidized bed reactor. *AIChE Symp. Ser.*, **90**(301), 92–104.

Senior, R.C. and Brereton, C.M.H. (1992) Modelling of circulating fluidized bed solids flow and distribution. *Chem. Eng. Sci.*, **47**, 281–296.

Sotudeh Garebaagh, R., Legros, R., Paris, J. and Chaoki, J. (1994) Process simulation of a circulating bed coal combustor, Abstract for *44th Canadian Chemical Engineering Conference* (ed. Svrcek, W.), Canadian Society for Chemical Engineering, Ottawa, pp. 141–142.

Sung, L.Y. (1995) NO_x and N_2O emissions from circulating fluidized bed combustion. MASc dissertation, Univ. of British Columbia, Vancouver.

Talukdar, J. and Basu, P. (1995) Modelling of nitric oxide emission from a circulating fluidized bed furnace, in *Preprints for Fluidization VIII* (eds C. Laguérie and J.F. Large), vol. 2, pp. 829–837.

Talukdar, J., Basu, P. and Joos, E. (1994) Sensitivity analysis of a performance predictive model for circulating fluidized bed boiler furnaces, in *Circulating Fluidized Bed Technology IV* (ed. A.A. Avidan), AIChE, New York, pp. 450–457.

Van Swaaij, W.P.M. (1978) The design of gas–solids fluid bed and related reactors. *A.C.S. Symp. Ser.* (eds D. Luss and V.W. Weekman), **72**, 193–222.

Weiss, V. and Fett, F.N. (1986) Modeling the decomposition of sodium bicarbonate in a circulating fluidized bed reactor, in *Circulating Fluidized Bed Technology* (ed. P. Basu), Pergamon, Toronto, pp. 167–172.

Weiss, V., Schöler, J. and Fett, F.N. (1988) Mathematical modelling of coal combustion in a circulating fluidized bed reactor, in *Circulating Fluidized Bed Technology II* (eds P. Basu and J.F. Large), Pergamon, Oxford, pp. 289–298.

Wen, C.Y. (1979) Chemical reaction in fluidized beds, in *Proceedings N.S.F. Workshop on Fluidization and Fluid-Particle Systems* (ed. H. Littman), Rensselaer Polytechnic Inst. Troy, NY, pp. 317–387.

Werther, J., Hartge, E.U. and Kruse, M. (1992) Gas mixing and interphase mass transfer in the circulating fluidized bed, in *Fluidization VII* (eds O.E. Potter and D.J. Nicklin), Engineering Foundation, New York, pp. 257–264.

White, C.C., Dry, R.J. and Potter, O.E. (1992) Modelling gas-mixing in a 9 cm diameter circulating fluidized bed, in *Fluidization VII* (eds O.E. Potter and D.J. Nicklin), Engineering Foundation, New York, pp. 265–273.

Wong, R., Pugsley, T. and Berruti, F. (1992) Modelling the axial voidage profile and flow structure in the riser of circulating fluidized beds. *Chem. Eng. Sci.*, **47**, 2301–2306.

Yerushalmi, J., Cankurt, N.T., Geldart, D. and Liss, B. (1978) Flow regimes in vertical gas–solid contact systems, *AIChE Symp. Ser.*, **74**(176), 1–13.

Zhang, L., Li, T.D., Zheng, Q.Y. and Lu, C.D. (1991) A general dynamic model for circulating fluidized bed combustion with wide particle size distributions, in *Proceedings 11th International Fluidized Bed Combustion Conference*, ASME, New York, pp. 1289–1294.

Zhang, W.T., Tung, Y. and Johnsson, F. (1991) Radial voidage profiles in fast fluid beds of different diameters. *Chem. Eng. Sci.*, **46**, 3045–3052.

Zhao, J. (1992) Nitric oxide emissions from circulating fluidized bed combustion. Ph.D. thesis, Univ. of British Columbia, Vancouver.

Zhou, J., Grace, J.R., Lim, C.J. and Brereton, C.M.H. (1995) Particle velocity profiles in a circulating fluidized bed riser of square cross-section. *Chem. Eng. Sci.*, **50**, 237–244.

16 Novel configurations and variants
YONG JIN, JING-XU ZHU AND ZHI-QING YU

16.1 Introduction

Compared with conventional bubbling and turbulent fluidized beds, circulating fluidized beds have many advantages including better gas–solids contacting and reduced backmixing (Lim *et al.*, 1995). However, due to the core-annulus structure, particle backmixing along the wall can still be significant. Gas–solids contacting is not optimal given the non-uniform distribution of gas and particle flow in the riser. Significant solids backmixing also occurs in the bottom solids-acceleration zone, often covering a significant portion of the riser. Backmixing can also occur at the riser exit if a strong constriction is used. For fast reactions such as fluid catalytic cracking where the intermediate is the product, uniform residence time distributions of gas and solids are very important. To further reduce backmixing and to improve gas–solid contact, internals can be installed inside CFBs to modify the flow pattern and many novel inlet and exit configurations can be employed. The first two sections of this chapter discuss the various available internals and novel geometry structures and their effects on gas and solids flow. The third section describes two special configurations where solids upflow and downflow are accommodated within a single vessel containing concentric upflow and downflow regions and the N-shape CFB loop.

Reversing the gas and solids flow from the upflow direction (riser) to the downflow direction (downer) has been shown to reduce axial gas and solids dispersion and to produce a more uniform radial flow structure (Zhu *et al.*, 1995). Extending current knowledge obtained in gas–solid systems to liquid–solids and gas–liquid–solid three-phase systems is expected to open a new horizon for applications of circulating fluidized bed technology and to lead to the developments of high efficiency liquid–solid and gas–liquid–solid reactors. Recent progress in the above two areas is covered in the last two sections of this chapter.

16.2 Internals

Non-uniform flow is an inherent characteristic of the gas–solids upflow riser fluidization systems. As discussed in Chapter 2, gas and particle velocities are higher and solids concentration lower in the centre, while gas and particle velocities are much lower and solids concentration much higher in the wall region, forming a core-annulus structure. This structure leads to two main problems: (1) inefficient gas–solids contact, and (2) significant gas and solids backmixing due to the radial non-uniform gas and solids flow structure and downflow in the wall region. To improve this non-uniform flow structure, internals can be introduced. Internals are usually aimed at redistributing gas and solids flow, in an effort to form a more uniform flow structure to improve interphase contact efficiency and to reduce gas and solids backmixing so as to increase the overall performance of the CFB reactors. Internals can also be installed in the riser bottom to enhance solids acceleration and to shorten the acceleration length. Quick acceleration of solids at the bottom is essential for fast reactions such as FCC to avoid large solids backmixing in the bottom dense region.

The various types of reported internals are summarized in Table 16.1. Davies and Graham (1988) swaged indentations 3.5 mm deep and 16 mm wide into the wall of a 0.152 m diameter riser to test their effect on pressure drop in the riser. In a 0.09 m diameter riser, Zheng *et al.* (1991b, 1992) investigated several internals including ring baffles, inverse cone and perforated plates with various arrangements of opening holes to compare their performance. Jiang *et al.* (1991) also installed ring baffles in a 0.1 m diameter riser to test their effect on chemical conversion. Gan *et al.* (1990) inserted a bluff-body (a half oval with parabolic curvature, flat end up, with its upper larger end occupying 50% of the riser cross-section) in the centre of a 0.14 m diameter riser to study its influence on the flow pattern, while van der Ham *et al.* (1991, 1993) installed regularly packed horizontal bars in a 0.126 × 0.126 cross-section riser as internal baffles. More recently, Salah *et al.* (1996) have tested the effect of ring size on radial voidage distribution in a 0.076 m diameter riser.

One key function of the baffles is to redistribute the radial gas and solids flow. Rings, swages and inverse cones scrape the downflowing solids away from the wall towards the centre, leading to a more uniform radial solids distribution. Using an optical fibre concentration probe, Zheng *et al.* (1991b, 1992) showed that the radial voidage distribution is much more uniform with ring baffles present (Figure 16.1). Jiang *et al.* (1991) measured more uniform radial gas distribution with ring baffles. However, Salah *et al.* (1996) have found that rings with smaller opening areas can make the radial voidage distribution less uniform under low gas velocity. Properly designed perforated plates help to redistribute the gas and solids flow as well. Zheng *et al.* (1990) found that a plate with larger orifices and more open area near the wall is very effective in flow redistribution. Addition of a narrow seam next to the wall further improved the performance. Gas preferably flows upwards in the annular region so that radial distribution becomes more uniform. The half oval bluff-body forces gas and solids to flow upwards in the narrow ring, reversing the radial flow distribution, with higher solids

NOVEL CONFIGURATIONS AND VARIANTS 527

Table 16.1 Configurations and characteristics of internals

Baffle type	Configuration	Main features reported	Riser size and open area	References
1. Swages		a. Decreased pressure drop	$D = 150$ mm 91%	Davies and Graham (1988)
2. Ring with large free area		a. Improved radial voidage distribution b. Decreased solids holdup c. Suppressed axial solids mixing d. Enhanced radial gas and solid mixing e. Improved gas–solids contact efficiency	$D = 90$ mm Open area not given	Zheng et al. (1991b, 1992)
3. Ring with small free area		a. Improved radial voidage distribution b. Increased pressure drop c. Increased solids holdup d. Enhanced radial gas and solid mixing e. Improved gas–solids contact efficiency f. Increased conversion of ozone decomposition reaction	$D = 102$ mm 56%	Jiang et al. (1991)
4. Perforated plates		a. Improved radial solids distribution	$D = 90$ mm 40%	Zheng et al. (1990)
5. Inverse cone		a. Improved radial solids distribution b. Decreased solids holdup	$D = 90$ mm 52–57%	Zheng et al. (1990)
6. Bluff-body		a. Increased pressure drop b. Decreased solids holdup c. Increased gas and particle velocities d. Improved radial voidage distribution e. Enhanced gas–solids contact efficiency f. Reversed radial flow pattern	$D = 140$ mm 49%	Gan et al. (1990)
7. Regular packed horizontal bars		a. Increased pressure drop b. Increased solids holdup c. Improved gas–solids contact efficiency	$D = 40$ mm 50% $D = 125$ mm 50%	van der Ham et al. (1991, 1994) van der Ham et al. (1993)

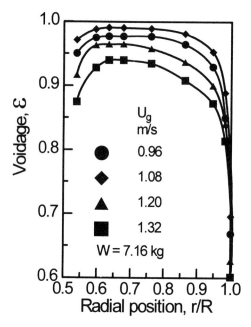

Figure 16.1 Effect of ring baffle on radial voidage profiles in the riser (FCC particles, $\rho_p = 930$ kg/m^3, $d_p = 46.7\,\mu$m, $D = 0.09$ m, $H = 10$ m, from Zheng et al., 1992, reproduced with permission).

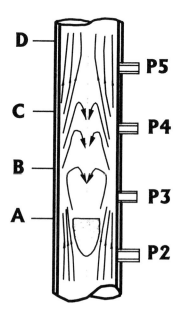

Figure 16.2 Effect of bluff-body on gas and solids flow in the riser (FCC particles, $\rho_p = 1545$ kg/m^3, $d_p = 59\,\mu$m, $U_g = 2$–10 m/s, $G_s = 30$–180 kg/m^{2}s, $D = 0.140$ m, $H = 11$ m, bluff-body 2.8 m away from the distributor, from Gan et al., 1990).

concentration and low gas and solids velocities in the core and dilute flow with higher gas and solids velocities in the wall region, above the baffle. Hence the dilute core and dense annulus flow pattern are disrupted as shown in Figure 16.2 (Gan *et al.*, 1990). As a consequence, gas–solids contacting is greatly improved.

Not all internals provide favourable gas and solids redistribution (Salah *et al.*, 1996). With a ring of 44% open area, Zheng *et al.* (1990) reported increased non-uniformity under certain operating conditions. For risers operating at high gas velocity and high solids flux, e.g. FCC units, particles mainly flow upwards at the wall so that installation of ring type baffles may actually induce particle downflow, leading to more radial flow segregation.

Installation of internal baffles can significantly affect the pressure distribution in the riser. Davies and Graham (1988) showed that swages evenly spaced along their circular riser wall caused the pressure drop in the fully developed region to be lower than for a bare tube. On the other hand, Jiang *et al.* (1991) measured an increase in pressure drop when four ring baffles with an open area of 56% were installed in their 0.1 m riser. Gan *et al.* (1990) experienced a localized high pressure reduction across their bluff-body baffle, coupled by significant solids acceleration. At the same time, internals can also change the average voidage. The results of Zheng *et al.* (1992) shown in Figure 16.1 indicate that ring-type baffles significantly reduce the average solids concentration. Gan *et al.* (1990) also found that their bluff-body baffle reduced the average solids concentration by scraping off superfluous particles next to the wall. On the other hand, Jiang *et al.* (1991) reported a higher local solids concentration in the presence of ring baffles. With the packed bar baffles, which take up large cross-sectional area (50%), van der Ham *et al.* (1993) measured 1.6 to 5 times higher solids holdup compared with an empty column. Although part of this increase was due to particles settling on top of the bars, at least some was due to the internals. It would appear that baffles with less flow restriction (e.g. less reduction in cross-sectional area) mainly increase the actual gas velocity, scraping the dense downflow solids away from the wall, leading to higher bed voidage and low pressure drop as observed by Davies and Graham (1988) and Zheng *et al.* (1991b, 1992), while baffles with a significant reduction of cross-sectional area constrict the solids flow, leading to lower bed voidage (e.g. Jiang *et al.*, 1991; van der Ham *et al.*, 1993). The bluff-body baffle seems to be an exception. It occupies 50% of the riser cross-section, significantly increases the gas and particle velocity and causes a higher pressure drop, but it does not increase solids holdup. This may be attributed to the smooth parabolic shape facing the gas and solids flow.

Most internals are likely to suppress axial solids backmixing. In their study of a ring-type baffle, Zheng *et al.* (1992) found that the axial solids dispersion is only half of that in an empty column. On the other hand, most internals appear to improve radial gas and solids mixing. Using FCC particles with

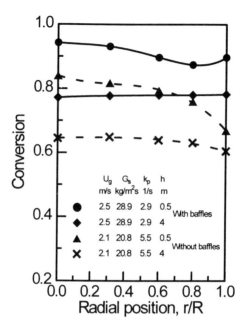

Figure 16.3 Effect of ring on radial ozone conversion in the riser (FCC particles, $\rho_p = 1500$ kg/m^3, $d_p = 89$ μm, $D = 0.102$ m, $H = 6.32$ m, ring inside diameter $= 0.076$ m, from Jiang et al., 1991, reproduced with permission of American Institute of Chemical Engineers, copyright © 1991 AIChE, all rights reserved).

a mean diameter of 89 μm impregnated with ferric oxide as catalysts for ozone decomposition, Jiang et al. (1991) observed a more uniform radial distribution of ozone concentration in the riser with ring baffles (Figure 16.3) and attributed this to higher radial gas and solids mixing. Internals can also enhance gas–solids contact through flow redistribution. van der Ham et al. (1991) showed that packed bar baffles enhanced the volumetric mass transfer coefficient. Internal baffles increase the turbulent intensity in the riser, leading to both intensified gas–solids contacting and increased lateral gas and solids mixing (Gan et al., 1990). Owing to the enhanced gas–solids contact efficiency, increased gas and solids radial mixing and suppressed axial mixing, CFB reactor performance is expected to be improved when internal baffles are present. Jiang et al. (1991) showed that ozone conversion is significantly higher when ring baffles were installed in a CFB reactor. Their results indicate that the increased catalyst holdup is mainly responsible for the increased conversion.

Installations of internals in commercial risers are scarce, due to two major concerns. Firstly, it is not known how to properly scale up the baffles and how much of the benefits reported for laboratory units can be extended to larger commercial risers. Secondly, erosion can be a major problem, especially for high velocity and/or high flux operations such as in FCC risers.

There are also other elements in the CFB riser, not installed as baffles to improve reactor performance, but necessary for some other purposes. One example is membrane heat transfer tubes mounted on the wall of circulating fluidized bed combustors. Lockhart *et al.* (1995) showed that solids concentration is much higher in the fin region than at the tube crests due to shielding by the adjacent half-tubes sandwiching the fin area. This high solids concentration at the wall enhances the core-annulus structure, resulting in reduced gas–solids contact and enhanced axial solids mixing, an undesirable situation. It is therefore very important to design the necessary elements in the riser in such a way that they improve the radial flow structure.

16.3 End configurations

16.3.1 Bottom sections

For circulating fluidized bed operation, it is essential to maintain continuous and smooth feeding of solids into the bottom of the riser. There are many variations of the bottom section (including both the riser bottom section and the bottom section of the solids return system) to accommodate solids return. The most commonly encountered bottom section structures and their characteristics are summarized in Table 16.2. The bottom section serves three purposes: to control the solids circulation rate, to deliver solids to the riser bottom and, for some configurations, to enhance solids acceleration. According to their operation mechanism, there are two primary types of solids control devices installed in the solids return line: mechanical valves and non-mechanical valves. Mechanical valves include the slide valve, butterfly valve, plug valve, screw feeder, etc., while non-mechanical valves include the L-valve, J-valve, V-valve, H-valve, internal nozzle, loop seal, sealpot, etc. Details are provided in Chapter 7. In terms of their solids delivering function, there are also two groups: those with and those without solids redistribution.

In bottom sections without solids redistribution (nos. 3, 4, 6, 8 and 11 in Table 16.2), solids are fed to the riser from one side, causing spatial non-uniform distribution of solids (Rhodes *et al.*, 1989; Senior, 1992). On the other hand, bottom sections with solids redistribution (nos. 1, 2, 5, 7 and 9 in Table 16.2) provide a better radial solids distribution in the bottom and can possibly benefit fast reactions for which reaction time is short so that initial uniform solids distribution becomes important. Dry *et al.* (1992) found that an enlarged riser bottom (no. 12 in Table 16.2) can promote radial solids mixing. Internal baffles can enhance solids redistribution at the riser bottom as well. When installed in the bottom section, internals such as a bluff-body (Gan *et al.*, 1990) and regular packed horizontal bars (van der Ham *et al.*, 1991, 1993) facilitate solids redistribution. Internals

Table 16.2 Configurations and characteristics of bottom inlet sections

Inlet type	Configuration	Main features reported	References
1. Inlets with internal nozzle		a. Easily adjustable solids feed rate (variable primary/auxiliary air ratio) b. Influences riser solids holdup	Naruse et al. (1991) Bai et al. (1992) Xia and Tung (1992)
2. Inlets with internal nozzle and narrowed bottom		a. Narrowed bottom to enhance solids acceleration b. Lower solids holdup in riser bottom	Gwyn (1994) Avidan (Chapter 13)
3. Loop seal, reversed V-valve or fluoseal		a. Steadier operation b. Better for Geldart Group B solids c. Limited solids circulation rate	Knowlton et al. (1988) Xia and Tung (1992)
4. L-valve		a. Operation easy to control b. Simple construction c. Good only for Group B solids d. Widely adjustable solids flow rate e. Operation strongly affected by geometry and standpipe design	Knowlton et al. (1988) Brereton and Grace (1994) Zheng et al. (1994)
5. J-valve		a. Very weak restrictive inlet b. Significant influence on riser solids holdup	Avidan and Yerushalmi (1982)
6. H-valve		a. Steadier operation b. Better for Geldart Group B solids c. Limited solids circulation rate	Yang et al. (1991a)
7. Venturi pipe inlet		a. Enhanced solids acceleration b. Limited solids circulation rate	Monceaux et al. (1986)

NOVEL CONFIGURATIONS AND VARIANTS

Table 16.2 Continued

Inlet type	Configuration	Main features reported	References
8. Inclined inlet with mechanical valve (slide valve, butterfly valve)		a. Fluidized or moving bed in the inclined pipe b. Easily adjustable solids circulation control by varying valve opening c. Less stable operation at large opening	Yerushalmi and Avidan (1985) Xia and Tung (1992)
9. Inclined inlet with concentric slide valve in the riser		a. Easily adjustable solids circulation rate by varying valve opening b. Large solids circulation rate possible c. Small pressure drop across the valve	Wei et al. (1994a)
10. Plug valve		a. Easily adjustable solids circulation rate by varying valve opening b. Large solids circulation rate possible c. Small pressure drop across the valve	Avidan (Chapter 13)
11. Screw feeder		a. Solids circulation rate not related to overall pressure balance in the circulation system b. Large solids circulation rate possible	Mori et al. (1991)
12. Enlarged bottom inlet		a. Good solids mixing b. More uniform radial solids distribution	Dry et al. (1992)

can also enhance solids acceleration in the riser bottom. For example, it has been a common practice to create a narrowed section above the vapour feed nozzle in modern CFB risers to increases the gas velocity, which in turn increases the particle velocity (no. 2 in Table 16.2). The Venturi pipe solids inlet (no. 7 in Table 16.2) serves a similar purpose. The bluff-body discussed above (see Table 16.1) significantly increases solids velocity by forcing particles to flow through the narrow gap between the body and the riser wall.

The bottom section configuration affects the hydrodynamics of the riser. Depending on their restriction to solids flow from the solids return system, solids inlets can be divided into two groups: weakly restrictive inlets and strongly restrictive inlets. Bai *et al.* (1992) showed significant differences in the axial bed voidage profile when the inlet is gradually changed from a strongly restrictive inlet (butterfly valve with small opening or internal nozzle inlet configuration with small auxiliary air/main air ratio) to a weakly restrictive inlet (fully open butterfly valve or nozzle inlet with large auxiliary air/main air ratio). For a fixed superficial gas velocity, reducing the solids inlet restriction by opening the butterfly valve results in a higher solids circulation rate and higher solids buildup in the riser bottom region (Figure 16.4). For a fixed superficial gas velocity, solids circulation rate and total solids inventory, a weakly restrictive inlet results in higher solids holdup in the riser bottom region than a strongly restrictive inlet (Figure 16.5). This is attributed to pressure buildup in the solids circulation system:

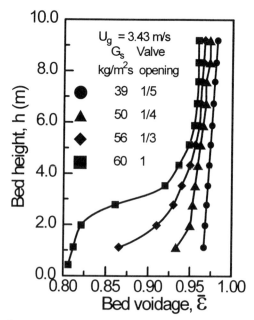

Figure 16.4 Effect of entrance structure on axial voidage profile (FCC particles, $\rho_p = 1545\,\text{kg/m}^3$, $d_p = 54\,\mu\text{m}$, $D = 0.140\,\text{m}$, $H = 10\,\text{m}$, from Bai *et al.*, 1991, reproduced with permission).

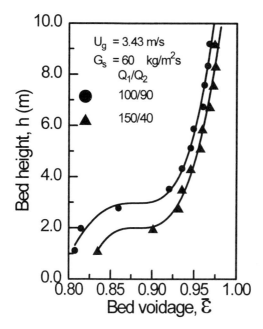

Figure 16.5 Effect of solids inlet restriction on axial voidage profile (FCC particles, $\rho_p = 1545\,\text{kg/m}^3$, $d_p = 54\,\mu\text{m}$, $D = 0.140\,\text{m}$, $H = 10\,\text{m}$, from Bai et al., 1992, reproduced with permission).

when the solids inventory is fixed, reducing the restriction decreases the pressure drop across the control valve, leading to a higher pressure head available from the solids return side (including downcomer and solids control valve). To keep the pressure in balance, solids holdup in the bottom of the riser must rise to increase the pressure head on the riser side (Bi and Zhu, 1993).

The riser bottom configuration can also influence the stability of operation. In general, a strongly restrictive solids inlet ensures stable operation due to a high pressure drop since minor disturbances in the system pressure distribution will not affect the solids flow rate. On the other hand, solids circulation can fluctuate more with a weakly restrictive inlet. As shown in Figure 16.5, Bai et al. (1992) found that increasing the total solids inventory significantly increases the solids buildup in the bottom of the riser with a weakly restrictive inlet while hardly changing the solids buildup in the same area with a strongly restrictive inlet. The geometric structure can affect the stability as well. Xia and Tung (1992) studied pressure wave transmission in riser bottoms equipped with a loop seal and an inclined pipe for solids feeding and found that the loop seal configuration gives much lower pressure wave velocity. The bottom with internal nozzle inlet configuration (no. 1 in Table 16.2) may also prevent apparent choking caused by insufficient pressure buildup in the solids return system. This type of choking, defined as standpipe-induced choking or type-B choking by Bi et al. (1993a), occurs when the pressure buildup in the solids return system is comparable with that in the riser and when gas starts to flow into the solids return system. For the main gas to flow into the solids return system in the bottom section with an internal nozzle inlet, it must overcome additional resistance due to the flow reversal in the annular ring outside the nozzle making this type of choking less likely. A screw feeder provides additional pressure head to the riser bottom, thereby stabilizing the system (Bi and Zhu, 1993).

Various bottom configurations used in commercial FCC units are discussed in Chapter 13.

16.3.2 Exit configuration

Riser exit structures can significantly affect flow conditions in the top section of a riser and sometimes even in the middle section (Jin et al., 1988). The structure and flow characteristics of several commonly used exits are outlined in Table 16.3. Exit structures can be divided into two types: abrupt exits (nos. 1, 2 and 6 in Table 16.3) and smooth exits (nos. 3, 4 and 5 in Table 16.3). Abrupt exits force the gas and solids flow leaving the riser to turn abruptly, while smooth exits provide a more gentle pathway for the gas and solids to change their flow direction more gradually. In general, abrupt exits create extra flow resistance, leading to increased solids holdup in the riser top, while smooth exits hardly affect the solids concentration in that region (Jin et al., 1988; Brereton and Grace, 1994). With high solids circulation rate and low gas velocity, Jin et al. (1988) showed that the exit effect can extend about 2/3 of the way down the riser from the top.

Since the solids density is normally about three orders of magnitudes larger than the gas density and in view of the particle size used in CFB systems, the relatively high inertia of the solids allows them to be very easily separated from the gas stream at the exit unless the change of gas flow direction is very smooth. Using a 140 mm diameter riser with FCC as solids material, Jin et al. (1988) studied both abrupt and smooth exit structures (nos. 2 and 3 in Table 16.3) and reported that an abrupt or restrictive exit creates a higher solids holdup at the top of the riser than a smooth or non-restrictive exit (see Figure 16.6). The pressure loss across the exit is higher for a more abrupt or stronger restrictive exit. Increasing solids circulation rate increases the pressure loss across the exit, resulting in a higher solids concentration at the riser top. Increasing the gas velocity, on the other hand, decreases the suspension density of the flow through the exit, leading to lower solids concentration and a weaker exit effect. Similar results have been reported by others in small-scale units (Hirama et al., 1992; Brereton and Grace, 1994; Zheng and Zhang, 1994) and in an industrial riser (Martin et al., 1992).

A restrictive exit can also lead to significant solids backmixing in the riser top region (Jin et al., 1988). This influence has been confirmed by Zheng and

Table 16.3 Configurations and characteristics of commonly used exit structures

Exit type	Configuration	Degree of restriction and main influence	References
1. L-shape exit (abrupt exit)		a. Medium strong restrictive exit b. Higher solids concentration at the riser top c. Violent turbulence and significant solids back-mixing in the riser exit region	Xia and Tung (1992) Zheng and Zhang (1994)
2. T-shape exit (abrupt exit)		a. Very strong restrictive exit b. Higher solids concentration at the riser top c. Violent turbulence and significant solids back-mixing in the riser exit region	Jin et al. (1988) Martin (1992) Brereton and Grace (1994) Zheng and Zhang (1994)
3. Right-angle bend exit with guiding baffle (smooth exit)		a. Weak restrictive exit b. No significantly high solids holdup at the region of exit c. No significant additional solids back-mixing activity	Jin et al. (1988) Zheng and Zhang (1994)
4. C-shape exit (smooth exit)		a. Weakest restrictive exit b. No significantly high solids holdup at the region of exit c. No significant additional solids back-mixing activity	Xia and Tung (1992) Brereton and Grace (1994)
5. Direct gas–solids separation exit (smooth exit)		a. Weak restrictive exit b. Insignificant influence on the bulk flow of gas and solids c. Same distribution of solid concentration in the exit region as in other parts of the CFB d. No additional back-mixing caused by exit e. Gas and solids separated efficiently	Qi et al. (1990)
6. Internal cyclone exit (abrupt exit)		a. Strong restrictive exit	Cao et al. (1991)

Zhang (1994) who found that using a more constricted exit increases the solids residence time. Both increased solids concentration and backmixing can be advantageous for some reactions such as combustion where increasing the residence time of solid fuel allows more efficient combustion and, while suspension density at the riser top enhances heat transfer in that region. To utilize this advantage, most CFB boilers have much smaller exit cross-section than the riser. Yang et al. (1991a) adopted an internal cyclone for

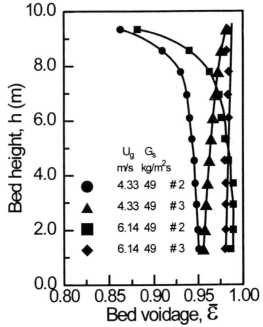

Figure 16.6 Effect of exit structure on axial voidage profile (FCC particles, $\rho_p = 1545 \, \text{kg/m}^3$, $d_p = 59 \, \mu\text{m}$, $D = 0.140 \, \text{m}$, $H = 11 \, \text{m}$, from Jin et al., 1988).

CFBC units. Since the exit area is much smaller than the riser cross-section, it acts as a strongly restrictive exit.

For fast reactions such as FCC, on the other hand, solids backmixing is undesirable so that a smooth exit structure would be preferable. However, most commercial FCC units employ a 90° turn exit with a cap that provides solids cushion to minimize erosion. To accommodate both considerations, Qi et al. (1989) proposed a smooth exit structure to eject gas and solids directly into gas–solids separators (no. 5 in Table 16.3). This arrangement has at least three significant advantages over conventional exits: (1) it causes very little restriction to the flow, (2) it allows immediate gas–solids separation after the solids leave the reactor, and (3) it reduces erosion.

16.4 Other bed configurations

Fast fluidization can also be achieved in a concentric circulating fluidized bed (Fusey et al., 1986). One option is to have solids upflow in the centre and downflow in the annular shell. This has been called an Internally Circulating Fluidized Bed (Milne et al., 1992a,b, 1994). Alternatively, one can have solids upflow in the annular shell and downflow in the centre, leading to what has been called the Integral Circulating Fluidized Bed (Liu et al., 1993a,b, Wang

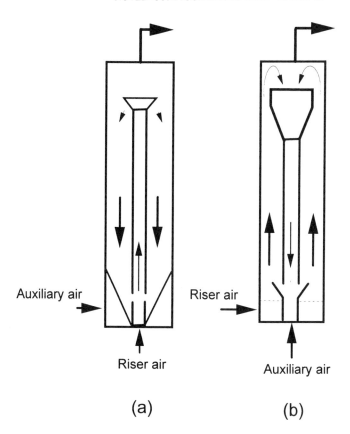

Figure 16.7 Schematics of *i*-ICFB, *o*-ICFB, and N-CFB (from Milne *et al.*, 1992a, and Wang *et al.*, 1993, reproduced with permission).

et al., 1993). These two options are shown in Figure 16.7. Since both have been abbreviated ICFB, the former will be identified as *i*-ICFB and the latter as *o*-ICFB with the *i* and *o* indicating the inside or outside location of the riser.

The *i*-ICFB originated from the spouted bed with a draft tube. To prevent spout gas from leaking through the annulus, causing non-uniform residence time distribution of gas, Milne *et al.* (1992b) sealed the bottom opening between the spout and annulus but left several small orifices for solids circulation. Solids enter the riser through these orifices and are entrained by the riser gas. A tee is located at the top of the riser to assist in disengaging solids from the riser gas stream. Auxiliary air is supplied to the annular region to keep the solids fluidized. Solids circulation rate is not independently controlled but is a function of the orifice size and the auxiliary gas velocity. Preliminary hydrodynamic studies by Milne *et al.* (1992a,b) and by earlier workers (Fusey *et al.*,1986) show stable operation of this system. One advantage is that a very high solids circulation rate (near 600 kg/m^{2}s) can be realized in this system (Milne et al., 1994). Milne et al. (1994) have shown that stable, isothermal operation at high temperature can also be achieved in an i-ICFB unit and suggested that it can be ideal for short contact time reactions where uniform residence time distribution and stable high temperature operation are needed, such as in pyrolysis.

o-ICFB has been developed as an alternative to the regular CFB systems to address the concerns of non-uniform solids feeding at the riser bottom and the energy requirements of a large amount of external solids separation and recirculation. There are several advantages: (1) simple structure; (2) high solids circulation rate; (3) uniform solids feeding. Solids first flow upward concurrently with gas in the annulus region, are separated by the multi-inlet cyclone at the top, then flow down the concentric central standpipe and are eventually fed back into the riser through a circular V-valve to ensure uniform feeding along the circumference of the inner tube. As for i-ICFB, the solids circulation rate is not independently controlled but varies with riser and auxiliary gas velocities. Liu et al. (1993a,b) and Wang et al. (1993) found that axial solids distributions in the riser of o-ICFB are very similar to those in regular CFB risers. The cross-sectional average solids holdup was somewhat higher than in regular CFB risers operated under similar conditions (Wang et al., 1993). Typical radial solids concentration profiles measured by Wang et al. (1993) show that a 'core-annulus' structure still exists inside the annular riser, with the dense layer at the outer wall slightly thicker than on the inner wall (Figure 16.8). Measuring at two vertical planes in the annular riser on the opposite sides of the central standpipe, Wang et al. (1993) found that the radial voidage distributions are symmetrical.

There are two other special arrangements for the CFB loop as shown in Figure 16.9, both labelled N-shaped circulating fluidized beds (N-CFB). In both situations, gas and particles first flow upwards in the first upflow section and then reverse direction to flow downwards concurrently in a downflow section. About 2/3 of the way down the downflow section, gas is separated from the solids stream to exit through a second upflow section (Zheng et al., 1989) or to flow into another CFB for further reactions (Qi et al., 1992). Zheng et al. (1989) developed their N-CFB system (Figure 16.9a) in an effort to make CFB combustion technology economical for small capacity boilers. By 'folding' the top portion of the riser downwards, the total height, and thus the cost of the boiler, can be reduced. Another reported advantage is that efficient inertial gas–solids separation can be achieved at the bottom connecting the downflow section and the second upflow section, which acts as a primary gas–solids separator. Qi et al. (1992) proposed a more complicated configuration (Figure 16.9b) where the flue gas withdrawn from the downflow section of the combustion loop is fed into another regular CFB loop for SO_2 dry-scrubbing, allowing

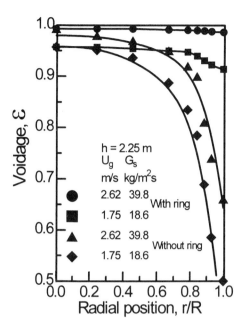

Figure 16.8 Radial voidage profile in o-ICFB (FCC particles, $\rho_p = 1500\,\text{kg/m}^3$, $d_p = 54\,\mu\text{m}$, $D = 0.118\,\text{m}$ equivalent, $H = 6.6\,\text{m}$, from Wang et al., 1993, reproduced with permission).

optimization of both combustion and dry-scrubbing. Some preliminary hydrodynamic studies have been carried out by both groups, showing that the flow structure in the first upflow section is similar to that in a regular CFB riser (see Chapter 2), while the flow pattern in the downflow section is similar to that reported for gas–solids downer reactors (see next section). One limitation of the N-CFB is that it can only be operated under very dilute conditions, since the static pressure head at the bottom of the riser provided by a dense solids bed in the downflow section would be too small to support a denser suspension in the riser (Bi and Zhu, 1993).

16.5 Gas–solids co-current downflow systems

In the last two decades, gas–solid co-current downflow circulating fluidized bed (CDCFB) or downer reactors have attracted the attention of many researchers given their significant advantages of short gas–solid contact time and uniform gas and solids residence time. These advantages are mainly due to the nature of gas and solids co-current downflow in the direction of gravity, rather than co-current upflow against gravity in riser reactors, which, while offering significant advantages over conventional bubbling fluidized bed (see Chapter 1), still suffer from severe gas and solids backmixing, non-uniform gas and solids flow structures and significant particle aggregation.

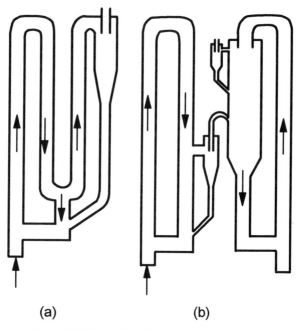

Figure 16.9 Schematics of the two types of N-CFB.

The first application of the downer reactor appeared to be the plasma ultrapyrolysis of coal in the 1960s and 1970s in the former USSR and Germany (Beiers et al., 1988; Brachold et al., 1993; Jin, 1994). The reason a downflow design was chosen seems to have been easy operation of the plasma torch rather than for better chemical reactor design. From the 1970s, Stone and Webster Engineering Corporation began to develop a new type of reactor, later referred to as the 'Quick Contact' (QC) reactor (Gartside, 1989), consisting of a solids–gas feed mixer, a downflow reactor section and a specially designed one-quarter-turn cyclone for ultra-fast gas–solids separation. The QC reactor offers very short residence times (~200 ms) and near plug flow condition. Downer reactors were also considered for the Fluid Catalytic Cracking (FCC) process (e.g. Murphy, 1992). Berg et al. (1989) proposed a downflow Ultra-Rapid Fluidized (URF) reactor, which has now been successfully applied to biomass pyrolysis (Graham et al., 1991). This reactor has recently been referred to (Bassi et al., 1994) as a Short Contact Time Fluidized Reactor (SCTFR).

16.5.1 Existing and potential applications of downflow systems

The downer reactor has distinct advantages for: (a) very fast reactions with the intermediate(s) as the desired product(s), where uniform and short contact time is essential to prevent over-reaction and to ensure good selectivity; (b) reactions where high gas–solid contact efficiency and low gas and solids backmixing are required; (c) reactions where a high solids-to-gas feed ratio is required; and (d) catalytic reactions with rapidly decaying catalysts.

Fluid catalytic cracking (FCC) appears to be a reaction for which downer reactors may eventually displace the current riser reactor. Well controlled reaction time (short and uniform) is essential to avoid overcracking and to ensure maximum yields of the desired products (e.g. gasoline). In current riser FCC units, the radial non-uniformity of gas and solids flow structure and the significant solids dispersion in the riser have caused over-cracking, resulting in lower gasoline recovery and higher yields of coke and light gases (Jin et al., 1990; Johnson et al., 1992). Murphy (1992) proposed a new FCC/heavy oil cracker unit that incorporates a downflow reactor and a riser regenerator. Catalyst flows from a hopper in a partially settled state, free of gas bubbles, into a number of parallel smaller diameter downflow reactors of length necessary to give the desired reaction time. The catalyst is then separated by a simple inertia separator before the product vapours are passed through high efficiency cyclones for final cleaning. Wei et al. (1994b) also proposed a new type of FCC downer reaction system with two-stage riser regeneration.

Mobil and Texaco have both patented downer reactors for the FCC process (Gross and Ramage, 1983; Gross, 1983; Niccum and Bunn, 1985). They claim uniform distribution of catalyst, decreased contact time of catalyst with the feed and reduced coking. The Texaco design (Niccum and Bunn, 1985) is sketched in Figure 16.10, which indicates that the required length of a downer cracker is much less than that of the riser (Figure 16.10), because of the shorter and more uniform catalyst residence time.

Application of downer reactor to ultrapyrolysis of coal under a hydrogen plasma stream started in the early 1970s with a 3 MW semi-commercial plant, built in Russia, and another 1.2 MW plant built in Germany (Jin, 1994). This process has been reported to lead to savings of up to 25% of the electricity consumption and to be able to largely reduce the adverse environmental impact of the CaC_2 process.

The downer reactors has also been used for the ultrapyrolysis of biomass (Freel et al., 1987; Graham et al., 1991), where various forms of biomass, including cellulose, wood, lignin and other fossil fuel feed-stocks, may be processed under extremely rapid pyrolysis conditions. In order to obtain the desired intermediate products, the reaction time must be controlled to the order of tens to hundreds of milliseconds and the product must be separated immediately after the reaction. To supply the heat required for the pyrolysis reaction, large amounts of sand 'THERMOFORS' are fed together with the reactant into the downer, thus taking advantage of the downer's ability to handle high solids/gas feed ratios. Further discussion regarding potential applications of downflow reactors is provided in Chapter 17.

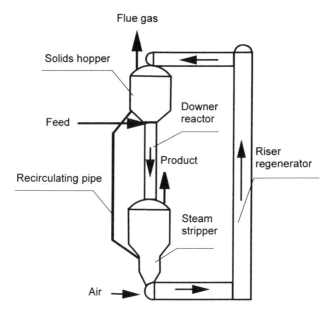

Figure 16.10 Texaco downer FCC design (Niccum and Bunn, 1985).

16.5.2 Typical structure of downers

A typical downer reactor consists of a vertical column with gas and solids distributors at the top and one or more gas–solids separator(s) at the bottom. For catalytic reactions or gas phase reactions with solids as heat carriers, solids are recirculated to the top of the downer after regeneration or re-heating. In a cold model laboratory downer reactor, gas–solids co-current downflow and solids circulation are usually achieved in a co-current downflow circulating fluidized bed (CDCFB), with solids separated from the gas stream at the bottom and carried back to the top of the downer via a riser.

The system used by the Tsinghua group is shown in Figure 16.11a. Starting from the dense fluidized bed storage tank at the bottom, solids are entrained upwards in a riser of diameter 140 mm and height 15 m. The gas–solids suspension is then transported to the top of the downer (140 mm in diameter and 11 m high), where gas and solids are redistributed through a gas–solids distributor as shown in Figure 16.11b. At the bottom of the downer, the gas–solids suspension enters a quick inertial separator (Figure 16.11c), where most entrained solids are recovered. The remaining solids are then separated from the gas stream in a secondary cyclone. Solids recovered by the quick internal separator and cyclone are returned to the dense fluidized bed storage tank. The solids circulation rate can be measured by deflecting collected solids into a measuring vessel.

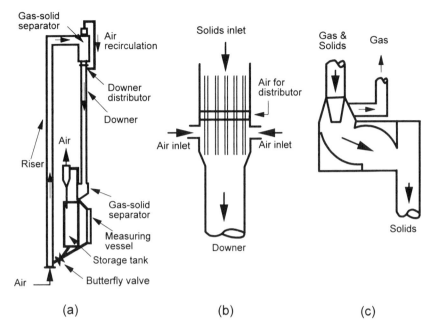

Figure 16.11 Schematic of the modified 140 mm downer CFB system in Tsinghua University.

Since solids acceleration in downers is caused by both gravity and drag, uniform solids distribution is as important as uniform gas distribution in downer distributor design. The gas and solids feeding system adopted at Tsinghua is sketched in Figure 16.11b. A fluidized bed is situated on top of the downer with many small diameter vertical distribution tubes to deliver solids (Gartside, 1989; Qi *et al.*, 1990; Aubert *et al.*, 1994). The main fluidization gas other than that required for solids feeding is introduced below the top distributor bed, and gas deflecting devices (Kim and Scader, 1983) may be installed to direct the gas flow downwards, in order not to disrupt the solids flow. The distributor bed is either semi-fluidized or kept around minimum fluidization; bubbles should be prevented since entrapped bubbles may block the distribution tubes causing feeding fluctuations. Small orifices may be drilled on the wall of the distributor tubes within the bed to improve solids distribution. The solids flowrate may be adjusted by the bed height and/or by the flowrate of fluidization air. For a system requiring higher solids flowrates, the diameter of the distribution tubes and the number and size of the orifices on the tubes are all expected to be larger. However, little research has been carried out on the feeding system so that no general design criteria are available. Developing improved gas and solids distribution methods is a priority area for downer reactors.

Downer reactors typically accommodate short contact time fast reactions, and especially fast reactions where an intermediate product is the desired product. Fast separation is therefore extremely critical and non-traditional quick gas–solids separators other than cyclones are needed to achieve quick separation. A novel one-quarter turn cyclone gas–solids separator has been patented by Stone & Webster (Gartside and Woebecke, 1982), for which a separation time of 30 ms and an efficiency of 98% were reported (Gartside, 1989). For the same purpose, a novel gas–solids quick separator as shown in Figure 16.11c was developed by the Tsinghua group in the 1980s (Qi et al., 1989). This is a simple inertial separator in which gas and solids suspension first pass through a specially designed nozzle and then impinge on a curved guiding plate with gradually increasing radius where more than 96% to 99% of the solids are separated from the gas phase within 0.05 to 0.3 s.

16.5.3 Hydrodynamics of downer

Hydrodynamic studies in downflow systems were pioneered by Shimizu et al. (1978) and Kim and Seader (1983). For both QC and SCTFR reactors, there has been no study of the hydrodynamics. Starting in the late 1980s, researchers at Tsinghua University (Qi et al., 1990; Bai et al., 1991: Yang et al., 1991b; Wang et al., 1992: Cao et al., 1994; Wei et al., 1994d, 1995; Zhu and Wei, 1995) have carried out a series of hydrodynamic studies in a 140 mm diameter and 5.8 m (recently changed to 11 m) high co-current downflow circulating fluidized bed (CDCFB) system. More recently, the research groups in the French Institute of Petroleum and the University of Western Ontario also published results from a 50 mm diameter downer (Aubert et al., 1994; Herbert et al., 1994; Roques et al., 1994).

(a) Three-section axial flow structure
Along the axis of downer reactors, there typically exist three distinct flow sections: first and second acceleration sections and a constant velocity section (Qi et al., 1990: Yang et al., 1991b and Wang et al., 1992). This axial flow structure is shown in Figure 16.12, and the flow characteristics of each section are summarized in Table 16.4. This three-section axial flow structure is consistent with what has been proposed by Kwauk (1964), based on his 'General Fluidization' theory.

Upon entering the downer, the gas immediately attains a velocity close to the superficial velocity, while the particle velocity is initially close to zero. Solid particles are first accelerated both by gravity and drag (Figure 16.12). As a consequence, pressure decreases continuously along the downer to compensate for the drag on the particles and the friction between the column wall and the gas–solid suspension (Figure 16.13). When the particle velocity reaches the gas velocity, the drag acting on the particles

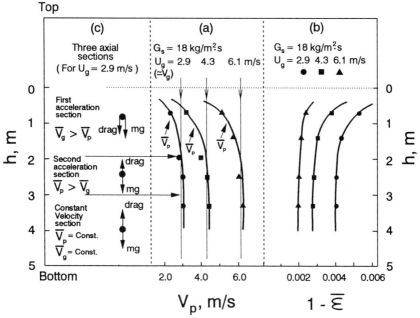

Figure 16.12 Axial flow structure in the downer (FCC particles, $\rho_p = 1545\,\text{kg/m}^3$, $d_p = 54\,\mu\text{m}$, $D = 0.140\,\text{m}$, $H = 5.8\,\text{m}$).

becomes zero and the pressure reaches a minimum. The section from the top to the position where particle velocity equals the gas velocity is the first acceleration section. In this section, particle velocity increases dramatically and so does the voidage (Figure 16.12).

Table 16.4 Characteristics of gas-particle flow structures in three different sections of downer reactors

	First acceleration section	Second acceleration section	Constant velocity section
Gas velocity, $\bar{V}_g$	Decreasing	Decreasing slightly	Constant
Particle velocity, $\bar{V}_p$	Increasing dramatically	Increasing at a reduced rate	Constant
Particle concentration, $1 - \bar{\varepsilon}$	Decreasing dramatically	Decreasing further	Constant
$\bar{V}_g$ relative to $\bar{V}_p$	$\bar{V}_g > \bar{V}_p$	$\bar{V}_g < \bar{V}_p$	$\bar{V}_g < \bar{V}_p$
Particle acceleration	Due to drag and gravity	Due to gravity only	No acceleration
Direction of drag	Downwards	Upwards	Upwards
Slip velocity, U_{slip}	Negative	Positive	Positive and constant
Absolute pressure, P	Decreasing	Increasing	Increasing linearly
Pressure gradient $\left(\dfrac{\partial P}{\partial h}\right)$	Negative	Positive	Positive
$\left(\dfrac{\partial^2 P}{\partial h^2}\right)$	Positive (large)	Positive (small)	Zero

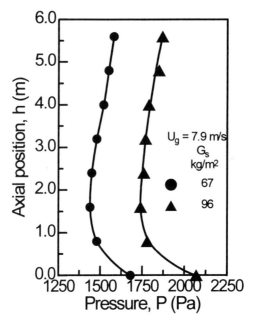

Figure 16.13 Axial distribution of pressure (FCC particles, $\rho_p = 1545 \text{ kg/m}^3$, $d_p = 54\,\mu\text{m}$, $D = 0.140\,\text{m}$, $H = 5.8\,\text{m}$, from Wang et al., 1992, reproduced with permission).

After acquiring the same velocity as the gas phase, solids are further accelerated by gravity while encountering drag in the upward direction exerted by the now slower-moving gas phase (Figure 16.12). The particle velocity increases until the slip velocity between the gas and particles reaches a value where the drag force counter-balances the gravitational force. This section is called the second acceleration section. In this section, particle velocity continues to increase, but at a smaller rate than in the first acceleration section, and solids holdup decreases accordingly (Figure 16.12). Pressure also increases gradually (Figure 16.13).

When the drag is sufficient to balance the gravitational force, particles are not accelerated further, and the remainder of the downer (excluding any exit effect) is named the constant velocity section. In this section, particles travel faster than gas, and both particle and gas velocities remain constant. Pressure increases linearly along the downer (Figure 16.13) and the pressure gradient becomes constant and is equal to the cross-sectional bed density, if wall friction is neglected.

The initial flow development length in downers seems to be much shorter than in risers (Zhu et al., 1995). For fast gas phase (catalytic or non-catalytic) reactions, a short flow development region is beneficial since axial solids dispersion is much greater and radial solids distribution less uniform in this region, resulting in undesirable gas and solids flow patterns and non-uniform gas and solids residence time distributions.

(b) Radial gas and solids flow structure

Typical radial profiles of gas and particle velocities measured in a downer (Bai et al., 1991; Cao et al., 1994) and in a riser (Yang et al., 1993), and the radial profile of solids concentration measured in both downers and risers (Wang et al., 1992) are given in Figure 16.14. Compared with the radial flow structure in the riser, the radial distributions of gas and particle velocities and solids concentration are all significantly more uniform in the downer. Solids concentration is seen to be relatively constant in the downer centre region until r/R reaches 0.8, where a dense ring with significantly higher solids concentration begins to develop. After reaching a maximum at $r/R \simeq 0.9$, solids concentration drops towards the wall. Gas and particle velocities also reach maxima in this dense ring region, with, however, much smaller peaks.

For the radial distribution of solids concentration in the downer, the local bed voidage is a function of r/R only, as in a riser (Bai et al., 1991). The following empirical correlation was presented by Bai et al. (1991)

$$\epsilon = \bar{\epsilon}^{30.62(1-r/R)\exp[-127.5(1-r/R)^2] + 22.8/(37.7-r/R)} \tag{16.1}$$

Radial distributions of local slip velocities in a riser (Yang et al., 1993) and a downer (Cao et al., 1994) are given in Figure 16.15. Although the gas and particle velocity profiles are quite different for the downer and the riser (Figure 16.14), the shapes of the two slip velocity profiles are similar, both increasing gradually with increasing r/R, reaching a maximum value near the wall at $r/R \simeq 0.95$ and then decreasing towards the wall. The distribution of slip velocity in the downer is similar to that of the radial solids concentration distribution, suggesting that a higher solids concentration leads to higher slip in downers. The slip velocity in the downer is seen to be significantly lower and its radial variation less than in the riser, indicating less particle aggregation in the downer.

Examination of the radial profiles of the gas and particle velocities and the solids concentration shown in Figure 16.14 indicates that the local particle velocity in the downer can be higher than the local gas velocity, with higher local solids concentration always corresponding to higher gas and particle velocities. This is contrary to the riser, where local particle velocity is always lower than local gas velocity and a higher local solids concentration always corresponds to lower gas and particle velocities.

Compared with riser reactors, downer reactors have a much more uniform radial gas–solids flow pattern, likely due to the change of the direction of gas flow from opposing to aiding gravity. The following mechanism is provided to explain the more favourable radial flow structure in the downer (Zhu and Wei, 1996): in both the downer and the riser, higher local solids aggregation

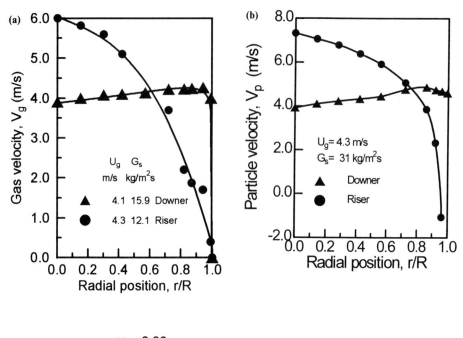
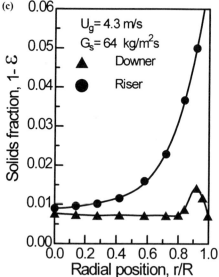

Figure 16.14 (a) Radial distributions of gas velocity in a downer and a riser; (b) radial distributions of particle velocity in the downer and the riser; (c) radial distributions of solids fraction in the downer and the riser (FCC particles, $\rho_p = 1545\,\text{kg/m}^3$, $d_p = 54\,\mu\text{m}$, $D = 0.140\,\text{m}$, $H = 5.8$ and $11\,\text{m}$, from Cao et al., 1994, from Wang et al., 1992, reproduced with permission, and from Yang et al., 1993, reproduced with permission of the American Institute of Chemical Engineers, copyright © 1993 AIChE, all rights reserved).

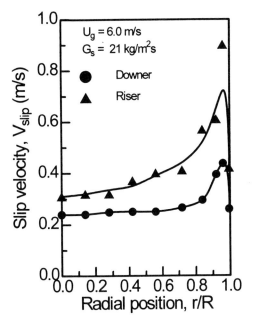

Figure 16.15 Gas–solid slip velocity profile in a downer and a riser (FCC particles, $\rho_p = 1550$ kg/m^3, $d_p = 54\,\mu$m, $D = 140$ mm, $H = 11$ m, from Cao et al., 1994, and from Yang et al., 1993, reproduced with permission of the American Institute of Chemical Engineers, copyright © 1993 AIChE, all rights reserved).

results in a reduction of the total drag/weight ratio. In the riser (where drag is the driving force for particle upflow), reduction of the drag/weight ratio due to more solids aggregation near the wall decreases the upwards particle velocity, which in turn increases the tendency for particle aggregation (Bi et al., 1993b). Increased particle aggregation then further reduces the drag/weight ratio and the local particle velocity, leading to steeper radial profiles for both gas and particle velocities. In the downer, on the other hand, the drag force is in the upwards direction. A reduction of the drag/weight ratio results in increased downwards particle velocity, which in turn leads to increased local gas velocity. On the other hand, increased local gas and particle velocities in the downer tend to reduce the extent of particle aggregation (Bi et al., 1993b), thereby increasing the drag. Therefore, the system stabilizes itself and a more uniform radial flow structure is present in the downer.

16.5.4 Gas and solids mixing

(a) Axial gas and solids mixing

Due to the more uniform radial distributions of gas and particle velocities and solids concentration, the axial gas and solids dispersion in the downer

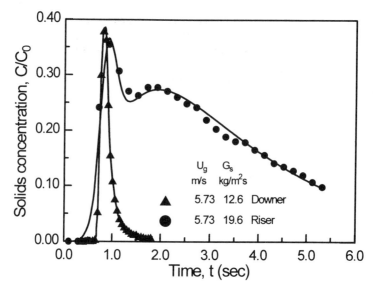

Figure 16.16 Typical particle residence time distribution in the downer and the riser (alumina particles, $\rho_p = 1710\,\text{kg/m}^3$, $d_p = 54\,\mu\text{m}$, $D = 140\,\text{mm}$, $H = 11\,\text{m}$, from Wei et al., 1994c).

are much smaller than in the riser. As shown in Figure 16.16, a phosphor particle tracing method, developed by Wei et al. (1994c,d), shows that the solids residence time distribution (RTD) in the downer is much narrower than in a riser operating under very similar conditions. The wide tail for the solids RTD in the riser is due to strong solids backmixing caused by particle aggregation and downflow along the wall. The tail for the solids RTD in the downer is much smaller, suggesting much reduced axial solids dispersion.

As a consequence, the axial solids Peclet number in the downer is reported (Wei et al., 1994d; Wei and Zhu, 1996) to be around 100, one to two orders of magnitude larger (Figure 16.17) than in the riser, where it is typically 1 to 9 (Zheng et al., 1992; Wei et al., 1994a). Figure 16.17 also shows that increasing the superficial gas velocity slightly increases the solids dispersion. The following correlation was recommended by Wei et al. (1994d) for the axial solids Peclet number:

$$Pe_a = \frac{8.93 \times 10^{-7}\,Re}{1 - \bar{\epsilon}} + 101 \qquad (16.2)$$

The axial gas dispersion in the downer is also small. Using hydrogen as the gas tracer, Wei et al. (1995) found that axial gas dispersion coefficients in downers are of order $0.01\,\text{m}^2/\text{s}$, one to two orders of magnitude less than the range of 0.4 to $1.0\,\text{m}^2/\text{s}$ reported for risers (Li and Wu, 1991). Given its extremely small magnitude, gas axial dispersion coefficients can only be measured at low superficial gas velocities (Wei et al., 1995).

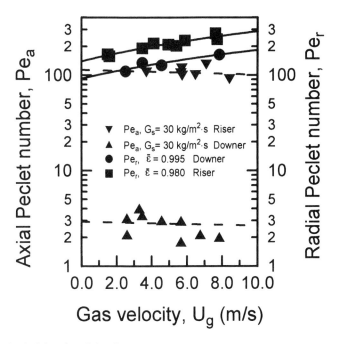

Figure 16.17 Axial and radial solids Peclet number in the downer and the riser (alumina particles, $\rho_p = 1710 \, kg/m^3$, $d_p = 54 \, \mu m$, $D = 140 \, mm$, $H = 11 \, m$, from Wei et al., 1994a, 1994d, reproduced with permission).

(b) Radial gas and solids mixing

The radial solids Peclet numbers, Pe_r, in the downer are comparable in magnitude with those in risers. Both increase with superficial gas velocity (Figure 16.17). The following correlation was obtained by Wei et al. (1994d) for the radial solids Peclet number in the downer:

$$Pe_r = 140 \, Re^{0.61}(1 - \bar{\epsilon})^{1.23} \tag{16.3}$$

Radial gas Peclet numbers in the downer measured by Wei et al. (1995) are of the order of 100–300, comparable with those obtained in risers (Adams, 1988; Amos et al., 1993). Radial gas dispersion in the downer is found to decrease with increasing superficial gas velocity.

Good radial mixing of gas and solids in both downer and riser reactors are essential to ensure good gas–solids contacting and therefore high productivity and selectivity. Significantly reduced axial gas and solids dispersions in the downer suggest that downer reactors operate under conditions much closer to plug flow than riser reactors. Excellent radial gas and solids mixing as well as significantly reduced axial gas and solids mixing in the downer reactor makes it a better candidate for short contact time reactions where intermediates are the desired product.

The large differences in hydrodynamic and mixing behaviour between risers and downers lie mainly in the different radial profiles of gas velocity, particle velocity and solids concentration. These differences in flow pattern are due to the direction of gas–solids flow, with or counter to gravity. Co-current gas–solid downflow provides some unique advantages for downer reactors. Zhu et al. (1995) recently provided a comprehensive review on downer reactors where more detailed experimental results and theoretical analyses can be found.

16.6 Liquid–solids and gas–liquid–solids systems

The liquid–solids circulating fluidized bed (LSCFB) and gas–liquid–solids circulating fluidized bed (GLSCFB) have received scant attention in the fluidization literature compared to gas–solids systems. Aside from two reports on applications of LSCFB to binary solids mixing (Di Felice et al., 1989) and fermentation (Pirozzi et al., 1989), most hydrodynamic studies have been carried out at Tsinghua University (Liang et al., 1993, 1994a,b, 1995a,b).

16.6.1 Liquid–solids (L–S) circulating fluidized bed

An important surfactant, linear alkylbenzene, is synthesized from benzene and l-dodecene with hydrogen fluoride (HF) as catalyst. However, HF is very corrosive and the disposal of fluorided neutralization products is a concern. An alternative catalyst is solid HY zeolite which, having many advantages over HF, deactivates quickly due to the deposition of gum-polymers onto the catalyst surface. To enable continuous regeneration of the solid catalyst, Jin et al. (1994) and Liang et al. (1994b, 1995a) adopted the concept of LSCFB. This called for hydrodynamic studies. LSCFB reactors could be useful not only for catalytic L–S reactions, but also for non-catalytic L–S reactions such as the manufacture of rutile and physical processes such as mineral ore dressing.

The cold model circulating fluidized bed apparatus used for L–S and G–L–S CFB studies by Liang et al. (1993, 1994a, 1995b) is shown in Figure 16.18. It is 140 mm in diameter and 3.0 m high. The solids circulation rate is controlled by the auxiliary liquid flow. For LSCFB, tap water was used as the liquid phase, while glass beads and silica gel particles of different sizes were used as the solids phase. Axial solids concentration distributions obtained by Liang et al. (1993) from pressure drop measurements are shown in Figure 16.19. At low superficial liquid velocities similar to those in conventional liquid–solid particulate fluidization, a dense bed region clearly exists at the bottom, with a freeboard devoid of solids above. Further increases of superficial liquid velocity cause significant particle entrainment, and recirculation of solids to the bottom of the riser becomes necessary.

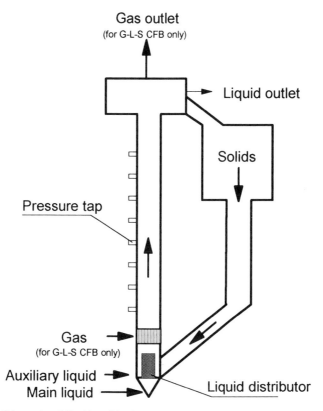

Figure 16.18 Schematic of liquid–solids CFB and gas–liquid–solids CFB apparatus (from Liang et al., 1994a).

Increasing liquid velocity also leads to more uniform axial solids concentration distributions until the axial solids holdup eventually becomes constant over the whole riser when the recirculation regime is reached. Between conventional particulate fluidization and the recirculation fluidization regimes lies a transition region where the dense bed surface becomes diffuse and gradually disappears. Increasing particle size and/or density makes the transition occur at higher liquid velocity. A regime map similar to that of Grace (1986) is proposed here for liquid–solids systems (Figure 16.20), giving the boundaries between the fixed bed and the particulate fluidization regimes and between the particulate and the recirculation fluidization regimes.

Figure 16.21 shows radial solids concentration profiles measured recently at Tsinghua. Contrary to the common belief that liquid–solids fluidized beds always exhibit uniform radial solids distribution (Kwauk and Zhuang, 1963), the radial solids distribution is not uniform when there is solids recirculation. Solids holdup is lower in the centre and increases towards the column

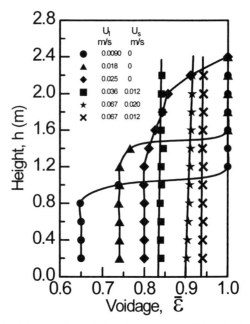

Figure 16.19 Axial voidage profiles at different operating conditions in the LSCFB (glass beads, $\rho_p = 2460 \, \text{kg/m}^3$, $d_p = 0.405 \, \mu\text{m}$, $D = 140 \, \text{mm}$, $H = 3 \, \text{m}$, from Liang et al., 1993).

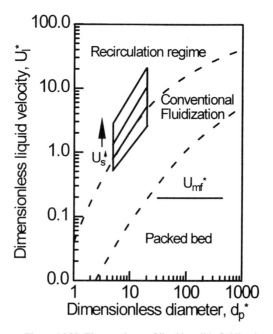

Figure 16.20 Flow regimes of liquid–solids fluidization.

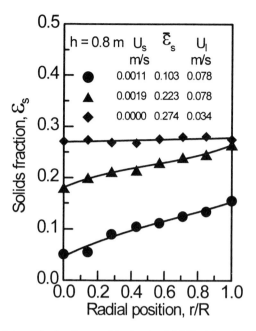

Figure 16.21 Radial solids fraction profiles in the LSCFB (glass beads, $\rho_p = 2460 \, \text{kg/m}^3$, $d_p = 0.405 \, \mu\text{m}$, $D = 140 \, \text{mm}$, $H = 3 \, \text{m}$).

wall. This radial non-uniformity increases with increasing liquid velocity and solids circulation rate. Figure 16.21 also shows that the voidage (i.e. the liquid phase holdup) in LSCFB decreases with decreasing liquid velocity and with increasing solids recirculation rate.

Apparent slip velocities obtained by Liang *et al.* (1993) are plotted against average solids holdup in Figure 16.22. With the Richardson and Zaki (1954) equation assumed to apply to the solids recirculating system and with the superficial liquid velocity replaced by the average slip velocity (Kwauk and Zhuang, 1963), the bed voidage may be calculated by:

$$U_t \epsilon^{n-1} = U_{slip} = \frac{U_l}{\epsilon} - \frac{U_s}{1-\epsilon} \tag{16.4}$$

Large differences exist between the predictions from equation 16.4 and the experimental results, suggesting that the Richardson and Zaki equation does not hold in LSCFB. This may be due to radial non-uniformity of the phase holdups in the LSCFB.

16.9.2 Gas–liquid–solid (G–L–S) three-phase circulating fluidized bed

The advantages of circulating fluidized beds suggest that CFB be applied to three-phase systems. The development of G–L–S CFB would open up new

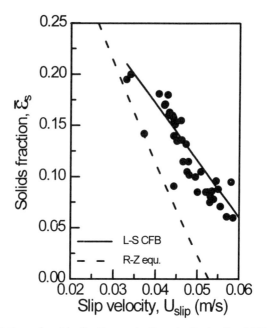

Figure 16.22 Relation of solids fraction and slip velocity in the LSCFB (glass beads, $\rho_p = 2460 \, \text{kg/m}^3$, $d_p = 0.405 \, \mu\text{m}$, $D = 140 \, \text{mm}$, $H = 3 \, \text{m}$, from Liang et al., 1993).

applications for circulating fluidized beds such as biochemical processes where intensified mass transfer and independent control of phase holdups as well as solids recirculation are required. Higher shear stress in CFB could promote biofilm renewal. (However, high shear can also be a disadvantage for some other biochemical processes such as mammalian cell processes.) Using the same experimental system as for LSCFB (Figure 16.18), Liang et al. (1994a, 1995b) studied the hydrodynamics of gas–liquid–solid circulating fluidized beds. Air and tap water were used as gas and liquid media to fluidize 0.405 mm glass beads. Local gas holdup was measured by an electrical conductivity probe. Liquid and solids phase holdups were derived from pressure drop measurements and the gas phase holdup.

The axial solids concentration distribution (Liang et al., 1995b) in a GLSCFB shown in Figure 16.23 has shapes similar to those of gas–solids CFB, i.e. S shape and exponential shape (Bai et al., 1992, see also Chapter 2). Increasing the superficial liquid velocity reduces both the solids holdup and its axial variation, as expected. However, increasing the superficial gas velocity increased the solids holdup, probably because increasing superficial gas velocity leads to increased gas holdup, which in turn reduces the effective buoyancy force exerted by the combined gas and liquid flow on the solid particles. The average solids holdup decreases with increasing superficial

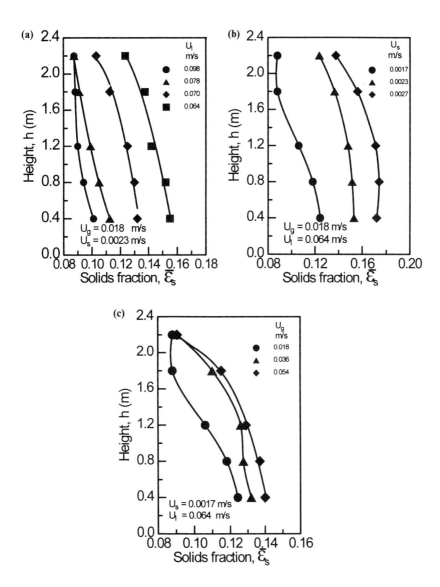

Figure 16.23 Axial solids fraction profiles at different operating conditions (glass beads, $\rho_p = 2460 \, \text{kg/m}^3$, $d_p = 0.405 \, \mu\text{m}$, $D = 140 \, \text{mm}$, $H = 3 \, \text{m}$, from Liang et al., 1995b, reproduced with permission of American Institute of Chemical Engineers, copyright © 1995 AIChE, all rights reserved).

liquid velocity and increases with increasing solids circulation rate. Figure 16.23 shows that the solids holdup in GLSCFB can be adjusted independently using both the gas and liquid velocities, a unique feature of circulating fluidized beds and a clear advantage of the G–L–S circulating fluidized bed as a three-phase reactor over conventional three-phase fluidized beds.

The general wake model for conventional three-phase fluidized beds (Bhatia and Epstein, 1974) was extended to three-phase circulating fluidized beds with the net solids flow now equal to U_s, instead of being zero. With this modification and following the details described by Bhatia and Epstein (1974), the average solids holdup in the G–L–S CFB has been calculated. The predictions agree well with the data reported in Figure 16.23.

Recent unpublished results of local gas holdup shown in Figure 16.24 indicate that the shape of the radial gas holdup distribution does not change significantly with the superficial velocities of gas, liquid and solids. Whether this is inherent for GLSCFB or column-dependent needs to be verified. The following correlation fits experimental results well:

$$\frac{\epsilon_g}{\bar{\epsilon}_g} = 1.58 - 0.092(r/R) - 1.2(r/R)^2 + 5.1(r/R)^3 \qquad (16.5)$$

The regime map proposed by Muroyama and Fan (1985) for a conventional three-phase fluidized bed has also been extended by the authors to include the recirculation regime as shown in Figure 16.25. With increases of liquid velocity from the dispersed bubble regime, significant particle entrainment starts and the bed enters the recirculation regime where continuous solids feeding to the bottom of the bed becomes necessary to maintain a fixed

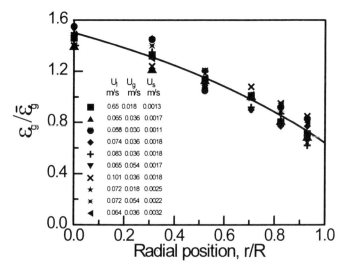

Figure 16.24 Radial relative gas fraction profiles in the GLSCFB (glass beads, $\rho_p = 2460 \, \text{kg/m}^3$, $d_p = 0.405 \, \mu\text{m}$, $D = 140 \, \text{mm}$, $H = 3 \, \text{m}$).

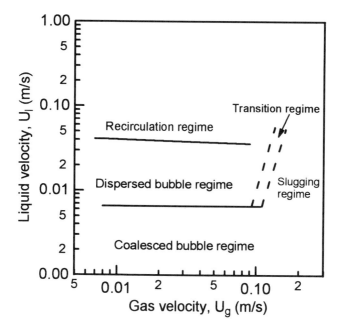

Figure 16.25 Flow regimes of gas–liquid–solid fluidization.

solids inventory in the bed. Within this recirculation regime, it may be possible to identify a circulating fluidization regime and a transport regime such as those in a gas–solids CFB (Liang et al. 1995b). More experimental results are needed to confirm the existence of these two separate regimes.

Acknowledgements

The authors would like to express their thanks to Drs F. Wei and W-G. Liang for their assistance in preparing this chapter.

Nomenclature

C/C_0	dimensionless tracer concentration
d_p	particle diameter, m
d_p^*	dimensionless particle diameter, $[d_p(\rho_l(\rho_s - \rho_l)/\mu_l)^{1/3}]$
D	downer/riser diameter, m
D_a	axial dispersion coefficient, m^2/s
D_r	radial dispersion coefficient, m^2/s
F_d	drag force on particle, N
G_S	solids circulation rate, kg/m^{2}s

h	vertical distance from top of downer or bottom of riser, m
H	downer or riser height, m
L	distance from solids tracer injection/stimulation point to sampling point, m
mg	gravitational force on particle, N
P	absolute pressure, Pa
Pe_a	axial solids Peclet number, $[= \bar{V}_p L/D_a]$
Pe_r	radial solids Peclet number, $[= \bar{V}_p L/D_r]$
Pe_{gr}	radial gas Peclet number, $[= U_g L/D_r]$
r	radial coordinate, m
R	downer/riser radius, m
Re	Reynolds number, $[= U_g \rho_g D/\mu]$
Q_1	primary air flow rate, m^3/s
Q_2	secondary air flow rate, m^3/s
t	time, s
U_g	superficial gas velocity, m/s
U_l	superficial liquid velocity, m/s
U_l^*	dimensionless liquid velocity, $[U_l(\rho_l^2/(\mu_l g(\rho_s - \rho_l)))^{1/3}]$
U_s	superficial solids velocity, m/s
U_{slip}	cross-section averaged slip velocity, m/s
U_t	single particle terminal velocity, m/s
V_g	local gas velocity, m/s
V_p	local particle velocity, m/s
V_{slip}	local slip velocity, m/s
$\bar{V}_g$	cross-section averaged gas velocity, m/s
$\bar{V}_p$	cross-section averaged particle velocity, m/s
W	solids inventory, kg
ϵ	local voidage
$\bar{\epsilon}$	cross-section averaged voidage
ϵ_g	local gas volume fraction
$\bar{\epsilon}_g$	cross-section averaged gas volume fraction
$\bar{\epsilon}_s$	cross-section averaged solids volume fraction
ρ_g	gas density, kg/m^3
ρ_l	liquid density, kg/m^3
ρ_p	particle density, kg/m^3
μ	gas viscosity, Pa·s
μ_l	liquid viscosity, Pa·s

References

Adams, C.K. (1988) Gas mixing in fast fluidized bed, in *Circulating Fluidized Bed Technology II* (eds P. Basu and J.F. Large), Pergamon Press, Oxford, pp. 299–306.

Amos, G., Rhodes, M.J. and Mineo, H. (1993) Gas mixing in gas–solids risers. *Chem. Eng. Sci.*, **48**, 943–949.

Arena, U., Cammarota, A. and Pistane, L. (1986) High velocity fluidization behavior of solids in a laboratory scale circulating fluidized bed, in *Circulating Fluidized Bed Technology* (ed. P. Basu), Pergamon Press, Toronto, pp. 119–125.

Aubert, E., Barreteau, D., Gauthier, T. and Pontier, R. (1994) Pressure profiles and slip velocities in a co-current downflow fluidized bed reactor, in *Circulating Fluidized Bed Technology IV* (ed. A.A. Avidan), AIChE, New York, pp. 403–405.

Avidan, A.A. and Yerushalmi, J. (1982) Bed expansion in high velocity fluidization. *Powder Technol.*, **32**, 223–232.

Bai, D.R., Jin, Y. and Yu, Z.Q. (1990) Particle acceleration length in a circulating fluidized bed. *Chem. Reaction Eng. & Tech.* (in Chinese), **6**(3), 34–39.

Bai, D-R., Jin, Y., Yu, Z-Q. and Gan, N-J. (1991) Radial profiles of local solid concentration and velocity in a concurrent downflow fast fluidized bed, in *Circulating Fluidized Bed Technology III* (eds. P. Basu, M. Horio and M. Hasatani), Pergamon Press, Oxford, pp. 157–162.

Bai, D-R., Jin Y., Yu, Z-Q. and Zhu, J-X. (1992) The axial distribution of the cross-sectionally averaged voidage in fast fluidized beds. *Powder Technol.*, **71**, 51–58.

Bassi, A.B., Briens, C.L. and Bergougnou, M.A. (1994) Short contact time fluidized reactors (SCTFRs), in *Circulating Fluidized Bed Technology IV* (ed. A.A. Avidan), AIChE, New York, pp. 15–19.

Beiers, H-G., Baumann, H., Bittner, D., Klein, J. and Juentgen, H. (1988) Pyrolysis of some gaseous and liquid hydrocarbons in hydrogen plasma. *Fuel*, **67**, 1012–1016.

Berg, D.A., Briens, C.L. and Bergougnou, M.A. (1989) Reactor development for the ultrapyrolysis reactor. *Can. J. Chem. Eng.*, **67**, 96–101.

Bhatia, V.K. and Epstein, N. (1974) Three phase fluidization: a generalized wake model, in *Fluidization and Its Applications* (ed. H. Angelino), Toulouse, France, pp. 380–392.

Bi, H-T. and Zhu, J-X. (1993) Static instability analysis of circulating fluidized beds and concept of high density risers. *AIChE J.*, **39**, 1272–1280.

Bi, H-T., Grace, J.R. and Zhu, J-X. (1993a) On type of choking in vertical pneumatic systems. *Int. J. Multiphase Flow*, **19**, 1077–1092.

Bi, H-T, Zhu, J-X., Jin, Y. and Yu, Z-Q. (1993b) Forms of particle aggregations in CFB, in *Proc. 6th National Conf. on Fluidization* (in Chinese), Wuhan, China, October 1993, pp. 162–167.

Brachold, H., Peuckerk, C. and Regner, H. (1993) Light-plasma-reactor for the production of acetylene from coke. *Chem. Ing. Tech.*, **63**, 293–297.

Brereton, C.M.H. and Grace, J.R. (1994) End effects in circulating fluidized bed hydrodynamics, in *Circulating Fluidized Bed Technology IV* (ed. A.A. Avidan), AIChE, New York, pp. 137–144.

Cao, B-L., Song, Z-X., Li, W-M., Zhang, J-Z., Gao, W-Y. and Pan, G-Z. (1991) Circulating fluidized bed boiler with an internal separator of three vertex chambers, in *Circulating Fluidized Bed Technology III* (eds P. Basu, M. Horio and M. Hasatani), Pergamon Press, Oxford, pp. 471–477.

Cao, C-S., Jin, Y., Yu, Z-Q. and Wang, Z-W. (1994) The gas–solids velocity profiles and slip phenomenon in a concurrent downflow circulating fluidized bed, in *Circulating Fluidized Bed Technology IV* (ed. A.A. Avidan), AIChE, New York, pp. 406–413.

Davies, C.E. and Graham, K.H. (1988) Pressure drop reduction by wall baffles in vertical pneumatic conveying tubes, in *CHEMECA '88, Australia's Bicentennial International Conference for the Process Industries*, Institute of Engineers, Australia, Sydney, Australia, August 1988, Vol. 2, pp. 644–651.

Di Felice, R., Gibilaro, L.G., Rapagna, S. and Foscolo, P.U. (1989) Particle mixing in a circulating liquid fluidized bed. *AIChE Symp. Ser.*, **85**(270), 32–36.

Dry, R.J. (1987) Radial particle size segregation in a fast-fluidized bed. *Powder Technol.*, **52**, 7–16.

Dry, R.J., White, R.B. and Close, R.C. (1992) The effect of gas inlet geometry on gas–solid contact efficiency in a circulating fluidized bed, in *Fluidization VII* (eds O.E. Potter and D.J. Nicklin), Engineering Foundation, New York, pp. 211–218.

Fan, L.S., Jean, R.H. and Kitano, K. (1987) On the operating regimes of concurrent upward gas–liquid–solid systems with liquid as the continuous phase. *Chem. Eng. Sci.*, **42**, 1853–1855.

Freel, B.A., Graham, R.G., Bergougnou, M.A., Overend, R.P. and Mok, L.K. (1987) The kinetics of the fast pyrolysis (ultrapyrolysis) of cellulose in a fast fluidized bed reactor. *AIChE Symp. Ser.*, **83**(255), 105–111.

Fusey, I., Lim, C.J. and Grace, J.R. (1986) Fast fluidization in a concentric circulating bed, in *Circulating Fluidized Bed Technology II* (ed. P. Basu), Pergamon Press, Oxford, pp. 409–416.

Gan, N., Jiang, D-Z., Bai, D-R., Jin, Y. and Yu, Z-Q. (1990) Concentration profiles in fast fluidized bed with bluff-body. *J. Chem. Eng. Chinese Univ.* (in Chinese), **3**(4), 273–277.

Gartside, R.J. (1989) QC – a new reaction system, in *Fluidization VI* (eds J.R. Grace, L.W. Shemilt and M.A. Bergougnou), Engineering Foundation, New York, pp. 25–32.

Gartside, R.J. and Woebecke, H.N. (1982) Solids feeding device and system. *U.S. Patent* 4,338,187.

Grace, J.R. (1986) Contacting modes and behaviour classification of gas–solid and other two-phase suspensions. *Can. J. Chem. Eng.*, **64**, 353–363.

Grace, J.R. (1990) High velocity fluidized bed reactors. *Chem. Eng. Sci.*, **45**, 1953–1966.

Graham, R.G., Freed, B.A. and Bergougnou, M.A. (1991) Scale-up and commercialization of rapid biomass pyrolysis for fuel and chemical production, in *Energy for Biomass and Wastes XIV* (ed. D.L. Klass), Inst. of Gas Technol., Chicago, pp. 1091–1104.

Gross, B. (1983) Heat balance in FCC process and apparatus with downflow reactor riser. *U.S. Patent* 4,411,773.

Gross, B. and Ramage, M.P. (1983) FCC reactor with a downflow reactor riser. *U.S. Patent* 4,385,985.

Gwyn, J.E. (1994) Entrance, exit, and wall effects on gas/particulate solids flow regimes, in *Circulating Fluidized Bed Technology IV* (ed. A.A. Avidan), AIChE, New York, pp. 679–684.

Harris, B.J., Davidson, J.F. and Xue, Y. (1994) Axial and radial variations of flow in circulating fluidized bed risers, in *Circulating Fluidized Bed Technology IV* (ed. A.A. Avidan), AIChE, New York, pp. 103–110.

Herbert, P.M., Gauthier, T.A., Briens, C.L. and Bergougnou, M.A. (1994) Application of fiber optic reflection probes to the measurement of local particle velocity and concentration in gas–solid flow. *Powder Technol.*, **80**, 243–252.

Hirama, M., Takeuchi, H. and Chiba, T. (1992) Regime classification of macroscopic gas–solid flow in a circulating fluidized bed riser. *Powder Technol.*, **70**, 215–222.

Horio, M. (1991) Hydrodynamics of circulating fluidization – present status and research needs, in *Circulating Fluidized Bed Technology III* (eds P. Basu, M. Horio and M. Hasatani), Pergamon Press, Oxford, pp. 3–14.

Hu, H-M., Wang, S-S. and Zhang, H-S. (1988) Study on the twin circulating fluidized bed reactors, in *Circulating Fluidized Bed Technology II* (eds P. Basu and J.F. Large), Pergamon Press, Oxford, pp. 555–563.

Ilias, S., Ying, S., Mathur, G.D. and Govind, R. (1988) Studies on swirling circulating fluidized beds, in *Circulating Fluidized Bed Technology II* (eds P. Basu and J.F. Large), Pergamon Press, Oxford, pp. 537–546.

Jiang, P. Bi, H-T., Jean, R.H. and Fan, L.S. (1991) Baffle effects on performance of catalytic circulating fluidized bed reactor. *AIChE J.*, **37**, 1392–1340.

Jin, Y. (1994) *Report on Visit to the Former USSR*, Tsinghua University, Beijing.

Jin, Y., Yu, Z-Q., Qi, C-M. and Bai, D-R. (1988) The influence of exit structure on the axial distribution of voidage in fast fluidized bed, in *Fluidization '88, Science and Technology* (eds. M. Kwauk and D. Kunii), Science Press, Beijing, pp. 165–173.

Jin, Y., Yu, Z-Q., Bai, D-R., Qi, C-M. and Zhong, X-X. (1990) Modelling of gas and solids two-phase flow for cocurrent downflow gas–solid suspensions. *Chem. Reaction Eng. & Technol.* (in Chinese), **6**(12), 17–23.

Jin, Y., Liang, W-G. and Yu, Z-Q. (1994) Synthesis of linear alkylbenzene using liquid-solids circulating fluidized bed reactors. *Chinese Patent* 94,105,710.0.

Johnson, T.E., Niccum, P.K., Raterman, F.M. and Schipper, P.H. (1992) FCC for the 1990s: new hardware developments. *AIChE Symp. Ser.*, **88**(291), 88–95.

Kim, J.M. and Seader, J.D. (1983) Pressure drop for cocurrent downflow of gas–solids suspensions. *AIChE J.*, **29**, 353–360.

Knowlton, T.M. (1988) Nonmechanical solids feed and recycle devices for circulating fluidized beds, in *Circulating Fluidized Bed Technology II* (eds. P. Basu and J.F. Large), Pergamon Press, Oxford, pp. 31–41.

Kwauk, M. (1964) Generalized fluidization II: Accelerative motion with steady profiles. *Scientia Sinica*, **13**, 1477–1492.

Kwauk, M. and Zhuang, Y-A. (1963) *Motion of Uniform Spheres and Fluid in Vertical Fluidized Systems*, Science Press, Beijing.
Lapidus, L. and Elgin, J.C. (1957) Mechanics of vertical-moving fluidized systems. *AIChE J.*, **3**, 63–68.
Lasch, B., Caram, H.S. and Chen, J.C. (1988) Solid circulation in cyclonic fluidized bed combustors, in *Circulating Fluidized Bed Technology II* (eds P. Basu and J.F. Large), Pergamon Press, Oxford, pp. 527–535.
Li, Y-C. and Wu, P. (1991) A study on axial gas mixing in a fast fluidized bed, in *Circulating Fluidized Bed Technology III* (eds P. Basu, M. Horio and M. Hasatani), Pergamon Press, Oxford, pp. 581–586.
Liang, W-G., Zhang, S-L., Yu, Z-Q., Jin, Y. and Wu, Q-W. (1993) Liquid–solids circulating fluidized bed (I) studies on the phase holdups and solid circulating rate and (II) studies on the apparent slip velocity and drag coefficient. *J. Chem. Ind. & Eng. China (Chinese Edition)*, **44**, 666–671 and 672–676.
Liang, W-G., Wu, Q-W., Yu, Z-Q., Jin, Y. and Wang, Z-W. (1994a) Hydrodynamics of gas–liquid–solid three-phase circulating fluidized bed, in *Fluidization '94, Science and Technology* (ed. Organizing Committee), Chemical Industry Press, Beijing, pp. 329–337.
Liang, W-G., Yu, Z-Q., Jin, Y., Wang, Z-W. and Wang, Y. (1994b) Synthesis of linear alkylbenzene in liquid–solid circulating fluidized bed reactor, in *Fluidization '94, Science and Technology* (ed. Organizing Committee), Chemical Industry Press, Beijing, pp. 345–352.
Liang, W-G., Jin, Y. and Yu, Z-Q. (1995a) Synthesis of linear alkylbenzene in a liquid–solid circulating fluidized bed reactor. *J. Chem. Tech. Biotech.*, **62**, 98–102.
Liang, W-G., Wu, Q-W., Yu, Z-Q., Jin, Y. and Bi, X-T. (1995b) Flow regimes of the three-phase circulating fluidized bed. *AIChE J.*, **41**, 267–271.
Lim, K.S., Zhu, J-X. and Grace, J.R. (1995) Hydrodynamics of gas fluidization. *Int. J. Multiphase Flow*, **21** (Suppl.), pp. 141–193.
Liu, D-J., Zheng, C-G., Li, H. and Kwauk, M. (1993a) A preliminary investigation on integral circulating fluidized bed. *Eng. Chem. Metall.* (in Chinese), **14**(1), 64–67.
Liu, D-J., Zheng, C-G., Li, H. and Kwauk, M. (1993b) Preliminary studies on integral circulating fluidized bed, in *Proceedings 6th National Conf. on Fluidization* (in Chinese), Wuhan, China, October 1993, pp. 63–66.
Lockhart, C., Zhu, J-X., Brereton, C., Lim, C.J. and Grace, J.R. (1995) Local heat transfer, solids concentration and erosion around membrane tubes in a cold model circulating fluidized bed. *Int. J. Heat Mass Transfer*, **38**, pp. 2403–2410.
Martin, M.P., Derouin, C., Tulier, P., Forissier, M., Wild, G. and Bernard, J.R. (1992) Catalytic cracking in riser reactors: core-annulus and elbow effects. *Chem. Eng. Sci.*, **47**, 2319–2324.
Mertes, T.S. and Rhodes, H.B. (1955) Liquid-particle behavior (Part I). *Chem. Eng. Prog.*, **51**, 429–432.
Milne, B.J., Berruti, F., Behie, L.A. and de Bruijn, T.J.W. (1992a) The internally circulating fluidized bed (ICFB): a novel solution to gas bypassing in spouted beds. *Can. J. Chem. Eng.*, **70**, 910–915.
Milne, B.J., Berruti, F. and Behie, L.A. (1992b) Solids circulation in an internally circulating fluidized bed (ICFB) reactor, in *Fluidization VII* (eds O.E. Potter and D.J. Nicklin), Engineering Foundation, New York, pp. 235–242.
Milne, B.J., Berruti, F., Behie, L.A. and de Bruijn, T.J.W. (1994) The hydrodynamics of the internally circulating fluidized bed at high temperature, in *Circulating Fluidized Bed Technology IV* (ed. A.A. Avidan), AIChE, New York, pp. 28–31.
Monceaux, L., Azzi, M., Molodtsof, Y. and Large, L.F. (1986) Overall and local characterization of flow regimes in a circulating fluidized bed, in *Circulating Fluidized Bed Technology II* (ed. P. Basu), Pergamon Press, Toronto, pp. 185–191.
Mori, S., Yan, Y., Kato, K., Matsubara, K. and Liu, D. (1991) Hydrodynamics of a circulating fluidized bed, in *Circulating Fluidized Bed Technology III* (eds P. Basu, M. Horio and M. Hasatani), Pergamon Press, Oxford, pp. 113–118.
Muroyama, K. and Fan, L.S. (1985) Fundamentals of gas–liquid–solid fluidization. *AIChE J.*, **31**, 1–34.
Murphy, J.R. (1993) Evolutionary design changes mark FCC process. *Oil & Gas J.*, May 18, 49–58.

Naruse, I., Kumita, M., Hattori, M. and Hasatani, M. (1991) Characteristics of flow behavior in riser of circulating fluidized bed with internal nozzle, in *Circulating Fluidized Bed Technology III* (eds P. Basu, M. Horio and M. Hasatani), Pergamon Press, Oxford, pp. 195–200.

Niccum, P.K. and Bunn, D.P. (1985) Catalytic Cracking System. *U.S. Patent*, 4,514,285.

Pirozzi, D., Gianfreda, L., Greco, G. and Massimilla, L. (1989) Development of a circulating fluidized bed fermentor: the hydrodynamic model for the system. *AIChE Symp. Ser.*, **85**(270), 101–110.

Qi, C-M., Jin, Y. and Yu, Z-Q. (1989) Investigation on the fast gas–solids separator for downer, *J. Petroleum Processing (Chinese Edition)*, **12**, pp. 51–56.

Qi, C-M., Yu Z-Q., Jin Y, Bai, D-R. and Yao, W-H. (1990) Hydrodynamics of cocurrent downwards fast fluidization (I) and (II). *J. Chem. Ind. & Eng. China (Chinese Edition)*, (3), 273–280 and 281–290.

Qi, C., Fregeau, J.R. and Farag, I.H. (1992) Hydrodynamics of the combustion loop of N-circulating fluidized bed combustion. *AIChE Symp. Ser.*, **88**(289), 26–32.

Rhodes, M.J., Hirama, T., Cerutti, G. and Geldart, D. (1989) Non-uniformity of solids flow in the risers of circulating fluidized beds, in *Fluidization VI* (eds J.R. Grace, L.W. Shemilt and M.A. Bergougnou), Engineering Foundation, New York, pp. 73–80.

Richardson, J.F. and Zaki, W.N. (1954) Sedimentation and fluidization: Part I. *Trans. Inst. Chem. Eng.*, **32**, 35–55.

Roques, Y., Gauthier, T., Pontier, R., Briens, C.L. and Bergougnou, M.A. (1994) Residence time distributions of solids in a gas–solids downflow transport reactor, in *Circulating Fluidized Bed Technology IV* (ed. A.A. Avidan), AIChE, New York, pp. 555–559.

Salah, H., Zhu, J-X., Zhou, Y-M. and Wei, F. (1996) Effect of internals on the hydrodynamics of circulating fluidized bed reactors, in *Preprints of 5th International Conference on Circulating Fluidized Beds*, Chinese Society of Particuology, Beijing, China, May 1996, pp. Eq. 1.1–1.6.

Senior, R.C. (1992) Circulating fluidized bed fluid and particle mechanics: modelling and experimental studies with application to combustion. Ph.D. Dissertation, University of British Columbia, Vancouver, Canada.

Shimizu, A., Echigo, R., Hasegawa, S. and Hishida, M. (1978) Experimental study of the pressure drop and the entry length of the gas–solid suspension flow in a circular tube. *Int. J. Multiphase Flow*, **4**, 53–64.

van der Ham, A.G.J., Prins, W. and van Swaaij, W.P.M. (1991) Hydrodynamics and mass transfer in a regularly packed circulating fluidized bed, in *Circulating Fluidized Bed Technology III* (eds. P. Basu, M. Horio and M. Hasatani), Pergamon Press, Oxford, pp. 605–611.

van der Ham, A.G.J., Prins, W. and van Swaaij, W.P.M. (1993) Hydrodynamics of a pilot-plant scale regularly packed circulating fluidized bed. *AIChE Symp. Ser.*, **89**(296), 53–72.

van der Ham, A.G.J., Prins, W. and van Swaaij, W.P.M. (1994) A small-scale regularly packed circulating fluidized bed, Part I: Hydrodynamics. *Powder Technol.*, **79**, 17–28.

Wang, X.S. and Gibbs, B.M. (1991) Hydrodynamics of circulating fluidized bed with secondary air injection, in *Circulating Fluidized Bed Technology III* (eds P. Basu, M. Horio and M. Hasatani), Pergamon Press, Oxford, pp. 225–230.

Wang, Z-L., Yao, J-Z., Liu, S-J., Li, H-Z. and Kwauk, M. (1993) Studies on the voidage distribution in an internally circulating fluidized bed, in *Proceedings 6th National Conf. on Fluidization* (in Chinese), Wuhan, China, October 1993, pp. 150–154.

Wang, Z-W., Bai, D-R. and Jin, Y. (1992) Hydrodynamics of cocurrent downflow circulating fluidized bed (CDCFB). *Powder Technol.*, **70**, 271–275.

Wei, F., Chen, W., Jin, Y. and Yu, Z-Q. (1994a) Studies on axial solids mixing behaviour in CFB riser, in *Proceedings 7th National Conference on Chemical Engineering* (in Chinese), Beijing, October 1994, pp. 350–353.

Wei, F., Jin, Y. and Yu, Z-Q. (1994b) Development of a new generation catalytic cracking reactor – gas–solid short contact catalytic cracking process. *Oil and Gas Processing* (in Chinese), **4**(1), 26–30.

Wei, F., Jin, Y., Yu, Z-Q., Gan, J. and Wang, Z-W. (1994c) Application of phosphor tracer technique to the measurement of solids RTD in circulating fluidized bed. *J. Chem. Ind. & Eng. China* (Chinese Edition), (2), 230–235.

Wei, F., Wang, Z-W., Jin, Y., Yu, Z-Q. and Chen, W. (1994d) Dispersion of lateral and axial solids in a cocurrent downflow circulating fluidized bed. *Powder Technol.*, **81**, 25–30.

Wei, F., Liu, J-Z., Jin, Y. and Yu, Z-Q. (1995) The gas mixing in cocurrent downflow circulating fluidized bed. *Chem. Eng. & Technol.*, **18**, pp. 59–62.

Wei, F. and Zhu, J-X. (1996) Effect of flow direction on axial solids dispersion in gas–solids cocurrent upflow and downflow systems. *Chem. Eng. J.*, accepted for publication.

Wilhelm, R.M. and Kwauk, M. (1948) Fluidization of solids particles. *Chem. Eng. Prog.*, **44**, 201–218.

Wu, S. and Alliston, M. (1993) Cold model testing of the effects of air proportions and reactor outlet geometry on solids behavior in a CFB, in *Proceedings 12th International Conference on Fluidized Bed Combustion* (ed. L.N. Rubow), ASME, New York, pp. 1003–1009.

Xia, Y. and Tung, Y. (1992) Effect of inlet and exit structures on fast fluidization, in *Selected Papers from Engineering Chemistry and Metallurgy* (English Translation) (eds M.-H. Mao and G. Xia), Science Press, Beijing, pp. 45–53.

Yang, L-D., Bao, Y-L., Zhang, Z-D., Zhao, M-G., Li, B-X., Wang, H.B., Wu, W-Y. and Lu, H-L. (1991a) H-valve and solid separator for circulating fluidized bed boilers, in *Fluidization '91, Science and Technology* (eds M. Kwauk and M. Hasatani), Science Press, Beijing, pp. 134–139.

Yang, Y-L., Jin, Y., Yu, Z-Q. and Wang, Z-W. (1991b) Particle flow-pattern in a dilute concurrent upflow and downflow circulating fluidized bed, in *Fluidization '91, Science and Technology* (eds M. Kwauk and M. Hasatani), Science Press, Beijing, pp. 66–75.

Yang, Y-L., Jin, Y., Yu, Z-Q., Wang Z-W. and Bai D-R. (1991c) The radial distribution of local particle velocity in a dilute circulating fluidized bed, in *Circulating Fluidized Bed Technology III* (eds P. Basu, M. Horio and M. Hasatani), Pergamon Press, Oxford, pp. 201–206.

Yang, Y-L., Jin, Y., Yu, Z-Q., Zhu, J-X. and Bi, H-T. (1993) Local slip behaviour in the circulating fluidized bed. *AIChE Symp. Ser.*, **89**(296), 81–90.

Yerushalmi, J. and Avidan, A.A. (1985) High-velocity fluidization, in *Fluidization*, 2nd edn (eds J.F. Davidson, R. Clift and D. Harrison), Academic Press, London, pp. 225–291.

Zheng, C-G., Tung, Y-G., Zhang, W-N. and Zhang, J-G. (1990) Impact of internals on radial distribution of solids in a circulating fluidized bed. *Engineering Chemistry & Metallurgy* (in Chinese), **11**(4), 296–302.

Zheng, C-G., Tung, Y-K., Xia, Y-S., Hun, B. and Kwauk, M. (1991a) Voidage redistribution by ring internals in fast fluidization, in *Fluidization '91, Science and Technology* (eds M. Kwauk and M. Hasatani), Science Press, Beijing, pp. 168–177.

Zheng, C-G., Tung, Y-K. and Xia, Y-S. (1991b) Internals for fast fluidized beds, in *Fluidization '91, Science and Technology* (eds M. Kwauk and M. Hasatani), Science Press, Beijing, pp. 413–413.

Zheng, C-G., Tung, Y-K., Li, H-Z. and Kwauk, M. (1992) Characteristics of fast fluidized beds with internals, in *Fluidization VII* (eds. O.E. Potter and D.J. Nicklin), Engineering Foundation, New York, pp. 275–284.

Zheng, Q-Y., Zhu, D-H., Feng, J-K. and Fan, J-H. (1989) Some aspects of hydrodynamics in N-shaped circulating fluidized bed combustor, in *Proceedings 10th International Conference on Fluidized Bed Combustion* (ed. A.M. Manaker), ASME, New York, Vol. 2, pp. 157–161.

Zheng, Q-Y. and Zhang, H. (1994) Experimental study of the effect of bed exits with different geometric structure of internal recycling of bed material in CFB boiler, in *Circulating Fluidized Bed Technology IV* (ed. A.A. Avidan), AIChE, New York, pp. 145–151.

Zheng, Q-Y., Ma, Z-W. and Wang, A-B. (1994) Experimental study of the flow pattern and flow behaviour of gas–solid two phase flow in L-valve, in *Circulating Fluidized Bed Technology IV* (ed. A.A. Avidan), AIChE, New York, pp. 246–252.

Zhou, J., Grace, J.R., Qin, S.Z., Brereton, C., Lim, C.J. and Zhu, J. (1994) Voidage profiles in a circulating fluidized bed of square cross-section. *Chem. Eng. Sci.*, **49**, 3217–3226.

Zhu, J-X. and Wei, F. (1996) Recent developments of downer reactors and other types of short contact reactors, in *Fluidization VIII* (eds C. Laguerie and J.F. Large), Tours, France, to be published.

Zhu, J-X., Yu, Z-Q., Jin, Y., Grace, J.R. and Issangya, A. (1995) Cocurrent downflow circulating fluidized bed (downer) reactors – a state of the art review. *Can. J. Chem. Eng.*, **73**, 662–677.

17 Future prospects
AMOS A. AVIDAN

17.1 Introduction

This chapter on future prospects of circulating fluid beds discusses three major topics. The introduction addresses the issue of why CFBs have gained commercial success in some applications, and why they have replaced low-velocity fluid beds (LFB), or fixed beds in many catalytic and non-catalytic applications. In some cases, CFB processes have been designed to solve the shortcomings of an LFB, only to be replaced eventually by improved, lower cost LFB processes. One such example, Fischer–Tropsch synthesis, is discussed briefly in section 17.2, and in more detail in Chapter 14. The third topic of this chapter addresses a new emerging brand of CFBs, the downflow short contact time contractor, or 'downer', discussed also in Chapter 16. This novel unit operation is gaining a commercial foothold, and may feature prominently in CFB's future prospects.

Circulating fluidized beds have made a large impact on the world economy, particularly in fluid catalytic cracking. In one broad sense, all fluidized beds exhibit some degree of circulation through such means as:

- entrainment from the dense bed to the freeboard, and from the freeboard back to bed;
- through cyclone diplegs;
- between two or more fluid beds.

If one accepts a narrow definition of circulating fluidized beds as one where there is no definite bed height and external circulation is required to maintain solids density, one can begin to differentiate between two broad types of fluidized beds:

- Low-velocity fluidized beds (LFB), which include bubbling and turbulent fluid beds.
- Circulating fluidized beds (CFB), such as FCC riser reactors.

This distinction is somewhat arbitrary. Consider, for example, a CFB boiler where solids entrainment rate is typically below $50 \text{ kg/m}^2\text{s}$, and the quintessential LFB turbulent fluid bed FCC regenerator, where the entrainment rate typically exceeds $100 \text{ kg/m}^2\text{s}$. The boiler is normally referred to as a CFB, while the higher flux regenerator is considered for current purposes as an LFB.

CFBs were usually developed to overcome shortcomings of LFBs, but in some cases, CFB operation is essential, and no LFB alternative is viable. The major example of the perfect CFB reactor application is the FCC riser reactor described in Chapter 13. Two examples of the evolution of LFBs and CFBs are shown in Table 17.1. In many cases CFBs were applied to solve shortcomings of LFBs, but later designers have found a simpler and less costly way to overcome these problems. For example:

Problem in LFB:	Poor contacting due to maldistribution, extensive backmixing and large bubble formation causing bypassing and poor gas–solids contact.
CFB solution:	High velocity operation at the upper reaches of the turbulent fluidization regime, in fast fluidization, or even in the dilute conveying regions.
Improved LFB solution:	Tailor and control particle size distribution in the Geldart A group, or AC group to maintain small bubbles; reactor internals to control solids and gas mixing; improved grid design.

The three examples provided in Table 17.2 indicate that a properly designed LFB can often be more attractive than a CFB unless very short contact times (as in FCC) are required. The well-designed LFB can then have equivalent or better performance than the CFB at a lower capital and operating cost.

Table 17.1 Some examples of CFB developments that have stood the test of time

Application	Early LFB	Some problems	CFB development	Status and future prospects
Fluid catalytic cracking (FCC) reactor	Long residence time dense bed	Development of zeolite catalyst makes dense bed obsolete (1960s)	Short residence time risers take advantage of zeolite catalyst (1970s)	Potential for ultra-short contact time CFB
Circulating fluid bed boilers	Bubbling bed boiler	Large number of feed points and poor performance limit application	Starting with Lurgi Alumina Calciners of the 1970s several versions of CFB combustors were developed and commercialized in the 1980s	Application in large-scale utility systems; potential for compact pressurized LFB combustors in the future?

Table 17.2 Some examples of CFBs developed to solve LFB problems and how the LFB was subsequently improved

Application	Early LFB	Some problems	CFB development	Subsequent improved LFB
Fischer–Tropsch synthesis	Hydrocol plant at Brownsville, Texas, 1952 (Squires, 1982)	Large particle size (Group B) caused large bubbles, low conversion	M.W. Kellogg and Sasol develop fast fluid bed Synthol process in 1955 using Group A solids (Squires, 1994; Steynberg et al., 1991)	Turbulent LFB, with proper design and fine Group AC catalyst shows better performance at a lower cost (Steynberg et al., 1991)
CFB regenerator for FCC	Bubbling and turbulent LFB	Bypassing, low efficiency, high NO_x emissions, high catalyst makeup rate	LFB designs developed in the 1970s in the USA and China (see Chapter 13)	Proper distribution of gas and solids, operation in partial combustion
Catalyst coolers for FCC regeneration	High velocity in upflow: 1940s; low velocity upflow: 1970s	Bed coils in LFB regenerator	Erosion, poor reliability, low heat transfer coefficient	Dense-phase downflow or backmixed catalyst coolers

17.2 Fischer–Tropsch synthesis

Fischer–Tropsch synthesis illustrates the prospects of CFB for catalytic applications. The Fischer–Tropsch (FT) reaction converts synthesis gas (carbon monoxide and hydrogen) to hydrocarbons and alcohols (see Chapter 14). FT has a long history, similar to catalytic cracking, and it has been commercialized, or tried on a large pilot-scale in several reactor types: tubular fixed-bed, recycle fixed-bed, CFB, LFB, and slurry. Unlike FCC, FT synthesis has only achieved site-specific commercial success till now. After World War II, it was commercialized in 1955 in South Africa in fixed and CFB reactors, and recently in an LFB reactor (Shell has recently constructed an FT plant in Malaysia).

The CFB FT Synthol reactor was invented by M.W. Kellogg, and developed in pilot plants in the late 1940s (Squires, 1994). The Synthol reactor (Figure 17.1) was scaled up from a 0.1 m M.W. Kellogg pilot plant to two 2.3 m commercial units at Sasol I. (Dry, 1981). A third reactor, of improved design, was added in 1960. Each reactor has a horizontal inlet pipe which is 1 m ID. Fresh feed (100 000 m³/h) and a similar flow of recycle gas pick up catalyst flowing down a straight standpipe. Catalyst flow is controlled by a slide valve. The expanded reaction zone contains heat exchange coils that remove 40% of the reaction heat. (Shell and tube heat exchanges were originally used but they tended to plug.) Approximately

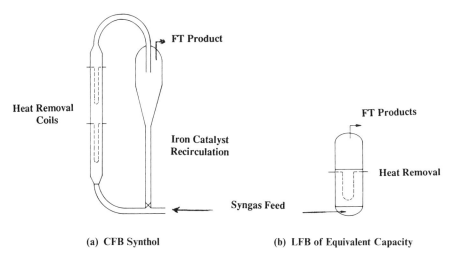

Figure 17.1 Sasol Fischer–Tropsch fluid bed reactor evolution from (a) CFB Synthol reactor at Sasol 2 and 3, to (b) equivalent capacity LFB reactor of the same or better performance (adapted from Silverman et al., 1986).

one-third of the catalyst inventory is in the reactor, while the rest is in the transfer lines and catalyst hopper. The exit standpipe from the reactor is curved, possibly to avoid excessive catalyst recycle (see Chapter 7). Another mechanism to control excessive solids backmixing is the use of proprietary internals. These prevent excessive temperature at the bottom of the reactor. The reactor pressure is raised to compensate for catalyst deactivation during the run. The riser voidage changes from 0.88 to 0.83, corresponding to the fast fluidization regime.

In 1974, the South African government decided to build a larger coal conversion complex, Sasol II at Secunda. This complex was designed to maximize fuel production, and includes only CFB Synthol reactors. Sasol III started up in the early 1980s. It is an almost exact replica of Sasol II. The Synthol reactors were scaled up to three-and-a-half times the fresh feed rate with the cooperation of the Badger Company. Many lessons learned in Sasol I, and further design improvements, were incorporated into the new design.

The importance of precise control of catalyst properties, including particle size distribution, was recognized at an early stage, in part thanks to a glass model. The catalyst is not microspherical and has a high density – properties that could promote severe segregation, large bubble formation, and erratic standpipe flow. It is pyrophoric and can become sticky because of wax formulation. The high velocity CFB eliminated bubbles – and it is probably this property that gave it an early advantage over the LFB. High pressure also helped improve fluidization properties, but the effect of fines was found to be crucial. A considerable fines fraction of particles smaller than 30 μm greatly improves flow properties. The Sasol investigators have termed this powder 'AC' on the Geldart powder classification. Fluidization aids (fine, divided, charcoal) have also been patented. Synthol reactors are unique – they solved a difficult fluid bed scale-up problem with ingenious hardware and modification of solids properties through continuous evolution. Yet at some point, it became apparent that with these improvements, the advantage of the better contact for a poor powder in a CFB may no longer be necessary.

The first commercial LFB FT reactor was developed by Hydrocarbon Research, Standard Oil, and others in the late 1940s (Squires, 1982). A large demonstration unit was started up in Brownsville, Texas in 1950. Two reactors, 5 m ID and 18 m tall, with a combined capacity of 360 000 tonnes per year were built. Each reactor contained about 200 tonnes of coarse (Group B) iron catalyst. The reactors were operated at similar conditions to the Synthol reactors, but at lower superficial gas velocities (around 0.2 m/s). Considerable difficulties were encountered, and a myriad of changes (including a change of plant ownership) were implemented. The difficulties were in two main areas: catalyst activity, and fluidization quality. Poor choice of catalyst properties led to large bubbles and loss of efficiency. This was solved by Synthol's CFB reactor, and the use of a Group A powder. Various improvements in Brownsville, including a reduction in catalyst size, ultimately solved the problem, but the plant was shut down in 1957 for economic reasons.

Recently Sasol has rediscovered the LFB (Silverman *et al.*, 1986). With an improved Group AC powder operation in the turbulent fluidization regime, and good internals and grid design, the LFB can process the same throughput as a CFB. (Catalyst inventory is the same though and since only one-third of the CFB catalyst contributes to a space velocity in the reactor, it can be assumed that the CFB still has better gas–solids contact.) The difference in size between the LFB and CFB Synthol reactor, for similar gas treating capacity is illustrated in Figure 17.1. From this figure it is easy to appreciate the cost savings inherent in the LFB design.

Interestingly, additional Sasol II-type CFB reactors were constructed in South Africa (The Mossel Bay Gas Project) in the late 1980s. One can attribute this to conservatism and wishing to proceed with the commercially proven, albeit, more expensive, CFB. It has, however, become clear to the Sasol developers that the LFB is the superior reactor for the Fischer–Tropsch process.

17.3 Ultra-short contact time fluid–particle reactors

One recently emerging novel type of CFBs is the ultra-short contact time reactor. The main feature of the short contact time fluidized reactor

(SCTFR) is its ability to contact solids and gas for average residence times of less than 1 second. Such short contact time is virtually impossible to achieve in an upflow CFB riser. For example, in recent years, some FCC risers have been designed with gas residence times as low as 1 s, while the corresponding solids residence time is still as high as 3 s (an average slip factor of 3) due to the existence of velocity and density profiles, and the action of gravity.

This shortcoming of upflow risers was recognized some time ago, and in this section we briefly describe major examples of SCTFR developments.

Professor Bergougnou and his colleagues at the University of Western Ontario have developed the 'downer' SCTFR concept and have proposed many potential applications in the past 15 years (Bassi et al., 1993). The main feature, common to most of their designs, is a jet mixing device to rapidly mix and contact two phases – a feed, which can be comprised of gas, a liquid and solids, and a hot particulate recycle stream. The original design used two opposing tangential jets, while later designs employed several impinging jets. The main application envisioned was rapid gasification of biomass. The process has been scaled up to 100 kg/h by Ensyn Engineers of Ottawa Ltd, and several commercial applications have been reported.

The SCTFR concept was applied in the early 1980s to the thermal cracking of heavy naphtha to produce light olefins by the Gulf Oil Corporation and Stone and Webster (Gartside, 1989). The process, which is illustrated schematically in Figure 17.2, was designed to contact hot inert solids (a Group A powder, such as FCC equilibrium catalyst) with vaporized naphtha. The mixing was accomplished in a co-current downflow mixing zone, with the cracking reaction taking place in a downflow contactor where the residence time is less than 0.2 s. An inertial separation device, with a separation efficiency as high as 98%, removed most of the coked solids from the cracked products. This was followed immediately by quench and a cyclonic separator to remove most of the remaining solids. The coked solids were regenerated in the regenerator vessel. A demonstration plant of this 'thermal regenerative cracking' or TRC process was constructed and operated for several years. While the changing economics for heavy naphtha cracking have prevented the commercialization of the TRC process, the system was proposed for other short contact time applications as the 'quick contact' or QC system (Gartside, 1989).

The SCTFR concept has appealed to FCC designers for some time. For example, Gross and Ramage (1983) describe a downflow SCTFR FCC system designed to take advantage of the high catalyst activity and superior selectivity of zeolite catalysts, while overcoming the disadvantages of upflow risers. Pilot-plant experiments have demonstrated the benefits of downflow. Bartholic (1991) patented an SCTFR FCC reactor shown schematically in Figure 17.3. Hot, high activity, FCC catalyst (containing over 40% zeolite) is introduced in downflow into a mixing chamber. In one embodiment, the FCC feed is atomized and introduced through horizontally mounted nozzles,

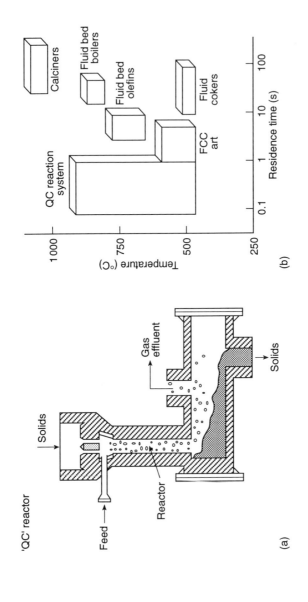

Figure 17.2 QC system (adapted from Gartside, 1989). (a) Schematic of the solids–feed mixer, downflow reactor and initial solids–gas separator. (b) Contact time temperature regime map, showing where QC systems fit (residence times below 1 s).

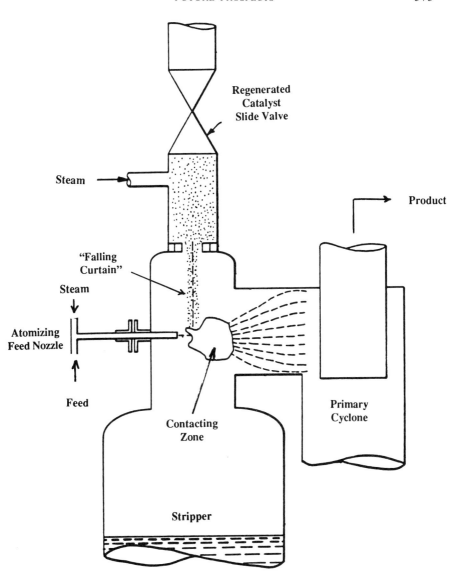

Figure 17.3 Schematic of a short contact time FCC reactor (adapted from Bartholic, 1991).

producing a horizontal spray. The atomized feed contacts the 'falling curtain' of hot catalyst, typically at higher temperatures (above 540°C), higher catalyst-to-oil ratio (over 10), and shorter contact time (less than 0.5 s) than in conventional cracking. The gas and solids velocities in the mixing zone are controlled so that most of the catalyst and the cracked product enter the horizontal conduit leading to the primary separation cyclone.

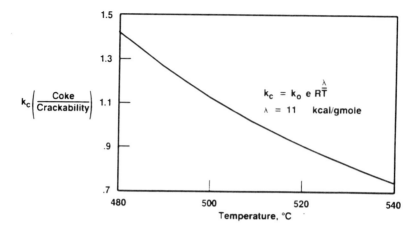

Figure 17.4 Increasing temperature reduces FCC coke selectivity (k_c = coke yield/crackability), hence providing potential for improved FCC operation at higher temperatures, higher catalyst activity, and shorter residence time than conventional FCC operation. Care must be taken to vaporize and mix the FCC feed perfectly with catalyst to achieve this benefit (Avidan and Shinnar, 1990).

This SCTFR system has been commercialized by the Coastal Corporation for FCC and for whole crude thermal cracking (Bartholic *et al.*, 1991). Additional commercial installations are planned. The SCTFR system is expected to provide improved selectivity for FCC. Coke selectivity is known to improve with increasing reactor temperature and decreasing contact time (Figure 17.4, Avidan and Shinnar, 1990), leading to increased heavy fuel oil conversion at the same overall coke yield. Reduced residence time is expected to further reduce hydrogen transfer activity and increase the yield of light olefins.

The short contact time fluidized reactor has already demonstrated significant advantages over conventional CFBs for some specific applications. The study of short contact time reactors, 'downers', and similar devices is not as extensive as of risers, but some progress is reported in Chapter 16. It is possible that it will find significant and increasing commercial application.

Nomenclature

CFB	circulating fluidized bed
FCC	fluid catalytic cracking
FT	Fischer–Tropsch
LFB	low-velocity fluidized bed
QC	quick contact
SCTFR	short contact time fluidized reactor
TRC	thermal regenerative cracking

References

Avidan, A.A. and Shinnar, R. (1990) Development of catalytic cracking technology. A lesson in chemical reactor design. *I&EC Res.* **29**, 931–947.

Bartholic, D.B. (1991) U.S. Patent 4,985,136.

Bartholic, D.B., Soudek, M. and Keim, M.R. (1991) The simplified approach to residual oil upgrading. Paper AM-91-46, NPRA Annual Meeting, March.

Bassi, A.S., Briens, C.L. and Bergougnou, M.A. (1993) Short contact time fluidized reactors (SCTFRs), in *Circulating Fluidized Bed Technology IV* (ed. A.A. Avidan), AIChE, New York.

Dry, M.E. (1981) The Fischer–Tropsch synthesis, in *Catalysis – Science and Technology* (eds Anderson, J.R. and Boudart, M.), Springer Verlag, Berlin, pp. 159–255.

Gartside, R. J. (1989) QC – A new reaction system, in *Fluidization VI* (eds J.R. Grace, L.W. Shemilt and M.A. Bergougnou), Engineering Foundation, New York, pp. 25–32.

Gross, B. and Ramage, M.P. (1993) U.S. Patent 4,385,985 (assigned to Mobil Oil Corporation).

Silverman, R.W., Thompson, A.H., Steynberg, A., Yukawa, Y. and Shingles, T. (1986) Development of a dense phase fluidized bed Fischer–Tropsch reactor, in *Fluidization V* (eds K. Ostergaard and A. Sorensen), Engineering Foundation, New York, pp. 441–448.

Squires, A.M. (1982) Contribution toward a history of fluidization, in *Proceedings Joint Meeting of Chemical Engineering Society of China and AIChE*, Beijing, Chemical Industry Press, pp. 322–353.

Squires, A.M. (1994) Origins of the fast fluid bed, in *Fast Fluidization* (ed. M. Kwauk), Academic Press, San Diego, pp. 4–38.

Steynberg, A.P., Shingles, T., Dry, M.E., Jager, B. and Yukawa, Y. (1991) Sasol commercial scale experience with LFB and CFB Fischer–Tropsch Reactors, in *Circulating Fluidized Bed Technology III* (eds P. Basu, M. Horio and M. Hasatani), Pergamon Press, Oxford, pp. 527–532.

Index

Acceleration effects 49–50, 190
Acceleration zone, see Bottom zone
Adsorption 88, 491
Advantages/disadvantages of circulating fluidized beds 4, 442, 461, 475
Agglomeration
 avoidance of 15
 in combustion systems 374–5
 in cyclones 446
 definition 16, 28
 in gasification 444
Aggregation, see Clusters
Annulus region
 gas mixing in 113
 solids motion 123, 263–9
 see also Wall layer thickness
Applications 4–6, 208
Asymmetry 47, 390, 482–3
Attrition 15, 16, 388, 500
Axial dispersion, see Gas mixing; Solids mixing

Backmixing, see Gas mixing; Solids mixing
Baffles
 bluff-body 526–32
 in cyclones 207–8
 in hydroforming process 490
 in MTG process 491
 in Synthol reactor 571
 ring-type 104, 304, 505, 508, 526–31
Blow-out velocity, see Transport velocity
Boilers 417–38
 ash disposal 400–401, 430
 bottom region in 42
 configuration 421–4, 431–3, 537–8
 gas mixing in 101, 103
 heat transfer surfaces 431
 refractory 425–6
 scale-up 430–34
 see also Combustion; Combustors; Emissions; Secondary air
Bottom zone 531–6
 in combustors 50
 gas mixing in 99, 109
 height of 22, 47, 62
 influence on riser hydrodynamics 534
 modeling 172, 507, 512–13
 nature of 34, 41–2
 particle motion 120–1
 pressure drop 49
 solids mixing 134, 531
 voidage distribution 43, 46, 54, 534
Boundary conditions 167–70
Bubbling beds
 bubbling regime 6, 11, 32
 chlorination of ilmenite 462
 as combustors 369, 372, 385–6
 distinguishing from circulating beds 22
 gas mixing in 96, 110
 heat transfer 370, 381
 models 26, 158, 378, 504, 514
 relative advantages/disadvantages 5, 370, 451
 scaling of 290–91, 451
 solids mixing in 120–1, 132–3, 139, 142
 stability 27
 in standpipes 219–20
Bypassing 21

Calcination 1, 89, 446–50
Capacitance measurements 29, 265, 317–21, 329–32, 343
Catalytic reactors 22–3, 49, 489–501
 see also Fischer–Tropsch synthesis; Fluid catalytic cracking; Reactor models
CFBC, see Combustion
Chemical conversion 21, 86
 effect of baffles 530
 in FCC 478
 measurements 21, 110
 predictions 510
Chemical reaction, see Chemical conversion; Reactor models
Choking 10, 14–15, 16, 535
Circulating fluidized bed combustion, see Combustion
Circulation rate 4, 22, 336, 495, 539–40
Clusters
 behavior 31, 285, 288
 contact time 280–82, 284–5
 definition 16
 dislodgement 304
 drag on 60
 frequency 125
 influence on heat transfer 270–77, 283, 286, 301

Clusters *cont'd*
  modeling 173
  motion of 55, 263
  occurrence 10, 23, 57, 123
  origin 27–9, 159, 171
  shape 30, 60
  size 29, 39, 61, 71, 125
  slip 57
  velocity 123, 267–9
  voidage 61, 280, 285
  wall coverage 265, 280, 296
Co-current downflow 3, 541–54, 573–6
  applications 542–4
  configuration 544–6
  distributor design 545
  hydrodynamics 546–51, 554
  mixing in 548–54
Co-current upward flow 2
Cold models 497
Collisions
  particle–particle 124, 134, 159–60, 161–2, 170, 173–6
  particle–wall 161–2, 167–70, 173–4
Combustion 369–411, 417–38
  in cyclones 105
  devolatilization 387, 389
  efficiency 371, 385, 398, 428, 429, 508
  flue gas recirculation 380–81
  fuels 417–18, 424–5
  history 5, 22
  load following 380–81, 384–5
  models 386, 508, 511, 515, 518–20
  residues 400–401, 430
  sintering 374–6
  stages 386
  temperature 373
  turndown 371, 379–83
  volatiles 386, 387, 516
  see also Boilers; Combustors; Emissions; NO_x emissions; Sulfur capture in CFBC
Combustors
  comparison with alternate types 1, 372–3
  comparison with FCC 377–9
  control of 379–81, 430
  cyclones 181, 433
  density profiles 43
  fuel feeding 356, 426
  heat transfer 302–4, 380
  instrumentation 321, 325, 334, 348–50
  operating conditions 87, 388, 421, 422
  pressure drops 49
  return system 214–15, 228–32
  segregation in 142
  temperature variation in 370, 380
  zones in 49–50
  see also Boilers; Combustion; Emissions; Fluid catalytic cracking
Concentration profiles
  in catalytic reactors 515–16
  in combustors 389–90, 515
Concentric circulating fluidized bed 538
Constitutive equations 160
Conversion, *see* Chemical conversion
Core-annulus structure
  analysis 50–61, 172
  in combustors 379, 389
  influence on gas mixing 99, 113
  interchange between core and annulus 173, 263
  occurrence 43, 46, 59, 121, 263, 379
  scaling law 70–72
  see also Models; Reactor models
Core radius, *see* Wall layer thickness
Corners 47, 516
Cyclones 181–210
  collection efficiency 194–202, 433, 446
  configurations 184–7
  design 493
  dipleg 221, 254–8
  in FCC 484–7
  gas mixing in 105
  one-quarter turn 546
  operating conditions 210
  particle motion in 187–91, 192–3
  pressure drop 189–90, 196–7
  reaction in 390
  scale-up 433
  stripper cyclone 485–6

Dense region, *see* Bottom zone
Density profiles, *see* Suspension density; Voidage
Diameter of column, *see* Scale-up
Dilute phase 16
Dilute-phase transport 1, 3, 12, 14–15, 57
  gas mixing in 100
  particle motion in 173
Dilute region 50
  height of 56
  particle motion in 121–4
  solids mixing in 134, 142
  voidage in 52
Diplegs, *see* Cyclones; Standpipes
Dispersion coefficients, *see* Gas mixing; Solids mixing
Distinguishing features of CFB 1–3, 441–2
Distributor 16, 431
Downcomer 22–3, 61, 64, 204
  see also Standpipes
Downer, *see* Co-current downflow
Downflow of particles
  influence on gas mixing 96, 100, 103, 110
  at walls 46, 379
  see also Co-current downflow; Wall layer thickness
Dry scrubbing 540

INDEX 581

Emissions
  from alumina calciners 447
  from combustors 370, 389, 392–4, 402–10, 418–20
  dioxins and furans 393
  from gasifiers 444
  metals 420, 444
  particulates 420, 428–30, 479
  see also Nitrous oxide emissions from CFBC; NO_x; Sulfur capture in CFBC
Energy minimization 50, 56–61
Entrainment 35
Erosion
  avoidance 241, 470, 493, 501
  of baffles 530
  of boilers 424
  of cyclones 493
  of heat transfer surfaces 304–5, 499
  in low-velocity fluidized beds 476
  models 166, 171
  of refractory 425–6, 461
  of trickle valve 258
Exchange rates, see Lateral solids motion; Mass transfer
Exit configuration 536–8
  in combustor 537–8
  in fluid catalytic cracker 486–7, 497, 538
  influence on density profiles 47–8
  influence on gas mixing 104
  influence on internal refluxing 43, 56, 125, 428, 493, 536
  influence on solids mixing 125, 536–7
  modeling 161
Experimental techniques 312–56
  circulation flux 336–7
  densitometry 325–9, 482
  heat transfer 292–3, 344–50
  particle flux measurement 332–6
  particle velocity measurement 337–44
  pressure measurement 314–17
  temperature measurement 347, 348, 350
  visualization 30, 313–14
  voidage 317
  see also Probes
External heat exchanger 230, 380, 381, 422, 433

Fast fluidization
  definition 16, 37
  gas mixing in 99–100
  regime 1, 2, 8–10, 15, 35
FCC, see Fluid catalytic cracking
Feed nozzles 479–82, 492, 498
Feeding of reactants
  distribution 86
  injection system in FCC 479
  number of feed points in CFBC 427, 431–2
  as sources of asymmetry 389, 390

Filters, see Separators (gas–solid)
Fines, see Particle size distribution
Fins 304–5
Fischer–Tropsch synthesis 87, 475, 498–500, 570–72
Fixed beds, 1, 11
Flow regimes 4, 6–15, 24
  in FCC regenerators 474
  in standpipes 218
  see also Transition velocities
Fluid catalytic cracking 446–87, 495–7
  afterburning 496
  catalysts 477, 478, 497
  density profiles 43
  in downflow reactors 543–4, 573–6
  feed composition 476
  feed nozzles 107, 470, 479
  history 5, 467, 495–6
  instrumentation 327
  maintenance 501
  operating conditions 87, 122, 477
  particle motion in 122
  pressure profiles 232
  products 466, 476
  regenerator 232, 473–6, 496
  return system 214–16, 227, 228, 232–5
  stripping 470, 485
  yields 476–7
Fluoseal, see Loop seal
Fouling
  in CFB boiler 424
  of distributor 455
  exploration in pilot plant 501
  of heat exchanger 499
  of optical probe 325
  of radiation-cooling chamber 452
Fragmentation, see Attrition
Friction 50, 72, 181, 193
Fully developed flow 34
Future prospects 463–4, 568–76

Gasification 369, 443–6, 463, 573
Gas–liquid–solid circulating fluidized bed 554, 557–61
Gas mixing 86–114
  axial 93–100, 490
  backmixing 22, 95
  bottom zone 109
  in combustors 389
  comparison with solids mixing 95–8, 132
  core region 113, 511
  in cyclones 105
  dispersion coefficients 92–3, 101
  importance of 86, 490
  lateral 101–3, 511
  models 109–13
Gas–solid contacting 490, 491, 526, 529

582 INDEX

Gas–solid reactions 6, 441–64
   see also Calcination; Combustion; Gasification; Ironmaking; Offgas treatment
Geldart powder classification 16
Granular temperature, see Kinetic theory
Gravity separation chamber 207–9
Grid, see Distributor

Heat transfer 261–305
   augmentation 304–5
   in combustor 370
   contact resistance 274, 280, 283, 296, 299
   gas convective 269–70
   heater length influence 272, 281, 296, 305
   measurement techniques 292–3, 344–50
   mechanisms 269–79, 283–4
   membrane wall 22, 302–5
   models 171, 279–85
   particle convective 269, 270–75, 283, 284, 285
   probes 294
   radiation 270, 283–4, 285–90, 300, 303, 348–50
History of circulating fluidized beds 5, 22, 467
Humidity, effect of 319
Hydrodynamic regimes, see Flow regimes
Hydrodynamics 21–75
   influence on heat transfer 263, 305
   modeling 149–76
   scaling 66–72
   see also Pressure fluctuations; Pressure profiles; Turbulence; Voidage

Incineration, see Combustion; Combustors
Inserts, see Baffles
Instability, see Stability
Instrumentation 312, 317–51, 355–6
Integral circulating fluidized bed 538–41
Interchange, see Mass transfer
Intermittency 46, 96
Internally circulating fluidized bed 538–41
Interparticle collisions, see Collisions
Inventory 61, 237, 238
Ironmaking 456–61

Jets, see Lateral gas injection; Plasma jet; Secondary air
J-valve, see Valves

Kinetic theory 124, 150, 159–60, 162–5

Laser Doppler anemometry 122–3, 163, 340–42
Lateral gas injection 106–8
   see also Secondary air

Lateral solids motion 51, 55, 72, 124, 263, 303
Limestone, see Sulfur capture in CFBC
Liquid–solid circulating fluidized bed 554–7
Loop seal 225, 230
   in combustors 373, 382, 421, 424
   definition 16
   feeding into 376, 427
   pressure balance 60, 230
   solids flow rate 64
   stability 535
Lower dense region, see Bottom zone
Low-velocity fluidized beds (LFB)
   comparison with CFB 475–6, 571–2
   distinguishing from circulating bed 3–5, 568–9
   in FCC technology 467, 474–6
   in Fischer–Tropsch synthesis 500
   modeling of 504
   see also Bubbling beds; Slug flow
L-valve
   aeration 243–8, 383
   in combustors 373, 382
   definition 16
   operation 243–4, 245–8
   pressure drop 230–32, 243, 244, 247
   principle of operation 242–3
   solids flow in 239–40

Maleic anhydride process 492, 497–8, 513
Mass transfer 21, 270, 275, 290
   effect of baffles 530
   interphase/inter-region 490, 509, 512, 513
   measurement 350–51
Membrane walls
   in combustors 372, 421, 424
   definition 16
   effect on hydrodynamics 104, 122, 281, 530–31
   heat transfer to 22, 302–5
   particle motion at 104, 122, 281
Methanol-to-gasoline (MTG) process 491
Methanol-to-olefins process 514
Minimum bubbling velocity 7, 12, 218
Minimum fluidization velocity 6, 12, 38, 245
Minimum slugging velocity 8, 12
Minimum transport velocity, see Transport velocity
Mixing, see Gas mixing; Solids mixing
Models
   axial dispersion 109–12, 127–33
   core-annulus 50–61, 110–13, 134, 136, 352
   dispersion 109–12, 127–33, 135
   freeboard 50, 378
   gas mixing 109–13

INDEX 583

heat transfer 158, 279–85, 300
hydrodynamic 149–76, 378
momentum balance 56–61
pressure profile 61–5
radial dispersion 135
solids mixing 124
two-fluid 25, 151–72
wall deposition 289–90
see also Reactor models
Moving packed bed flow 218–19
Multi-solid fluidized bed 142

Nitrous oxide emissions from CFBC 402, 407–9, 410, 419–20, 515
Non-mechanical solids flow devices 240–54, 495
  automatic flow mode 249–54
  in combustors 382
  valve mode 242–8, 471
  see also L-valve; Valves
NO_x emissions
  control via secondary air injection 105–6, 373
  effect of operating conditions 373, 405–6
  levels 371, 418–19
  predictive models 508, 511, 515
  reactions 402–7
  from regenerators 474
N-shaped circulating fluidized bed 540–42

Offgas treatment 452–6, 461
Onset velocities, *see* Transition velocities
Optical fiber probes
  flow structure determination 29, 30, 38, 46
  local particle concentration measurement 321–5, 356, 526
  particle velocity measurement 29, 340
  viewing of interior 314

Packed beds, *see* Fixed Beds; Moving packed bed flow
Particle motion 120–6
Particle properties 15, 478
Particle shape 15
Particle size distribution
  cyclones 183–4, 199, 206
  effect on capacitance measurements 319
  effect on hydrodynamics 160, 174
  FCC catalyst 478
  fines content 478–9, 490, 491, 571–2
  importance of 571
  specifications 15
Particle tracking 166, 353–5
Perturbation theory, *see* Stability, analysis
Phase diagrams 35, 37, 61
Photography *see* Video photography; Visualization

Pilot plants 491, 501
Plasma jet 542, 543
Plumes 389, 390
Pneumatic transport, *see* Dilute-phase transport
Pollutants, *see* Emissions
Power generation, *see* Combustion; Gasification
Pre-reduction of iron ore, *see* Ironmaking
Pressure
  effect on heat transfer 284, 299
Pressure drop
  across cyclone 64, 189, 195–6, 203–5
  across downcomer 64
  effect of baffles 529
  influence on bottom zone 22
  across lamella separator 209
  across loop seal 230
  across riser 64, 529
  across standpipe 225–7
  across valves 64, 225, 231, 244
Pressure fluctuations 9, 34–5, 315–16
Pressure profiles
  axial 42–50
  around circulation loop 61–5
  in FCC systems 232, 486–7
  in return system 221
Pressurized circulating fluidized beds 369, 397, 411, 434–7, 444
Probes
  capacitance 29, 265, 317–21, 343
  heat transfer 294, 344–7
  impact 336, 343
  isokinetic sampling 332–4
  suction 332–6
  for visualizing interior 313–14
  zirconia oxygen 355
  see also Optical fiber probes
Pyrolysis 542, 543, 573, 576

Radial flux, *see* Lateral solids motion
Radial mixing, *see* Gas mixing; Solids mixing
Reactor models 490, 504–21
  bubbling beds 504
  cluster/gas 516–17
  co-existing upflow/downflow 517
  core-annulus 508–15
  fundamental 518
  Monte Carlo 517–18
  one-dimensional 506–8
  transient 519
  turbulent fluidization regime 505
Regenerator, *see* Fluid catalytic cracking
Regime diagrams 10–14, 560–61
Regimes, *see* Flow regimes
Residence time distributions, *see* Gas mixing; Solids mixing

584   INDEX

Return systems 214
Ring-type baffle, see Baffles
Riser 16, 492
Roasting of ores 6, 22, 449–52
Rotary kiln 462–3
Roughness 104, 174
 in cyclones 194
 effect on clusters 265, 282
 effect on heat transfer 262, 304

Sampling
 gas 92, 355
 solids 121–2, 332–6, 355
Saturation carrying capacity 36, 237
Scale-up 113
 of baffles 529
 of boilers 430–34
 of catalytic reactors 489–91
 effect on gas mixing 103–4
 effect on heat transfer 262, 300, 303
 effect on hydrodynamics 65–6, 266
 effect on solids mixing 141
 of FCC units 479, 496
 see also Scaling relations
Scaling relations 65–72, 433, 491
 hydrodynamic 65–72, 290–91
 thermal 291–2
 validation 72–5, 290–91
Seal 236, 256
 dense phase 225
 seal pot 249–50
 see also Loop seal
Secondary air 17
 in combustors 373, 427
 effect on gas mixing 105
 effect on suspension density 47
 penetration 424, 431
 see also Lateral gas injection
Segregation of solids 17, 119, 139–4, 164
Selectivity 88, 499
Separation efficiency
 in cyclones 193–203, 433, 446
 in gravity separation chambers 210
Separators (gas–solid)
 baghouse 428
 electrostatic precipitator 428
 hot ceramic filters 437, 443
 impact 230, 373, 428, 484–5, 546
 impingement 230, 373, 428, 484–5, 546
 inertial 230, 373, 428, 484–5, 546
 lamella 207–9
 see also Cyclones; Gravity separation chamber
Settling chambers, see Gravity separation chamber
Short contact time fluidized reactor 542, 546, 572–6
Sintering, see Agglomeration

Slide valve
 in FCC units 232, 471, 494–5
 pressure drop 227
 in Synthol reactor 492, 494, 500, 570
Slip factor 484, 493, 498, 513
Slip velocity 58, 162, 549
Slug flow 7, 11, 32, 235
Smelting, see Ironmaking; Offgas treatment
Solids circulation rate, see Circulation rate
Solids downflow, see Downflow of particles
Solids hold-up, see Suspension density; Voidage
Solids mixing 119–44
 axial 121, 126–34
 comparison with gas mixing 95–8, 132
 lateral 123, 134–8, 531
 models 124
Solids transport, see Dilute-phase transport
Solids volume fraction, see Suspension density; Voidage
Sorbents, see Sulfur capture in CFBC
Splash zone 50, 142, 514
Stability
 analysis 27
 flow in cyclone 186–7
 of riser operation 535
Standpipes 214–58, 494
 aeration 227–8, 470, 473, 494
 angled 232–5, 470, 494
 bridging in 499
 design 500
 flow regimes 218–20
 history 214
 maximum pressure 238
 maximum solids flowrate 238–40, 245
 overflow 220–22
 pressure drop 217, 220, 223, 494
 stripping in 470
 underflow 220–22
Stick-slip flow 219, 245
Strands, see Clusters
Streamers, see Clusters
Streaming flow 220
Suction probe, see Sampling
Sulfur capture in CFBC 394–402
 effect of temperature 373
 mechanism 371, 394–6
 predictive model 508
 sorbents 398, 420, 427
 sorbent utilization 397–402, 428, 429
Superficial gas velocity 4, 17, 267, 388
Surface roughness, see Roughness
Suspension density
 axial profile 377–9, 380, 520
 effect of 302–3
 radial profiles 482–3
 use in CFBC control 383
 see also Voidage

Swirl 47, 161
Synthol reactor, *see* Fischer–Tropsch synthesis

Temperature, effect of
  on heat transfer 261, 296–9
  in L-valve 248
Temperature profiles
  in CFB combustors 384
  near membrane walls 301
  use in determining solids mixing 134, 135–8
Terminal settling velocity 38, 465
Thermal boundary layer 270, 278, 282, 285, 300
Thermal cracking, *see* Pyrolysis
Three-phase operation, *see* Gas–liquid–solid circulating fluidized bed
Tomography 327, 329–32
Tracers
  gas 88–92, 354, 490, 498, 514
  solids 127–30, 134, 135, 341, 351–3, 490
Transition velocities 6–10, 34–5, 37
Transport disengagement height (TDH) 17, 36, 43, 47
Transport velocity 9–10, 12–13, 35, 467, 474
Trickle valve 255–8
Turbulence
  in cyclones 105, 196
  data 60
  gas–particle interaction 124
  gas phase 163, 171, 174–6
  influence on flow patterns 162
  influence on gas mixing 86, 101–3, 105, 133
  influence on solids mixing 55, 123, 133
  modeling 171
  solids phase 164, 165
Turbulent fluidization
  in FCC regenerators 473
  gas mixing in 96, 110
  in MTG process 491
  properties 39–40, 46
  reactor 505
  regime 8–9, 11, 32, 474

Upper dilute zone, *see* Dilute region

Valves 225, 237
  butterfly 534
  H-valve 531
  J-valve 16, 230, 242
  N-valve 251
  plug valve 471
  V-valve 16, 251–2
  see also L-valve; Non-mechanical solids flow devices; Slide valve; Trickle valve
Velocities
  solids 43, 122–3
  see also Clusters; Superficial gas velocity
Vibration 499
Video photography 123, 263, 265, 288
Viscosity
  effective 160
  effect of gas viscosity 33
Visualization 30, 313–14, 343
Voidage
  axial profiles 23, 24, 35–7, 61–2
  definition 17
  radial profiles 43–6, 121, 526
  ranges 4
  see also Suspension density
V-valve, *see* Valves

Wall coverage 265, 280, 287–8, 296, 303
Wall effects 32, 34, 43, 159
  see also Roughness
Wall friction, *see* Friction
Wall layer thickness 47, 113, 263, 379
Wall region, *see* Annulus region
Wastage, *see* Erosion
Waterwalls, *see* Membrane walls
Wear, *see* Erosion
Windbox 431
Wing walls 304–5, 424, 431

Zirconia cell probes, *see* Probes